AF802973

Microorganisms for Sustainability

Volume 5

Series editor
Naveen Kumar Arora, Environmental Microbiology, School for Environmental Science, Babasaheb Bhimrao Ambedkar University, Lucknow, Uttar Pradesh, India

More information about this series at http://www.springer.com/series/14379

Dilfuza Egamberdieva • Parvaiz Ahmad
Editors

Plant Microbiome: Stress Response

Springer

Editors
Dilfuza Egamberdieva
Faculty of Biology
National University of Uzbekistan
Tashkent, Uzbekistan

Parvaiz Ahmad
Department of Botany and Microbiology
King Saud University
Riyadh, Riyadh, Saudi Arabia

ISSN 2512-1901 ISSN 2512-1898 (electronic)
Microorganisms for Sustainability
ISBN 978-981-10-5513-3 ISBN 978-981-10-5514-0 (eBook)
https://doi.org/10.1007/978-981-10-5514-0

Library of Congress Control Number: 2017963993

Printed on acid-free paper

This Springer imprint is published by Springer Nature
The registered company is Springer Nature Singapore Pte Ltd.
The registered company address is: 152 Beach Road, #21-01/04 Gateway East, Singapore 189721, Singapore

Foreword

Abiotic and biotic stresses affect the productivity and quality of agroecosystems around the globe. Indiscriminate use of chemicals to enhance the crop productivity has further aggravated the problem. The agroecosystems not only have become more and more polluted but also have become stressed due to the unhealthy practices. The use of chemicals such as pesticides and fertilizers have made the soil saline, polluted water bodies, and also resulted in the loss of biodiversity besides being dangerous to humans and other organisms. Several plant beneficial microbes have been affected by the uncontrolled use of harmful chemicals in fields. Although these chemicals resulted in the green revolution, scientists, agriculturists, and governments now know the adverse effects of the chemical-induced green revolution. The problem has been further worsened due to the impact of climate change resulting from urbanization, industrialization, and the ever-increasing human population. One of the major challenges faced by mankind is to feed the swelling human population. Food and nutritional security is one of the biggest tasks of the present time. The task becomes even more challenging because the sustainability and diversity of the agroecosystems have to be maintained. The solution is the use of green approaches in agriculture and lesser reliance on the harmful chemicals. Plant beneficial microbes are the symbiotic partners of almost all the plant species known and can be exploited to enhance the quality and quantity of food production. These microorganisms also known as plant growth-promoting microbes are the only alternatives for the harmful chemicals. Already, the use of biofertilizers and biopesticides is being encouraged around the globe. However, we still need to understand the mechanisms using latest biotechnological tools to properly utilize these useful microbes in enhancing the productivity and maintaining the health of the plant. Researchers working in the area of agriculture microbiology and biotechnology know that the next "green revolution" which will provide healthy and balanced diet simultaneously maintaining the sustainability of agroecosystems will be through the use of these plant beneficial microbes.

Plant beneficial microbes interact with the plant and help it in its survival and ability to fight the biotic and abiotic stresses. These microbes also known as "plant microbiome" provide nutrients to the plant and also combat phytopathogens by diverse mechanisms. It is essential to understand the mechanisms of interactions between the plant and its microbiome so as to exploit it for providing nutrition; fighting abiotic stresses such as salinity, temperature, and drought; or even controlling the phytopathogens. The plant microbiome plays an even more important role in hostile conditions. After going through the book *Plant Microbiome: Stress Response*, I would like to congratulate the editors, Dr. Dilfuza Egamberdieva and Dr. Parvaiz Ahmad, for covering the diverse aspects of plant-microbe interactions particularly in relation to biotic and abiotic stresses. The editors are renowned scientists in the domain area of this volume and have also roped in contributions from eminent researchers from around the globe working in the area of plant-microbe interactions in relation to environmental and biotic stresses. Chapters included in the volume provide the fundamental knowledge of the diversity of plant microbiome and its role in combating the stresses in plants. The work presented in the volume also throws light on how the plant microbiome is important for the survival and how it elicits the stress tolerance responses in the plant particularly in hostile conditions. The tome also provides the inputs related to the utilization of useful plant-associated microbes in enhancing the productivity and yields of stressed ecosystems. With ever-increasing mouths to feed, it is very important that we use the marginalized and stressed agroecosystems for better productivity. The book covers the aspects of utilization of plant microbiome to enhance the productivity of stressed agroecosystems in a holistic and sustainable manner. Latest tools and techniques of molecular biology and metabolomics to study the intricacies of plant-microbe interactions at the molecular and biochemical level have been explained and used. The work presented in the volume also invokes the thought for future directions of research in the area of utilization of plant microbiome for enhancing yields of hostile and marginalized agroecosystems. The effort, reviews, and research put in the volume are up to the mark and need of the hour. The book will be very useful for the scholars, students, and researchers working in the field of plant-microbe interactions particularly related to stress response and management. The future of mankind is in using such green approaches so as to make our platter free of toxics and full of nutrients and for food security of each and every human living on this planet. I once again congratulate the editors and the contributors for this timely volume with futuristic and sustainable approach.

Professor, School for Environmental Science
BBA University
Lucknow, UP, India

Naveen Kumar Arora

About the Series Editor

Prof. Naveen Kumar Arora PhD in microbiology, Professor in School for Environmental Sciences, Babasaheb Bhimrao Ambedkar University (a Central University), Lucknow, Uttar Pradesh, India, is a renowned researcher in the field of environmental microbiology and biotechnology. His specific area of research is rhizosphere biology and Plant Growth Promoting Rhizobacteria (PGPRs). He has more than 50 research papers published in premium international journals and several articles published in magazines and dailies. He is editor of three books, published by Springer. He is member of a number of national and international societies and reviewer of several international journals. He has delivered lectures in conferences and seminars around the globe. He has a long-standing interest in teaching at the postgraduate level and is involved in teaching courses in bacteriology, microbial physiology, environmental microbiology, agriculture microbiology, and industrial microbiology. He has been advisor to 57 postgraduate and 8 doctoral students. Recently he was awarded for excellence in research by the Honorable Governor of Uttar Pradesh. Although an academician and researcher by profession, he has a huge obsession for wildlife and its conservation and has authored a book, *Splendid Wilds*. He has a dedicated website, www.naveenarora.co.in, for the cause of wildlife and environment conservation.

Contents

About the Editors

Dr. Dilfuza Egamberdieva graduated in biology from the National University of Uzbekistan in 1993. She received her PhD in agricultural sciences from the Humboldt University of Berlin, Germany, in 2000 and conducted her first postdoctoral research at the Helsinki University of Finland in 2001. In 2003, she conducted her second postdoctoral research in the Department of Soil Science and Plant Nutrition, University of Florence, Italy. In 2006, she continued her research activities in the Department of Biological Sciences of Manchester Metropolitan University and, in 2007, in the Institute of Biology, Leiden University of Netherlands. She returned to Uzbekistan in 2009 and established a research laboratory in the National University of Uzbekistan, where the main focus of research is directed towards an understanding of the structural and functional organization of plant-associated microbial communities. For her best academic achievements, she received numerous fellowships and awards including UNESCO-L'ORÉAL Fellowship for Women in Science (2006), IUBMB Young Scientist Award (2006), ASM Morrison Rogosa Award (2006), UNESCO-MAB Award (2005), and TWAS-TWOWS-SCOPUS Young Women Research Award (2009), and in 2012 she was awarded the TWAS Prize in Agricultural Sciences for her innovative contributions to the study of plant-microbe interactions under environmental stress and their potential to improve crop production. In 2014, she was elected as member of the Global Young Academy, and in the same year she was awarded the Alexander Von Humboldt Fellowship and began her research as senior scientist at the Leibniz Centre for Agricultural Landscape Research (ZALF) in Germany. She has over 80 peer-reviewed publications in international journals and 40 chapters.

Dr. Parvaiz Ahmad is Senior Assistant Professor in the Department of Botany at Sri Pratap College, Srinagar, Jammu and Kashmir, India. Presently Dr. Ahmad is visiting scientist at Department of Botany and Microbiology, King Saud University, Riyadh, Saudi Arabia. Dr. Ahmad completed his postgraduate studies in botany in 2000 from Jamia Hamdard, New Delhi, India. After receiving a doctorate from the Indian Institute of Technology (IIT), Delhi, India, he joined the International Centre for Genetic Engineering and Biotechnology, New Delhi, in 2007. His main research areas are stress physiology, molecular biology, and microbiology. He has published more than 60 research papers in peer-reviewed journals and 40 book chapters. He is also editor of 19 books (one with Studium Press (India) Pvt. Ltd., New Delhi, India; nine with Springer, NY, USA; five with Elsevier, USA; and four with John Wiley& Sons Ltd). He is a recipient of the Junior Research Fellowship and Senior Research Fellowship by CSIR, New Delhi, India. Dr. Parvaiz has been awarded the Young Scientist Award under the Fast Track Scheme in 2007 by the Department of Science and Technology (DST), Government of India. Dr. Parvaiz is actively engaged in studying the molecular and physio-biochemical responses of different agricultural and horticultural plants under environmental stress.

Chapter 1
Diversity, Functions, and Stress Responses of Soil Microorganisms

Shyam Narain Pandey, Murtaza Abid, and Mirza Mohammad Abid Ali Khan

Abstracts Plant-associated soil microorganisms colonize the rhizosphere, face many stresses, and play the significant role in the functioning of plants by influencing their growth and metabolism. Microorganisms are beneficial to decompose organic matter, for mineralization, and for the availability of plant nutrients in the soil. They also maintain the soil ecosystem and biogeochemical cycle. The soil microflora and microfauna cause synergistic or antagonistic effects on plants and face various biotic and abiotic stresses in the rhizosphere. Soil microbes make a gene pool involved in microbes-plant interactions. The main categories of soil microorganisms are bacteria, algae, protozoa, fungi, viruses, and multicellular animal parasites. The activities of soil microorganisms are influenced by interactions between soil physicochemical properties and environmental conditions. Bacteria are present in all types of soil and play their roles in atmospheric nitrogen fixation. In this review, the stress conditions in the rhizosphere, diversity of microorganisms, and their role in increasing soil fertility have been emphasized. Most of the microflora and microfauna are pathogenic in nature, but their positive interactions into the soil (in the rhizosphere) are very significant. Therefore, review of recent studies on diversity, stresses, and functions of soil microorganisms are described in this article.

Keywords Soil microbes • Biology • Enzymes • Plant interaction • Abiotic stress

S.N. Pandey
Department of Botany, University of Lucknow, Lucknow, Uttar Pradesh, India

M. Abid
Department of Biochemistry, King Georges, Medical University, Lucknow, Uttar Pradesh, India

M.M. Abid Ali Khan (✉)
Department of Botany, Shia P.G. College, Associated with University of Lucknow, Lucknow, Uttar Pradesh, India
e-mail: mmabidalikhan265@gmail.com

D. Egamberdieva, P. Ahmad (eds.), *Plant Microbiome: Stress Response*, Microorganisms for Sustainability 5, https://doi.org/10.1007/978-981-10-5514-0_1

1.1 Introduction

The collective communities of plant-associated microorganisms are referred to as the plant microbiome. Soil microbiology is the study of small microscopic organisms in the soil and their functions and mode of effects on soil properties. Soil microorganisms are involved in various biotic and abiotic stress conditions in the rhizosphere (Newman et al. 2016). The soil is the large sink of organic carbon; the soil microorganisms transform the organic material (Doi et al. 2006; Yadav and Pandey 2015) and nutrient for plants (Gul and Whalen 2016; Kamble and Baath 2016). Plants fix carbon into the soil through its exudates, which includes sugars, amino acids, flavonoids, aliphatic acids, and fatty acids that attract beneficial microorganisms while repelling and killing harmful ones (Janga et al. 2017; Jimtha et al. 2017).

Soil microorganisms are vital to the cycle of life on earth. They play important role in the soil to support ecosystem (Hiinninghaus et al. 2017) as well as fertility (Brady 2001; Bhat 2013). The microorganisms are microscopic (< 0.1 mm in diameter) with variable groups, shapes, and sizes (Smith et al. 1992; Wall and Virginia 1999; Neufeld and Mohn 2005). These are of significance in the soil biological processes that are so critical for the animal (Du and Liu 2012) and plant life (Lavelle and Spain 2001). The organic residues from various sources added into the soil are devoured by soil microorganisms to form dark-colored, complex organic compound humus (Leckie et al. 2004). In the synthesis process, humus residues are decomposed by various types of microorganisms (Torsvik and Ovreas 2002), yielding carbon dioxide and water. Also, nutrient elements held in the organic residues are released through mineralization (Brady and Weil 2008; Verma and Pandey 2016) in forms that are available for growing plants (Griffith et al. 1994; Yadav and Pandey 2015). The soil environment is one of the most complex biological communities on earth and niches to an even larger share of biodiversity than tropical forests (Tongway and Ludwig 2005). Beijerinck (1901) developed "enrichment culture technique" which isolated many soil microorganisms including aerobic nitrogen-fixing bacterium (symbiotically growing *Azotobacterchrococcum*), *Lactobacillus* species, *Acetobacter*, green algae and the yeast *Schizosaccharomyces octosporus*, and many others. Beijerinck also isolated *Bacillus radicicola* (later called *Rhizobium*) which is responsible to nodulation in legumes for nitrogen fixation. The *Thiobacillus denitrificans*, the first identified sulfate-reducing bacterium, use sulfate as an electron acceptor instead of oxygen. The *Beggiatoa*, a sulfur-oxidizing bacterium, use inorganic H_2S as an energy source and CO_2 as a carbon source (chemoautotrophs).

1.2 Stress Responses of Plant Microbiome

A large diversity of soil microorganisms are involved in various stress conditions in the rhizosphere (Lugtenberg and Kamilova 2009; Qiang et al. 2012). Soil microorganisms interact with various stresses (biotic and abiotic) through their diverse genetic compositions (Newman et al. 2016; Koch et al. 2001). They face stresses of nutrients (Miransari 2013), temperature and water potential (Moore-Kucera and Dick 2008; Herron et al. 2010), etc. A wide diversity of soil microorganisms react to various stresses, and different stress response systems of microorganisms interact with each other and play important role in the virulence of pathogenic organisms (Koele et al. 2009; Ayalaja and Pedro 2012). In a report, the bacterial cell wall of *Escherichia coli* is involved in the maintenance, adaptation, and protection of bacteria (Ayalaja and Pedro 2012) against stress conditions. The soil stress conditions affect soil fertility through microbial-activity-mediated decomposition, mineralization, and nitrogen fixation processes (Panikov et al. 2006; Borken and Matzner 2009; Kaiser and Kalbitz 2012). The moisture and temperature stress highly affect the activity of soil microorganisms (Panikov et al. 2006). The cell wall component of fungi responds to changes in the soil environment by making a cell surface epitope to protect the cell from the host immune response (Latge 2017). Soil microorganisms alter their metabolic responses against drought stress conditions. In response to stress conditions, soil microorganisms alter protein's nature through mutations in the genetic system (Hartmann et al. 2017). The bacteria and fungi produce antibiotic metabolites against pathogenic microbes (Hoffmeister and Keller 2007). In response to soil environmental stresses, microorganisms have various adaptations through physiological acclimation mechanisms, to survive. Soil microorganisms influence growth and biochemical responses of plants through various activities (Mendes et al. 2013).

1.2.1 Abiotic Stresses

Soil abiotic factors such as freezing/thawing, drying/rewetting, water percolation, oxic/anoxic conditions alter physical environment of the soil and affect microbial population and activities (Borken and Matzner 2009).The rhizosphere microbiome plays an important role for plants to survive under extreme climatic conditions (Jorquera et al. 2012).The soil microorganisms were shown to support plant growth in osmotic drought stress and halotolerant bacterial strain (Siddikee et al. 2010; Berg et al. 2013) in saline soil (Upadhyay et al. 2009). The dominant bacterial genus *Bacillus* has been isolated under saline stress conditions (Upadhyay et al. 2009). The activity of soil microorganisms is variable at different development stages of plant root growth with specific time and space (Barret et al. 2011). Microbial genes are involved in nutrient acquisition, stress responses, and secretions in the rhizosphere (Rainey 1999). A greater variation in moisture and temperature highly affect microbial population and functions (Panikov et al. 2006; Brady and Weil 2008). Grazing affects the composition and diversity of plant species, increases soil carbon loss, and affects soilborne microbial community (Kowalchuk et al. 2002; Klumpp et al. 2009). Drought stress affects soil microbial biomass and enzyme activities in the rhizosphere (Sanaullah et al. 2011). Tillage effects on soil microbial activity and carbon dynamics during decomposition process have been reported by White and

Charls (2009). Moisture stress in soil limited organic matter inputs due to which carbon stocks in dry land soil area tend to about half of that soils in moist environmental condition (IPCC 2006). Rainfall events triggered moisture fluctuations that cause stress, which strongly affects soil microbe's respirations and nutrient mineralization. The decrease in moisture content in soil decomposers activity reduced microfauna generally undergoing stress sooner than bacteria and fungi (Manzoni et al. 2012). Bacteria are more sensitive than fungi. Polycyclic aromatic hydrocarbons indirectly affect microbial community in legumes, which decreases nitrogen fixation in the root (Kawasaki et al. 2012). Some abiotic factors, such as pH and toxic metals, largely affect the microbial population as well as plant growth (Raudales et al. 2009; Singh et al. 2015).

1.2.2 Biotic Stresses

The rhizosphere microbiomes help plants to survive under extreme stress conditions (Jorquera et al. 2012). Soil microorganisms in the rhizosphere protect and promote the plant growth from various pathogens through activities such as biofertilization, stimulation of plant root growth, rhizoremediation, control of abiotic stresses, and disease control (Kogel et al. 2006; Lugtenberg and Kamilova 2009; Qiang et al. 2012). The main mechanism by which rhizosphere microorganisms ward off plant pathogens is antibiosis (Lugtenberg and Kamilova 2009). Along with bacteria in the rhizosphere, fungi are prolific producers of antibiotic metabolites (Hoffmeister and Keller 2007; Druzhinina et al. 2011). Most biocontrol strains (bacteria and fungi) produce more than one antibiotic compound with overlapping or different degrees of antimicrobial activity. The soil microorganisms were shown to support plant growth in various soil conditions (Upadhyay et al. 2009; Berg et al. 2013). The activity of soil microorganisms is variable at different development stages of plant root growth in specific time and space (Barret et al. 2011). In the rhizosphere, various microbial groups constitute a gene pool that protects microorganisms from extreme stress conditions (Rainey 1999). A multitude of small RNAs (sRNAs, 18–25 not in length) that accumulate in plant tissues have functions in regulating plant responses to various microbes during infections (Ferrer and Voinnet 2009). The microbe-plant root associations are also variable at the different depth. Most of the algae are dominated over the surface layer, but some blue-green algae are found in association with roots that produce organic chemicals. A large group of soil microorganisms in soil protect plant roots from various pathogens before and during primary infection (Mela et al. 2011; Pieterse 2012). Some species of *Trichoderma* produce antimicrobial compounds (Vyas and Mathus 2002; Druzhinina et al. 2011). Some bacterial and fungal groups produce more than one antibacterial compounds (Reader et al. 2005).

1.3 Environment and Soil Microorganisms

Human alteration of the global environment has triggered the sixth major extinction event in the history of life and caused widespread changes in the global distribution of organisms. These changes in biodiversity alter ecosystem processes (Young and Crawford 2004) and change the resilience of ecosystems to environmental change (Bardgett and Putten 2014; Romero-Olivares et al. 2017). The activities of microorganisms have a central role in the global fluxes of the biogenic gases (such as carbon dioxide, methane, and nitrous oxide) (Singh et al. 2010). The mechanisms by which microorganisms regulate terrestrial greenhouse gas flux are not very clear because these involve complex interactions that occur between microorganisms (Watt et al. 2006a, b) and other biotic and abiotic factors (Sharma et al. 2008). Simulated climate change affects microorganisms, nematode, density, and biodiversity in cold climate condition soil. The principal source of organic carbon is planted residues (Kuzyakov 2010). Soil carbon stocks manifest losses due to decomposition of soil organic matter through the action of soil fauna and microbes (Lal 2004) and physical export by leaching and erosion (Panagos et al. 2015). Moisture stress in soil (IPCC 2006) limited organic matter inputs at very low level in arid and semiarid regions (Tongway and Ludwig 2005). Soil organic matter is a significant scale to evaluate soil fertility, because it is the sink for nutrients and medium for nutrient cycling (Kamble and Baath 2016). The change in climatic conditions directly influences moisture content in soil organic matter and, consequently, affects the physical, chemical, and biological behavior of the soil (Uphoff et al. 2006). Carbon compounds oxidizing bacteria (methanotrophs) require trace metal copper for oxidation of methane. Therefore, they accumulate copper in their body (Nicolas et al. 2015) and have enormous potential in bioremediation of copper in soil (Ma et al. 2009) and biotransformations of chemicals (Hakemian and Rosenzweig 2007) and bioenergy (Nicolas et al. 2015). Methane monooxygenases are nature's primary biological mechanism for suppressing atmospheric levels of methane, a potent greenhouse gas (Nicolas et al. 2015). Soil microbial communities mediate critical ecosystem carbon and nutrient cycles (Waldrop and Firestone 2006). The coupled cycling of nutrients (nitrogen, phosphorus, sulfur, etc.) governs numerous ecosystem processes, including carbon sequestration in soil and vegetation (Tongway and Ludwig 1996; Penman et al. 2010). The climate change influences the nematode density, microbial biomass, and nutrients availability in soil (Hui et al. 2016). The soil management practices entailing deforestation, conversion from perennial to annual plant species, the heavy grazing effects of soil carbon, and changes in soil temperature alter the composition of microbial diversity (Schmidt et al. 2011).

1.4 Classification of Soil Microorganisms

The soil microorganisms can broadly classify into two groups.

1.4.1 Soil Flora

Soil microflora present in the soil are classified as (1) bacteria, (2) fungi, (3) actinomycetes, and (4) algae. The bacterial group again classified in A is heterotrophic (symbiotic and nonsymbiotic nitrogen fixers, ammonifier, cellulose decomposers, denitrifiers) and B autotrophic (*Nitrosomonas*, *Nitrobacter*, sulfur oxidizers, etc.). The heterotrophic soil microorganisms obtain their energy and carbon from decomposition of soil organic matter. These organisms dominated over the autotrophs (Zhou et al. 2017). They include the protozoa, nematodes, and most of the bacteria, fungi, and actinomycetes. The autotrophs obtain their energy from solar energy (photoautotrophs) and from the oxidation of inorganic elements nitrogen, sulfur, and iron (chemoautotrophs). Fungi are the second largest fraction of the microbial biomass after bacteria in most well-aerated soil. The fungal population ranges from 2×10^4 to 1×10^6 fungal propagules (part of the fungus per gram of the soil); usually more than half of the fungal biomass is of basidiomycetous fungi alone. The population of an aerobic bacterial group is much higher than anaerobic bacteria. The *Bacillus* and *Pseudomonas* bacteria decompose complex organic compounds. The coryneforms, the nocardioforms, and the true filamentous bacteria or actinomycetes are important components of soil microbial community (Aneja et al. 2008).

1.4.2 Soil Bacteria

It is a single very small (4–5 μm in length) prokaryotic cell (Ingham 2009) that increases rapidly in numbers. In soil, bacteria exist as mats, clumps, and filaments in colonies. While bacteria may be small, they make up both the largest number and biomass of any soil microorganisms (Zhou et al. 2017; Garland 1997). Bacteria are present in all kinds of soil, but their population decreases in lower horizons of soil (Brady and Weil 2008; Doi et al. 2007). Their mode of nutrition is either heterotrophic or autotrophic. Bacteria are very significant in soil (Smith et al. 1999) and play a vital role in controlling nutrient availability to plant root (Kar et al. 2017). The autotrophs obtain their energy from the oxidation of inorganic elements such as nitrogen, sulfur, and iron. They obtain most of the carbon from carbon dioxide (Sylvia et al. 2005). Bacteria participate in oxidation as well as reduction reactions in the soil. A significant process performed by bacteria is nitrogen fixation, a biochemical process of combination of atmospheric nitrogen with hydrogen to form ammonia which is then incorporated into organic nitrogen compounds utilized by plants. Bacteria not only synthesize the new organic compound but also decomposed them to recycle (Islam 2008). The activities of bacteria are maximum at 20–40 °C; the temperature extremes kill the bacteria. Exchangeable calcium and alkaline pH (6–8) are best for bacterial growth. There are about 200 bacterial genera and about 4000 different bacterial species found even in a single sample of soil, but less than 1% of the species are vulnerable (Aneja et al. 2008). Bacteria *Azotobacter* produces growth-stimulating chemical substances which help in improving soil fertility (Young and Crawford 2004). The microorganisms such as *Rhizobium* and *Glomus* species have been shown to play a role in reducing diseases in soil (Avis 2008).

1.4.3 Actinomycetes

These are the broad group of Gram-positive, filamentous, resemble molds and are similar to bacteria. Actinomycetes are most prominent in moist soils high in humus (Macfadyen 1963). They release nutrients in the soil solution as the result of their metabolic processes. In association with some leguminous plants, actinomycetes fix nitrogen available to plants. The actinomycetes group help in the aggregation of soil particles by secreting sticky exude containing polysaccharides. The association between actinomycetes and plant roots is called "actinorhizal." These filamentous bacteria fix nitrogen in nodules (Zhou et al. 2017). Nitrogen fixation occurs in symbiotic vesicles which maintain anaerobic conditions necessary for nitrogenase. The nodules formed by *Alnus* (family Betulaceae), *Ceanothus* (family Rhamnaceae), etc. have nodules as large as baseballs and that of Casuarina (family Casuarinaceae) is of soccer ball size. The genus *Frankia* is well-known actinomycetes that forms nodules in several non-leguminous trees and shrubs belonging to eight families of angiosperms (Aneja et al. 2008). Actinorhizal nodules are pink in color when cut open; this color is due to anthocyanins, not by leghemoglobin (as occurs in legumes). Actinomycetes are of great importance in decomposing soil organic matter and help in mineralization of nutrients (Howell and Kenzie 2017). They produce sources of novel bioactive metabolites and antibiotics (Bibb 2005; Subramani and Aalbersberg 2012; Chanthasena and Nantapong 2016). There are more than 2200 known microbial secondary metabolites, 70% of which are produced by actinomycetes (Berdy 2005). Antifungal activities of actinomycetes of soil origin have been reported by Augustine et al. (2004). Selvameenal et al. (2009) isolated an actinomycetes strain (*Streptomyces hygroscopicus* subsp. *ossamyceticus*) from the desert soil, which produces a yellow color pigment with antibiotic activity. Diversity and properties of acidophilic actinomycetes from rhizosphere beneficial to plants has been isolated (Poomthongdee et al. 2015) from soil using acidified media of pH 5.5. Several strains of actinomycetes are isolated from saline soil (Deshmukh and Vidhale 2015), wetland soil (George et al. 2011), and marine, brackish, and terrestrial sediments of Samal Island (Parungao et al. 2007). The various *Streptomyces* species from the soil with antibacterial activity have been identified (Jeya et al. 2013; Shetty et al. 2014). Phylogenetic analysis of the genus *Actinomyces* based on 16S rRNA gene sequences has been reported by Newman et al. (2016).

1.4.4 Soil Fungi

Soil fungi are the largest and ancient organisms in the world. The fungus *Armillaria* (a basidiomycetous fungus), a facultative parasite of many roots, has been reported by Smith et al. (1992) as one of the largest and oldest living organisms that occupy a minimum of 15 hectares, weigh in excess of 1000 kg, and have remained genetically stable for more than 1500 years. Fungi are microscopic, single-celled to multicellular organisms that are usually growing with long threadlike structure called the hyphae. Some soil fungi are free-living; some occur in symbiotic relationship (mycorrhizae) with plant roots. In soil, most fungal species are *Deuteromycetes* (anaerobic fungi), *Zygomycetes* (especially *Mucorales*), and *Ascomycetes* and some belonging to *Chytridiomycetes* and *Oomycetes*. Fungi dominated mostly on the upper soil profile. They play important roles in water dynamics, nutrient cycling, and diseases suppression. In soil, over 690 species of 170 genera have been identified (Brady 2001). Many genera of molds are found in soils; most dominant are *Penicillium*, *Mucor*, *Fusarium*, and *Aspergillus* (Pandey et al. 2017). All fungal species lacking chlorophyll pigments therefore depend on their carbon and energy requirement on the organic matter in the soil. Fungi are responsible for decomposition in terrestrial ecosystem as they degrade cellulose, hemicelluloses, pectins, starch, and lignin, the component of plant cell walls. Soil fungi can be grouped into three groups based on their mode of nutrition – saprophytic, symbiotic, and parasitic fungi. The saprophytic fungi decompose dead organic matter and immobilize nutrients (nitrogen, phosphorus, etc.) and proteins in the soil. Some fungal group associated with plant roots symbiotically (mycorrhizal fungi).

1.4.5 Mycorrhizae

An important fungal group, mutually beneficial (symbiotic) in association with higher plant roots in soil, is called "mycorrhizae." Mycorrhizae increase the surface area in the vicinity of plant roots to the immobilized availability of nutrients while getting their energy sources in the form of carbohydrates from plant roots. Soil moisture and topography affect population composition of root-associated fungal species (Kohout and Tedersoo 2017). There are mainly two groups of mycorrhizal associations, *ectomycorrhiza* and *endomycorrhiza*. The ectomycorrhiza group includes a large number of fungal species which cover the external surface of plant roots with a fungal mantle; they penetrate the roots and develop around the cortical cells but do not penetrate these cell walls. The endomycorrhizal group, the most important and wide spread of which are called vesicular-arbuscular mycorrhizae (VAM), invade and penetrate the cortical cell walls and form hyphal masses within the cells. The arbuscular are branched fingerlike hyphae probably having an absorptive function (Lambert et al. 1979).

Endomycorrhizae with 89 identified species of fungi in soils from the tropical to arctic region forming VAM (Lal 1987). The VAM fungi have very wide host range which includes angiosperm species of almost all the families. The structures called vesicles act as storage organs for the plant nutrients and other products (Zhou et al. 2017).

1.4.6 Nematodes

Nematodes are microscopic animals (2000 μm long) that are called threadworms or eelworms, found in all types of soil. More than 1000 species of soil nematodes are known. They are free-living in soil and feed on bacteria and fungi. The parasitic fungi harm plant growth. Mostly, nematodes are saprophytes that live on dead organic matter; some are predatory on other nematodes, bacteria, protozoa, algae, etc. Some species of the genus *Heterodera* can invade the root of plants. They cause serious damage, especially to vegetable crops. The Mermithidae nematodes which may be 20 cm long are very common in tropical soils, being parasites of some arthropods such as locusts. Plant parasitic nematodes cause great economic losses in crop plants (Koenning et al. 1999; Kashaija et al. 2004). Nematodes can move through the soil where a film of moisture surrounds the soil particles. They live in spaces between soils particles filled with water. They enter into the dormant stage in hot and dry conditions, and as soon as water becomes available, they spring back to activity. Nematode-suppressive effects of marigold (*Tagetes* spp.) produce essential oil that cause allelopathic effect, which suppresses nematodes population in the soil (Hooks et al. 2010). The activity of nematodes may also be suppressed by some bacterial groups by their antagonistic effects (Sturz and Kimpinski 2004). The actinomycetes in organic matter of marigold-rich soil and soil solarization reduce population of soil nematodes (Oka et al. 2007).

1.4.7 Protozoa

Protozoa are the simplest form of animal life, small, single-celled, and microscopic (5–500n μm). They are considerably larger than bacteria, primarily feed on bacteria and other protozoa, and affect the activity of microorganisms, rate of mineralization, and nitrogen fixation in soil (Griffiths 1994; Wolters 2001). Mostly, amoebas and flagellated protozoa require water for their locomotion and feeding; their activity is limited to the water-filled pore space in soil. They can withstand drying of the soil and other adverse conditions (Hoorman 2011) by forming resistant cysts. The number ranges from few hundreds to several hundred thousands per gram of moist soil rich in organic matter (Aneja et al. 2008).

1.5 Functions of Microorganisms

Various groups of microorganisms are parasites on plants, because they can cause many soilborne diseases responsible to the great loss of plant produce, but a large group of them are necessary to maintain soil fertility by their various functions. Some important functions of soil microorganisms are the following.

1.5.1 Mineralization

The release of organically bound nitrogen to inorganic forms ($NH_4^+$and NO_3^-) is termed mineralization. Soil microorganisms release nutrients after decomposition of organic matter (Brady 1995). When microorganisms decompose organic matter, they use the carbon and nutrients in the organic matter for their growth. They release excess nutrients into the soil where they can be taken up by plants (Sharma 2006). If the organic matter has a low nutrient content, microorganisms will take nutrients from the soil to meet their requirements. For example, applying organic matter with carbon to nitrogen ratios lower than 22:1 to soil generally increases mineral nitrogen in the soil. In contrast, applying organic matter with carbon to nitrogen ratios higher than 22:1 generally results in microorganisms taking up mineral nitrogen from soil (Hoyle et al. 2011). Due to the action of soil microorganisms, organically bound forms of nitrogen, sulfur, and phosphorus are made available to the plant roots in the soil (Singh and Bahel 1993). The plant proteins also succumb to microbial decay, yielding carbon dioxide and water (Stotzky 1997) and also amino acids such as glycerine and cysteine. In turn, the nitrogen and sulfur in these amino acids are further broken down, eventually yielding simple inorganic ions such as NH_4^+, NO_3^-, and SO_4^{2-}. When food for microorganisms exhausted in the soil, the number of microorganisms decrease. The dead microorganisms are attacked by living microbes in the soil. The process further releases carbon dioxide, water, and mineral nutrients (nitrogen, phosphorus, sulfur, and trace elements) into the soil.

1.5.2 Fixing Atmospheric Nitrogen

Nitrogen can be added to the soil by some microbes that "fix" it from the atmosphere and can then be released back to the atmosphere by other microorganisms. Symbiotic nitrogen fixation is a significant source of nitrogen content in soil made available through the process of nitrogen fixation by microbial activities (Unkovich 2003). In the symbiosis, rhizobia or bradyrhizobia fix nitrogen gas from the atmosphere and make it available to the legume. In exchange, they receive carbon from the legume. The symbiosis is highly specific, and particular species of rhizobia and bradyrhizobia are required for each legume. The elemental nitrogen ($N = N$) in the atmosphere cannot be used directly by higher plants. The area of the agricultural field contains only about 3.3 mg of nitrogen as compared to the air above that area which contains about 300,000 mg of nitrogen (Brady and Weil 2008). Atmospheric nitrogen (80% N) is a limitless source of nitrogen not used by plants in elemental form (Beijerinck 1901). Bacteria along with cyanobacteria (blue-green algae) and certain actinomycetes are important to capture gaseous nitrogen (Brady 2001) and have it fixed and made available to be absorbed by plant roots. In leguminous plants, the nodule-forming organisms such as free-fixing bacteria of several kinds and some actinomycetes have the ability to fix nitrogen (Tilman and Downing 1994; Sharma et al. 2008). Much of the nitrogen added to the soil in organic combination is subjected to simplification first to amino compounds, then to ammonium (NH_4^+) ion, and finally to nitrate (NO_3^-). Microorganisms incorporate ammonium nitrogen into organic forms in their bodies.

1.5.3 *Nitrification*

Nitrification is a process of enzymatic oxidation of ammonia to nitrates by several microorganisms present in the soil. Firstly, nitrate (NO_2^-) is produced by organisms *Nitrosomonas*; secondly, further oxidation by another organism *Nitrobacter* produces nitrate (NO_3^-).

Step 1 Activity of *Nitrosomonas*

$$\underset{\text{Ammonium}}{NH_4^+} + 1^{1/2}O \xrightarrow{\text{Nitrosomonas}} \underset{\text{Nitrite}}{NO_2^-} + 2H^+ + H_2O + \text{Energy}(275\,\text{kJ})$$

Step 2 Activity of *Nitrobacter*

$$\underset{\text{Nitrite}}{NO_2^-} \xrightarrow{[o]} \underset{\text{Nitrate}}{NO_3^-} + \text{Energy}(76\,\text{kJ})$$

If nitrite is not converted into nitrate by bacteria, nitrite is toxic to plants and mammals (Brady, 2001). Nitrifying microorganisms are very sensitive to their surrounding conditions. Nitrification can take place only if there is source ammonia to be oxidized (Broadbent et al. 1982). Factors such as high C/N ratio of residues, which prevent the release of ammonia, also prevent nitrification. Soil aeration and good soil drainage are needed to provide the oxygen for the nitrification process. The temperature most favorable for nitrification is 25–35 °C; nitrification is slow in cool soil (ASA 1980). Nitrification is also retarded by both low and high moisture content in the soil. Soil with low pH also influences the nitrification which adversely depends on the concentration of base-forming cations. Pesticides in soil affect the activity of nitrifying organisms, which inhibit nitrification (Burns and Hardy 1975). The negatively charged colloids in the soil are not absorbed by the negatively charged nitrate ions. Consequently, they are leaching from the soil's lower horizons.

1.5.4 Biological Nitrogen Fixation

The symbiosis of legumes and bacteria of the genus *Rhizobium* biologically fixes the atmospheric nitrogen in soil. The specific bacterial group infects the root hairs, and cortical cells induce nodule formation (Thilakrathana and Raizada 2017). The mutual benefits here are that the host plant provides carbohydrates for energy to bacteria and the bacteria reciprocate by supplying the plant with fixed nitrogen compounds. Therefore, the association is symbiotic. *Rhizobium* is host specific, for example, *Rhizobium trifold species infect* clovers, and *Rhizobium phaseoli* inoculates *Phaseolus vulgaris* (beans) and other specific bacteria-host plant relationship.

It is the biochemical process by which elemental nitrogen is combined into organic forms. The process is carried out by several species of bacteria (most of them are associated with roots of legumes), a few actinomycetes and blue-green algae. The overall effect of the process is to reduce nitrogen gas to ammonia.

$$N_2 + 8H + 6e \xrightarrow[\text{[Fe,Mo]}]{\text{Nitrogenase}} 2NH_3 + H_2$$

Ammonia is, in turn, combined with organic acids to form amino acids and ultimately proteins.

$$NH_3 + \text{Organic acids} \rightarrow \text{Amino acids} \rightarrow \text{Proteins}$$

The nitrogen fixation is a very important biological process on the earth, next to photosynthesis.

1.5.5 Denitrification

The facultative anaerobic microbes are involved in the reduction of nitrate nitrogen to gaseous compounds. The process is facilitated by specific reductase enzyme at different steps of denitrification (Leffelaar 1986). The stages NO, N_2O, and N_2 can be released to the atmosphere. The oxygen atoms become incorporated into the bodies of the anaerobic bacteria. There are non-microbial processes by which nitrogen may be reduced in soils to gaseous forms. If nitrogen may be reduced in soils, losses by denitrification may be very high in flooded soils and losses by denitrification may be very high due to poor aeration (Patrick 1982).

1.5.6 Sulfur Immobilization

Immobilization of inorganic form of sulfur by soil microbes occurs when low-sulfur organic materials are added to the soils (Brady 2001). The mechanism is same as in the case of nitrogen mineralization. The energy-rich organic residues stimulate microbial growth (Smith and Goodman 1999), and the sulfate is synthesized into the microbial cell. After death, inorganic sulfate appears in the soil solution. During the microbial breakdown of organic materials, several sulfur-containing gases are formed such as hydrogen sulfide (H_2S), carbon disulfide (CS_2), carbonyl sulfide (COS), and mercaptan (CH_3SH). All are more prominent in anaerobic soils. Most of the others are formed from the microbial decomposition of sulfur-containing amino acids methionine and cysteine (Stipanuk, 1986). These gases can be adsorbed by soil colloids, but some escape to the atmosphere where they undergo chemical changes and eventually return to the soil (Zhou et al. 2017).

1.6 Conclusion

Microflora and microfauna in soil, along with other biological sources are vital to the cycle of life on earth. The soil microbes-plant interactions show synergistic and antagonistic effects to each other. They face many biotic and abiotic stresses, struggle for food and space, help in mineralization, and return carbon dioxide to the atmosphere where it is recycled by higher plants. Soil microorganisms form humus organic constituent, so vital in improving soil conditions. During decomposition of organic residues, soil microorganisms release essential plant nutrients (nitrogen, phosphorus, sulfur, etc.) in inorganic forms available for plant root absorption. Fungi, actinomycetes, and bacteria are decay organisms, while bacteria and algae play special roles in providing essential nutrients, especially nitrogen, through the processes of nitrogen fixation. A large group of microorganisms, parasites, adversely affects plants. The study and assessment of microbial activities and their output in the soil are key factors in agricultural soil management.

Acknowledgment I acknowledge Dr. Shalini Srivastava, Department of Botany, University of Lucknow, Lucknow, for the suggestions and providing reference materials during the preparation of the article.

References

Aneja KR, Pray J, Raman J (2008) Soil microbiology. In: Basic and applied microbiology. New Age International Publishers, New Delhi, pp 379–400

ASA (1980) Nitrification inhibitors-potentials and limitations, ASA special publication no. 38. American Society of Agronomy/Soil Science Society of America, Midison

Augustine SK, Bhavsar SP, Baserisalehi M, Kapadnis BP (2004) Isolation and characterization of antifungal activity of an actinomycete of soil origin. Indian J Exp Biol 42:928–932
Avis PG (2008) Ectomycorrhizal fungal communities in two North American oak forests respond to nitrogen addition. New Phytol 179(2):472–483
Ayalaja CF, Pedro A (2012) Stress response in microbiology. In: Jose M. Rcguena (ed) Caister Academic Press, Madrid
Bardgett RD, Putten WH (2014) Belowground biodiversity and ecosystem functioning. Nature 515:505–511
Barret M, Morrissey JP, O'Gara F (2011) Functional genomics analysis of plant growth-promoting rhizobacterial traits involved in rhizosphere competence. Biol Fertil Soils 47:729–747
Beijerinck MW (1901) UberoligonitrophileMikroben.zbl.Backt 7:561–582
Berdy J (2005) Bioactive microbial metabolites. J Antibiot (Tokyo) 58:1–26
Berg G, Alvi M, Schmidt C, Zachow C, Egamberdieva D, Kamilova F, Lugtenberg B (2013) Biocontrol and osmoprotection for plants under salinated conditions. In: de Bruijn FJ (ed) Molecular microbiology, ecology of the rhizosphere. Wiley, Blackwell. https://doi.org/10.1002/9781118297674.ch55
Bhat AK (2013) Preserving microbial diversity of soil ecosystem: a key to sustainable productivity. Int J Curr Microbiol App Sci 2(8):85–101
Bibb MJ (2005) Regulation of secondary metabolism in Streptomycetes. Curr Opin Microbiol 8:208–215
Borken W, Matzner E (2009) Reappraisal of drying and wetting effects on C and N mineralization and fluxes in soils. Glob Chang Biol 15:808–824
Brady NC (1995) The nature and properties of soils, 10th edn. Prentice-Hall, New Delhi
Brady NC (2001) The nature and properties of soils, 10th edn. Prentice-Hall, New Delhi
Brady NC, Weil RR (2008) The nature and properties of soils, 14th edn. Printice Hall, Upper Saddle River
Broadbent FE, Nakashima T, Chang GY (1982) Estimation of nitrogen fixation by Isotope dilution in field and green house experiments. Agron J 74:625–628
Burns RC, Hardy RWF (1975) Nitrogen fixation in bacteria and higher plants. Springer, Berlin
Chanthasena P, Nantapong N (2016) Biodiversity of antimicrobial-producing actinomycetes strains isolated from dry Dipterocarp forest soil in Northeast Thailand. Braz Arch Biol Technol 59:145–151
Deshmukh AA, Vidhale NN (2015) Characterization of novel antibacterial Actinomycetes strain N8 from saline soil of Vidarbha region. Int J Life Sci Biotechnol Pharma Res 4(1):22–25
Doi T, Matsumoto H, Oshita N, Takemoto Y, Shinada T (2006) Microflora analysis on hydrogen fermentation of feedstock wastes. J Water Waste 48:784–790
Doi T, Hagiwara Y, Abe J, Morita S (2007) Analysis of rhizosphere bacteria of rice cultivated in Andisol lowland and upland fields using molecular biological methods. Plant Roots 1:66–74
Druzhinina IS, Seidl-Seiboth V, Herrera-Estrella A et al (2011) Trichoderma: the genomics of opportunistic success. Nat Rev Microbiol 9:896
Du L, Liu W (2012) Occurrence, fate and ecotoxicity of antibiotics in agroecosystems. A review. Agron Sustain Dev 32:309–327
Ferrer VR, Voinnet O (2009) Roles of plant small RNAs in biotic stress responses. Annu Rev Plant Biol 60:485–510
Garland JL (1997) Analysis and interpretation of community-level physiological profile in microbial ecology. FEMS Microb Ecol 24:289–300
George M, George G, Hatha AM (2011) Diversity and antibacterial activity of actinomycetes from wetland soil. South Pac J Nat Appl Sci 28(1):52–57
Griffith P, Shiah FK, Gloersen K, Ducklow HW, Fletcher M (1994) Activity and distribution of attached bacteria in Chesapeake Bay. Mar Ecol Prog Ser 108:1–10
Griffiths BS (1994) Microbial-feeding nematodes and protozoa in soil: their effects on microbial activity and nitrogen mineralization in decomposition hotspots and the rhizosphere. Plant Soil 164:25–33

Gul S, Whalen JK (2016) Biochemical cycling of nitrogen and phosphorus in biochar-amended soils. Soil Biol Biochem 103:1–15

Hakemian AS, Rosenzweig AC (2007) The biochemistry of methane oxidation. Annu Rev Biochem 76:223–241

Hartmann FE, Vallet AS, McDonald BA, Croll D (2017) A fungal wheat pathogen evolved host specialization by extensive chromosomal rearrangements. ISME J 11:1189–1204. https://doi.org/10.1038/15mej.2016.196

Herron PM, Gage DJ, Cardon ZG (2010) Microscale water potential gradients visualized in soil around plant root tips using microbiosensors. Plant Cell Environ 33:199–210

Hiinninghaus M, Koller R, Kramer S, marhan S, Ellen K, Bonkowski M (2017) Changes in bacterial community composition and soil respiration indicate rapid successions of protest grazers during mineralization of maize crop residues. Pedobiologia 62:1–8

Hoffmeister D, Keller NP (2007) Natural products of filamentous fungi: enzymes, genes and their regulation. Nat Prod Rep 24:393–416

Hooks CRR, Wang KH, Ploeg A, McSorley R (2010) Using marigold (*Tagetes* spp.) as a cover crop to protect crops from plant parasitic nematodes. Appl Soil Ecol 46:307–320

Hoorman JJ (2011) The role of soil protozoa and nematodes. Agriculture and Natural resources, Ohio State University, Columbus, pp 1–4

Howell DM, Kenzie DM (2017) Using bioavailable nutrients and microbial dynamics to assess soil type and placement depth in reclamation. Appl Soil Ecol 116:87–95

Hoyle FC, Baldock JA, Murphy DV (2011) Soil organic carbon-role in rainfed farming systems with particular reference to Australian conditions. Springer, Dordrecht

Hui D, Deng Q, Tian H, Luo Y (2016) Climate change and carbon sequestration in forest ecosystems. In: Suzuki T, Lackner M (eds) Handbook of climate change migration and application. Springer, Berlin, pp 555–594

Ingham ER (2009) Soil biology primer, chapter 4: soil fungus. Soil and Water Conservation Society, Ankeny, pp 22–23

IPCC (2006) IPCC guidelines for national greenhouse gas inventories. In: Egghleston HS, Bunendia L, Miwa K, Ngara T and Tanabe K (eds). Tokyo

Islam KR (2008) Lecture on soil physics, personal collection of K. Islam. Ohio State University, School of natural resources, Columbus, Ohio

Janga MR, Raoof MA, Ulaganthan K (2017) Effective biocontrol of *Fusarium* wilt in castor (*Ricinuscommunis* L.) with Bacillus sp., in pot experiment. Rhizosphere 3(1):50–52

Jeya K, Kiruthika K, Veerapagu M (2013) Isolation of antibiotic producing Streptomyces sp. from soil of Perambalur district and a study on the antibacterial activity against clinical pathogens. Int J Pharm Tech Res 5:1207–1211

Jimtha CJ, Jishma P, Shreelekha S, Chitra S, Radhekrishnan EK (2017) Antifungal properties prodigiosin producing rhizospheric serratia sp. Rhizosphere 3(1):105–108

Jorquera MA, Shaharoona B, Nadeem SM, de La Luz Mora M, Crowley DE (2012) Plant growth-promoting rhizobacteria associated with ancient clones of creosote bush (*Larrea tridentata*). Microb Ecol 64:1008–1017

Kaiser K, Kalbitz K (2012) Cycling downwards- dissolved organic matter in soils. Soil Biol Biochem 52:29–32

Kamble PN, Baath E (2016) Comparison of fungal and bacterial growth after alleviating induced N-limitation in soil. Soil Biol Biochem 103:97–105

Kar G, Hilger D, Schoenau JJ, Peak D (2017) Effect of plant growth and time on phosphorus speciation in a manure-amended prairie soil under controlled conditions. Rhizosphere. https://doi.org/10.1016/j.rhisph.2017.05.004

Kashaija IN, Mcintyre BD, Ssali H, Kizito F (2004) Spatial distribution of roots, nematode populations and root necrosis in highland banana in Uganda. Nematology 6:7–12

Kawasaki A, Watson ER, andKertesz MA (2012) Indirect effect of polycyclic aromatic hydrocarbon contamination on microbial communities in legume and grass rhizospheres. Plant Soil 358:169–182

Klumpp K, Fontaine S, Attard E, Le Roux X, Gleixner G, Soussana JF (2009) Grazing triggers soil carbon loss by altering plant roots and their control on soil microbial community. J Ecol 97:876–885

Koch B, Worm J, Jonson LE, Hojberg O, Nymbroe O (2001) Carbon limitation induces sigma (S)-dependent gene expression in *Pseudomonas fluorescens* in soil. Appl Environ Microbiol 67:3363–3370

Koele N, Turpanlt MP, Hildebrand EE, Uroz S, Klett EF (2009) Interaction between mycorrhizal fungi and mycorrhizosphere bacteria during mineral weathering. Soil Biol Biochem 9:1935–1942

Koenning SR, Overstreet C, Noling JW, Donald PA, Becker JO, Fortnum BA (1999) Survey of crop losses in response to phytoparasitic nematodes in the United States in 1999. Suppl J Nematol 31:587–618

Kogel KH, Franken P, Huckelhoven R (2006) Endophyte or parasite-what decides? Curr Opin Plant Biol 9:358–363

Kohout P, Tedersoo L (2017) Effect of soil moisture on root-associated fungal communities of *Erica dominans in* Drakensberg mountains in South Africa. Micorrhiza 27:1–10

Kowalchuk G, Buma DS, de Boer W, Klinkhamer PGL, van Veen JA (2002) Effects of above-ground plant species composition and diversity on the diversity of soil borne microorganisms. Anton Leeuw Int J Gen Mol Microbiol 81:509–520

Kuzyakov Y (2010) Primary effects: interactions between living and dead organic matter. Soil Biol Biochem 42(9):1363–1371

Lal R (1987) Tropical ecology and physical Edaphology. Wiley, New York

Lal R (2004) Soil carbon sequestration impacts on global climate change and global security. Science 403:1623–1627

Lambert DH, Baker DE, Cole Jr H (1979) The role of mycorrhizae in the interactions of phosphorus with zinc, copper, and other elements. Soil Sci Soc Amer 43:976–980

Latge JP (2017) Immune evasion: face changing in the fungal opera. Nat Microbiol 2:16266–16269

Lavelle P, Spain A (2001) Soil Ecology. Kluwer Academic Publishers, Dordrecht

Leckie SE, Prescott CE, Grayston SJ, Neufeld JD, Mohn WW (2004) Comparison of chloroform fumigation extraction, phospholipid fatty acid and DNA methods to determine microbial biomass in forest humus. Soil BiolBiochem 36:532–536

Leffelaar PA (1986) Dynamics of anaerobiosis, denitrification and water in a soil aggregate. Soil Sci 142:352–366

Lugtenberg B, Kamilova F (2009) Plant growth promoting rhizobacteria. Annu Rev Microbiol 63:541–558

Ma Z, Jacobson FE, Geidroc DP (2009) Coordination chemistry of bacterial metal transport and sensing. Chem Rev 109:4644–4681

MacFadyen A (1963) Soil Organisms. In: Doekson J, Drift JV (eds). North-Holland Public Co., Amsterdam

Manzoni S, Schimel JP, Porporato A (2012) Responses of soil microbial communities to water stress: results from a meta-analysis. Ecology 93(4):930–938

Mela F, Fritsche K, de Boer W, van Veen JA, de Graaff LH, van den Berg M, andLeveau JHJ (2011) Dual transcriptional profiling of a bacterial/fungal confrontation: *Collimonas fungivorans* versus *Aspergillus niger*. ISME J 5:1494–1504

Mendes R, Garbeva P, Raaijmakers JM (2013) The rhizosphere microbiome: significance of plant beneficial, plant pathogenic, and human pathogenic microorganisms. FEMS Microbiol Rev 37(5):634–663

Miransari M (2013) Soil microbes and the availability of soil nutrients. Acta Physiol Plant 35(11):3075–3084

Moore-Kucera J, Dick RP (2008) PLFA profiling of microbial community structure and seasonal shifts in soils of a Douglas- fir chronosequence. Microb Ecol 55(3):500–511

Neufeld JD, Mohn WW (2005) Unexpectedly high bacterial diversity in arctic tundra relative to boreal forest soils, revealed by serial analysis of ribosomal sequence tags. Appl Environ Microbiol 71:5710–5718

Newman MM, Hoilett N, Lorenz N, Dick RP, Liles MR, Ramsier C, Kloepper JW (2016) Glyphosate effects on soil rhizosphere-associated bacterial communities. Sci Total Environ 543:155–160

Nicolas V, Platsaki S, Basle A, Allen SJ, Peterson NG, Crombie AT, Deninson C (2015) A four helix bundle stores copper for methane oxidation. Nature 525:140–143

Oka Y, Shapira N, Fine P (2007) Control of root-knot nematodes in organic farming systems by organic amendments and soil solarization. Crop Prot 26:1556–1565

Panagos P, Borrell P, Meusburger K, Alewell C, Lugato E, Montanarella L (2015) Estimating the soil erosion cover-management factor at European scale. Land Use Policy 48:38–50

Pandey P, Chauhan D, Pandey B (2017) Isolation, identification and seasonal distribution of soil fungi in Achanakmar Bilaspur. Indian J Sci Res 13(2):51–56

Panikov NS, Flanagan PW, Oechel WC, Mastepanov MA, Christenson TR (2006) Microbial activity in soils frozen to below -39 degrees C. Soil BiolBiochem 38:3520

Parungao MM, Maceda EBG, Villano MAF (2007) Screening of antibiotic-producing actinomycetes from marine, brackish and terrestrial sediments of Samal Island. Philipp J Res Sci Comput Eng 4(3):29–38

Patrick WH Jr (1982) Nitrogen transformations in submerged soils. In: Stevenson FJ (ed) Nitrogen in agricultural soils, Agronomy series no. 22. American Society of Agronomy/Soil Science Society of America, Midison

Penman TD, Pike DA, Webb JK, Shine R (2010) Predicting the impact of climate change on Australia's most endangered snake, *Hoplocephalus bungaroides*. Diver Distribut 16:109–118

Pieterse CMJ (2012) Prime time for transgenerational defense. Plant Physiol 158:545

Poomthongdee N, Daungmal K, Panthom-aree W (2015) Acidophilic actinomycetes from rhizosphere soil: diversity and properties beneficial to plants. J Antibiot 68:106–114. https://doi.org/10.1111/1574-6976.12028

Qiang X, Weiss M, Kogel KH, Schafer P (2012) *Piriformospora indica and* mutualistic basidiomycete with an exceptionally large plant host range. Mol Plant Pathol 13:508–518

Rainey PB (1999) Adaptation of *Pseudomonas fluorescens* to the plant rhizosphere. Environ Microbiol 1:243–257

Raudales RE, Stone E, McSpadden Gardener BB (2009) Seed treatment with 2, 4 diacetylphloroglucinol-producing pseudomonads improves crop health in low pH soils by altering pattern of nutrients uptake. Phytology 99:506–511

Reader JS, Ordoukhanian PT, Kim JG, de Crecy-Lagard V, Hwang I, Farrand S, Schimmel P (2005) Major biocontrol of plant tumors targets tRNA synthetase. Science 309:1533

Romero-Olivares L, Allison SD, Treseder KK (2017) Soil microbes and their response to experimental warming over time: a meta-analysis of field studies. Soil Biol Biochem 107:32–40

Sanaullah M, Blagodatskaya E, Chabbi A, Rumpel C, Kuzyakov Y (2011) Drought effects on microbial biomass and enzyme activities in rhizosphere of grasses depend on plant community composition. Appl Soil Ecol 48:38–44

Schmidt MWI, Torn MS, Abivens S, Dittmar T, Guggenberger G, Janssens IA, Kleber M, Kögel-Knabner I, Lehmann J, Manning DA, Nannipieri P, Rasse DP, Weiner S, Trumbore SE (2011) Persistence of soil organic matter as an ecosystem property. Nature 478:49–56

Selvameenal L, Radhakrishnan M, Balagurunathan R (2009) Antibiotic pigment from desert soil actinomycetes; biological activity, purification and chemical screening. Indian J Pharm 71(5):499–504

Sharma CP (2006) Plant Micronutrients, 1st edn. Science, Enfield, pp 5–15

Sharma N, Sudarshan Y, Sharma R, Singh G (2008) RAPD analysis of soil microbial diversity in Western Rajasthan. Curr Sci 94(8):1058–1061

Shetty PR, Buddana SK, Tatipamula VB, Naga YVV, Ahmad J (2014) Production of polypeptide antibiotic from Streptomyces parvulus and its antibacterial activity. Braz J Microbiol 45(1):303–312

Siddikee M, Chauhan P, Anandham R, Han GH, Sa T (2010) Isolation, characterization and use for plant growth promotion under salt stress, of ACC deaminase-producing halotolerant bacteria derived from coastal soil. J Microbiol Biotechnol 20:1577–1584

Singh A, Bahel GS (1993) Phosphate equilibria in soils in relation to added phosphorus, Sesbania aculeate incorporation and cropping- a study of solubility relationship. J Indian Soc Soil Sci 41:233–237

Singh BK, Bardgett RD, Smith P, Dave S (2010) Microorganisms and climate change: Terrestrial feedbacks and mitigation options. Nat Rev Microbiol 8:779–790

Singh K, Pandey SN, Mishra A (2015) Preference of heavy metals accumulation, tolerance limit and biochemical responses of *Eichhornia crassipes* (Mart.) exposed to industrial waste water. Int J Curr Res 7(1):11818–11822

Smith KP, Goodman RM (1999) Host variation for interactions with beneficial plant-associated microbes. Annu Rev Phytopathol 37:473–491

Smith ML, Bruhn JN, Anderson JB (1992) The fungus Armillaria bulbosa is among the largest and oldest living organism. Nature 356:428–431

Stipanuk MH (1986) Metabolism of sulfur-containing amino acids. Annu Rev Nutr 6:179–209

Stotzky G (1997) Soil as an environment for microbial life. In: JDV E (ed) Modern soil microbiology. Trevors JT and Wellington/Marcel Dekker, New York, pp 20–28

Sturz AV, Kimpinski J (2004) Endoroot bacteria derived from marigold (*Tagetes* spp.) can decrease soil population densities of root-lesion nematodes in the potato root zone. Plant Soil 262:241–249

Subramani R, Aalbersberg W (2012) Marine actinomycetes: an ongoing source of novel bioactive metabolites. Microbiol Res 167:571–580

Sylvia DM, Hartel PG, FuhrmannJJ JDA (2005) In: David MS (ed) Principals and applications of microbiology, 2nd edn. Pearson Printice hall, Upper Saddle River

Thilakrathana MS, Raizada MN (2017) A meta-analysis of the effectiveness of diverse rhizobia inoculants on soybean traits under field conditions. Soil BiolBiochem 105:177–196

Tilman D, Dowing JA (1994) Biodiversity and stability in grasslands. Nature 367:363–365

Tongway D, Ludwig J (1996) Rehabilitation of semiarid landscapes in Australia, restoring productive soil patches. Restora Ecol 4:388–397

Tongway DJ, Ludwig JA (2005) Heterogeneity in arid and semi-arid lands. In: Lovett GM, Turner MG, Jones GG, Weathers KC (eds) Ecosystem function in heterogeneous landscapes. Springer, New York, pp 189–205

Torsvik V, Ovreas I (2002) Microbial diversity and function in soil: from genes to ecosystems. Curr Opin Microbiol 5:240–245

Unkovich (2003) Symbiotic nitrogen fixation and fertilizers in Australian agriculture. In: Proceedings of the 12th Australian nitrogen fixation conference, Glenelg

Upadhyay SK, Singh DP, Saikia R (2009) Genetic diversity of plant growth promoting rhizobacteria isolated from rhizospheric soil of wheat under saline condition. Curr Microbiol 59:489–496

Uphoff N, Ball AS, Fernandes E, Herren H, Husson O, Laing M, Palm C, Pretty J, Sanchez P, Sanginga N, Thies J (2006) Biological approaches to sustainable soil systems. CRC Press/ Francis & Taylor, London

Verma P, Pandey SN (2016) Effect of integrated nutrient management in alluvial soil on growth and biochemical responses of radish. J Biol Chem Res 33(1):34–39

Vyas RK, Mathus K (2002) Trichoderma spp. in cumin rhizosphere and their potential in suppression of wilt. Indian Phytopathol 55:455–457

Waldrop MP, Firestone MK (2006) Response of microbial community composition and function to soil climate change. MicrobEcol 52:716–724

Wall DH, Virginia RA (1999) Controls on soil biodiversity: insights from extreme environments. Appl Soil Ecol 13:137–150

Watt M, Kirkegaard JA, Passioura J (2006a) Rhizosphere biology and crop productivity- a review. Aust J Soil Res 44:299–317

Watt M, Silk W, Passioura J (2006b) Rates of root and organism growth, soil conditions and temporal and spatial development of the rhizosphere. Ann Bot 97:839–855

White PM, Charls WR (2009) Tillage effects on microbial and carbon dynamics during plant residue decomposition. Soil Sci Soc Am J 73:138–145. https://doi:10.2136/sssaj2007.0384

Wolters V (2001) Biodiversity of soil animal and its function. Eur J Soil Biol 37:221–227

Yadav A, Pandey SN (2015) Effect of integrated nutrient management on the growth, biochemical constituents and yield of tomato (*Lycopersicon esculentum* Mill.) J Biol Chem Research 32(2):835–838

Young IM, Crawford JW (2004) Interactions and self-organization in the soil-microbe complex. Science 304:1634–1637

Zhou X, Guo Z, Chen C, Jia Z (2017) Soil microbial community structure and diversity are largely influenced by soil pH and nutrient quality in 78-year-old tree plantations. Biogeosciences 14:2101–2111

Chapter 2
Harnessing the Plant Microbiome for Improved Abiotic Stress Tolerance

Syed Sarfraz Hussain, Samina Mehnaz, and Kadambot H.M. Siddique

Abstract The benefits of the green revolution in agriculture are over because current agricultural productivity has touched its limits of effectiveness in increasing plant yield. This problem is complicated by shrinking farmland, high labour costs and biotic and abiotic stresses. In fact, global agriculture and increased production would depend on the application and utilisation of microorganisms of agricultural importance, which will serve as an alternative strategy for higher crop productivity in the future. Efficient microbes play a key role in integrated management practices such as biotic and abiotic stresses and nutrient management to reduce chemical use and improve cultivar performance. On the other hand, high food demand and ever-increasing population increase pressure and urgency of how to exploit the microbiome for high crop yields and reduced losses caused by environmental stresses. This chapter highlights the importance of the designer plant microbiome, a strategy that may provide an effective and sustainable increase in crop yield and ultimately leads to food security by efficiently tackling biotic and abiotic stresses.

Keywords Microbiome • Plant nutrition • Abiotic stress

S.S. Hussain (✉)
School of Agriculture, Food & Wine, Waite Campus, University of Adelaide, Adelaide, SA, Australia

Department of Biological Sciences, Forman Christian college (A Chartered University), Lahore, Pakistan
e-mail: syedsarfraz_2000@yahoo.com

S. Mehnaz
Department of Biological Sciences, Forman Christian college (A Chartered University), Lahore, Pakistan

K.H.M. Siddique
The UWA Institute of Agriculture, The University of Western Australia, Crawley, WA, Australia

D. Egamberdieva, P. Ahmad (eds.), *Plant Microbiome: Stress Response*, Microorganisms for Sustainability 5, https://doi.org/10.1007/978-981-10-5514-0_2

2.1 Introduction: An Overview of Abiotic Stresses in Plants

A plethora of abiotic stresses affects crop plants including drought, extreme temperatures, salinity, nutrient deficiency and flooding which are expected to intensify due to climatic changes (Hussain et al. 2012; Timmusk et al. 2013, 2014; Rolli et al. 2015). Abiotic stresses represent a significant threat to agricultural productivity (Capell et al. 2004). A global water shortage due to significant climatic changes is the leading cause of these abiotic stresses. Drought is the most significant abiotic stress, adversely affecting the productivity and distribution of crop plants worldwide (Hussain et al. 2012; Marasco et al. 2013). For example, drought alone effects up to 45% of the global agricultural land, characterised by high human population (38%) with increased food demands (Bartels and Hussain 2008; Hussain et al. 2012). Plant growth and development face deleterious effects even with short-term water imbalance. Plants exhibit a plethora of responses at physiological, metabolic and molecular levels to survive or tolerate adverse conditions which include stomatal closure, increased aquaporin and H^+-pyrophosphatase activity and accumulation of a variety of compatible solutes (Bartels and Sunkar 2005; Shinozaki and Yamaguchi-Shinozaki 2007; Marasco et al. 2013; Hussain et al. 2016). Several studies have revealed overall effects of drought on plant growth and development; however, it is difficult to understand the damage caused to plants at the cellular and molecular levels under water deficit conditions (Zhu 2002; Chaitanya et al. 2003; Chaves and Oliveira 2004).

Salinity limits agricultural production in arid and semiarid areas, characterised by low annual precipitation, where agriculture is dependent on irrigation (Agrawal et al. 2013). Increased salt ion concentrations (such as Na^+ and Cl^-, but also others including Ca^{2+}, K^+, CO_3^{2-}, NO_3^-, SO_4^{2-}) in soil reduce water uptake by roots which ultimately results in the accumulation of toxic salt ions within plant cells (Tester and Davenport 2003). Plants have the ability to tackle this problem such as low Na+ concentration by actively maintaining translocation into vacuolar compartments via ATP-dependent ion pumps. However, excessive NaCl in irrigation water results in osmotic stress (Zhang et al. 2009a, b). Any misbalance in intracellular ion homeostasis leads to the damaging effects, for example, cell signalling pathways including those that lead to the synthesis of osmotically active metabolites, specific proteins, nutritional disorders, assimilation, membrane disorganisation, reduced cell division and expansion, genotoxicity and certain free radical scavenging enzymes that control ion and water flux (Zhang et al. 2009a, b) which ultimately impaired optimal growth and development.

Abiotic stresses such as drought, salinity and extreme temp are often interconnected and induce a similar set of plant responses by activating the same or interacting pathways (Shinozaki and Yamaguch-Shinozaki 2000; Seki et al. 2001, 2002; Kreps et al. 2002). A general response such as compatible solute accumulation and the synthesis of stress proteins and antioxidants at cellular level in many crop plants have been reported for all these stresses (Cushman and Bohnert 2000; Bartels and Sunkar 2005; Chinnusamy et al. 2005; Bartels and Hussain 2008; Hussain et al. 2011, 2012).

Efforts to enhance plant performance under abiotic stress have met with little success due to an incomplete understanding of the stress tolerance mechanisms in plants. Several groups have studied the complex mechanisms involved in stress response and adaptation—such as stress signalling, readjusting metabolism and reprogramming gene expression—to improve stress tolerance (Ma et al. 2011; Marasco et al. 2016; Thao and Tran 2016). However, for agricultural and environmental sustainability, the development of stress-tolerant plants is a viable approach, which seems imperative to fulfil the growing demands for quality food (Castiglioni et al. 2008). However, current breeding methods lack suitable methodological means to manage crop production in stress environment (Ashraf and Foolad 2007). In contrast, genetic engineering of crop plants can play a major role in developing stress-tolerant plants. Combining transgenic approaches with current breeding methods can be used to develop enhanced stress tolerance of crop plants (Capell et al. 2004). Current transgenic approaches aim to transfer to the target plant one or several different genes involved in several pathways including regulatory transcription factors, compatible solutes/osmoprotectants (proline, glycine betaine, polyamines) and proteins (LEA, heat shock, aquaporin) for generating stress-tolerant plants (Wang et al. 2003; Vinocur and Altman 2005; Valliyodan and Nguyen 2006; Bhatnagar-Mathur et al. 2007; Kathuria et al. 2007; Sreenivasulu et al. 2007; Marasco et al. 2016; Thao and Tran 2016). The bottleneck of transgenic approaches has been and continues to be the identification of key genes and their use in transgenic crops with improved stress tolerance without sacrificing yield (Bartels and Hussain 2008).

The last century has witnessed several significant, diverse and unexpected discoveries related to the plant-associated microbiome by molecular and omics tools combined with novel microscopic techniques (Mendes et al. 2011; Bulgarelli et al. 2012; Lundberg et al. 2012; Bhattacharyya et al. 2016; Berg et al. 2016; Timmusk et al. 2017). A wide range of agriculturally important microbiomes has been extensively exploited for increased growth and disease management in plants. It is expected that plant-associated beneficial microbiomes can significantly contribute to alleviating abiotic stresses using a variety of mechanisms (Hayat et al. 2010; Mapelli et al. 2013; Vejan et al. 2016). The sustainability of crop plants challenged by environmental stresses becomes more important and needs nonconventional solutions such as the use of microbiomes (Schaeppi and Bulgarelli 2015). Strengthening microbial traits beneficial to plants, the environment or both offers a promising avenue for the development of sustainable future agriculture. Microbial collection and utilisation can serve as a valuable tool and key determinants in managing plant health and productivity under an array of biotic and abiotic stresses (Celebi et al. 2010; Mengual et al. 2014; Rolli et al. 2015; Berg et al. 2016; Marasco et al. 2016). The identification, characterisation and utilisation of beneficial microbiomes which enhance abiotic stress tolerance in plants would help to sustain the next generation in agriculture worldwide (Jorquera et al. 2012; Nadeem et al. 2014). Diverse mechanisms which these microbes use to confer stress have been reviewed elsewhere (Lugtenberg and Kamilova 2009; Yang et al. 2009; Grover et al. 2010; Zelicourt et al. 2013; Nadeem et al. 2014). In this chapter, we will highlight advantages of the plant-associated microbiome approach, in particular, increasing plant tolerance to different abiotic stresses, which pose a serious threat to global crop productivity.

2.2 Exploring the Plant-Associated Microbiome for Improving Abiotic Stress Tolerance in Plants

2.2.1 Drought Stress

Recent data have revealed that the plant-associated microbiome can influence several plant traits including growth and biotic and abiotic stress tolerance (Mendes et al. 2011; Lau and Lennon 2012; Marasco et al. 2012, 2013; Bainard et al. 2013; Sugiyama et al. 2013; Berg et al. 2014; Rolli et al. 2015; Panke-Buisse et al. 2015). Drought stress represents a serious threat to agriculture worldwide. The contribution of the plant-associated microbiome to plant adaptation to drought stress is poorly understood. Rolli et al. (2015) tested in vivo eight isolates, over 510 strains, for their ability to support grapevine and *Arabidopsis* growth under drought stress; they demonstrated that plant growth-promoting activity is stress dependent and not a per se feature of the strains. Similarly, a pepper plant inoculated with selected strains under irrigated and drought conditions exhibited a stress-dependent plant growth-promoting pattern by increasing shoot and leaf biomass and shoot length and enhancing photosynthesis in drought-challenged grapevine, with a profound positive effect on drought-sensitive rootstock. Overall, these results indicate that the tested bacteria significantly contributed to plant adaptation to drought via stress-induced plant growth promotion. Certain PGPR, such as *Achromobacter piechaudii* ARV8, enhance drought stress tolerance in pepper and tomato by 1-aminocyclopro pane-1-carboxylate (ACC) deaminase. The mechanisms which render drought stress tolerance in plants remain largely speculative. However, it is possible that the breakdown of plant ACC by bacterial ACC will inhibit ethylene synthesis which ultimately reduces plant stress and enables normal plant growth (Glick et al. 2007; Arshad et al. 2008; Duan et al. 2009; Yang et al. 2009). Another study highlighted the positive influence of bacterial priming on wheat seedlings under drought stress (Timmusk et al. 2014, 2017); this method increased plant biomass by 78% and improved photosynthesis fivefold under severe drought. Furthermore, three of seven volatiles from bacterially primed drought-stressed wheat seedlings have been used to assess plant performance under drought stress in early stages of stress development (Timmusk et al. 2014). Wheat inoculated with *Burkholderia phytofirmans* PsJN had an increased photosynthesis, better water use efficiency, and high chlorophyll content and grain yield than the control under water deficit in the field conditions (Naveed et al. 2014a). Similarly, maize inoculated with both *B. phytofirmans* and *Enterobacter* sp. FD17 performed better compared to controls (Naveed et al. 2014b). Three bacterial strains isolated from extremely water-stressed soil, viz. *Pseudomonas putida*, *Pseudomonas* sp. and *Bacillus megaterium*, stimulated plant growth under drought conditions (Marulanda et al. 2009). Similarly, Sandhya et al. (2009) reported that inoculation of sunflower seedlings with *Pseudomonas* sp. strain GAP-45 enhanced survival and plant biomass under drought stress. It is possible that inoculated bacteria can efficiently colonise the root-adhering soil resulting in stable soil aggregates and ultimately enhanced stress tolerance. In a similar study, maize plants inoculated with *Pseudomonas* strain GAP-45 showed increased compatible solutes and antioxidant under water deficit conditions (Sandhya et al. 2010). In tomato, grapevine, olive and pepper plants, microbes isolated from roots of plants growing under extreme dry conditions improved the growth of another host species under similar growth conditions (Marasco et al. 2013). This stress-resistance solution strategy has the potential to save time, effort and costs. Kohler et al. (2008) inoculated lettuce with *Pseudomonas mendocina* and arbuscular mycorrhizal fungi (*Glomus intraradices* or *G. mosseae*) which resulted in antioxidant catalase activity under severe drought conditions pointing to possible use of microbes in alleviation of oxidative stress. Similarly, the accumulation of 14-3-3 protein along with glutathione and ascorbate has played important roles in maintaining plant metabolic functions and conferring protection under drought conditions. Lavender plants inoculated with *Glomus intraradices* and *Glomus* sp. strain accumulated these compounds and exhibited high drought tolerance by improving water contents, root biomass and N and P contents (Porcel et al. 2006; Marulanda et al. 2007). Plant growth-promoting bacteria have improved growth in sunflower, pea, sorghum, tomato, pepper, rice, common bean and lettuce under drought conditions (Alami et al. 2000; Creus et al. 2004; Mayak et al. 2004; Dodd et al. 2005; Cho et al. 2006; Marquez et al. 2007; Figueiredo et al. 2008; Arshad et al. 2008; Kohler et al. 2008; Sandhya et al. 2009; Kim et al. 2013; Perez-Montano et al. 2014; Marasco et al. 2016).

2.2.2 *Salinity Stress*

Extreme climatic conditions and the misuse of agricultural land over the past few decades have led to high salinity, which is a limiting factor to global crop productivity (Wicke et al. 2011). Several approaches, in addition to molecular technologies, have been implicated for addressing salinity such as soil reclamation and management practices. However, these methods are expensive and not always practical and sustainable for controlling salinity. In contrast, the use of natural plant growth-promoting bacteria as inoculants for crop plants growing on salt-affected land is gaining momentum (Tiwari et al. 2011; Shabala et al. 2013; Paul and Lade 2014; Qin et al. 2014; Ruiz et al. 2015). A growing body of research has shown that microbial communities increase productivity and improve plant health following adverse environmental stresses (Berendsen et al. 2012; Zuppinger-Dingley et al. 2014; Sloan and Lebeis 2015).

It is proposed that microbes inhabiting sites exposed to frequent stress conditions develop adaptive tolerant traits and are potential candidates as plant growth promoters under stress conditions (Yang et al. 2016b). Halotolerant microbes thrive under soil salinity stress and express traits to help plants to survive high salinity. Upadhyay et al. (2009) isolated 130 rhizobacterial strains from wheat plants sown under saline conditions and showed that 24 isolates tolerated relatively high levels (8%) of NaCl stress. The authors attributed this tolerance to different genes, hormones and proteins such as *nif*H, indole-3-acetic acid (IAA), siderophores and gibberellin. Similarly, halotolerant bacterial strains isolated from Korea enhanced plant growth under salinity stress by reducing ethylene production via ACC deaminase activity (Siddikee et al. 2010). The availability of new halotolerant diazotrophic bacteria, with traits such as IAA, phosphorus solubilisation and ACC deaminase activity, isolated from roots of *Salicornia brachiate* (extreme halophyte) represents other potential candidates (Jha et al. 2012). Arora et al. (2014) demonstrated that 17 of 20 bacteria isolated from halophytes and other salt-tolerant plant species happily grew in 7.5% NaCl in culture and two of these grew in 10% NaCl. Plant-associated microbiomes have improved growth in canola, pepper, tomato, bean, wheat and lettuce (Yildirim and Taylor 2005; Barassi et al. 2006; Upadhyay et al. 2009; Ali et al. 2014; Leite et al. 2014; Zhao et al. 2016).

There are reports that the involvement of arbuscular mycorrhizal fungi (AMF) has increased host plant tolerance to salinity stress. Co-inoculation of AMF plants with *Glomus* sp. has increased growth in saline soils possibly due to increased phosphate and decreased Na^+ concentration in shoots compared to uninoculated controls (Giri and Mukerji 2004). AMF treatment has improved salt tolerance in maize, mungbean, clover, tomato and cucumber due to P acquisition, improved osmoregulation by proline accumulation and reduced NaCl concentration (Jindal et al. 1993; Al-Karaki et al. 2001; Feng et al. 2002; Ben Khaled et al. 2003; Grover et al. 2010; Velazquez-Hernandez et al. 2011). However, research on the ability of bacterial and AM species to induce protective proteins and osmoprotectants is needed. The above reports suggest that plants under stress may readily recruit diverse bacterial strains with broad implications for plants grown under salt stress. This phenomenon has been collectively termed induced systemic tolerance (Yang et al. 2009).

2.2.3 Extreme Temperature Stress (Low and High)

The Intergovernmental Panel on Climate Change (IPCC: 2007) reported that global temperatures are predicted to increase by 1.8–3.6 °C by the end of this century due to climate changes. High temperatures are a major obstacle in crop production as well as microbial colonisation, which results in major cellular damage such as protein degradation and aggregation. All organisms respond to high temperature by producing a specific group of polypeptides known as heat shock proteins (HSPs). Stress adaptation in microorganisms represents a complex multilevel regulatory process that may involve several genes (Srivastava et al. 2008), such that microbes develop different adaptation strategies to combat the stress. Certain microbes perform better at high temperatures, and these microbes may be important for crop plants under high temperature (Yang et al. 2016a). Srivastava et al. (2008) isolated *P. putida* strain NBR10987, which exhibited thermotolerance in the drought-stressed rhizosphere of chickpea and was attributed to the stress sigma factor δs overexpression and thick biofilm formation. Certain bacterial strains combat stress by producing exopolysaccharides (EPS) which possess unique water holding and cementing characteristics and play vital roles in stress tolerance by water retention and biofilm formation. Sorghum seedlings inoculated with *Pseudomonas* AKM-P6 strain had improved tolerance to heat stress through enhanced physiological and metabolic performance indicating a unique interaction of inducible proteins in heat tolerance using microbes (Ali et al. 2009).

Low-temperature stress is an important limiting factor to crop productivity because it adversely affects plant growth and development. Grapevines inoculated with *B. phytofirmans* PsJN increased tolerance to low nonfreezing temperatures and resistance to grey mould. Similarly, endophyte inoculation resulted in higher and faster accumulation of stress-related proteins and metabolites, which lead to more effective resistance to low temperature, indicating a positive priming effect on plants (Theocharis et al. 2012). Similarly, Barka et al. (2006) noted that grapevine roots inoculated with *B. phytofirmans* PsJN resulted in better root growth, higher plant biomass and increased physiological activity at low temperature (4 °C). Further analysis revealed that bacterised plantlets significantly increased proline, starch and phenolic levels compared with uninoculated control plantlets, which enhanced grapevine plantlets to tolerate low temperature. Low temperature usually inhibits soybean symbiotic activities (nodule infection and nitrogen fixation), but inoculation of soybean with both *Bradyrhizobium japonicum* and *Serratia proteamaculans* resulted in faster growth at 15 °C (Zhang et al. 1995, 1996). Switchgrass inoculated with *B. phytofirmans* PsJN had enhanced growth under glasshouse conditions (Kim et al. 2012). According to Mishra et al. (2009), wheat seedlings inoculated with *Pseudomonas* sp. strain PPERs23 highly improved root and shoot lengths resulting in dry root/shoot biomass and total phenolics, chlorophyll and amino acid contents. Furthermore, inoculated wheat seedlings had enhanced physiologically available iron, anthocyanins, proline, protein and relative water contents and reduced Na^+/K^+ ratio and electrolyte leakage, resulting in enhanced cold tolerance (Mishra et al. 2009). Many studies have explored several bacterial strains for enhanced cold stress tolerance in plants (Selvakumar et al. 2008a, b, 2009, 2010a, b). It is apparent from the above studies that *B. phytofirmans* PsJN has a wide host spectrum, which includes grapevines, maize, soybean, sorghum, wheat and switchgrass with promising results under different abiotic stresses.

2.2.4 Heavy Metal Stress

Heavy metal contamination due to increased industrialisation has recently received attention because heavy metals cannot be degraded (Kidd et al. 2009; Ma et al. 2011; Rajkumar et al. 2012). Various physiochemical and biological techniques developed to remove contaminants have failed due to being expensive, environmentally unsafe and unacceptable by the public (Boopathy 2000; Vidali 2001; Doble and Kumar 2005). Phytoremediation using plants to eliminate soil contaminants is cost-effective and environmentally friendly with high public acceptance technology (Hadi and Bano 2010; Beskoski et al. 2011; Fester et al. 2014; Arslan et al. 2015). Another viable and promising alternative is the application of plant-associated microbiomes whereby microbial activities in the rhizosphere increase plant metal uptake by several ways like altering mobility and bioavailability of metals (Rajkumar et al. 2010; Ma et al. 2011; Aafi et al. 2012; Yang et al. 2012). Several plant growth-promoting substances, such as plant growth hormones (IAA, cytokinins and gibberellins), siderophores and ACC deaminase, are produced by plant-associated microbiomes to improve plant growth in heavy metal-contaminated soils (Babu and Reddy 2011; Luo et al. 2011, 2012; Wang et al. 2011; Bisht et al. 2014; Kukla et al. 2014; Waqas et al. 2015; Ijaz et al. 2016; Santoyo et al. 2016). High soil contamination could reduce plant growth including root growth and expansion mainly due to oxidative stress, which limits the rate of phytoremediation (Gerhardt et al. 2009; Hu et al. 2016). The lack of nutrients and reduced microbial density also limit phytoremediation (Gerhardt et al. 2009). Common heavy metals include manganese (Mn), cadmium (Cd), lead (Pb), chromium (Cr), zinc (Zn), aluminium (Al) and copper (Cu). Some metalloids also show toxicity such as antimony (Sb) and arsenic (As) (Duruibe et al. 2007; Park 2010; Wuana and Okieimen 2011; Pandey 2012).

Rhizosphere bacteria deserves close attention among the microbes involved in phytoremediation (Arora et al. 2005) as these can directly improve process efficiency by altering soil pH and oxidation/reduction reactions (Khan et al. 2009; Kidd et al. 2009; Uroz et al. 2009; Wenzel 2009; Rajkumar et al. 2010; Afzal et al. 2011; Ma et al. 2011). *Microbacterium* sp. G16 and *Pseudomonas fluorescens* G10 significantly increased the solubility of lead (Pb) in *Brassica napus* compared with uninoculated controls and were mainly attributed to IAA, siderophores, ACC deaminase and phosphate solubilisation (Sheng et al. 2008). Similarly, co-inoculation of *Zea mays* with *Azotobacter chrococcum* or *Rhizobium leguminosarum* improved plant growth and biomass in Pb-contaminated soil (Hadi and Bano 2010; Hussain et al. 2013). Several endophyte genera like *Bacillus* sp., *Serratia*, *Enterobacter*, *Burkholderia* sp., *Agrobacterium* and others have increased the phytoremediation rate and biomass production in metal-contaminated soils (Wani et al. 2008; Kumar et al. 2009; Mastretta et al. 2009; Luo et al. 2012; Nonnoi et al. 2012; Afzal et al. 2014; Glick 2014, 2015; Hardoim et al. 2015; Ijaz et al. 2016; Singh et al. 2016; Zheng et al. 2016; Feng et al. 2017).

Moreover, mycorrhizal fungi play significant role in phytoremediation due to hyperaccumulators of heavy metals with heavy metal tolerance (Zarei et al. 2010; Orłowska et al. 2011).

2.2.5 Nutrient Deficiency Stresses

Beneficial microbes can be used to enhance the sustainability of current agricultural systems. Members of the rhizosphere microbiome are playing significant roles in plant nutrient management (Adhya et al. 2015). Well-known examples include nitrogen-fixing rhizobia and mycorrhizal fungi involved in phosphorus uptake (Hawkins et al. 2000; Richardson et al. 2009; Miransari 2011). Plants usually get nutrients from the rhizosphere and from the phyllosphere (Turner et al. 2013). Plant nutrient management requires optimal use of soil, water, atmospheric factors and NPK fertilisers (Miao et al. 2011), along with a beneficial microbiome to help improve nutrient use efficiency. A plethora of research is available on the usefulness of symbionts such as mycorrhizal fungi for channelling nutrients and minerals such as phosphorus, water and other essential macro- and microelements from soil to growing plants (Gianinazzi et al. 2010; Adeleke et al. 2012; Johnson and Graham 2013; Salvioli et al. 2016) and for modelling and improved soil structure and aggregates (Miller and Jastrow 2000) in crops such as cereals, pulses, fruits and oilseeds to meet their nutritional requirements (Jeffries and Barea 2001; Johnson et al. 2012; Salvioli and Bonfante 2013). Apart from *Rhizobium* and *Bradyrhizobium*, several other bacterial endophytes have been reported to establish symbiosis with plants for bioavailable nitrogen fixation in unspecialised host tissues even in the absence of nodules (Zehr et al. 2003; Gaby and Buckley 2011; Guimaraes et al. 2012; Santi et al. 2013). For example, *Cyanobacteria* are in symbiotic association with a range of plants from different clads, such as gunnera, cycads and lichens, and form heterocysts suitable for biological nitrogen fixation (BNF) with nitrogenase (Berman-Frank et al. 2003; Santi et al. 2013). Another study revealed that 74 and 77 of 102 bacteria associated with sugarcane roots successfully fix nitrogen and solubilise phosphorus, respectively (Leite et al. 2014). Similarly, analysis of the cowpea rhizosphere using 16S rRNA sequencing revealed that *Burkholderia* and *Achromobacter* species along with *Rhizobium* and *Bradyrhizobium* can nodulate cowpea and support BNF (Guimaraes et al. 2012). Some reports have indicated that algal genera such as *Anabaena*, *Aphanocapsa* and *Phormidium* can fix atmospheric nitrogen in paddy fields (Shridhar 2012; Hasan 2013).

Considering the importance of essential plant nutrients, it would be logical to discover bacterial species that affect macro- and micronutrient uptake in plant species under different deficient and toxic conditions (Leveau et al. 2010; Mapelli et al. 2012). Microbiomes can also facilitate the uptake of several trace elements such as iron (Zhang et al. 2009a, b; Marschner et al. 2011; Shirley et al. 2011) and calcium (Lee et al. 2010). Collectively, members of the plant microbiome play essential roles in degrading organic compounds which are required not only for their survival but also for plant growth in nutrient-poor soils (Leveau et al. 2010; Mapelli et al. 2012; Turner et al. 2013; Bhattacharyya et al. 2015).

2.2.6 *Establishing a Functional Plant Microbiome in Agriculture*

It is important to understand microbe–microbe and plant–microbe interactions to generate/develop a beneficial soil microbiome. However, it is unknown whether such beneficial microbial communities would be stable in agricultural soils. Under natural conditions, two factors, i.e. soil type and plant roots, usually determine the composition and association of microbial communities with plant roots. The influence of soil type and plant roots on the rhizomicrobiome has been reviewed extensively (Berg and Smalla 2009; Philippot et al. 2013; Bulgarelli et al. 2013, 2015; Lareen et al. 2016). Physiochemical properties of soils have a direct influence on specific microbes and plant root exudates (Hamel et al. 2005; Dumbrell et al. 2010) whereby soil type mostly determines the soil biome and plant root exudates tend to establish a favourable rhizobiome. Collectively, soil type and plant species are important players which determine the composition of rhizosphere and recruit diverse microbial communities for the establishment of a favourable rhizobiome to increase crop yields and reduce losses to biotic and abiotic stresses (Bulgarelli et al. 2012, 2015; Peiffer et al. 2013; Philippot et al. 2013; Schlaeppi et al. 2014; Tkacz et al. 2015; Lebeis et al. 2015; Yeoh et al. 2016). These factors significantly contribute to the selective enrichment of beneficial microbes in the rhizobiome, which may help to identify heritable traits to improve plant health and productivity (Tkacz and Poole 2015). Consequently, this mechanistic approach has the potential to create a microbiome that can improve plant traits following species or genotype-driven selection in the composition of rhizobiome structure as revealed in maize, barley, potato, *Arabidopsis*, *Brassica rapa* and sugarcane (Rasche et al. 2006; Bulgarelli et al. 2012, 2015; Lundberg et al. 2012; Peiffer et al. 2013; Lebeis et al. 2015; Panke-Buisse et al. 2015; Yeoh et al. 2016).

2.3 Customised Adjustment of the Plant Microbiome: A Revolution in Progress

Recent studies have highlighted the potential of customised or synthetic microbial communities to reap maximum benefits in crop production in terms of plant growth, yield and resistance to abiotic and biotic challenges (Mendes et al. 2011; Lau and Lennon 2012; Berendsen et al. 2012; Bainard et al. 2013; Bulgarelli et al. 2015; Lebeis 2015). Using the plant microbiome in crop production is not a new concept. The plant microbiome is a key determinant of plant health and productivity (Berendsen et al. 2012; Ziegler et al. 2013; Chaparro et al. 2014) and has received considerable attention in recent years (Lebeis et al. 2012; Bulgarelli et al. 2013). Manipulation of the plant microbiome can increase tolerance to biotic and abiotic stresses (Barka et al. 2006; Jha et al. 2012; Jorquera et al. 2012; Berg et al. 2013), increase agricultural production (Yang et al. 2009; Bakker et al. 2012; Turner et al. 2013), reduce chemical inputs (Adesemoye et al. 2009; Adesemoye and Egamberdieva 2013; Jha et al. 2015) and reduce greenhouse gas emissions (Singh et al. 2010), resulting in more sustainable agricultural productivity. This is vital for sustaining the ever-growing global population. Furthermore, identified naturally occurring beneficial microbes are now being used in agriculture for significant improvement of crop plant performance (Zolla et al. 2013; Nadeem et al. 2014).

Despite the fact that the richness of species and diversity of microbial communities recruited in plant microbiomes are mostly unknown, assembling a specific trait-associated microbiome is critical into new plant hosts for the development of improved production systems. There is ample evidence that many molecules, microbes, plant species and mechanisms support the establishment of a rhizobiome with the potential to play significant roles in enhanced plant productivity in the future (Berendsen et al. 2012; Miller and Oldroyd 2012; Bakker et al. 2013; Oldroyd 2013; Qiu et al. 2014; Zhang et al. 2015). Some strategies have been worked out to reshape the rhizobiome and redirect microbial activity by bringing about change in root exudates using conventional and modern breeding approaches (Bakker et al. 2012). Efforts to develop PGPB and/or PGPF consortia by mimicking or partially reconstructing the plant microbiome/rhizobiome are in progress. Tomato plants inoculated with these PGP consortia (*Bacillus amyloliquefaciens* IN937a, *Bacillus pumilus* T4, AMF *Glomus intraradices*) in greenhouse conditions resulted in full yield with 30% fewer inputs (Adesemoye et al. 2009). Similarly, Atieno et al. (2012) reported increased biomass in two soybean cultivars after inoculation with *B. japonicum* 532C, RCR3407 and *B. subtilus* MIB600. In another study, co-inoculation of soybean with *B. japonicum* E109 and *Bacillus amyloliquefaciens* LL2012 indirectly improved soybean nodulation efficiency. Phytohormones produced by *Bacillus amyloliquefaciens* LL2012 helped to improve nodulation efficiency in *B. japonicum* E109 (Masciarelli et al. 2014). Mengual et al. (2014) employed a consortium of *B. megaterium*, *Enterobacter* sp., *Bacillus thuringiensis* and *Bacillus* sp. along with composted sugar beet residues on *Lavandula dentata* L. to help restore soils by increasing phosphorus bioavailability, soil nitrogen fixation and foliar NPK contents. Hence, the success of a rational design of a plant microbiome depends on several factors including smart integration of all players in the system. In this context, genetic diversity of the local soil microbiome can help to improve and stabilise the effects of microbial inoculants. Therefore, it is recommended that microbiome profiling be implemented for the determination, monitoring and targeted application of microbial inoculants under field conditions.

2.4 Conclusion and Future Perspectives

The growing body of research relating to the plant microbiome is bringing into focus its importance for plant health, growth and productivity. While most research findings are preliminary, intensive research is required to unravel the intricacies of this highly complex phenomenon to understand microbe community dynamics and communication to exploit this largely untapped resource. Opportunities for exploiting the plant microbiome for raising crops are numerous and diverse. Plant-associated microbes would play a significant role in stress management in plants and provide excellent models for understanding stress tolerance mechanisms. Another strategy would be to generate transgenic plants harbouring beneficial genes from microbes, similar to transgenic plants harbouring ACC deaminase gene from bacteria. However, considering the timeframe and other issues involved in the development of transgenic plants, it would be more cost-effective and environmentally friendly to develop easy-to-handle microbial inoculants to alleviate abiotic stresses.

While several studies have shown significant improvements to stress tolerance using PGPM to crops under field conditions (Celebi et al. 2010; Mengual et al. 2014; Rolli et al. 2015), others have revealed inconsistent or negative results (Nadeem et al. 2014). One promising strategy for a stable beneficial outcome is to use a microbial consortium in the field to tailor the rhizobiome to respond to specific biotic and abiotic stresses without compromising plant growth and productivity (Trabelsi and Mhamdi 2013). Therefore, the mechanisms by which microbes confer stress tolerance to their hosts need further research to develop suitable microbial consortia for ready-to-use formulations under different biotic and abiotic stresses. However, this will require concerted efforts at interdisciplinary levels from microbiologists, molecular biologists, plant physiologists, plant breeders, soil scientists and agronomists. Recent developments in this field provide opportunities to understand how the microbe–microbe and plant–microbe interactions mediate the functional relationship between different players.

References

Aafi NE, Brhada F, Dary M, Maltouf AF, Pajuelo E (2012) Rhizostabilization of metals in soils using Lupinus luteus inoculated with the metal resistant rhizobacterium Serratia sp. MSMC 541. Int J Phytoremediation 14:26174

Adeleke RA, Cloete TE, Bertrand A, Khasa DP (2012) Iron ore weathering potentials of ectomycorrhizal plants. Mycorrhiza 22:535–544

Adesemoye AO, Egamberdieva D (2013) Beneficial effects of plant growth-promoting rhizobacteria on improved crop production: prospects for developing economies. In: Maheshwari DK, Saraf M, Aeron A (eds) Bacteria in agrobiology: crop productivity. Springer, Berlin, pp 45–63

Adesemoye AO, Torbert HA, Kloepper JW (2009) Plant growth promoting rhizobacteria allow reduced application rates of chemical fertilizers. Microb Ecol 58:921–929

Adhya TK, Kumar N, Reddy G, Podile AR, Bee H, Bindiya S (2015) Microbial mobilization of soil phosphorus and sustainable P management in agricultural soils. Curr Sci 108:1280–1287

Afzal M, Yousaf S, Reichenauer TG, Kuffner M, Sessitsch A (2011) Soil type affects plant colonization, activity and catabolic gene expression of inoculated bacterial strains during phytoremediation of diesel. J Hazard Mater 186:1568–1575

Afzal M, Khan QM, Sessitsch A (2014) Endophytic bacteria: prospects and applications for the phytoremediation of organic pollutants. Chemosphere 117:232–242

Agrawal AA, Johnson MTJ, Hastings AP, Maron JL (2013) A field experiment demonstrating plant life-history evolution and its co-evolutionary feedback to seed predator populations. Am Nat 181:S35–S45

Alami Y, Achouak W, Marol C, Heulin T (2000) Rhizosphere soil aggregation and plant growth promotion of sunflowers by exopolysaccharide producing *Rhizobium* sp. strain isolated from sunflower roots. Appl Environ Microbiol 66:3393–3398

Ali Sk Z, Sandhya V, Grover M, Kishore N, Rao LV, Venkateswarlu B (2009) *Pseudomonas* sp. strain AKM-P6 enhances tolerance of sorghum seedlings to elevated temperatures. Biol Fertil Soils 46:45–55

Ali S, Duan J, Charles TC, Glick BR (2014) A bioinformatics approach to the determination of genes involved in endophytic behavior in *Burkholderia* sp. J Theor Biol 343:193–198

Al-Karaki GN, Ammad R, Rusan M (2001) Response of two tomato cultivars differing in salt tolerance to inoculation with mycorrhizal fungi under salt stress. Mycorrhiza 11:43–47

Arora DK, Saikia R, Dwievdi R, Smith D (2005) Current status, strategy and future prospects of microbial resource collections. Curr Sci 89:488–495

Arora S, Patel PN, Vanza MJ, Rao GG (2014) Isolation and characterization of endophytic bacteria colonizing halophyte and other salt tolerant plant species from coastal Gujarat. Afr J Microbiol Res 8:1779–1788

Arshad M, Sharoona B, Mahmood T (2008) Inoculation with Pseudomonas spp. containing ACC deaminase partially eliminate the effects of drought stress on growth, yield and ripening of pea (*Pisum sativum* L.) Pedosphere 18:611–620

Arslan M, Imran A, Khan QM, Afzal M (2015) Plant–bacteria partnerships for the remediation of persistent organic pollutants. Environ Sci Pollut Res:1–15

Ashraf M, Foolad MR (2007) Roles of glycine betaine and proline in improving plant abiotic stress resistance. Environ Exp Bot 59:206–216

Atieno M, Herrmann L, Okalebo R, Lesueur D (2012) Efficiency of different formulations of *Bradyrhizobium japonicum* and effect of co-inoculation of *Bacillus subtilis* with two different strains of *Bradyrhizobium japonicum*. World J Microbol Biotechnol 28:2541–2550

Babu AG, Reddy S (2011) Dual inoculation of arbuscular mycorrhizal and phosphate solubilizing fungi contributes in maintenance of plant health in fly ash ponds. Water Air Soil Pollut 219:3–10

Bainard LD, Koch AM, Gordon AM, Klironomos JN (2013) Growth response of crops to soil microbial communities from conventional monocropping and tree-based intercropping systems. Plant Soil 363:345–356

Bakker MG, Manter DK, Sheflin AM, Weir TL, Vivanco JM (2012) Harnessing the rhizosphere microbiome through plant breeding and agricultural management. Plant Soil 360:1–13

Bakker PA, Berendsen RL, Doornbos RF, Wintermans PC, Pieterse CM (2013) The rhizosphere is revisited: root microbiomics. Front Plant Sci 4:165

Barassi CA, Ayrault G, Creus CM, Sueldo RJ, Sobero MT (2006) Seed inoculation with *Azospirillum* mitigates NaCl effects on lettuce. Sci Hortic (Amsterdam) 109:8–14

Barka EA, Nowak J, Clement C (2006) Enhancement of chilling resistance of inoculated grapevine plantlets with a plant growth-promoting rhizobacterium, *Burkholderia phytofirmans* strain PsJN. Appl Environ Microbiol 72:7246–7252

Bartels D, Hussain SS (2008) Current status and implications of engineering drought tolerance in plants using transgenic approaches. CAB Rev Persp Agr Vet Sci Nutri Natu Sci 3:020

Bartels D, Sunkar R (2005) Drought and salt tolerance in plants. Crit Rev Plant Sci 24:23–58

Ben Khaled L, Gomez AM, Ourraqi EM, Oihabi A (2003) Physiological and biochemical responses to salt stress of mycorrhized and/or nodulated clover seedlings (*Trifolium alexandrinum* L.) Agronomie 23:571–580

Berendsen RL, Pieterse CMJ, Bakker P (2012) The rhizosphere microbiome and plant health. Trends Plant Sci 17:478–486

Berg G, Smalla K (2009) Plant species and soil type cooperatively shape the structure and function of microbial communities in the rhizosphere. FEMS Microbiol Ecol 68:1–13

Berg G, Zachow C, Müller H, Philipps J, Tilcher R (2013) Next-generation bio-products sowing the seeds of success for sustainable agriculture. Agronomy 3:648–656
Berg G, Grube M, Scholter M, Smalla K (2014) Unraveling the plant microbiome: looking back and future perspectives. Front Microbiol 5:148
Berg G, Rybakova D, Grube M, Koberl M (2016) The plant microbiome explored: implications for experimental botany. J Exp Bot 67:995–1002
Berman-Frank I, Lundgren P, Falkowski P (2003) Nitrogen fixation and photosynthetic oxygen evolution in *Cyanobacteria*. Res Microbiol 154:157–164
Beskoski VP, Gojgic-Cvijovic G, Milic J, Ilic M, Miletic S, Solevic T, Vrvic MM (2011) Ex situ bioremediation of a soil contaminated by mazut (heavy residual fuel oil), a field experiment. Chemosphere 83:34–40
Bhatnagar-Mathur P, Vadez V, Shrama KK (2007) Transgenic approaches for abiotic stress tolerance in plants: retrospect and prospects. Plant Cell Rep 27:411–424
Bhattacharyya PN, Sarmah SR, Dutta P, Tanti AJ (2015) Emergence in mapping microbial diversity in tea (*Camellia sinensis* L.) soil of Assam, North-East India: a novel approach. Eur J Biotechnol Biosci 3:20–25
Bhattacharyya PN, Goswani MP, Bhattacharyya LH (2016) Perspective of beneficial microbes in agriculture under changing climatic scenario: a review. J Phytol 8:26–41
Bisht S, Pandey P, Kaur G, Aggarwal H, Sood A, Sharma S, Kumar V, Bisht NS (2014) Utilization of endophytic strain Bacillus sp. SBER3 for biodegradation of polyaromatic hydrocarbons (PAH) in soil model system. Eur J Soil Biol 60:67–76
Boopathy R (2000) Factors limiting bioremediation technologies. Bioresour Technol 74:63–67
Bulgarelli D, Rott M, Schlaeppi K, Ver L, van Themaat Ahmadinejad N, Assenza F, Rauf P, Huettel B, Reinhardt R, Schmelzer E, Peplies J, Gloeckner FO, Amann R, Eickhorst T, Schulze-Lefert P (2012) Revealing structure and assembly cues for *Arabidopsis* root-inhabiting bacterial microbiota. Nature 488:91–95
Bulgarelli D, Schlaeppi K, Spaepen S, Ver Loren van Themaat E, Schulze-Lefert P (2013) Structure and functions of the bacterial microbiota of plants. Annu Rev Plant Biol 64:807–838
Bulgarelli D, Garrido-Oter R, Munch PC, Weiman A, Droge J, Pan Y, McHardy AC, Schulze-Lefert P (2015) Structure and function of the bacterial root microbiota in wild and domesticated barley. Cell Host Microbe 17:392–403
Capell T, Bassie L, Christou P (2004) Modulation of the polyamine biosynthetic pathway in transgenic rice confers tolerance to drought stress. Proc Natl Acad Sci U S A 101:9909–9914
Castiglioni P, Warner D, Bensen RJ, Anstrom DC, Harrison J, Stoecker M, Abad M, Kumar G, Salvador S, D'ordine R, Navarro S, Back S, Fernandes M, Targolli J, Dasgupta S, Bonin C (2008) Bacterial RNA chaperones confer abiotic stress tolerance in plants and improved grain yield in maize under water limited conditions. Plant Physiol 147:446–455
Celebi SZ, Demir S, Celebi R, Durak ED, Yilmaz IH (2010) The effect of arbuscular mycorrhizal fungi (AMF) applications on the silage maize (*Zea mays* L.) yield in different irrigation regimes. Eur J Soil Biol 46:302–305
Chaitanya KV, Sundar D, Jutur PP, Reddy RA (2003) Water stress effects on photosynthesis in different mulberry cultivars. Plant Growth Regul 40:75–80
Chaparro JM, Badri DV, Vivanco JM (2014) Rhizosphere microbiome assemblage is affected by plant development. ISMI J 8:790–803
Chaves MM, Oliveira MM (2004) Mechanisms underlying plant resistance to water deficits: prospects for water-saving agriculture. J Exp Bot 55:2365–2384
Chinnusamy V, Jagendorf A, Zhu JK (2005) Understanding and improving salt tolerance in plants. Crop Sci 45:437–438
Cho K, Toler H, Lee J, Ownley B, Stutz JC, Moore JL, Auge RM (2006) Mycorrhizal symbiosis and response of sorghum plants to combined drought and salinity stresses. J Plant Physiol 163:517–528
Creus CM, Sueldo RJ, Barassi CA (2004) Water relations and yield in *Azospirillum*-inoculated wheat exposed to drought in the field. Can J Bot 82(2):273–281

Cushman JC, Bohnert HJ (2000) Genomic approaches to plant stress tolerance. Curr Opin Plant Biol 3:117–124

de Zelicourt A, Al-Yousif M, Hirt H (2013) Rhizosphere microbes as essential partners for plant stress tolerance. Mol Plant 6:242–245

Doble M, Kumar A (2005) Biotreatment of industrial effluents. Elsevier, Butterworth-Heinemann, Oxford, pp 1–5

Dodd IC, Belimov AA, Sobeih WY, Safronova VI, Grierson D, Davies WJ (2005) Will modifying plant ethylene status improve plant productivity in water-limited environments? 4th international crop science congress. http://www.cropscience.org.au/icsc2004/poster/1/3/4/510_doddicref.htm. Accessed 17 June 2007

Duan J, Muller KM, Charles TC, Vesley S, Glick BR (2009) 1-aminocyclopropane-1-carboxylate (ACC) deaminase genes in rhizobia from southern Saskatchewan. Microb Ecol 57:423

Dumbrell AJ, Nelson M, Helgason T, Dytham C, Fitter AH (2010) Relative roles of niche and neutral processes in structuring a soil microbial community. ISME J 4:337–345

Duruibe JO, Ogwuegbu MOC, Egwurugwu JN (2007) Heavy metal pollution and human biotoxic effects. Int J Phys Sci 2:112–118

Feng G, Zhang FS, Li XL, Tian CY, Tang C, Renegal Z (2002) Improved tolerance of maize plants to salt stress by *arbuscular mycorrhiza* is related to higher accumulation of leaf P-concentration of soluble sugars in roots. Mycorrhiza 12:185–190

Feng NX, Yu J, Zhao HM, Cheng YT, Mo CH, Cai QY, Li YW, Li H, Wng MH (2017) Efficient phytoremediation of organic contaminants in soils using plant-endophyte partnerships. Sci Total Environ 583:352–368

Fester T, Giebler J, Wick LY, Schlosser D, Kästner M (2014) Plant–microbe interactions as drivers of ecosystem functions relevant for the biodegradation of organic contaminants. Curr Opin Biotechnol 27:168–175

Figueiredo MVB, Burity HA, Martinez CR, Chanway CP (2008) Alleviation of drought stress in common bean (*Phaseolus vulgaris* L.) by co-inoculation with *Paenibacillus polymyxa* and *Rhizobium tropici*. Appl Soil Ecol 40:182–188

Gaby JC, Buckley DH (2011) A global census of nitrogenase diversity. Environ Microbiol 13:1790–1799

Gerhardt KE, Huang XD, Glick BR, Greenberg BM (2009) Phytoremediation and rhizoremediation of organic soil contaminants: potential and challenges. Plant Sci 176:20–30

Gianinazzi S, Gollotte A, Binet MN, van Tuinen D, Redecker D, Wipf D (2010) Agroecology: the key role of *arbuscular mycorrhizae* in ecosystem services. Mycorrhiza 20:519–530

Giri B, Mukerji KG (2004) Mycorrhizal inoculant alleviate salt stress in *Sesbania aegyptiaca* and *Sesbania grandiflora* under field conditions: evidence for reduced sodium and improved magnesium uptake. Mycorrhiza 14:307–312

Glick BR (2014) Bacteria with ACC deaminase can promote plant growth and help to feed the world. Microbiol Res 169:30–39

Glick BR (2015) Phytoremediation. Beneficial plant-bacterial interactions. Springer International Publishing, Switzerland, pp 191–221

Glick BR, Cheng Z, Czarny J, Duan J (2007) Promotion of plant growth by ACC deaminase producing soil bacteria. Eur J Plant Pathol 119:329–339

Grover M, Ali SZ, Sandhya V, Rasul A, Venkateswarlu B (2010) Role of microorganisms in adaptation of agriculture crops to abiotic stresses. World J Microbiol Biotechnol 27:1231–1240

Guimaraes AA, Jaramillo PMD, RSA N, Florentino LA, Silva KB, de Souza Moreira FM (2012) Genetic and symbiotic diversity of nitrogen-fixing bacteria isolated from agricultural soils in the western Amazon by using cowpea as the trap plant. Appl Environ Microbiol 78:6726–6733

Hadi F, Bano A (2010) Effect of diazotrophs (*Rhizobium* and *Azotobacter*) on growth of maize (Zea mays L.) and accumulation of Lead (Pb) in different plant parts. Pak J Bot 42:4363–4370

Hamel C, Vujanovic V, Jeannotte R, Nakano-Hylander A, St-Arnaud M (2005) Negative feedback on perennial crop: fusarium crown and root rot of asparagus is related to changes in soil microbial community structure. Plant Soil 268:75–87

Hardoim PR, van Overbeek LS, Berg G, Pirtilla AM, Compant S, Campisano A, Doring M, Sessitsch A (2015) The hidden world within plants: ecological and evolutionary considerations for defining functioning of microbial endophytes. Microbiol Mol Biol Rev 79:293–320

Hasan MA (2013) Investigation on the nitrogen fixing *Cyanobacteria* (BGA) in rice fields of North-West region of Bangladesh. J Environ Sci Nat Resour 6:253–259

Hawkins HJ, Johansen A, George E (2000) Uptake and transport of organic and inorganic nitrogen by arbuscular mycorrhizal fungi. Plant Soil 226:275–285

Hayat R, Ali S, Amara U, Khalid R, Ahmed I (2010) Soil beneficial bacteria and their role in plant growth promotion: a review. Ann Microbiol 60:579–598

Hu S, Gu H, Cui C, Ji R (2016) Toxicity of combined chromium (VI) and phenanthrene pollution on the seed germination, stem lengths, and fresh weights of higher plants. Environ Sci Pollut Res 23:15227–15235

Hussain SS, Iqbal MT, Arif MA, Amjad M (2011) Beyond osmolytes and transcription factors: drought tolerance in plants via protective proteins and aquaporins. Biol Plant 55:401–413

Hussain SS, Raza H, Afzal I, Kayani MA (2012) Transgenic plants for abiotic stress tolerance: current status. Arch Agron Soil Sci 58:693–721

Hussain A, Abbas N, Arshad F, Akram M, Khan ZI, Ahmad K, Mansha M, Mirzaei F (2013) Effects of diverse doses of Lead (Pb) on different growth attributes of Zea-Mays L. Agric Sci 4:262–265

Hussain SS, Asif MA, Sornaraj P, Ali M, Shi BJ (2016) Towards integration of system based approach for understanding drought stress in plants. In: Ahmad P, Rasool SP (eds) Water stress and crop plants: a sustainable approach. Elsevier, Atlanta, pp 227–247., (Chapter 16

Ijaz A, Imran A, Ul Haq MA, Khan QM, Afzal M (2016) Phytoremediation: recent advances in plant-endophytic synergistic interactions. Plant Soil 405:179–195

IPCC (2007) Climate change 2007. The physical science basis. Contribution of working group I to the fourth assessment report of the intergovernmental panel on climate change. Cambridge University Press, Cambridge, NY

Jeffries P, Barea JM (2001) *Arbuscular mycorrhiza*: a key component of sustainable plant-soil ecosystems. In: Hock B (ed) The mycota (vol. IX: fungal associations). Springer, Berlin

Jha B, Gontia I, Hartmann A (2012) The roots of the halophyte *Salicornia brachiata* are a source of new halotolerant diazotrophic bacteria with plant growth-promoting potential. Plant Soil 356:265–277

Jha P, Panwar J, Jha PN (2015) Secondary plant metabolites and root exudates: guiding tools for polychlorinated biphenyl biodegradation. Int J Environ Sci Technol 12:789–802

Jindal V, Atwal A, Sekhon BS, Rattan S, Singh R (1993) Effect of vesicular-arbuscular mycorrhiza on metabolism of moong plants under salinity. Plant Physiol Biochem 31:475–481

Johnson NC, Graham JH (2013) The continuum concept remains a useful framework for studying mycorrhizal functioning. Plant Soil 363:411–419

Johnson D, Martin F, Cairney JWG, Anderson IC (2012) The importance of individuals: intraspecific diversity of mycorrhizal plants and fungi in ecosystems. New Phytol 194:614–628

Jorquera MA, Shaharoona B, Nadeem SM, de la Luz Mora M, Crowley DE (2012) Plant growth-promoting rhizobacteria associated with ancient clones of creosote bush (*Larrea tridentata*). Microb Ecol 64:1008–1017

Kathuria H, Giri J, Tyagi H, Tyagi AK (2007) Advances in transgenic rice biotechnology. Crit Rev Plant Sci 26:65–103

Khan A, Jilani G, Akhtar MS, Naqvi SMS, Rasheed M (2009) Phosphorus solubilizing bacteria: occurrence, mechanisms and their role in crop production. J Agric Biol Sci 1:48–58

Kidd P, Barcelo J, Bernal MP, Navari-Izzo F, Poschenrieder C, Shilev S, Clemente R, Monterroso C (2009) Trace element behavior at the root-soil interface: implications in phytoremediation. Environ Exp Bot 67:243–259

Kim S, Lowman S, Hou G, Nowak J, Flinn B, Mei C (2012) Growth promotion and colonization of switchgrass (*Panicum virgatum*) cv. Alamo by bacterial endophyte *Burkholderia phytofirmans* strain PsJN. Biotechnol Bioguels 5:37

Kim YC, Glick B, Bashan Y, Ryu CM (2013) Enhancement of plant drought tolerance by microbes. In: Aroca R (ed) Plant responses to drought stress. Springer Verlag, Berlin

Kohler J, Hernandez JA, Caravaca F, Rolden A (2008) Plant growth promoting rhizobacteria and arbuscular mycorrhizal fungi modify alleviation biochemical mechanisms in water stressed plants. Funct Plant Biol 35:141–151

Kreps JA, Wu Y, Chang HS, Zhu T, Wang X, Harper JF (2002) Transcriptome changes for *Arabidopsis* in response to salt, osmotic and cold stress. Plant Physiol 130:2129–2141

Kukla M, Płociniczak T, Piotrowska-Seget Z (2014) Diversity of endophytic bacteria in *Lolium perenne* and their potential to degrade petroleum hydrocarbons and promote plant growth. Chemosphere 117:40–46

Kumar KV, Srivastava S, Singh N, Behl HM (2009) Role of metal resistant plant growth promoting bacteria in ameliorating fly ash to the growth of *Brassica juncea*. J Hazard Mater 170:51–57

Lareen A, Burton F, Schäfer P (2016) Plant root-microbe communication in shaping root microbiomes. Plant Mol Biol 90:575–587

Lau JA, Lennon JT (2012) Rapid responses of soil microorganisms improve plant fitness in novel environments. Proc Natl Acad Sci U S A 109:14058–14062

Lebeis SL (2015) Greater than the sum of their parts: characterizing plant microbiomes at the community level. Curr Opin Plant Biol 24:82–86

Lebeis SL, Rott M, Dangl JL, Schulze-Lefert P (2012) Culturing a plant microbiome community at the cross-Rhodes. New Phytol 196:341–344

Lebeis SL, Paredes SH, Lundberg DS, Breakfield N, Gehring J, McDonald M, Malfatti S, Glavina del Rio T, Jones CD, Tringe SG, Dangl JL (2015) Salicylic acid modulates colonization of the root microbiome by specific bacterial taxa. Science 349:860–864

Lee SW, Ahn PI, Sy S, Lee SY, Seo MW, Kim S, Sy P, Lee YH, Kang S (2010) *Pseudomonas* sp. LSW25R antagonistic to plant pathogens promoted plant growth and reduced blossom red rot of tomato roots in a hydroponic system. Eur J Plant Pathol 126:1–11

Leite MCBS, de Farias ARB, Freire FJ, Andreote FD, Sobral JK, Freire MBGS (2014) Isolation, bioprospecting and diversity of salt-tolerant bacteria associated with sugarcane in soils of Pernambuco, Brazil. Rev Bras Eng Agrıc Amb 18:S73–S79

Leveau JHJ, Uroz S, de Boer W (2010) The bacterial genus Collimonas: mycophagy, weathering and other adaptive solutions to life in oligotrophic soil environments. Environ Microbiol 12:281–292

Lugtenberg B, Kamilova F (2009) Plant-growth-promoting rhizobacteria. Annu Rev Microbiol 63:541–556

Lundberg DS, Lebeis SL, Paredes SH, Yourstone S, Gehring J, Malfatti S, Tremblay J, Engelbrektson A, Kunin V, del Rio TG, Edgar RC, Eickhorst T, Ley RE, Hugenholtz P, Tringe SG, Dangl JL (2012) Defining the core *Arabidopsis thaliana* root microbiome. Nature 488:86–90

Luo SL, Chen L, Chen JI, Xiao X, Xu TY, Wan Y, Rao C, Liu CB, Liu Y, Lai C, Zeng GM (2011) Analysis and characterization of cultivable heavy metal-resistant bacterial endophytes isolated from Cd-hyperaccumulator *Solanum nigrum* L. and their potential use for phytoremediation. Chemosphere 85:1130–1138

Luo S, Xu T, Chen L, Chen J, Rao C, Xiao X, Wan Y, Zeng G, Long F, Liu C, Liu Y (2012) Endophyte-assisted promotion of biomass production and metal-uptake of energy crop sweet sorghum by plant-growth-promoting endophyte *Bacillus* sp. SLS18. Appl Microbiol Biotechnol 93:1745–1753

Ma Y, Prasad MNV, Rajkumar M, Freitas H (2011) Plant growth promoting rhizobacteria and endophytes accelerate phytoremediation of metalliferous soils. Biotechnol Adv 29:248–258

Mapelli F, Marasco R, Rolli E, Cappitelli F, Daffonchio D, Borin S (2012) Mineral-microbe interactions: biotechnological potential of bio-weathering. J Biotechnol 157:473–481

Mapelli F, Marasco R, Rolli E, Barbato M, Cherif H, Guesmi A, Ouzari I, Daffonchio D, Borin S (2013) Potential for plant growth promotion of rhizobacteria associated with Salicornia growing in Tunisian hypersaline soils. Biomed Res 2013:248078

Marasco R, Rolli E, Attoumi B, Vigani G, Mapelli F, Borin S, Daffonchio D (2012) A drought resistance-promoting microbiome is selected by root system under desert farming. PLoS One 7:e48479

Marasco R, Rolli E, Vigani G, Borin S, Sorlini C, Ouzari H, Zocchi G, Daffonchio D (2013) Are drought-resistance promoting bacteria cross-compatible with different plant models? Plant Signal Behav 8:e26741

Marasco R, Mapelli F, Rolli E, Mosqueira MJ, Fusi M, Bariselli P, Reddy M, Cherif A, Tsiamis G, Borin S, Daffonchio D (2016) Salicornia strobilacea (synonym of Halocnemum strobilaceum) growth under different tidal regimes selects rhizosphere bacteria capable of promoting plant growth. Front Microbiol 7:1286

Marquez LM, Redman RS, Rodriguez RJ, Roosinck MJ (2007) A virus in a fungus in a plant: three-way symbiosis required for thermal tolerance. Science 315:513–515

Marschner P, Crowley D, Rengel Z (2011) Rhizosphere interactions between microorganisms and plants govern iron and phosphorus acquisition along the root axis–model and research methods. Soil Biol Biochem 43:883–894

Marulanda A, Porcel R, Barea JM, Azcon R (2007) Drought tolerance and antioxidant activities in lavender plants colonized by native drought tolerant or drought sensitive *Glomus* species. Microb Ecol 54:543–552

Marulanda A, Barea JM, Azcón R (2009) Stimulation of plant growth and drought tolerance by native microorganisms (AM fungi and bacteria) from dry environments: mechanisms related to bacterial effectiveness. J Plant Growth Regul 28:115–124

Masciarelli O, Llanes A, Luna V (2014) A new PGPR co-inoculated with *Bradyrhizobium japonicum* enhances soybean nodulation. Microbiol Res 169:609–615

Mastretta C, Taghavi S, Van Der Lelie D, Mengoni A, Galardi F, Gonnelli C, Barac T, Boulet J, Weyens N, Vangronsveld J (2009) Endophytic bacteria from seeds of *Nicotiana tabacum* can reduce cadmium phytotoxicity. Int J Phytoremediation 11:251–267

Mayak S, Tirosh T, Glick BR (2004) Plant growth promoting bacteria that confer resistance to water stress in tomato and pepper. Plant Sci 166:525–530

Mendes R, Kruijt M, de Bruijn I, Dekkers E, van der Voort M, Schneider JH, Piceno YM, DeSantis TZ, Andersen GL, Bakker PA, Raaijmakers JM (2011) Deciphering the rhizosphere microbiome for disease-suppressive bacteria. Science 332:1097–1100

Mengual C, Schoebitz M, Azcón R, Roldán A (2014) Microbial inoculants and organic amendment improves plant establishment and soil rehabilitation under semiarid conditions. J Environ Manag 134:1–7

Miao B, Stewart BA, Zhang F (2011) Long-term experiments for sustainable nutrient management in China. A review. Agron Sustain Dev 31:397–414

Miller RM, Jastrow JD (2000) Mycorrhizal fungi influence soil structure. In: Kapulnik Y, Douds DD Jr (eds) *Arbuscular mycorrhizas*: physiology and function. Kluwer Academic Publishers, London, pp 3–18

Miller JB, Oldroyd GD (2012) The role of diffusible signals in the establishment of rhizobial and mycorrhizal symbioses. In: Perotto S, Baluška F (eds) Signaling and communication in plant symbiosis. Springer, Berlin, pp 1–30

Miransari M (2011) Hyperaccumulators, arbuscular mycorrhizal fungi and stress of heavy metals. Biotechnol Adv 29:645–653

Mishra P, Mishra S, Selvakumar G, Kundub S, Gupta HS (2009) Enhanced soybean (*Glycine max* L) plant growth and nodulation by *Bradyrhizobium japonicum*-SB1 in presence of *Bacillus thuringiensis*-KR1. Acta Agric Scand Sect B Soil Plant Sci 59:189–196

Nadeem SM, Ahmad M, Zahir ZA, Javaid A, Ashraf M (2014) The role of mycorrhizae and plant growth promoting rhizobacteria (PGPR) in improving crop productivity under stressful environments. Biotechnol Adv 32:429–448

Naveed M, Hussain B, Zahir A, Mitter B, Sessitsch A (2014a) Drought stress amelioration in wheat through inoculation with Burkholderia phytofirmans strain PsJN. Plant Growth Regul 73:121–131

Naveed M, Mitter B, Reichenauer TG, Wieczorek K, Sessitsch A (2014b) Increased drought stress resilience of maize through endophytic colonization by *Burkholderia phytofirmans* PsJN and *enterobacter* sp. FD17. Environ Exp Bot 97:30–39

Nonnoi F, Chinnaswamy A, García de la Torre VS, Coba de la Peña T, Lucas MM, Pueyo JJ (2012) Metal tolerance of rhizobial strains isolated from nodules of herbaceous legumes *Medicago* sp. and *Trifolium* sp. growing in mercury-contaminated soils. Appl Soil Ecol 61:49–59

Oldroyd GED (2013) Speak, friend, and enter: signaling systems that promote beneficial symbiotic associations in plants. Nat Rev Microbiol 11:252–263

Orłowska E, Przybyłowicz W, Orlowski D, Turnau K, Mesjasz-Przybyłowicz J (2011) The effect of mycorrhiza on the growth and elemental composition of Ni-hyperaccumulating plant *Berkheya coddii* Roessler. Environ Pollut 159:3730–3738

Pandey VC (2012) Phytoremediation of heavy metals from fly ash pond by *Azolla caroliniana*. Ecotoxicol Environ Saf 82:8–12

Panke-Buisse K, Poole A, Goodrich J, Ley R, Kao-Kniffin J (2015) Selection on soil microbiomes reveals reproducible impacts on plant function. ISME J 9:980–989

Park JD (2010) Heavy metal poisoning. Hanyang Med Rev 30:319–325

Paul D, Lade H (2014) Plant-growth-promoting rhizobacteria to improve crop growth in saline soils: a review. Agron Sustain Dev 34:737–752

Peiffer JA, Spor A, Koren O, Jin Z, Tringe SG, Dangl JL, Buckler ES, Ley RE (2013) Diversity and heritability of the maize rhizosphere microbiome under field conditions. Proc Natl Acad Sci U S A 110:6548–6553

Perez-Montano F, Alías-Villegas C, Bellogin RA, del Cerro P, Espuny MR, Jimenez-Guerrero I, Lopez-Baena FJ, Ollero FJ, Cubo T (2014) Plant growth promotion in cereal and leguminous agricultural important plants: from microorganism capacities to crop production. Microbiol Res 169:325–336

Philippot L, Raaijmakers JM, Lemanceau P, van der Putten WH (2013) Going back to the roots: the microbial ecology of the rhizosphere. Nat Rev Microbiol 11:789–799

Porcel R, Aroca R, Cano C, Bago A, Ruiz-Lozano JM (2006) Identification of a gene from the arbuscular mycorrhizal fungus *Glomus intraradices* encoding for a 14–3-3 protein that is upregulated by drought stress during the AM symbiosis. Microb Ecol 52:575–582

Qin S, Zhang YJ, Yuan B, Xu PY, Xing K, Wang J, Jiang JH (2014) Isolation of ACC deaminase-producing habitat-adapted symbiotic bacteria associated with halophyte *Limonium sinense* (Girard) Kuntze and evaluating their plant growth promoting activity under salt stress. Plant Soil 374:753–766

Qiu M, Li S, Zhou X, Cui X, Vivanco J, Zhang N, Shen Q, Zhang R (2014) De-coupling of root-microbiome associations followed by antagonist inoculation improves rhizosphere soil suppressiveness. Biol Fertil Soils 50:217–224

Rajkumar M, Ae N, Prasad MNV, Freitas H (2010) Potential of siderophore-produing bacteria for improving heavy metal phytoextraction. Trends Biotechnol 28:142–149

Rajkumar M, Sandhya S, Prasad MNV, Freitas H (2012) Perspectives of plant-associated microbes in heavy metal phytoremediation. Biotechnol Adv 30:1562–1574

Rasche F, Velvis H, Zachow C, Berg G, van Elsas JD, Sessitsch A (2006) Impact of transgenic potatoes expressing antibacterial agents on bacterial endophytes is comparable with the effects of plant genotype, soil type and pathogen infection. J Appl Ecol 43:555–566

Richardson AE, Barea JM, McNeill AM, Prigent-Combaret C (2009) Acquisition of phosphorus and nitrogen in the rhizosphere and plant growth promotion by microorganisms. Plant Soil 321:305–339

Rolli E, Marasco R, Vigani G, Ettoumi B, Mapelli F, Deangelis ML, Gandolfi C, Casati E, Previtali S, Gerbino R, Cei FP, Borin S, Sorlini C, Zocchi G, Daffonchio D (2015) Improved resistance to drought in prompted by the root-associated microbiome as a water stress dependent trait. Environ Microbiol 17:316–331

Ruiz KB, Biondi S, Martínez EA, Orsini F, Antognoni F, Jacobsen SE (2015) Quinoa – a model crop for understanding salt-tolerance mechanisms in halophytes. Plant Biosystems 150:357–371

Salvioli A, Bonfante P (2013) Systems biology and "omics" tools: a cooperation for next-generation mycorrhizal studies. Plant Sci 203:107–114

Salvioli A, Ghignone S, Novero M, Navazio L, Venice F, Bagnaresi P, Bonfante P (2016) Symbiosis with an endobacterium increases the fitness of a mycorrhizal fungus, raising its bioenergetic potential. ISME J 10:130–144

Sandhya V, Ali SZ, Grover M, Kishore N, Venkateswarlu B (2009) Pseudomonas sp. strain P45 protects sunflowers seedlings from drought stress through improved soil structure. J Oilseed Res 26:600–601

Sandhya V, Ali SZ, Grover M, Reddy G, Venkateswaralu B (2010) Effect of plant growth promoting Pseudomonas spp. On compatible solutes, antioxidant status and plant growth of maize under drought stress. Plant Growth Regul 62:21–30

Santi C, Bogusz D, Franche C (2013) Biological nitrogen fixation in non-legume plants. Ann Bot 111:743–767

Santoyo G, Moreno-Hagelsieb G, Orozco-Mosqueda MC, Glick BR (2016) Plant growth-promoting bacterial endophytes. Microbiol Res 183:92–99

Schaeppi K, Bulgarelli D (2015) The plant microbiome at work. Mol Plant-Microbe Interact 28:212–217

Schlaeppi K, Dombrowski N, Oter RG, Ver Loren van Themaat E, Schulze-Lefert P (2014) Quantitative divergence of the bacterial root microbiota in *Arabidopsis thaliana* relatives. Proc Natl Acad Sci U S A 111:585–592

Seki M, Narusaka M, Abe M, Kasuga M, Yamaguchi-Shinozaki K, Carninci P, Hayashizaki Y, Shinozaki K (2001) Monitoring the expression pattern of 1300 *Arabidopsis* under drought and cold stresses by using a full length cDNA microarray. Plant Cell 13:62–72

Seki M, Narusaka M, Ishida J, Nanjo T, Fujita M, Oono Y, Kamiya A, Nakajima M, Enju A, Sakurai T, Satou M, Akiyama K, Taji T, Yamaguchi-Shinozaki K, Carninci P, Kawai J, Hayashizaki Y, Shinozaki K (2002) Monitoring the expression profiles of 7000 *Arabidopsis* genes under drought, cold, and high salinity stresses using full length cDNA microarray. Plant J 31:279–292

Selvakumar G, Kundu S, Joshi P, Nazim S, Gupta AD, Mishra PK, Gupta HS (2008a) Characterization of a cold-tolerant plant growth-promoting bacterium *Pantoea dispersa* 1A isolated from a sub-alpine soil in the north western Indian Himalayas. World J Microbiol Biotechnol 24:955–960

Selvakumar G, Mohan M, Kundu S, Gupta AD, Joshi P, Nazim S, Gupta HS (2008b) Cold tolerance and plant growth promotion potential of *Serratia marcescens* strain SRM (MTCC 8708) isolated from flowers of summer squash (*Cucurbita pepo*). Lett Appl Microbiol 46:171–175

Selvakumar G, Joshi P, Nazim S, Mishra PK, Bisht JK, Gupta HS (2009) Phosphate solubilization and growth promotion by *Pseudomonas fragi* CS11RH1 (MTCC 8984) a psychrotolerant bacterium isolated from a high altitude Himalayan rhizosphere. Biologia 64:239245

Selvakumar G, Joshi P, Suyal P, Mishra PK, Joshi GK, Bisht JK, Bhatt JC, Gupta HS (2010a) *Pseudomonas lurida* M2RH3 (MTCC 9245), a psychrotolerant bacterium from the Uttarakhand Himalayas, solubilizes phosphate and promotes wheat seedling growth. World J Microbiol Biotechnol 5:1129–1135

Selvakumar G, Kundu S, Joshi P, Nazim S, Gupta AD, Gupta HS (2010b) Growth promotion of wheat seedlings by *Exiguobacterium acetylicum* 1P (MTCC 8707) a cold tolerant bacterial strain from the Uttarakhand Himalayas. Ind J Microbiol 50:50–56

Shabala S, Hariadi Y, Jacobsen SE (2013) Genotypic difference in salinity tolerance in quinoa is determined by differential control of xylem Na^+ loading and stomatal density. J Plant Physiol 170:906–914

Sheng X, He L, Wang Q, Ye H, Jiang C (2008) Effects of inoculation of biosurfactant-producing *Bacillus* sp. J119 on plant growth and cadmium uptake in a cadmium-amended soil. J Hazard Mater 155:17–22

Shinozaki K, Yamaguchi-Shinozaki K (2007) Gene networks involved in drought stress response and tolerance. J Exp Bot 58:221–227

Shinozaki K, Yamaguch-Shinozaki K (2000) Molecular responses to dehydration and low temperature: difference and cross talk between two stress signaling pathways. Curr Opin Plant Biol 3:217–223

Shirley M, Avoscan L, Bernuad E, Vansuyt G, Lemanceau P (2011) Comparison of iron acquisition from Fe-pyoverdine by strategy I and strategy II plants. Botany 89:731–735

Shridhar BS (2012) Review: nitrogen fixing microorganisms. Int J Microbiol Res 3:46–52

Siddikee MA, Chauhan PS, Anandham R, Han GH, Sa T (2010) Isolation, characterization and use for plant growth promotion under salt stress, of ACC deaminase producing halotolerant bacteria derived from coastal soil. J Microbiol Biotechnol 20:1577–1584

Singh BK, Bardgett RD, Smith P, Reay DS (2010) Microorganisms and climate change: terrestrial feedbacks and mitigation options. Nat Rev Microbiol 8:779–790

Singh B, Kaur T, Kaur S, Manhas RK, Kaur A (2016) Insecticidal potential of an endophytic Cladosporiumvelox against *Spodoptera litura* mediated through inhibition of alpha glycosidases. Pestic Biochem Physiol 131:46–52

Sloan SS, Lebeis S (2015) Exercising influence: distinct biotic interactions shape root microbiome. Curr Opin Plant Biol 26:32–36

Sreenivasulu N, Sopory SK, Kavi Kishor PB (2007) Deciphering the regulatory mechanisms of abiotic stress tolerance in plants by genomic approaches. Gene 388:1–13

Srivastava S, Yadav A, Seem K, Mishra S, Chaudhary V, Srivastava CS (2008) Effect of high temperature on *Pseudomonas putida* NBRI0987 biofilm formation and expression of stress sigma factor RpoS. Curr Microbiol 56:453–457

Sugiyama A, Bakker MG, Badri DV, Manter DK, Vivanco JM (2013) Relationships between *Arabidopsis* genotype-specific biomass accumulation and associated soil microbial communities. Botany-Botanique 91:123126

Tester M, Davenport R (2003) Na^+ tolerance and Na^+ transport in higher plants. Ann Bot 91:503–527

Thao NP, Tran LS (2016) Enhancement of plant productivity in the post genomic era. Curr Genomics 17:295–296

Theocharis A, Bordiec S, Fernandez O, Paquis S, Dhondt-Cordelier S, Baillieul F, Clement C, Barka EA (2012) *Burkholderia phytofirmans* PsJN primes *Vitis vinifera* L. and confers a better tolerance to low nonfreezing temperatures. Mol Plant-Microbe Interact 25:241–249

Timmusk S, Timmusk K, Behers L (2013) Rhizobacterial plant drought stress tolerance enhancement: towards sustainable water resource management and food security. J Food Security 1:6–9

Timmusk S, El-Daim IA, Cpolovici L, Tanilas T, Kannaste A, Behers L, Nevo E, Seisenbaeva G, Stenstrom E, Niinemets U (2014) Drought-tolerance of wheat improved y rhizosphere bacteria from harsh environments: enhanced biomass production and reduced emissions of stress volatiles. PLoS One 9:e96086

Timmusk S, Behers L, Muthoni J, Aronsson AC (2017) Perspectives and challenges of microbe application for crop improvement. Front Plant Sci 8:49

Tiwari S, Singh P, Tiwari R, Meena KK, Yandigeri M, Singh DP, Arora DK (2011) Salt-tolerant rhizobacteria-mediated induced tolerance in wheat (*Triticum aestivum* L.) and chemical diversity in rhizosphere enhance plant growth. Biology Fert Soils 47:907–916

Tkacz A, Poole P (2015) Role of root microbiota in plant productivity. J Exp Bot 66:2167–2175

Tkacz A, Cheema J, Chandra G, Grant A, Poole PS (2015) Stability and succession of the rhizosphere microbiota depends upon plant type and soil composition. ISME J 9:2349–2359

Trabelsi D, Mhamdi R (2013) Microbial inoculants and their impact on soil microbial communities: a review. Biomed Res 2013:e863240

Turner TR, James EK, Poole PS (2013) The plant microbiome. Genome Biol 14:209

Upadhyay SK, Singh DP, Saikia R (2009) Genetic diversity of plant growth promoting rhizobacteria from rhizospheric soil of wheat under saline conditions. Curr Microbiol 59:489–496

Uroz S, Dessaux Y, Oger P (2009) Quorum sensing and quorum quenching: the yin and yang of bacterial communication. Chembiochem 10:205–216

Valliyodan B, Nguyen H (2006) Understanding regulatory networks and engineering for enhanced drought tolerance in plants. Curr Opin Plant Biol 9:1–7

Vejan P, Abdullah R, Khadiran T, Ismail S, Nusrulhaq Boyce A (2016) Role of plant growth-promoting rhizobacteria in agricultural sustainability: a review. Molecules 21:573

Velazquez-Hernandez ML, Baizabal-Aguirre VM, Cruz-Vazquez F, Trejo-Contreras MJ, Fuentes-Ramırez LE, Bravo-Patino A, Valdez-Alarcon JJ (2011) Gluconacetobacter *diazotrophicus levansucrase* is involved in tolerance to NaCl, sucrose and desiccation, and in biofilm formation. Arch Microbiol 193:137–149

Vidali M (2001) Bioremediation: an overview. Pure Appl Chem 73:1163–1172

Vinocur B, Altman A (2005) Recent advances in engineering plant tolerance to abiotic stress: achievements and limitations. Curr Opin Biotechnol 16:123–132

Wang WX, Vinocur B, Altman A (2003) Plant responses to drought, salinity and extreme temperatures: towards genetic engineering for stress tolerance. Planta 219:1–14

Wang HG, Zhnag XZ, Li H, He HB, Fang CX, Zhang AJ, Li QS, Chen RS, Guo XK, Lin HF, Wu LK, Lin S, Chen T, Lin RY, Peng XX, Lin WX (2011) Characterization of metaproteomics in crop rhizospheric soil. J Proteome Res 10:932–940

Wani PA, Khan MS, Zaidi A (2008) Effect of metal tolerant plant growth-promoting Rhizobium on the performance of pea grown in metal-amended soil. Arch Environ Contam Toxicol 55:33–42

Waqas M, Khan AL, Hamayun M, Shahzad R, Kim YH, Choi KS, Lee IJ (2015) Endophytic infection alleviates biotic stress in sunflower through regulation of defense hormones, antioxidants and functional amino acids. Eur J Plant Pathol 141:803–824

Wenzel WW (2009) Rhizosphere processes and management in plant-assisted bioremediation (phytoremediation) of soils. Plant Soil 321:385–408

Wicke B, Smeets E, Dornburg V, Vashev B, Gaiser T, Turkenburg W, Faij A (2011) The global technical and economic potential of bioenergy from salt-affected soils. Energy Environ Sci 4:2669–2681

Wuana RA, Okieimen FE (2011) Heavy metals in contaminated soil: a review of sources, chemistry, risks and best available strategies for bioremediation. ISRN Ecol 2011:402647

Yang J, Kloepper JW, Ryu CM (2009) Rhizosphere bacteria help plants tolerate abiotic stress. Trends Plant Sci 14:1–4

Yang Q, Tu S, Wang G, Liao X, Yan X (2012) Effectiveness of applying arsenate reducing bacteria to enhance arsenic removal from polluted soils by *Pteris vittata* L. Int J Phytoremediation 14:89–99

Yang AZ, Akhtar SS, Amjad M, Iqbal S, Jacobsen SE (2016a) Growth and physiological responses of quinoa to drought and temperature stress. J Agron Crop Sci 202:421–589

Yang AZ, Akhtar SS, Iqbal S, Amjad M, Naveed M, Zahir ZA, Jacobsen SE (2016b) Enhancing salt tolerance in quinoa by halotolerant bacterial inoculum. Funct Plant Biol 43:632–642

Yeoh YK, Paungfoo-Lonhienne C, Dennis PG, Robinson N, Ragan MA, Schmidt S, Hugenholtz P (2016) The core root microbiome of sugarcanes cultivated under varying nitrogen fertilizer application. Environ Microbiol 18:1338–1135

Yildirim E, Taylor AG (2005) Effect of biological treatments on growth of bean plans under salt stress. Ann Rep Bean Improv Coop 48:176–177

Zarei M, Wubet T, Schäfer SH, Savaghebi GR, Jouzani GS, Nekouei MK (2010) Molecular diversity of arbuscular mycorrhizal fungi in relation to soil chemical properties and heavy metal contamination. Environ Pollut 158:2757–2765

Zehr JP, Jenkins BD, Short SM, Steward GF (2003) Nitrogenase gene diversity and microbial community structure: a cross-system comparison. Environ Microbiol 5:539–554

Zhang F, Lynch DH, Smith DL (1995) Impact of low root temperatures in soybean [*Glycine max* (L.) Merr.] on nodulation and nitrogen fixation. Environ Exp Bot 35:279–285

Zhang F, Dashti N, Hynes R, Smith DL (1996) Plant growth promoting rhizobacteria and soybean [*Glycine max* (L.) Merr.] nodulation and nitrogen fixation at suboptimal root zone temperatures. Ann Bot 77:453–460

Zhang H, Dong H, Li W, Sun Y, Chen S, Kong X (2009a) Increased glycine betaine synthesis and salinity tolerance in *AhCMO* transgenic cotton lines. Mol Breed 23:289–298

Zhang H, Sun Y, Xie X, Kim MS, Dowd SE, Pare PW (2009b) A soil bacterium regulates plant acquisition of iron via deficiency-inducible mechanisms. Plant J 58:568–577

Zhang Z, Le Velly M, Rhind SM, Kyle CE, Hough RL, Duff EI, McKenzie C (2015) A study on temporal trends and estimates of fate of Bisphenol A in agricultural soils after sewage sludge amendment. Sci Total Environ 515–516:1–11

Zhao S, Zhou N, Zhao ZY, Zhang K, Wu GH, Tian CY (2016) Isolation of endophytic plant growth-promoting bacteria associated with the halophyte Salicornia europea and evaluating their promoting activity under salt stress. Curr Microbiol 73:574–581

Zheng YK, Qiao XG, Miao CP, Liu K, Chen YW, Xu LH, Zhao LX (2016) Diversity, distribution and biotechnological potential of endophytic fungi. Ann Microbiol 66:529–542

Zhu JK (2002) Salt and drought stress signal transduction in plants. Annu Rev Plant Biol 53:247–273

Ziegler M, Engel M, Welzl G, Schloter M (2013) Development of a simple root model to study the effects of single exudates on the development of bacterial community structure. J Microbiol Methods 94:30–36

Zolla G, Badri DV, Bakker MG, Manter DK, Vivanco JM (2013) Soil microbiomes vary in their ability to confer drought tolerance to *Arabidopsis*. Appl Soil Ecol 68:1–9

Zuppinger-Dingley D, Schmid B, Petermann JS, Yadav V, De Deyn GB, Flynn DF (2014) Selection for niche differentiation in plant communities increases biodiversity effects. Nature 515:108–111

Chapter 3
Plant Growth Promotion and Biocontrol Mediated by Plant-Associated Bacteria

Miguel A. Matilla and Tino Krell

Abstract The rhizosphere, defined as the volume of soil under the physical, chemical and biological influences of plant roots, is a region of enormous microbial diversity and activity. This microbial activity is essential for plant nutrition and health since it favours the uptake of nutrients by the plant and offers resistance against a wide range of plant pathogens. Bacteria are the main microbial representatives in the rhizosphere, and plant growth-promoting rhizobacteria (PGPR) stimulate plant growth by multiple mechanisms. In this chapter, we present an overview of the strategies employed by PGPR to exert their beneficial effects on the colonized plants. The direct effects of PGPR on plant growth are mainly derived from their capacity to improve the nutritional status of plants and the production of phytohormones. Alternatively, beneficial rhizospheric bacteria can also promote plant health by protecting plants against pathogens mainly through the induction of systemic resistance and the production of exoenzymes and multiple antagonistic metabolites. Here, special attention has been given to the biosynthesis and biological activities of bioactive volatiles, non-ribosomal peptides and polyketides by PGPR. Finally, the promising use of PGPR-based products as sustainable agricultural practices is discussed.

Keywords Plant growth-promoting rhizobacteria • Phytohormones • Biocontrol • Induced systemic resistance • Bioactive secondary metabolites

M.A. Matilla (✉) • T. Krell
Department of Environmental Protection, Estación Experimental del Zaidín, Consejo Superior de Investigaciones Científicas, Granada, Spain
e-mail: miguel.matilla@eez.csic.es

D. Egamberdieva, P. Ahmad (eds.), *Plant Microbiome: Stress Response*, Microorganisms for Sustainability 5, https://doi.org/10.1007/978-981-10-5514-0_3

3.1 Introduction

The term rhizosphere was first introduced by Lorenz Hiltner (1904) as the area of soil surrounding the plant root system that supports high levels of microbial activity. Currently, the rhizosphere is considered as the volume of soil that is under the physical, chemical and biological influences of plant roots, including root tissues colonized by the microorganisms (Lugtenberg and Bloemberg 2004; Lugtenberg and Kamilova 2009). It was estimated that plants release up to 40% of the photosynthetically fixed carbon through root exudates, which leads to changes in the biochemical and physical properties of the surrounding soil (Hutsch et al. 2000; Walker et al. 2003; Bais et al. 2006; Newmann and Römheld 2007; Uren 2007; Badri and Vivanco 2009). Indeed, the substantial amount of nutrients released (e.g. sugars, vitamins, organic acids and sugars, among others) supports the emergence of a complex community of organisms, including bacteria, fungi, nematodes, protozoa and algae; of which some have a major impact on plant health, growth and development (Bais et al. 2006; Perry et al. 2007; Lugtenberg and Kamilova 2009; Raaijmakers et al. 2009; Bulgarelli et al. 2013; Vacheron et al. 2013; Huang et al. 2014). Among them, bacteria (rhizobacteria) are considered one of the main inhabitants of the rhizosphere (Berendsen et al. 2012). Importantly, many rhizobacteria can colonize plant roots and form biofilms associated to the root system (Danhorn and Fuqua 2007; Matilla et al. 2011a). Furthermore, rhizosphere-associated bacteria have been shown to be more versatile in the metabolization and solubilization of nutrients as compared to bacteria from bulk soils, which converts them into key players for improving soil fertility (Hayat et al. 2010).

Multiple studies have demonstrated the influence of plant species and their developmental stages on the rhizosphere microbial communities (Graner et al. 2003; Milling et al. 2004; Mougel et al. 2006; Micallef et al. 2009; Inceoglu et al. 2013). Thus, plants select and attract specific microbes and, therefore, alter the composition and diversity of the root-associated microbiome in a plant-specific manner (Houlden et al. 2008; Badri et al. 2009, 2013). The microbial diversity found in the rhizosphere is extraordinary since 1 g of soil can contain more than ten billion bacteria belonging to thousands of different species and, in general, the total number of bacteria in the rhizosphere are two to three orders of magnitude superior to the corresponding values derived from the analysis of bulk soil (Roselló-Mora and Amann 2001; Roesch et al. 2007; Chaparro et al. 2013a; Reinhold-Hurek et al. 2015). However, plant properties (i.e. genotype, growth stage, nutritional status) and plant root exudates exert a strong selective pressure on the rhizosphere bacterial composition that finally results in a decreased bacterial diversity and in the shaping of the structure of the bacterial community (Marilley et al. 1999; Bais et al. 2006; Haichard et al. 2008; Berg and Smalla 2009; Doornbos et al. 2012; Bulgarelli et al. 2013; Chaparro et al. 2013a; Li et al. 2014a). In fact, several studies propose the existence of a co-evolution process between plants and their associated bacteria, which has resulted in a high degree of host specificity (Burdon and Thrall 2009; Raaijmakers et al. 2009; Bakker et al. 2014). In accordance with this, transgenic plants with altered root exudate composition were used to confirm the alterations in root-associated bacterial communities (Andreote et al. 2008; Oliver et al. 2008; Badri et al. 2009; Aira et al. 2010). In this chapter, we focus on the recent progress made to unravel the interactions between plants and rhizosphere bacteria and, more specifically, in the plant growth-promoting and biocontrol properties of root-associated bacteria, including their capacity to synthesize bioactive secondary metabolites.

3.2 Plant Growth and Health Mediated by Beneficial Rhizobacteria

Bacteria isolated from a particular rhizosphere can be classified as harmful, beneficial or neutral with respect to the plant life cycle (Somers et al. 2004; Antoun and Prevost 2006; Berendensen et al. 2012). Within this heterogeneous bacterial community, plant growth-promoting rhizobacteria (PGPR) have been demonstrated to increase growth and productivity of many plants, including a significant number of commercial species (Gray and Smith 2005; Van Loon 2007; Berg 2009; Lugtenberg and Kamilova 2009; Duta and Podile 2010; Hayat et al. 2010; Saharan and Nehra 2011; Bakker et al. 2013; Drogue et al. 2013; Mendes et al. 2013; Reed and Glick 2013; Ahemad and Kribet 2014; Prathap and Ranjitha-Kumari 2015). Thus, PGPR are responsible for causing multiple beneficial pleiotropic effects on plants, besides having strong incidence on plant transcriptome (Srivastava et al. 2012; Vacheron et al. 2013). In order to develop these beneficial properties, PGPR must be capable of competing with other rhizospheric microorganisms for nutrients secreted by the plant roots, besides being efficient in the colonization of the root tissues (Parray et al. 2016).

During evolution, plants have established interactions with a broad range of PGPR. The study of the bacterial rhizobiome is rapidly progressing mainly due to the fast development of metagenomics and massive genome-sequencing strategies, state-of-the-art techniques that are facilitating the identification of novel bacterial determinants and mechanisms involved in biocontrol and plant growth promotion. Beneficial root-associated bacteria are known to directly affect plant growth and development by multiple mechanisms, including phosphate solubilization, N_2 fixation as well as the production of phytohormones and different volatiles (Lugtemberg and Kamilova 2009; Richardson et al. 2009; Bakker et al. 2013; Prathap and Ranjitha-Kumari 2015). Likewise, PGPR can also promote plant growth indirectly by eliciting plant defence responses (Matilla et al. 2010; Pieterse et al. 2014), reducing susceptibility to plant diseases (e.g. production of antibiotics) (Gross and Loper 2009; Raaijmakers et al. 2009; Matilla et al. 2012, 2015, 2016b, 2017; Hellberg et al. 2015; Chowdhury et al. 2015a; Mousa and Raizada 2015) or competing with plant pathogens for nutrients and niche (Lugtemberg and Kamilova 2009; Ahmed and Holsmtrön 2014) (Fig. 3.1). Additionally, some PGPR can help plants to overcome drought and saline stresses and to increase the capacity of plants to sequester heavy

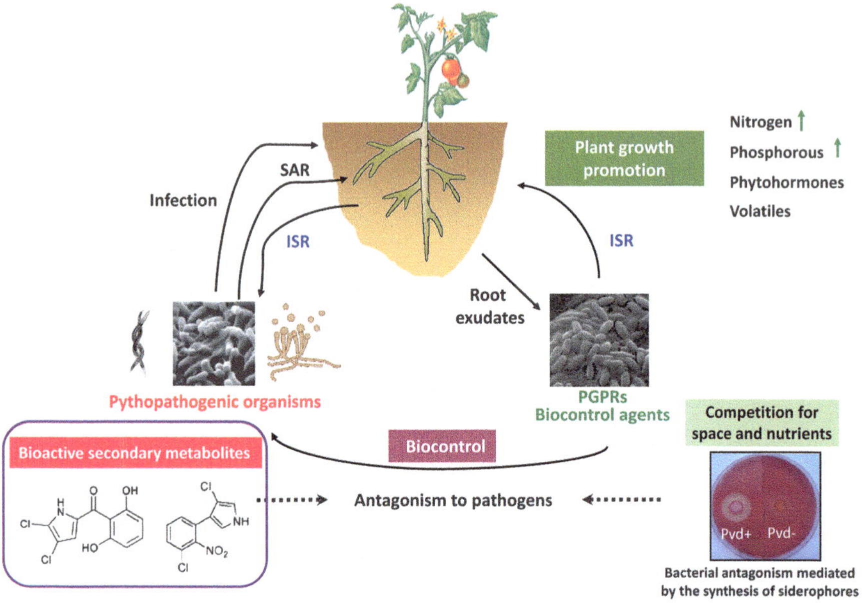

Fig. 3.1 Direct and indirect mechanisms of plant growth promotion and biocontrol mediated by plant growth-promoting rhizobacteria. *ISR* induced systemic resistance, *SAR* systemic acquired resistance, *Pvd* siderophore pyoverdine

metals (Lugtemberg and Kamilova 2009; Richardson et al. 2009; Bakker et al. 2013; Prathap and Ranjitha-Kumari 2015). Based on these beneficial effects derived from plant-PGPR interactions, a significant number of PGPR-based products have been commercialized in recent years, and these have proven to be efficient biofertilizers, phytostimulators and biopesticides (Table 3.1). Among them, bacterial strains belonging to the *Bacillus*, *Pseudomonas, Serratia* and *Azospirillum* genera, which are equipped with multiple mechanisms for biocontrol of phytopathogens and plant growth promotion, are within the most exploited biopesticides and biofertilizers (Berg 2009; Bashan et al. 2014; Ahirwar 2015; Vejan et al. 2016).

3.2.1 Root Exudates Regulate PGPR Functions

Plant roots release large amounts of organic and inorganic compounds through secretion, diffusion and cell lysis. These compounds include carbohydrates, polysaccharides, amino acids, organic acids and fatty acids, among others, which are responsible for the generation of a wide and flexible barrier of chemicals in the rhizosphere (Uren 2007; Jones et al. 2009). Plant root exudates are passively released, and in general, the rate of exudation was shown to be higher at root tips, root hairs and regions of emergence of primary and secondary roots (Neumann y

Table 3.1 Commercial biofertilizers based on plant growth-promoting rhizobacteria

Product name	Bacterial composition	Product properties	Company/patent
Nitrocode AZ+	*Azospirillum* spp.	Plant growth promotion due to the production of phytohormones and the fixation of atmospheric nitrogen	Agrocode
BioNem	*Bacillus firmus* GVB-126	Suppression of diseases caused by plant-pathogenic nematodes	Bayer CropScience
Green Releaf	*Bacillus licheniformis* SB3086	Biocontrol of plant pathogenic fungi through the antibiotic agents and hydrolytic enzymes	Novozymes Biologicals, Inc
YieldShield	*Bacillus pumilus* strain GB34	Inducion of systemic resistance against *Rhizotonia* and *Fusarium*	Bayer CropScience
Serenade ASO®	*Bacillus subtilis* QST713	Fungicide active against phytopathogenic fungi and oomycete such as *Pythium, Rhizoctonia, Fusarium* and *Phytophthora*	Bayer CropScience
FZB24®	*Bacillus subtilis* FZB24	Induction of sytemic resistance, promotion of plant growth and inhibition of soil-borne pathogens	ABITEP
Easy Start® TE-Max	*Bacillus subtilis* E4-CDX	Efficient colonizer of grass roots. Inhibits diseases caused by phytopathogenic fungi	COMPO Expert GmbH
BioBoost®	*Delftia acidovorans*	Positively impact on plant root growth due to the production of phytohormones and the oxydation of sulphur	BrettYoung
NH	*Paenibacillus polymyxa* AC-1	Plant growth promotion due to the production of phytohormones and bioactive secondary metabolites	Green Biotech Company Ltd
Cedomon®	*Pseudomonas chlororaphis*	Biopesticide effective against seed-borne diseases in barley and oats seeds	Bio-Agri AB
BlightBan® A506	*Pseudomonas fluorescens* A506	Biocontrol of fire blight on pome fruits and suppression of frost damage on economically important crops	NuFarm
Fosfogel®	*Pseudomonas putida* BIRD-1	Promotes plant rooting and growth mainly due to the synthesis of indole acetic acid and phosphatases	Bio-Iliberis R&D
Fungikiller®	*Pseudomonas fluorescens* BIRD-2	Inhibits the growth of phytopathogenic fungi and oomycete from the *Fusarium, Rhizoctonia, Pythium* and *Phytophthora* genera	Bio-Iliberis R&D
EVL Coating	PGPR consortia	Plant growth promotion due to the increase uptake of mineral nutrients	EVL INC
VitaSoil®	PGPR consortia	Plant growth promotion due to the increase uptake of mineral nutrients	Symborg
Bioshield™	*Serratia entomophila*	Control of grass grub larvae	Ballance Agri-Nutrients
Rhizostar®	*Serratia plymuthica* HRO-C48	Fungicide to control *Verticillium* wilt	EU Patent 98124694.5
MSU97	*Serratia marcescens* MSU97	Biocontrol agent to protect plants from oomycete pathogens	Patent WO 2002091825 A2

Römheld 2007; Uren 2007; Jones et al. 2009). The released compounds, in addition to their use as nutrients by rhizospheric microorganisms, can act as signal molecules to initiate diverse physical and chemical interactions between plant roots and rhizospheric bacteria (Wen et al. 2007; De la Peña et al. 2008; Hawes et al. 2012; Baetz and Martinoia 2013; Huang et al. 2014). As a result, through the secretion of metabolites by roots, plants can modulate bacterial gene expression, including the expression of genes involved in the establishment of bacteria-plant interactions and encoding other plant-beneficial traits (Mark et al. 2005; Matilla et al. 2007a; Fan et al. 2012; Vacheron et al. 2013).

The chemical properties and composition of plant root exudates vary according to plant species and age but are also largely influenced by multiple biotic and abiotic factors (Uren 2007; Lesuffleur and Liquet 2010; Badri and Vivanco 2009; Matilla et al. 2010). Thus, several studies have shown that PGPR can elicit changes in the root metabolism, leading to changes in root exudation patterns (Matilla et al. 2010; Vacheron et al. 2013). On the other hand, plants can drive and shape the composition of their microbiome depending on the chemical composition of their root exudates (Berendensen et al. 2012; Chaparro et al. 2013b). As an example, under N-limiting conditions, legume roots secrete increasing amounts of flavones and flavonols that favour the initiation of the legume-rhizobia symbiosis (Abdel-Lateif et al. 2012). Furthermore, components of plant root exudates can be sensed and metabolized by PGPR and serve as chemoattractants to facilitate plant root colonization (Reyes-Darias et al. 2015; Yuan et al. 2015; Corral-Lugo et al. 2016; Webb et al. 2016).

3.2.2 *Improvement of Plant Nutritional Status by PGPR*

3.2.2.1 Phosphorous

Phosphorous is a major growth-limiting nutrient that is mainly present in soils as insoluble inorganic phosphate, thereby limiting the levels of soluble phosphate accessible to plants (Rodríguez et al. 2006; Yang et al. 2009). The ability of some PGPR to solubilize phosphate is an important trait for promoting plant growth, and plant-associated bacteria belonging to the *Pseudomonas*, *Bacillus* or *Rhizobium* genera are efficient phosphate solubilizers (Rodríguez and Fraga 1999; Rodríguez et al. 2006; Rosas et al. 2006; Ramaekers et al. 2010; Glick 2012; Roca et al. 2013).

PGPR can use various strategies to solubilize phosphate, but it is well recognized that the primary mechanism consists in the production of low molecular weight organic acids. The production and secretion of these organic acids either decrease the pH of the soil or chelate mineral ions, finally resulting in phosphate solubilization (Glick 2012; Khan et al. 2014). Gluconic and 2-ketogluconic acids are referred as some of the most frequent organic acids involved in phosphate solubilization by Gram-negative PGPR (Sashidhar and Podile 2010; Roca et al. 2013; Khan et al. 2014). However, other organic acids such as lactic, isovaleric, isobutyric, acetic, glycolic, oxalic, malonic and succinic are also produced by multiple phosphate-solubilizing bacteria (Rodríguez and Fraga 1999; Ahemad and Kribet 2014; Khan et al. 2014). Alternatively, PGPR can release organic and inorganic phosphorus from soil compounds by the secretion of enzymes such as phosphatases, phytases, phosphonatases and lyases. Among these extracellular enzymes are acid phosphatases, which are widely distributed within plant-associated bacteria, and their action is the main mechanism for the enzymatic phosphate solubilization (Rossolini et al. 1998; Vassilev et al. 2006; Mohammadi 2012; Roca et al. 2013; Khan et al. 2014).

3.2.2.2 Nitrogen

Most biological N_2 fixation processes are being carried out by the activity of molybdenum nitrogenases found in diazotrophic (nitrogen fixing) bacteria (Bishop and Jorerger 1990; Glick 2012). Among them, rhizobia form root nodules in leguminous plants and are responsible for most of the bacterial-mediated N_2 fixation (Glick 2012; Maróti and Kondorosi 2014). However, in addition to rhizobia, non-symbiotic bacteria are also able to fix N_2 and make it accessible to plants, but their contribution to the total biologically accessible N_2 for plants is under debate (Steenhoudt and Vanderleyden 2000; Somers et al. 2004; Glick 2012). Among these non-symbiotic N_2-fixing bacteria, free-living strains from the *Azospirillum* genus are the best characterized, but their plant growth-promoting properties have been mainly associated to the production of phytohormones and to the increase in the uptake of mineral nutrients by plant roots (Steenhoudt and Vanderleyden 2000; Lugtenberg and Kamilova 2009).

3.2.2.3 Iron

The availability of this essential micronutrient for living organisms in soils is limited (Ma 2005; Aguado-Santacruz et al. 2012). However, in response to iron deficiency, bacteria can secrete low molecular mass iron-chelating molecules called siderophores. Siderophores are capable of scavenging ferric ions, the most common form of iron in nature, forming stable siderophore-Fe^{+3} complexes. These complexes are subsequently recognized by specific transporters at the bacterial surface in a process that finally results in the reduction of Fe^{+3} and the release of soluble Fe^{+2} into the cytoplasm (Schalk et al. 2011; Ahmed and Holsmtrön 2014; Llamas et al. 2014). Currently, more than 500 different siderophores have been identified (Hider and Kong 2010), and among them, pyoverdines are the main siderophores produced by plant growth-promoting fluorescent pseudomonads (Visca et al. 2007; Matilla et al. 2007b, 2011b; Llamas et al. 2014).

Siderophores produced by PGPR can exert beneficial effects on plant growth by different direct and indirect mechanisms (Lemanceau et al. 2009; Gamalero and Glick 2011). Direct beneficial effects include the improvement of the iron nutritional status of the plant and the subsequent promotion of plant growth. Although the exact mechanisms of this process are currently unknown, it has been hypothesized that bacterial siderophores can chelate Fe^{+3} from soils and make it accessible to phytosiderophores (Glick 2012; Ahmed and Holsmtrön 2014; Saha et al. 2016). On the other hand, it has been shown that Fe^{+3}-pyoverdine complexes can be incorporated by the plant, finally resulting in an increase iron content in the plant tissues (Vansuyt et al. 2007; Ahmed and Holsmtrön 2014).

Indirect plant growth promotion mediated by bacterial siderophores mainly results from their capacity to reduce the availability of iron to phytopathogens (Lugtenberg and Kamilova 2009; Yu et al. 2011; Schenk et al. 2012; Saha et al. 2013; Ahmed and Holsmtrön 2014). Importantly, the perception of bacterial siderophores by plants has been proven to activate plant defences by triggering induced systemic resistance (ISR) (Höfte and Bakker 2007; De Vleesschauwer and Höfte 2009; Aznar and Dellagi 2015). However, the elicitation of the plant immune responses by bacterial siderophores has been recently shown to be dependent on the plant iron status (Trapet et al. 2016).

3.2.3 Phytohormone Production

Plant-associated bacteria can also benefit plants through the synthesis of phytohormones or the modulation of the plant hormonal balance (Karadeniz et al. 2006; Kloepper et al. 2007; Lugtenberg and Kamilova 2009; Glick 2012). The phytohormones ethylene, auxins, abscisic acid, cytokinins, or gibberellins control plant growth and development. Conversely, ethylene as well as jasmonic and salicylic acid are stress-essential regulators of plant immunity since they are responsible for creating a central signalling backbone that specifically coordinates defence responses against phytopathogens (Bari and Jones 2009; Pieterse et al. 2012; Denancé et al. 2013; Naseem et al. 2015; Shigenaga and Arqueso 2016). In this chapter, the role of phytohormones produced by PGPR in the physiology of the plant is briefly reviewed.

3.2.3.1 Ethylene

The hormone ethylene (ET), among other effects, inhibits root elongation and auxin transport (Matilla-Vázquez and Matilla 2014). Importantly, it has been shown that some plant-pathogenic bacteria produce ET, and although its role in the development of the disease remains unclear, experimental data suggest that it may act as a virulence factor *in planta* (Weingart et al. 2001; Matilla-Vázquez and Matilla 2014). Alternatively, PGPR can act as sink for the ET precursor, 1-aminocyclopropane-1-carboxylic acid (ACC), consequently lowering ET levels in roots and concomitantly increasing root length and plant growth (Glick et al. 2007; Matilla-Vázquez and Matilla 2014; Singh et al. 2015). Thus, ACC exuded by roots and seeds can be actively taken up by plant-associated rhizobacteria and subsequently cleaved to ammonia and α-ketobutyrate by the bacterial ACC deaminase (ACCd) (Glick et al. 2007). The ACCd-encoding genes are widespread within PGPR bacteria, and ACC deaminase activities have been found in a multitude of PGPR strains (Singh et al. 2015). As a consequence, PGPR with ACCd activity have been shown to possess the capacity to reduce ET production in the colonized roots and promote plant growth, mainly in response to multiple biotic and abiotic stresses (Glick et al. 2007; Gamalero and Glick 2012, 2015; Matilla-Vázquez and Matilla 2014; Singh et al. 2015). Recently, it was shown that the biocontrol properties of *Pseudomonas putida* strain UW4 against pine wilt syndrome, a disease caused by the nematode *Bursaphelenchus xylophilus,* were dependent on the ACC deaminase activity of UW4 (Nascimento et al. 2013).

3.2.3.2 Auxins

Auxins regulate multiple plant functions but, in general, affect plant growth and development by stimulating plant cell division, elongation and differentiation. At the root level, auxins can promote root elongation, formation of lateral roots and root hairs (Spaepen and Vanderleyden 2011; Glick 2012; Vacheron et al. 2013; Duca et al. 2014). It has been estimated that around 80% of the rhizospheric bacteria possess the ability to synthesize and secrete auxins, mainly indole-3-acetic acid (IAA) (Spaepen and Vanderleyden 2011; Glick 2012). It is currently accepted that IAA production by PGPR increases root length and surface, thereby resulting in an increased uptake of soil nutrients and plant growth (Spaepen et al. 2007; Lugtenberg and Kamilova 2009; Spaepen and Varderleyden 2011). However, the production of IAA by bacterial pathogens has been associated with phytopathogenesis. The current proposed mechanism for this increased plant susceptibility to IAA-producing bacterial pathogens results from the IAA-mediated decrease in plant cell wall integrity and from the inhibition of programmed cell death (Spaepen and Varderleyden 2011; Duca et al. 2014).

Although tryptophan-independent pathways of IAA synthesis have been demonstrated, tryptophan is the main precursor for the biosynthesis of IAA in bacteria, and five different tryptophan-dependent biosynthetic pathways have been described (Spaepen and Varderleyden 2011; Patten et al. 2013; Vacheron et al. 2013) (Table 3.2). Tryptophan is a component of root exudates, and it can be taken up by bacterial cells to promote IAA synthesis (Kamilova et al. 2006; Glick 2012). Importantly, it has recently been shown that the plant growth-promoting rhizobacterium *Bacillus amyloliquefaciens* SQR9 increases around fourfold the tryptophan secretion by cucumber plant roots. This increased exudation of tryptophan resulted in an enhanced production of IAA by SWR9 and in the stimulation of total root surface area (Liu et al. 2016). Interestingly, auxin-signalling pathways and lateral root growth in *Arabidopsis* can be elicited by non-IAA-producing PGPR, probably by altering IAA distribution in the root system (Contesto et al. 2010). In accordance with this, the plant-associated bacterium *Pseudomonas putida* 1290 exhibits chemotaxis towards IAA, and it can use the phytohormone as carbon and energy source

Table 3.2 Biosynthesis of indole-3-acetic acid in plant-associated bacteria

Biosynthetic pathway	Biosynthetic intermediates	Producing bacteria[a]
Indole-3-acetamide pathway	Indole-3-acetamide	*Agrobacterium tumefaciens*, *Erwinia herbicola*, *Dickeya dadantii* 3937
		Pseudomonas fluorescens, *Pseudomonas savastanoi*, *Pseudomonas syringae*, *Pantoea agglomerans*, *Ralstonia solanacearum*, *Rhizobium* sp., *Bradyrhizobium* sp.
Indole-3-pyruvate pathway	Indole-3-piruvic acid, indole-3-acetaldehyde	*Agrobacterium brasilense*, *Azospirillum brasilense*, *Azospirillum lipoferum*, *Bacillus subtilis*, *Enterobacter cloacae*, *Paenibacillus polymyxa*, *Pantoea agglomerans*, *Pseudomonas putida*, *Rhizobium* sp., *Serratia plymuthica*
Tryptamine pathway	Tryptamine, indole-3-acetic aldehyde	*Azospirillum brasilense*, *Bacillus cereus*
Indole-3-acetonitrile	Indole-3-acetaldoxime, indole-3-acetonitrile	*Agrobacterium tumefaciens*, *Bacillus amyloliquefaciens*, *Pseudomonas syringae*, *Pseudomonas fluorescens*, *Rhizobium* sp.
Tryptophan side-chain oxidase	Indole-3-acetaldehyde	*Pseudomonas fluorescens*
Tryptophan-independent pathway	Unknown	*Azospirillum brasilense*

[a]Sources: Spaepen and Vanderleyden (2011) and Duca et al. (2014)

(Scott et al. 2013). Multiple bacterial IAA degraders have been described, and their IAA-degradation properties may help to modulate different aspects of plant physiology, leading to an improvement of their survival in the rhizosphere (Scott et al. 2013; Duca et al. 2014). Thus, IAA degradation by the plant-associated bacterium *Burkholderia phytofirmans* positively influences the colonization of the rhizosphere by the strain and also its plant growth-promoting properties (Zúñiga et al. 2013).

3.2.3.3 Cytokinins

Cytokinins (CKs) regulate multiple processes in plants, including the promotion of cell division and the modulation of cell growth and differentiation (Frébort et al. 2011; Ha et al. 2012). The production of CKs, especially zeatin, has been reported in various plant-associated bacteria, including PGPR from the *Pseudomonas* and *Bacillus* genera (García de Salome et al. 2001; Arkhipova et al. 2005; Frébort et al. 2011; Glick 2012). Although the production of CKs in phytopathogens has been associated with plant invasion and virulence (Frébort et al. 2011), their mechanisms of action in PGPR are less known. However, it has been speculated that CKs produced by beneficial plant-associated bacteria may become part of the plant CKs pool, subsequently influencing plant growth and development (Glick 2012). Alternatively, it was recently shown that CKs, synergistically with salicylic acid (SA), induce the expression of defence genes in plants (Jiang et al. 2013; Naseem et al. 2015). In accordance with this observation, the production of CKs by the PGPR *Pseudomonas fluorescens* G20-18 has been associated with its ability to protect *Arabidopsis* plants against the phytopathogen *Pseudomonas syringae*. Thus, mutants of *P. fluorescens* G20-18 defective in CKs synthesis exhibited reduced biocontrol properties *in planta*. The authors also showed that the activation of plant resistance mediated by G20-18 required the functional perception of bacterial-derived cytokinins by the plant (Großkinsky et al. 2016).

3.2.3.4 Abscisic Acid

The phytohormone abscisic acid (ABA) regulates multiple aspects of plant growth and development, including plant responses to different environmental stresses such as cold, salinity and desiccation (Finkelstein 2013). Several plant-associated bacteria have been shown to produce ABA and to increase the levels of the phytohormone *in planta* (Frankenberger and Arshad 1995; Perrig et al. 2007; Cohen et al. 2008, 2015; Sgroy et al. 2009; Salomon et al. 2014). This bacterial-mediated increase in the ABA concentration was shown to alleviate multiple environmental stresses, finally resulting in plant growth promotion (Cohen et al. 2008, 2015; Salomon et al. 2014). Alternatively, some beneficial rhizobacteria were shown to metabolize ABA and to decrease its concentration (or spatial distribution) *in planta* (Hartung et al. 1996; Zhang et al. 2008; Jiang et al. 2012; Belimov et al. 2014). This reduction in the levels of the phytohormone correlated with an increase in the chlorophyll content, photosynthetic efficiency and plant growth promotion (Zhang et al. 2008; Jiang et al. 2012; Belimov et al. 2014).

It is also becoming apparent that ABA is an important factor involved in modulating plant defences. Thus, plant mutants impaired in ABA biosynthesis or insensitive to the perception of the hormone are more resistant to pathogens compared to wild-type plants (Adie et al. 2007; Asselbergh et al. 2008; Xu et al. 2013). Furthermore, the addition of exogenous ABA prevents accumulation of SA and results in the suppression of the resistance against phytopathogens in different plant species (Mohr and Cahill 2003; Koga et al. 2004). This phenomenon is based on the antagonistic action of ABA in SA-mediated signalling processes, therefore blocking the induction of systemic resistance (SAR) (Yasuda et al. 2008; de Torres et al. 2009; Jiang et al. 2010; Xu et al. 2013). In accordance with this, several phytopathogens have been shown to activate ABA synthesis in their host plants in order to promote virulence (de Torres-Zabala et al. 2007, 2009). It can be hypothesized that the degradation of ABA by PGPR may result in the activation of the plant-inducible defence responses.

3.2.4 Induction of Systemic Resistance

In the early 1990s, initial evidences indicating that PGPR can stimulate plant immune system were provided (Alström 1991; Van Peer et al. 1991; Wei et al. 1991). Since then, multiple studies (mainly involving strains from the *Pseudomonas*, *Bacillus* and *Serratia* genera) have reported the potential of PGPR to induce resistance locally and systemically in plants (De Vleesschauwer and Höfte 2009; Pieterse et al. 2014). The phenomenon of PGPR-mediated induced systemic resistance (ISR) requires the efficient colonization of plant roots by the inducing bacteria (Raaijmakers et al. 1995; Lugtenberg and Kamilova 2009) and is characterized by the systemic activation of plant defences. As a result, ISR-triggering PGPR prime the plants to respond faster and stronger to the attack of a broad range of attacking species, including bacteria, fungi, oomycetes, viruses and insects (Verhagen et al. 2004; Conrath et al. 2006; Berendsen et al. 2012; Walters et al. 2013; Pieterse et al. 2014). For the onset of the systemic immunity, PGPR must produce elicitors, named microbe-associated molecular patterns (MAMPs), that stimulate pattern-recognition receptors (PRRs) of plants. These elicitors include proteins, lipids, carbohydrates and small molecules (Boller and Felix 2009; De Vleesschauwer and Höfte 2009; Pieterse et al. 2014), and some of them were found to act redundantly triggering ISR (Meziane et al. 2005; Pieterse et al. 2014). As observed in PGPR, plant pathogens also produce elicitors (named pathogen-associated molecular patterns or PAMPs) that can also be recognized by the plant, finally resulting in an enhanced systemic defence capacity towards a broad spectrum of phytopathogens. This pathogen-mediated induced resistance is known as systemic acquired resistance (SAR) and is characterized by an increase in the levels of SA and the expression of pathogenesis-related (PR) encoding genes in the triggered plants (Vlot et al. 2009; Spoel and Dong 2012; Pieterse et al. 2014). Although ISR and SAR are phenotypically similar, ISR is SA-independent, does not involve the accumulation of PR proteins and requires intact jasmonic acid (JA) and ET signalling pathways (van Loon et al. 1998; Matilla-Vázquez and Matilla 2014; Pieterse et al. 2014). However, some exceptions include PGPR that require the SA-dependent pathway in order to induce plant immunity (Matilla et al. 2010; van de Mortel et al. 2012; Pieterse et al. 2014). Overall, ISR and SAR provide protection against a broad range of pathogens, and both strategies reduce the fitness costs of having plant defences constitutively activated (Conrath et al. 2006; Pieterse et al. 2014).

3.2.5 Production of Lytic Exoenzymes

Fungal cell wall (CW) is mainly composed of chitin, glucans and proteins (Peberdy 1990) and a number of PGPR produce exoenzymes such as proteases, chitinases, cellulases and β-1,3-glucanases (Chernin and Chet 2002; Nagpure et al. 2014; Parray et al. 2016). Among them, chitinases are ubiquitous lytic enzymes responsible for the hydrolyzation of the β-(1–4)-glycosidic bonds of the polymer chitin (Nagpure et al. 2014; Mousa and Raizada 2015). Chitinases are one of the most characterized antifungal enzymes, and chitinase-producing plant- and soil-associated bacteria, specially belonging to *Bacillus*, *Pseudomonas* and *Streptomyces* genera, have been shown to be effective against plant-pathogenic fungi (Quecine et al. 2008; Nagpure et al. 2014; Mousa and Raizada 2015). Importantly, the exoskeletons of insects, arthropods and nematodes consist largely of chitin, and chitinase-producing PGPR are potential biocontrol agents acting as insecticides and nematicides (Tripathi et al. 2002; Nagpure et al. 2014). Besides the secretion of chitinases, the synthesis of additional lytic enzymes such as proteases and β-1,3-glucanases was shown to be essential for the biocontrol properties of certain PGPR since these exoenzymes have been demonstrated to play a role in CW degradation of plant pathogenic fungi and oomycetes (Dunne et al. 1998; Kamensky et al. 2003; Compant et al. 2005; Glick 2012; Li et al. 2015). Importantly, the synthesis of lytic enzymes is highly controlled, and the two component systems GacA/GacS and GrrA/GrrS were among the most characterized regulatory systems (Heeb and Haas 2001; Haas and Keel 2003; Ovadis et al. 2004; Workentine et al. 2009; Kim et al. 2011).

3.3 Synthesis of Volatiles and Secondary Metabolites by PGPR

As stated above, plant growth promotion mediated by PGPR can be linked to their capacity to protect plants against pathogens, and PGPR are an extraordinary source of antimicrobial, antiviral and nematicide compounds. Thus, plant-associated bacteria produce multiple bioactive volatile compounds and secondary metabolites. The latter can be classified into different groups based on their biosynthetic origin and chemical structure and mainly include non-ribosomal peptides (NRP), polyketides (PK), terpenoids and bacteriocins (Haas and Défago 2005; Gross and Loper 2009; Pidot et al. 2014; Chowdhury et al. 2015a; Mousa and Raizada 2015; Venugopalan and Srivastava 2015). The scope of this section is to describe the diversity of bioactive volatiles and secondary metabolites produced by plant-associated bacteria that have a role in the biocontrol of plant diseases.

3.3.1 Production of Hydrogen Cyanide

Hydrogen cyanide (HCN) is a volatile compound that acts as a strong inhibitor of many metalloenzymes, besides chelating and inactivating multiple metals in soils (Blumer and Haas 2000; Brandl et al. 2008). HCN can be synthesized by bacteria, algae, fungi, plants and insects as a mean to avoid predation or competition, mainly due to its broad spectrum of antibiotic activity (Blumer and Haas 2000; Haas and Défago 2005; Gross and Loper 2009). The synthesis of HCN in cyanogenic bacteria occurs through the oxidation of glycine mediated by the flavoenzyme HCN synthase (Blumer and Haas 2000). HCN biosynthesis in PGPR has been mostly investigated in fluorescent pseudomonads (Haas and Keel 2003; Haas and Défago 2005; Gross and Loper 2009), and it was estimated that around 50% of rhizosphere-isolated pseudomonads are able to produce HCN *in vitro* (Bakker and Schippers 1987). Interestingly, a phylogenetic analysis of 30 plant-associated fluorescent pseudomonads showed no evidence for an acquisition of HCN biosynthetic gene clusters (*hcnABC*) through horizontal gene transfer (Frapolli et al. 2012).

The first experimental evidence of HCN in plant biocontrol was reported in the late 1980s using the PGPR *Pseudomonas protegens* CHA0 (Voisard et al. 1989). Since then, the characterization of mutants defective in the synthesis of cyanide (or the transfer of HCN biosynthetic gene clusters from cyanogenic to non-cyanogenic bacteria) allowed to demonstrate that cyanogenesis in PGPR is an important trait for the efficient biocontrol of plant diseases caused by phytopathogenic fungi, oomycetes and nematodes (Voisard et al. 1989; Flaishman et al. 1996; Nandi et al. 2015; Zdor 2015). Interestingly, the simultaneous production of HCN and the polyketide 2,4-diacetylphloroglucinol (DAPG) resulted in an increased potential for the biological control of the bacterial canker of tomato (Lanteigne et al. 2012). Although the use of cyanogenic bacteria for the biocontrol of plant diseases has been demonstrated, several studies also showed that HCN inhibits plant growth (Rudrappa et al. 2008; Grossmann 2010; Zdor 2015). Therefore, although the biochemical and cellular mechanisms of HCN on plant growth suppression remain to be identified, the use of cyanogenic biopesticides/biofertilizers may require further consideration when optimizing crop growth and yield.

3.3.2 Volatile Organic Compounds

Plant-associated bacterial strains are able to produce and actively release a broad range of volatile organic compounds (VOCs). VOCs are molecules with low molecular weight (<300 Da), low boiling points and high vapour pressure, and so they can freely cross though the cellular membranes and diffuse into the surrounding environment (Wheatley 2002; Effmert et al. 2012; Audrain et al. 2015; Kanchiswamy et al. 2015; Chung et al. 2016). Currently, more than 1500 different VOCs have been isolated from 450 bacterial and fungal strains (Lemfack et al. 2014). Based on their structure, bacterial VOCs are currently classified into six chemical groups, including acids, hydrocarbons, ketones/alcohols, nitrogen-containing compounds, sulphur compounds and terpenes (Audrain et al. 2015). However, the biosynthetic routes for most of these VOCs remain unknown, but an interconnection between primary and secondary metabolisms for the biosynthesis of some of them has been suggested (Dudareva et al. 2013; Audrain et al. 2015; Kanchiswamy et al. 2015).

Many of the identified bacterial VOCs are produced by soil- and plant-associated bacteria (Effmert et al. 2012; Audrain et al. 2015; Kanchiswamy et al. 2015). Thus, more than 120 different VOCs were identified within 26 *Streptomyces* species (Schöller et al. 2002; Dickschat et al. 2005; Lemfack et al. 2014). In addition, rhizobacterial strains belonging to *Bacillus*, *Enterobacter*, *Paenibacillus*, *Pseudomonas*, *Serratia* and *Stenotrophomonas* genera were found to produce a high diversity of VOCs (Ryu et al. 2003; Kai et al. 2007; Vespermann et al. 2007; Farag et al. 2013; Lemfack et al. 2014 Audrain et al. 2015; Kanchiswamy et al. 2015). Among them, one particularly interesting PGPR strain is *Serratia plymuthica* 4Rx13, which has been shown to produce more than 100 chemically different VOCs with multiple biological properties. Interestingly, the bicyclic octadiene sodorifen contributes to around 50% of the total content of VOCs produced by 4Rx13, and recent data suggest that its biosynthesis derives from the metabolism of terpenes (Kai et al. 2010; Domik et al. 2016).

The function of most of the currently identified bacterial VOCs remains elusive, but it has been shown that they can (1) either positively or negatively influence the growth of plants, bacteria or fungi, (2) stimulate plant defences or (3) act as interspecies chemical signals between bacteria, plants, fungi and nematodes (Ryu et al. 2003; Wenke et al. 2012; Farag et al. 2013; Audrain et al. 2015; Kanchiswamy et al. 2015; Chung et al. 2016). Interestingly, recent research showed that VOCs emitted from pathogen-resistant plants induce resistance against phytopathogens in susceptible plants (Quintana-Rodríguez et al. 2015). It is currently unknown whether plants discriminate between VOCs produced by beneficial and phytopathogenic microorganisms, and further research is needed to shed light into how VOCs of different sources are perceived by plant cells.

VOCs emitted by PGPR were shown to enhance plant growth, modulate the root system architecture or regulate processes such as cell expansion and photosynthetic efficiency (Ryu et al. 2003; Zhang et al. 2007; Xie et al. 2009; Gutierrez-Luna et al. 2010; Blom et al. 2011; Bailly and Weisskopf 2012; Groenhagen et al. 2013; Kanchiswamy et al. 2015). The VOCs acetoin and 2,3-butanediol were originally isolated from the PGPR *Bacillus subtilis* GB03 and *B. amyloliquefaciens* IN937a, and identified as the first bacterial-derived VOCs to play a role in plant growth and development (Ryu et al. 2003). Subsequent studies showed that acetoin and 2,3-butanediol induce ISR in *Arabidopsis* against *Pectobacerium carotovorum* subsp. *carotovorum* and *Pseudomonas syringae*. For the triggering of these immune responses, the activation of ET-dependent signalling pathways was required (Ryu et al. 2004; Rudrappa et al. 2010). Recent data also showed that the production of 2,3-butanediol by the PGPR *Enterobacter aerogenes* increases protection of maize seedlings against phytopathogenic fungi (D'Alessandro et al. 2014). Additionally, the role of long-chain VOCs produced by the PGPR *Paenibacillus polymyxa* E681 in triggering ISR was also evidenced (Lee et al. 2012). Alternatively, bacterial VOCs such as 2,3-butanediol were shown to benefit plant health conferring tolerance to abiotic stresses such as drought, salinity and heavy metals (Cho et al. 2008; Bitas et al. 2013; Farag et al. 2013). On the other hand, certain VOCs produced by strains from the *Burkholderia*, *Pseudomonas* and *Serratia* genera were demonstrated to drastically inhibit plant growth (Vespermann et al. 2007; Kai et al. 2009, 2010; Bailly and Weisskopf 2012; Bitas et al. 2013; Kanchiswamy et al. 2015). It has been suggested that bacterial pathogens can use VOCs to suppress plant immune responses (Blom et al. 2011).

The production of VOCs by PGPR was shown to protect plants against a wide range of pathogens. Thus, VOCs produced by *Serratia plymuthica* and *Pseudomonas fluorescens* were found to inhibit the growth of *Agrobacerium tumefaciens* and *A. vitis in vitro* (Dandurishvili et al. 2011). Additionally, the VOCs 2,3-butanediol and 2,3-butanedione produced by *Bacillus subtilis* repressed the expression of virulence factor-encoding genes in the phytopathogenic bacterium *P. carotovorum* subsp. *carotovorum* (Chung et al. 2016). However, particular attention to the antagonist properties of bacterial VOCs has been given to those volatiles showing inhibitory effects against plant-pathogenic fungi (Kanchiswamy et al. 2015). VOCs produced by plant-associated biocontrol strains belonging to *Serratia*, *Pseudomonas*, *Bacillus* and *Burkholderia* genera inhibit the growth of multiple plant pathogenic fungi and oomycete, including *Rhizoctonia solani, Colletotrichum gloeosporioides, Sclerotinia sclerotiorum, Phytophthora infestans* and *Fusarium oxysporum* (Wheatley 2002; Fernando et al. 2005; Vespermann et al. 2007; Kai et al. 2010; Elshafie et al. 2012; Groenhagen et al. 2013; De Vrieze et al. 2015).

3.3.3 Secondary Metabolites: Non-ribosomal Peptides and Polyketides

Non-ribosomal peptides (NRPs) and polyketides (PKs) constitute the main secondary metabolites produced by PGPR (Haas and Défago 2005; Gross and Loper 2009; Pidot et al. 2014; Chowdhury et al. 2015a; Mousa and Raizada 2015; Venugopalan and Srivastava 2015). The structural diversity of NRPs and PKs is highly remarkable, even though current estimations suggest that only 1% of the total number of secondary metabolites have been discovered (Fischbach and Walsh 2009). This chemical diversity is reflected in their broad range of biological and pharmacological properties since these metabolites can act as pigments, siderophores, antibacterial, antifungal, nematicides, antitumour and immunosuppressants (Sattely et al. 2008; Gross and Loper 2009; Hertweck 2009; Mousa and Raizada 2015).

NRPs constitutes a large group of natural products which are synthesized by large multifunctional enzymes, the non-ribosomal peptide synthetases (NRPS). Although recent *in silico* analyses showed that non-modular NRPS are frequently found in the bacterial genomes (Wang et al. 2014), in general, NRPS are organized into sets of repeated catalytic units called modules (Sattely et al. 2008; Miller and Gulick 2016). Each of these modules is responsible of the addition of a single amino acid during the biosynthetic process. The minimal NRPS module required for a single monomer addition consists of three domains. The adenylation domain (A) is responsible for selecting and loading amino acids onto the adjacent peptide carrier protein (PCP), where the growing chain remains attached via a thioester bond. The condensation domain (C) catalyzes the formation of the peptide bond between the amino acid loaded on the PCP domain of the same module and the peptidyl intermediate from the preceding module (Fig. 3.2). Additionally, the minimal A-C-PCP module can also include additional domains such as methyltranferases or epimerases (Sattely et al. 2008; Strieker et al. 2010; Gulick 2016; Miller and Gulick 2016).

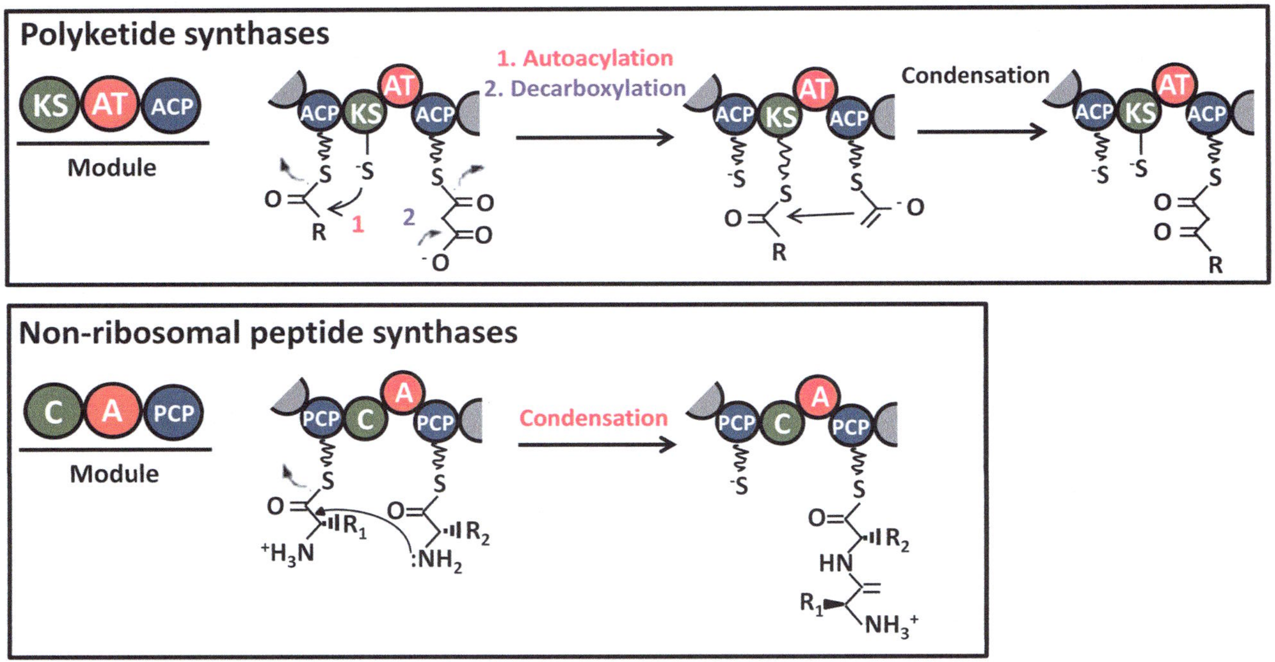

Fig. 3.2 Biosynthesis of non-ribosomal peptides and polyketides. The activity of ketosynthase (KS) and condensation (C) domains in the synthesis of polyketides and non-ribosomal peptides, respectively. C and KS domains catalyze the formation of amide and carbon-carbon bonds, respectively, between the up- and downstream substrates attached on the peptide carrier protein (PCP) and acyl carrier protein (ACP) domains. *AT* acyltransferase domain, *A* adenylation domain (Adapted from Sattely et al. 2008)

NRPs can be linked to fatty acids generating linear or cyclic lipopeptides with a broad range of antagonistic bioactivities against a wide range of plant-pathogenic bacteria, fungi and oomycetes (Ongena et al. 2007a; Raaijmakers et al. 2010). The mechanism of action of lipopeptides mainly results from their ability to bind to lipid layers, subsequently interfering with the integrity and permeability of membranes (Ongena et al. 2007a; Raaijmakers et al. 2010; Hamley 2015). However, non-ribosomal lipopeptides such as surfactin and fengycin were also found to act as ISR elicitors (Ongena et al. 2007b). Among the lipopeptides produced by PGPR, those produced by *Pseudomonas* and *Bacillus* strains are the best characterized (Raaijmakers et al. 2010; Chowdhury et al. 2015a).

The other main group of bacterial-derived secondary metabolites, polyketides, are synthesized by polyketide synthases (PKS) (Hertweck 2009; Soares-Gomes et al. 2013; Till and Race 2014; Helfrich and Piel 2016). Although type II and III non-modular PKS are present in bacteria, most of the known bacterial polyketides are produced by multimodular type I PKS (Hertweck 2009). As described above for NRPS, type I PKS are large multifunctional enzymes in which the catalytic domains are covalently fused and organized in modules; each of which is responsible of one elongation cycle of the growing polyketide. The minimal functional module consists of an acyl carrier protein (ACP) and domains with acyltransferase (AT) and ketosynthase (KS) activities. During each elongation cycle, AT domains add new extender units (generally malonyl-CoA or acetyl-CoA) onto the ACP domains, and the KS domains catalyze the condensation reactions, while the growing polyketide is covalently attached to the ACP domain (Fig. 3.2). Furthermore, KS-AT-ACP modules can be supplemented with domains that exert ketoreductase, dehydratase and enoylreductase activities (Sattely et al. 2008; Hertweck 2009; Till and Race 2014; Helfrich and Piel 2016). Importantly, bacteria can also produce NRP/polyketide hybrids, which are secondary metabolites that are synthesized by type I PKS-NRPS hybrids (Du and Shen 2001; Fisch 2013).

Polyketides, NRPs and polyketide/NRP hybrids can also suffer post-assembly modifications mediated by different tailoring enzymes (Fig. 3.3). These chemical modifications include hydroxylations, halogenations, glycosylations or alkylations and are generally important for the biological activities of the resulting molecules (Sattely et al. 2008; Hertweck 2009; Till and Race 2014; Helfrich and Piel 2016) (Fig. 3.3).

The recent analysis of 2478 bacterial genomes resulted in the identification of 2976 NRPS/PKS gene clusters of which 48%, 15.9% and 36.1% are NRPS, PKS and hybrid NRPS/PKS biosynthetic clusters, respectively (Wang et al. 2014). Importantly, some bacteria can devote around 10% of their genomes to the synthesis of secondary metabolites (Udwary et al. 2007; Nett et al. 2009; Chowdhury et al. 2015a), and the recent progress in bacterial genomics, bioinformatics and analytical techniques has evidenced the potential of plant-associated bacteria to produce multiple NRPs and PKs (Gross and Loper 2009; Wang et al. 2014; Chowdhury et al. 2015a; Helfrich and Piel 2016; Mousa and Raizada 2015; Weber et al. 2015; Matilla et al. 2016a, 2017). The synthesis of multiple bioactive secondary metabolites by PGPR has been associated with their plant colonization fitness and their capacity to

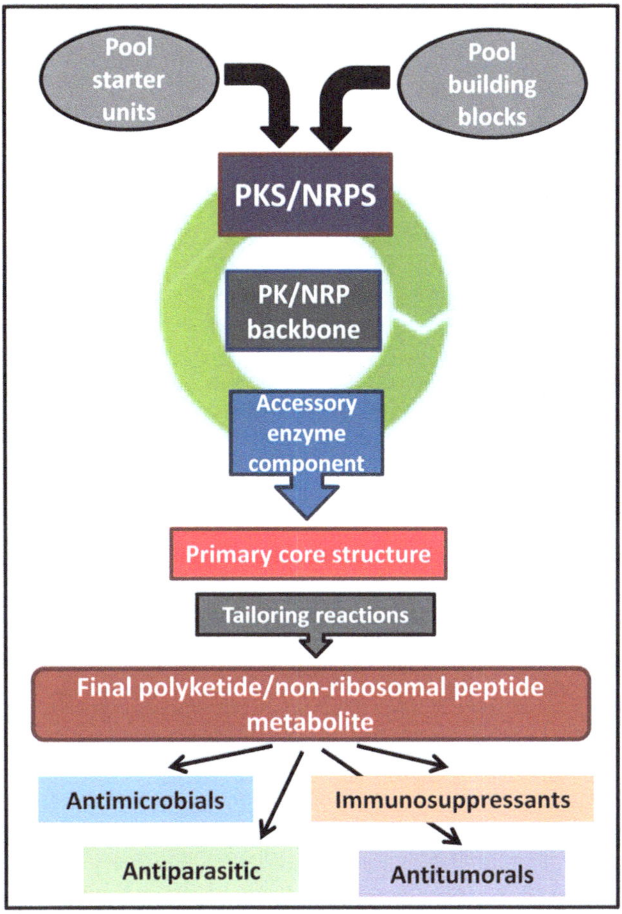

Fig. 3.3 Overview of the biosynthetic process of non-ribosomal peptides and polyketides mediated by polyketide synthases (PKS) and non-ribosomal peptide synthases (NRPS) (Adapted from Hertweck 2009)

protect plants against phytopathogens (De Vleesschauwer and Höfte 2007; Gross and Loper 2009; Scherlach et al. 2013; Chowdhury et al. 2015a; Mousa and Raizada 2015). Indeed, NRPS and PKS gene clusters with roles in the biocontrol against phytopathogens have been identified in bacterial strains isolated from disease suppressive soils (Mendes et al. 2011; Van Der Voort et al. 2015). In accordance with this, the production of bioactive NRPs and PKs by PGPR has been detected during plant colonization *in vivo* (Ongena et al. 2007a; Nihorimbere et al. 2012; Debois et al. 2014; Chowdhury et al. 2015b) and, in some cases, their biosynthesis *in planta* was increased in the presence of the target phytopathogens (Li et al. 2014b; Chowdhury et al. 2015b). Although multiple PGPR from different genera have been shown to produce NRPs and PKs, strains from the *Pseudomonas*, *Serratia* or

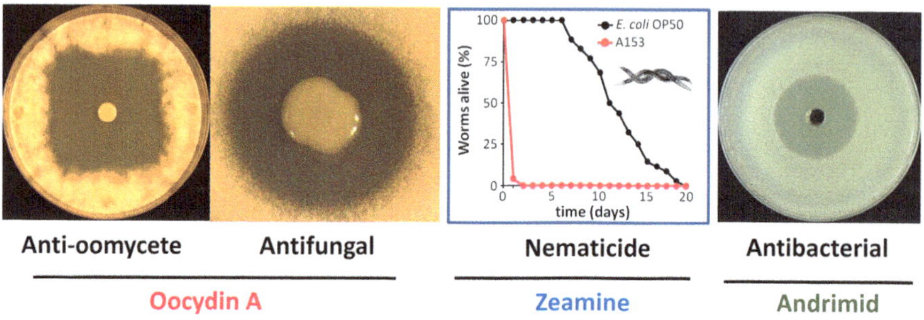

Fig. 3.4 Broad range of biological activities identified in the plant growth-promoting rhizobacteria *Serratia plymuthica* A153. The strain A153 produces the polyketide, oocydin A and the hybrid NRP/polyketides, zeamine and andrimid, compounds that exhibit multiple antimicrobial and nematicidal properties

Bacillus genera are prolific producers of these secondary metabolites (De Vleesschauwer and Höfte 2007; Gross and Loper 2009; Raaijmakers et al. 2010; Chowdhury et al. 2015a; Mousa and Raizada 2015). Among them, the NRP bacillibactin (Chen et al. 2009a), bacillomycin D (Chowdhury et al. 2015b), iturin (Nihorimbere et al. 2012) and fengycin (Nihorimbere et al. 2012); the PK oocydin A (Matilla et al. 2012, 2015), bacillaene (Chen et al. 2006), difficidin (Chen et al. 2009b) and mupirocin (Gurney and Thomas 2011); and the hybrid NRP-polyketides rhizoxin (Partida-Martínez and Hertweck 2007), andrimid (Matilla et al. 2016a) and zeamine (Hellberg et al. 2015) are some examples of secondary metabolites synthesized by PGPR. These natural products show a broad spectrum of biological activities, and they have been shown to target plant-pathogenic bacteria, fungi, oomycete, nematodes or even being involved in triggering ISR (De Vleesschauwer and Höfte 2007; Gross and Loper 2009; Scherlach et al. 2013; Chowdhury et al. 2015a; Mousa and Raizada 2015). Some specific plant-associated bacteria have been demonstrated to be highly prolific in the production of bioactive secondary metabolites, including the PGPR *Bacillus amyloliquefaciens* FZB42 (Chowdhury et al. 2015a), *Pseudomonas fluorescens* Pf-5 (Loper et al. 2007) and *Serratia plymuthica* A153 (Hellberg et al. 2015; Matilla et al. 2012, 2015, 2016a, b) (Fig. 3.4).

One of the most investigated antibiotics with roles in plant biocontrol is 2,4-diacetylphloroglucinol (DAPG), a phenolic PK that is produced by many plant-associated fluorescent pseudomonads (Haas and Defago 2005; Rezzonico et al. 2007; Gross and Loper 2009). DAPG has been shown to trigger ISR (Iavicoli et al. 2003; Weller et al. 2012) and to inhibit the growth of fungal and oomycete phytopathogens such as *Fusarium oxysporum*, *Rhizoctonia solani* and *Pythium ultimum* (Keel et al. 1992; de Souza et al. 2003; Rezzonico et al. 2007). Furthermore, its activity against plant-pathogenic bacteria (Keel et al. 1992) and nematodes (Cronin et al. 1997; Meyer et al. 2009) have been also demonstrated. DAPG production has been shown *in vivo* in the rhizosphere (Haas and Keel 2003) and, in accordance with this, DAPG-producing *Pseudomonas* species have been associated with the natural suppressiveness of several soils against multiple fungal plant pathogens (Weller 2007;

Frapolli et al. 2010; Kyselková and Moënne-Loccoz 2012). Thus, mutants defective in DAPG synthesis exhibit reduced biocontrol properties than the parental strains (Vincent et al. 1991; Fenton et al. 1992; Keel et al. 1992; Cronin et al. 1997). The 8-kb DAPG gene cluster (*pHGFACBDE*) encodes proteins responsible for the biosynthesis, regulation, degradation and secretion of DAPG (Gross and Loper 2009; Moynihan et al. 2009), and phylogenetic analyses determined that DAPG synthesis is an ancestral trait within the producing strains (Moynihan et al. 2009). Although the target of DAPG has not been identified, intact membrane function and cell homeostasis correlated with the sensitivity to DAPG (Kwak et al. 2011). Recently, new DAPG analogues have been chemically synthesized, and some of them were found to have improved *in vitro* and *in vivo* activities – opening new strategies for the rational design of novel DAPG-derived antibiotics (Gong et al. 2016).

3.4 Future Perspectives

The continuous increase in the world total population (9 billion by 2050) and the imminent global climate change are starting to challenge the sustainability of our current food supplies and energy resources. Furthermore, plant pests, currently responsible for up to 40% of the losses in the main crops worldwide (Oerke and Dehne 2004; Glare et al. 2012), will continue hampering plant growth and agricultural production (Miller et al. 2009; Velivelli et al. 2014). A significant body of evidence shows that synthetic pesticides and fertilizers represent important risks to the environment and human health. In consequence, governments are starting to restrict the use of some synthetic agrochemicals in order to promote more sustainable agricultural practices (Pimental et al. 2005; Glare et al. 2012), and ecologically friendly alternative management strategies to chemical fertilizers and pesticides are urgently needed.

As described in this chapter, beneficial plant-associated bacteria can promote plant growth directly acting as biofertilizers and phytostimulators or indirectly suppressing phytopathogens (biopesticides) (Fig. 3.1). Consequently, beneficial rhizospheric microorganisms are one of the most promising strategies for the rational management of crops, an aspect that is reflected in the constant increase in the commercialization of microbial-based agronomical products (Table 3.1). At present, the global market for biofertilizers is estimated at USD 591 million, with an expected annual rise of 13% in sales (https://www.mordorintelligence.com/). Alternatively, the use biopesticides for the rational control of plant diseases currently represents around 4% of the overall pesticide market (Thakore 2006; Lehr 2010), with a global market estimated at USD 3 billion in 2017 (Velivelli et al. 2014). However, this significant increase in the global market of biofertilizers and biopesticides continues to be potentially limited due to (1) the discrepancies between countries in the regulation of microbial-based products; (2) the elevated costs for the discovery, development and registration of new products (mainly in the case of biopesticides); (3) the currently inconsistent field performance of biopesticides; (4) the difficulties in biopesticide delivery to plant cultures; and (5) the frequent need for their reapplication (Glare et al. 2012). Overcoming these obstacles will encourage the use of biofertilizers and biopesticides as sustainable crop management practices. In view of this, new molecular technologies and screening methods are being employed for the development of a new generation of microbial inoculants with increased plant promotion and biocontrol properties (Glare et al. 2012; Velivelli et al. 2014).

Acknowledgements Miguel A. Matilla was supported by the Spanish Ministry of Economy and Competitiveness Postdoctoral Research Program, Juan de la Cierva (JCI-2012-11815). The Tino Krell laboratory is supported by FEDER funds and Fondo Social Europeo through grants from the Junta de Andalucía (grant CVI-7335) and the Spanish Ministry for Economy and Competitiveness (grant BIO2013-42297). We thank Angel J. Matilla (Department of Plant Physiology, University of Santiago de Compostela, Spain) for the critical review of this chapter and for his advice on the phytohormone section.

References

Abdel-Lateif K, Bogusz D, Hocher V (2012) The role of flavonoids in the establishment of plant roots endosymbioses with arbuscular mycorrhiza fungi, rhizobia and *Frankia* bacteria. Plant Signal Behav 7:636–641

Adie BA, Pérez-Pérez J, Pérez-Pérez MM, Godoy M, Sánchez-Serrano JI, Schmelz EA (2007) ABA is an essential signal for plant resistance to pathogens affecting JA biosynthesis and the activation of defenses in Arabidopsis. Plant Cell 19:1665–1681

Aguado-Santacruz GAA, Moreno-Gómez BA, Jiménez-Francisco BB, García-Moya EB, Preciado-Ortiz RE (2012) Impact of the microbial siderophores and phytosiderophores on the iron assimilation by plants: a synthesis. Rev Fitotec Mex 35:9–21

Ahemad M, Kribet M (2014) Mechanisms and applications of plant growth promoting rhizobacteria: current perspective. J King Saud Univ-Sci 26:1–20

Ahirwar N (2015) PGPR current and future prospects for development of sustainable agriculture. J Microbiol Biotechnol 7:96–102

Ahmed E, Homlströn SJM (2014) Siderophores in environmental research: roles and applications. Microb Biotechnol 7:196–208

Aira M, Gómez-Brandón M, Lazcano C, Baath E, Domínguez E (2010) Plant genotype strongly modifies the structure and growth of maize rhizosphere microbial communities. Soil Biol Biochem 42:2276–2281

Alström S (1991) Induction of disease resistance in common bean susceptible to halo blight bacterial pathogen after seed bacterization with rhizosphere pseudomonads. J Gen Appl Microbiol 37:495–501

Andreote F, Mendes R, Dini-Andreote F, Rossetto PB, Labate CA, Pizzirani-Kleiner AA, Van Elsas JD, Azevedo JL, Araújo WL (2008) Transgenic tobacco revealing altered bacterial diversity in the rhizosphere during early plant development. Antonie Van Leeuwenhoek 93:415–424

Antoun H, Prevost D (2006) Ecology of plant growth promoting rhizobacteria. In: Siddiqui ZAS (ed) PGPR: biocontrol and biofertilization. Springer, Dordrecht, pp 1–38

Arkhipova TN, Veselov SU, Melentiev AI, Martynenko EV, Kudoyarova GR (2005) Ability of bacterium *Bacillus subtilis* to produce cytokinins and to influence the growth and endogenous hormone content of lettuce plants. Plant Soil 272:201–209

Asselbergh B, Achuo AE, Hofte M, Van Gijsegem F (2008) Abscisic acid deficiency leads to rapid activation of tomato defense responses upon infection with *Erwinia chrysanthemi*. Mol Plant Pathol 9:11–24

Audrain B, Farag MA, Ryu CM, Ghigo JM (2015) Role of bacterial volatile compounds in bacterial biology. FEMS Microbiol Rev 39:222–233
Aznar A, Dellagi A (2015) New insights into the role of siderophores as triggers of plant immunity: what can we learn from animals? J Exp Bot 66:3001–3010
Badri DV, Vivanco JM (2009) Regulation and function of root exudates. Plant Cell Environ 32:666–681
Badri DV, Quintana N, El Kassis EG, Kim HK, Choi YH, Sugiyama A, Verpoorte R, Martinoia E, Manter DK, Vivanco JM (2009) An ABC transporter mutation alters root exudation of phytochemicals that provoke an overhaul of natural soil microbiota. Plant Physiol 151:2006–2017
Badri DV, Chaparro JM, Zhang R, Shen Q, Vivanco JM (2013) Application of natural blends of phytochemicals derived from the root exudates of Arabidopsis to the soil reveal that phenolic-related compounds predominantly modulate the soil microbiome. J Biol Chem 288:4502–4512
Baez U, Martinoia E (2013) Root exudates: the hidden part of plant defense. Trends Plant Sci 19:90–97
Bailly A, Weisskopf L (2012) The modulating effect of bacterial volatiles on plant growth: current knowledge and future challenges. Plant Signal Behav 7:79–85
Bais HP, Weir TL, Perry LG, Gilroy S, Vivanco JM (2006) The role of root exudates in rhizosphere interactions with plants and other organisms. Annu Rev Plant Biol 57:233–266
Bakker AW, Schippers B (1987) Microbial cyanide production in the rhizosphere in relation to potato yield reduction and *Pseudomonas* SPP-mediated plant growth-stimulation. Soil Biol Biochem 19:451–457
Bakker PA, Berendsen RL, Doornbos RL, Wintermans PC, Pieterse CM (2013) The rhizosphere revisited: root microbiomics. Front Plant Sci 165:1–7
Bakker MG, Schlatter DC, Otto-Hanson L, Kinkel LL (2014) Diffuse symbioses: roles of plant-plant, plant-microbe and microbe-microbe interactions in structuring the soil microbiome. Mol Ecol 23:1571–1583
Bari R, Jone JDG (2009) Role of plant hormones in plant defence responses. Plant Mol Biol 69:473–488
Bashan Y, de-Bashan LE, Prabhu SR, Hernández JP (2014) Advances in plant growth-promoting bacterial inoculant technology: formulations and practical perspectives (1998–2013). A review. Plant Soil 378:1–33
Belimov AA, Dodd IC, Safronova VI, Dumova VA, Shaposhnikov AI, Ladatko AG, Davies WJ (2014) Abscisic acid metabolizing rhizobacteria decrease ABA concentrations *in planta* and alter plant growth. Plant Physiol Biochem 74:84–91
Berendsen RL, Pieterse CMJ, Bakker PAHM (2012) The rhizosphere microbiome and plant health. Trends Plant Sci 17:478–486
Berg G (2009) Plant-microbe interactions promoting plant growth and health: perspectives for controlled use of microorganisms in agriculture. App Microbiol Biotech 14:11–18
Berg G, Smalla K (2009) Plant species and soil type cooperatively shape the structure and function of microbial communities in the rhizosphere. FEMS Microbiol Ecol 68:1–13
Bishop PE, Jorerger RD (1990) Genetics and molecular biology of an alternative nitrogen fixation system. Plan Mol Biol 41:109–125
Bitas V, Kim HS, Bennett JW, Kang S (2013) Sniffing on microbes: diverse roles of microbial volatile organic compounds in plant health. Mol Plant-Microbe Interact 26:835–843
Blom D, Fabbri C, Connor E, Schiestl F, Klauser D, Boller T et al (2011) Production of plant growth modulating volatiles is widespread among rhizosphere bacteria and strongly depends on culture conditions. Environ Microbiol 13:3047–3058
Blumer C, Haas D (2000) Mechanism, regulation, and ecological role of bacterial cyanide biosynthesis. Arch Microbiol 173:170–177
Boller T, Felix G (2009) A renaissance of elicitors: perception of microbe-associated molecular patterns and danger signals by pattern-recognition receptors. Annu Rev Plant Biol 60:379–406
Brandl H, Lehmann S, Faramarzi MA, Martinell D (2008) Biomobilization of silver, gold, and platinum from solid waste materials by HCN-forming microorganisms. Hydrometallurgy 94:14–17

Bulgarelli D, Schlaeppi K, Spaeten S, Ver Loren Van Themaat E, Schulze-Lefert P (2013) Structure and functions of the bacterial microbiota of plants. Annu Rev Plant Biol 64:807–838

Burdon JJ, Thrall PH (2009) Coevolution of plants and their pathogens in natural habitats. Science 324:755–756

Chaparro JM, Badri DV, Bakker MG, Sugiyama A, Manter DK, Vivanco JM (2013a) Root exudation of phytochemicals in *Arabidopsis* follows specific patterns that are developmentally programmed and correlate with soil microbial functions. PLoS One 8:e55731

Chaparro JM, Badri DV, Vivanco JM (2013b) Rhizosphere microbiome assemblage is affected by plant development. ISME J 8:790–803

Chen XH, Vater J, Piel J, Franke P, Scholz R, Schneider K, Koumoutsi A, Hitzeroth G, Grammel N, Strittmatter AW, Gottschalk G, Süssmuth RD, Borriss R (2006) Structural and functional characterization of three polyketide synthase gene clusters in *Bacillus amyloliquefaciens* FZB 42. J Bacteriol 188:4024–4036

Chen XH, Koumoutsi A, Scholz R, Borriss R (2009a) More than anticipated – production of antibiotics and other secondary metabolites by *Bacillus amyloliquefaciens* FZB42. J Mol Microbiol Biotechnol 16:14–24

Chen XH, Scholz R, Borriss M, Junge H, Mögel G, Kunz S, Borriss R (2009b) Difficidin and bacilysin produced by plant-associated *Bacillus amyloliquefaciens* are efficient in controlling fire blight disease. J Biotechnol 140:38–44

Chernin L, Chet I (2002) Microbial enzymes in biocontrol of plant pathogens and pests. In: Burns RG, Dick RP (eds) Enzymes in the environment: activity, ecology, and applications. Marcel Dekker, New York, pp 171–225

Cho SM, Kang BR, Han SH, Anderson AJ, Park JY, Lee YH, Cho BH, Yang KY, Ryu CM, Kim YC (2008) 2R,3R-butanediol, a bacterial volatile produced by *Pseudomonas chlororaphis* O6, is involved in induction of systemic tolerance to drought in *Arabidopsis thaliana*. Mol Plant-Microbe Interact 21:1067–1075

Chowdhury SP, Hartmann A, Gao X, Borriss R (2015a) Biocontrol mechanism by root-associated Bacillus amyloliquefaciens FZB42 – a review. Front Microbiol 6:780

Chowdhury SP, Uhl J, Grosch R, Alquéres S, Pittroff S, Dietel K, Schmitt-Kopplin P, Borriss R, Hartmann A (2015b) Cyclic Lipopeptides of *Bacillus amyloliquefaciens* subsp. *plantarum* colonizing the lettuce rhizosphere enhance plant defense responses toward the bottom rot pathogen Rhizoctonia solani. Mol Plant-Microbe Interact 28:984–995

Chung J, Song GC, Ryu CM (2016) Sweet scents from good bacteria: case studies on bacterial volatile compounds for plant growth and immunity. Plant Mol Biol 90:677–687

Cohen AC, Bottini R, Piccoli P (2008) *Azospirillum brasilense* Sp 245 produces ABA in chemically-defined culture medium and increases ABA content in *Arabidopsis* plants. Plant Growth Regul 54:97–103

Cohen AC, Bottini R, Pontin M, Berli FJ, Moreno D, Boccanlandro H, Travaglia CN, Piccoli PN (2015) *Azospirillum brasilense* ameliorates the response of *Arabidopsis thaliana* to drought mainly via enhancement of ABA levels. Physiol Plant 153:79–90

Compant S, Duffy B, Nowak J, Clément C, Barka EA (2005) Use of plant growth-promoting bacteria for biocontrol of plant diseases: principles, mechanisms of action, and future prospects. Appl Environ Microbiol 71:4951–4959

Conrath U, Beckers GJM, Flors V, García-Agustín P, Jakab G et al (2006) Priming: getting ready for battle. Mol Plant-Microbe Interact 19:1062–1071

Contesto C, Milesi S, Mantelin S, Zancarini A, Desbrosses G, Varoquaux F (2010) The auxin-signaling pathway is required for the lateral root response of *Arabidopsis* to the rhizobacterium *Phyllobacterium brassicacearum*. Planta 232:1455–1470

Corral-Lugo A, De la Torre J, Matilla MA, Fernández M, Morel B, Espinosa-Urgel M, Krell T (2016) Assessment of the contribution of chemoreceptor-based signalling to biofilm formation. Environ Microbiol 18:3355–3372

Cronin D, Moenne-Loccoz Y, Fenton A, Dunne C, Dowling DN, O'gara F (1997) Role of 2,4-Diacetylphloroglucinol in the Interactions of the biocontrol Pseudomonad strain F113 with the potato cyst nematode *Globodera rostochiensis*. Appl Environ Microbiol 63:1357–1361

D'Alessandro M, Erb M, Ton J, Brandenburg A, Karlen D, Zopfi J, Turlings TC (2014) Volatiles produced by soil-borne endophytic bacteria increase plant pathogen resistance and affect tritrophic interactions. Plant Cell Environ 37:813–826

Dandurishvili N, Toklikishvili N, Ovadis M, Eliashvili P, Giorgobiani N, Keshelava R, Tediashvili M, Vainstein A, Khmel I, Szegedi E, Chernin L (2011) Broad-range antagonistic rhizobacteria *Pseudomonas fluorescens* and *Serratia plymuthica* suppress *Agrobacterium* crown gall tumours on tomato plants. J Appl Microbiol 110:341–352

Danhorn T, Fuqua C (2007) Biofilm formation by plant-associated bacteria. Annu Rev Microbiol 61:401–422

De García Salamone IE, Hynes RK, Nelson LM (2001) Cytokinin production by plant growth promoting rhizobacteria and selected mutants. Can J Microbiol 47:404–411

De la Peña C, Lei Z, Watson BS, Sumner LW, Vivanco JM (2008) Root-microbe communication through protein secretion. J Biol Chem 283:25247–25255

de Souza JT, Arnould C, Deulvot C, Lemanceau P, Gianinazzi-Pearson V, Raaijmakers JM (2003) Effect of 2,4-diacetylphloroglucinol on *Pythium*: cellular responses and variation in sensitivity among propagules and species. Phytopathology 93:966–975

de Torres Zabala M, Bennett MH, Truman WH, Grant MR (2009) Antagonism between salicylic and abscisic acid reflects early host pathogen conflict and moulds plant defence responses. Plant J 59:375–386

de Torres-Zabala M, Truman W, Bennett MH, Lafforgue G, Mansfield JW, Rodriguez Egea P, Bogre L, Grant M (2007) *Pseudomonas syringae* pv. *tomato* hijacks the Arabidopsis abscisic acid signalling pathway to cause disease. EMBO J 26:1434–1443

De Vleesschauwer D, Höfte M (2007) Using Serratia plymuthica to control fungal pathogens of plants. CAB Rev Perspect Agric Vet Sci Nutr Nat Resour 2:1–12

De Vleesschauwer D, Höfte M (2009) Rhizobacteria-induced systemic resistance. Adv Bot Res 51:223–281

De Vrieze M, Pandey P, Bucheli TD, Varadarajan AR, Ahrens CH, Weisskopf L, Bailly A (2015) Volatile organic compounds from native potato-associated *Pseudomonas* as potential anti-oomycete agents. Front Microbiol 6:1295

Debois D, Jourdan E, Smargiasso N, Thonart P, De Pauw E, Ongena M (2014) Spatiotemporal monitoring of the antibiome secreted by *Bacillus* biofilms on plant roots using MALDI mass spectrometry imaging. Anal Chem 86:4431–4438

Denancé N, Sánchez-Vallet A, Goffner D, Molina A (2013) Disease resistance or growth: the role of plant hormones in balancing immune responses and fitness costs. Front Plant Sci 4:155

Dickschat JS, Martens R, Brinkhoff T, Simon M, Schulz S (2005) Volatiles released by a *Streptomyces* species isolated from the North Sea. Chem Biodivers 2:837–865

Domik D, Thürmer A, Weise T, Brandt W, Daniel R, Piechulla B (2016) A terpene synthase is involved in the synthesis of the volatile organic compound sodorifen of *Serratia plymuthica* 4Rx13. Front Microbiol 7:737

Doornbos L, Van Loon L, Bakker PAHM (2012) Impact of root exudates and plant defense signaling on bacterial communities in the rhizosphere. A review. Agron Sustain Dev 32:227–243

Drogue B, Combes-Meynet E, Moënne-Loccoz Y, Wisniewski-Dyé F, Prigent-Combaret C (2013) Control of the cooperation between plant growth-promoting rhizobacteria and crops by rhizosphere signals. de Bruijn, F.J. Molecular microbial ecology of the rhizosphere. 1 and 2. Wiley, Hoboken, 281–294

Du L, Shen B (2001) Biosynthesis of hybrid peptide-polyketide natural products. Curr Opin Drug Discov Devel 4:215–228

Duca D, Lorv J, Patten CL, Rose D, Glick BR (2014) Indole-3-acetic acid in plant-microbe interactions. Antonie Van Leeuwenhoek 106:85–125

Dudareva N, Klempien A, Muhlemann JK, Kaplan I (2013) Biosynthesis, function and metabolic engineering of plant volatile organic compounds. New Phytol 198:16–32

Dunne C, Moënne-Loccoz Y, McCarthy J, Higgins P, Powell J, Dowling DN et al (1998) Combining proteolytic and phloroglucinol-producing bacteria for improved biocontrol of *Pythium*-mediated damping-off of sugar beet. Plant Pathol 47:299–307

Dutta S, Podile AR (2010) Plant growth promoting rhizobacteria (PGPR): the bugs to debug the root zone. Crit Rev Microbiol 36:232–244

Effmert U, Kalderas JJ, Warnke R, Piechulla B (2012) Volatile mediated interactions between bacteria and fungi in the soil. J Chem Ecol 36:665–703

Elshafie HS, Camele I, Racioppi R, Scrano L, Iacobellis NS, Bufo SA (2012) In vitro antifungal activity of *Burkholderia gladioli* pv. *agaricicola* against some phytopathogenic fungi. Int J Mol Sci 13:16291–16302

Fan B, Carvalhais LC, Becker A, Fedoseyenko D, von Wirén N, Borriss R (2012) Transcriptomic profiling of *Bacillus amyloliquefaciens* FZB42 in response to maize root exudates. BMC Microbiol 12:116

Farag MA, Zhang H, Ryu CM (2013) Dynamic chemical communication between plants and bacteria through airborne signals: induced resistance by bacterial volatiles. J Chem Ecol 39:1007–1018

Fenton AM, Stephens PM, Crowley J, O'Callaghan M, O'Gara F (1992) Exploitation of gene(s) involved in 2,4-diacetylphloroglucinol biosynthesis to confer a new biocontrol capability to a *Pseudomonas* strain. Appl Environ Microbiol 58:3873–3878

Fernando WGD, Ramarathnam R, Krishnamoorthy AS, Savchuk SC (2005) Identification and use of potential bacterial organic antifungal volatiles in biocontrol. Soil Biol Biochem 37:955–964

Finkelstein R (2013) Abscisic acid synthesis and response. Arabidopsis Book 11:e0166

Fisch KM (2013) Biosynthesis of natural products by microbial iterative hybrid PKS–NRPS. RSC Adv 3:18228–18247

Fischbach MA, Walsh CT (2009) Antibiotics for emerging pathogens. Science 325:1089–1093

Flaishman M, Eyal Z, Zilberstein A, Voisard C, Haas D (1996) Suppression of *Septoria tritici* leaf blotch and leaf and rust of wheat by recombinant cyanide-producing strains of *Pseudomonas putida*. Mol Plant-Microbe Interact 9:642–645

Frankenberger WT, Arshad M (1995) Phytohormones in soils: production and function. Marcel Dekker, New York

Frapolli M, Défago G, Moënne-Loccoz Y (2010) Denaturing gradient gel electrophoretic analysis of dominant 2,4-diacetylphloroglucinol biosynthetic *phlD* alleles in fluorescent *Pseudomonas* from soils suppressive or conducive to black root rot of tobacco. Soil Biol Biochem 42:649–656

Frapolli M, Pothier JF, Défago G, Moënne Y (2012) Evolutionary history of synthesis pathway genes for phloroglucinol and cyanide antimicrobials in plant-associated fluorescent pseudomonads. Mol Phylogen Evol 63:877–890

Frébort I, Kowalska M, Hluska T, Frébortová J, Galuszka P (2011) Evolution of cytokinin biosynthesis and degradation. J Exp Bot 62:2431–2452

Gamalero E, Glick BR (2011) Mechanisms used by plant growth-promoting bacteria. In: Maheshwari DK (ed) Bacteria in agrobiology: plant nutrient management. Springer, Berlin, pp 17–46

Gamalero E, Glick BR (2012) Ethylene and abiotic stress tolerance in plants. In: Ahmad P, Prassad MNV (eds) Environmental adaptations and stress tolerance on plants in the era of climatic change. Springer, New York, pp 395–412

Gamalero E, Glick BR (2015) Bacterial modulation of plant ethylene levels. Plant Physiol 169:13–22

Glare T, Caradus J, Gelernter W, Jackson T, Keyhani N, Köhl J, Marrone P, Morin L, Stewart A (2012) Have biopesticides come of age? Trends Biotechnol 30:250–258

Glick BR (2012) Plant growth promoting-rhizobacteria: mechanisms and applications. Hindawi Publishing Corporation, Scientifica: 963401

Glick BR, Cheng Z, Czarny J, Duan J (2007) Promotion of plant growth by ACC deaminase-producing soil bacteria. Eur J Plant Pathol 119:329–339

Gong L, Tan H, Chen F, Li T, Zhu J, Jian Q, Yuan D, Xu L, Hu W, Jiang Y, Duan X (2016) Novel synthesized 2, 4-DAPG analogues: antifungal activity, mechanism and toxicology. Sci Rep 6:32266

Graner G, Persson P, Meijer J, Alstrom S (2003) A study on microbial diversity in different cultivars of *Brassica napus* in relation to its wilt pathogen, *Verticillium longisporum*. FEMS Microbiol Lett 29:269–276

Gray EJ, Smith DL (2005) Intracellular and extracellular PGPR: commonalities and distinctions in the plant-bacterium signaling processes. Soil Biol Biochem 37:395–412

Groenhagen U, Baumgartner R, Bailly A, Gardiner A, Eberl L, Schulz S, Weisskopf L (2013) Production of bioactive volatiles by *different Burkholderia ambifaria* strains. J Chem Ecol 39:892–906

Gross H, Loper JE (2009) Genomics of secondary metabolite production by *Pseudomonas* spp. Nat Prod Rep 26:1408–1446

Großkinsky DK, Tafner R, Moreno MV, Stenglein SA, García de Salamone IE, Nelson LM, Novák O, Strnad M, van der Graaff E, Roitsch T (2016) Cytokinin production by *Pseudomonas fluorescens* G2018 determines biocontrol activity against *Pseudomonas syringae* in Arabidopsis. Sci Rep 6:23310

Grossmann K (2010) Auxin herbicides: current status of mechanism and mode of action. Pest Manag Sci 66:113–120

Gulick AM (2016) Structural insight into the necessary conformational changes of modular nonribosomal peptides synthetases. Curr Opin Chem Biol 35:89–96

Gurney R, Thomas CM (2011) Mupirocin: biosynthesis, special features and applications of an antibiotic from a gram-negative bacterium. Appl Microbiol Biotechnol 90:11–21

Gutierrez-Luna FM, Lopez-Bucio J, Tamirano-Hernandez J, Valencia-Cantero E, de la Cruz HR, Andias-Rodriguez L (2010) Plant growth-promoting rhizobacteria modulate root system architecture in *Arabidopsis thaliana* through volatile organic compound emission. Symbiosis 51:75–83

Ha S, Vankova R, Yamaguchi-Shinozaki K, Shinozaki K, Tran LS (2012) Cytokinins: metabolism and function in plant adaptation to environmental stresses. Trends Plant Sci 17:172–179

Haas D, Defago G (2005) Biological control of soil-borne pathogens by fluorescent pseudomonads. Nat Rev Microbiol 3:307–319

Haas D, Keel C (2003) Regulation of antibiotic production in root-colonizing *Pseudomonas spp* and relevance for biological control of plant disease. Annu Rev Phytopathol 41:117–153

Haichar FE, Marol C, Berge O, Rangel-Castro JI, Prosser JI, Balesdent J, Heulin T, Achouak W (2008) Plant host habitat and root exudates shape soil bacterial community structure. ISME J 2:1221–1230

Hamley IW (2015) Lipopeptides: from self-assembly to bioactivity. Chem Commun 51:8574–8583

Hartung W, Sauter A, Turner NC, Fillery I, Heilmeier H (1996) Abscisic acid in soils: what is its function and which factors and mechanisms influence its concentration? Plant Soil 184:105–110

Hawes MC, Curlango-Rivera G, Xiong Z, Kessler JO (2012) Roles of root border cells in plant defense and regulation of rhizosphere microbial population by extracellular DNA "trapping". Plant Soil 355:1–16

Hayat R, Ali S, Amara U, Khalid R, Ahmed I (2010) Soil beneficial bacteria and their role in plant growth promotion: a review. Ann Microbiol 60:579–598

Heeb S, Haas D (2001) Regulatory roles of the GacS/GacA two-component system in plant-associated and other Gram-negative bacteria. Mol Plant-Microbe Interact 14:1351–1363

Helfrich EJ, Piel J (2016) Biosynthesis of polyketides by *trans*-AT polyketide synthases. Nat Prod Rep 33:231–316

Hellberg JE, Matilla MA, Salmond GP (2015) The broad-spectrum antibiotic, zeamine, kills the nematode worm *Caenorhabditis elegans*. Front Microbiol 6:137

Hertweck C (2009) The biosynthetic logic of polyketide diversity. Angew Chem Int Ed Engl 48:4688–4716

Hider RC, Kong X (2010) Chemistry and biology of siderophores. Nat Prod Rep 27:637–657

Hiltner L (1904) Uber neuere Erfahrungen und Probleme auf dem Gebiete der Bodenbakteriologie unter besonderden berucksichtigung und Brache. Arb Dtsch Landwirtsch Gesellschaft 98:59–78

Höfte M, Bakker PAHM (2007) Competition for iron and induced systemic resistance by siderophores of plant growth-promoting rhizobacteria. In: Varma A, Chincholkar SB (eds) Microbial siderophores. Springer, Berlin, pp 121–134

Houlden A, Timms-Wilson TM, Day MJ, Bailey MJ (2008) Influence of plant developmental stage on microbial community structure and activity in the rhizosphere of three field crops. FEMS Microbiol Ecol 65:193–201

Huang X-F, Chaparro JM, Reardon KF, Zhang R, Shen Q, Vivanco JM (2014) Rhizosphere interactions: root exudates, microbes, and microbial communities. Botany 92:267–275

Hutsch BW, Augustin J, Merbach W (2000) Plant rhizodeposition an important source for carbon turnover in soils. J Plant Nutr Soil Sci 165:397e407

Iavicoli A, Boutet E, Buchala A, Métraux JP (2003) Induced systemic resistance in *Arabidopsis thaliana* in response to root inoculation with *Pseudomonas fluorescens* CHA0. Mol Plant-Microbe Interact 16:851–858

Inceoglu Ö, Overbeek LS, Salles JF, Elsas JD (2013) Normal operating range of bacterial communities in soil used for potato cropping. Appl Environ Microbiol 79:1160–1170

Jiang CJ, Shimono M, Sugano S, Kojima M, Yazawa K, Yoshida R et al (2010) Abscisic acid interacts antagonistically with salicylic acid signaling pathway in rice- *Magnaporthe grisea* interaction. Mol Plant-Microbe Interact 23:791–798

Jiang F, Chen L, Belimov AA, Shaposhnikov AI, Gong F, Meng X, Hartung W, Jeschke DW, Davies WJ, Dodd IC (2012) Multiple impacts of the plant growth-promoting rhizobacterium *Variovorax paradoxus* 5C-2 on nutrient and ABA relations of *Pisum sativum*. J Exp Bot 63:6421–6430

Jiang CJ, Shimono M, Sugano S, Kojima M, Liu X, Inoue H, Sakakibara H, Takatsuji H (2013) Cytokinins act synergistically with salicylic acid to activate defense gene expression in rice. Mol Plant-Microbe Interact 26:287–296

Jones DL, Nguyen C, Finlay RD (2009) Carbon flow in the rhizosphere: carbon trading at the soil-root interface. Plant Soil 321:5–33

Kai M, Effmert U, Berg G, Piechulla B (2007) Volatiles of bacterial antagonists inhibit mycelial growth of the plant pathogen *Rhizoctonia solani*. Arch Microbiol 187:351–360

Kai M, Haustein M, Molina F, Petri A, Scholz B, Piechulla B (2009) Bacterial volatiles and their action potential. Appl Microbiol Biotech 81:1001–1012

Kai M, Crespo E, Cristescu SM, Harren FJ, Francke W, Piechulla B (2010) *Serratia odorifera*: analysis of volatile emission and biological impact of volatile compounds on *Arabidopsis thaliana*. Appl Microbiol Biotechnol 88:965–976

Kamensky M, Ovadis M, Chet I, Chernin L (2003) Soil-borne strain IC14 of *Serratia plymuthica* with multiple mechanisms of antifungal activity provides biocontrol of *Botrytis cinerea* and *Sclerotinia sclerotiorum* diseases. Soil Biol Biochem 35:323–331

Kamilova F, Kravchenko LV, Shaposhnikov AI, Azarova T, Makarova N, Lugtenberg B (2006) Organic acids, sugars and L-tryptophane in exudates of vegetables growing on stonewool and their effects on activities of rhizosphere bacteria. Mol Plant-Microbe Interact 9:250–256

Kanchiswamy CN, Malnoy M, Maffei ME (2015) Chemical diversity of microbial volatiles and their potential for plant growth and productivity. Front Plant Sci 6:151

Karadeniz A, Topcuoglu SF, Inan S (2006) Auxin, gibberellin, cytokinin and abscisic acid production in some bacteria. World J Microbiol Biotechnol 22:1061–1064

Keel C, Schenider U, Maurhofer M, Voisard C, Laville J, Burger U, Wirthener P, Haas D, Défago G (1992) Suppression of root diseases by *Pseudomonas fluorescens* CHA0: importance of the bacterial secondary metabolite 2,4-diacetylphloroglucinol. Mol. Plant Microbe Interact 5:4–13

Khan MS, Zaidi A, Ahmad E (2014) Mechanism of phosphate solubilization and physiological functions of phosphate-solubilizing microorganisms. In: Khan et al (eds) Phosphate solubilizing microorganisms. Springer, Cham, pp 31–62

Kim YC, Leveau J, Gardener BB, Pierson EA, Pierson LS, Ryu C-M (2011) The multifactorial basis for plant health promotion by plant-associated bacteria. Appl Environ Microbiol 77:1548–1555

Kloepper JW, Gutiérrez-Estrada A, McInroy JA (2007) Photoperiod regulated elicitation of growth promotion but not induced resistance by plant growth-promoting rhizobacteria. Can J Microbiol 53:159–167

Koga H, Dohi K, Mori M (2004) Abscisic acid and low temperatures suppress the whole plant-specific resistance reaction of rice plants to the infection of *Magnaporthe grisea*. Physiol Mol Plant Pathol 65:3–9

Kwak YS, Han S, Thomashow LS, Rice JT, Paulitz TC, Kim D, Weller DM (2011) *Saccharomyces cerevisiae* genome-wide mutant screen for sensitivity to 2,4-diacetylphloroglucinol, an antibiotic produced by *Pseudomonas fluorescens*. Appl Environ Microbiol 77:1770–1776

Kyselková M, Moënne-Loccoz Y (2012) *Pseudomonas* and other microbes in disease-suppressive soils. In: Lichtfouse E (ed) Sustainable agriculture reviews: organic fertilisation, soil quality and human health, vol 9. Springer, Dordrecht, pp 93–140

Lanteigne C, Gadkar VJ, Wallon T, Novinscak A, Filion M (2012) Production of DAPG and HCN by *Pseudomonas* sp. LBUM300 contributes to the biological control of bacterial canker of tomato. Phytopathology 102:967–973

Lee B, Farag MA, Park HB, Kloepper JW, Lee SH, Ryu CM (2012) Induced resistance by a long-chain bacterial volatile: elicitation of plant systemic defense by a C13 volatile produced by *Paenibacillus polymyxa*. PLoS One 7:e48744

Lehr P (2010) Biopesticides: the global market. Report code CHM029B. BCC Research, London

Lemanceau P, Expert D, Gaymard F, Bakker P, Briat JF (2009) Role of iron in plant-microbe interactions. Plant Innate Immun 51:491–549

Lemfack MC, Nickel J, Dunkel M, Preissner R, Piechulla B (2014) mVOC: a database of microbial volatiles. Nucleic Acids Res 42:D744–D748

Lesuffleur F, Liquet JBC (2010) Characterisation of root amino acid exudation in white clover (*Trifolium repens* L.) Plant Soil 333:191–201

Li X, Rui J, Mao Y, Yannarell A, Mackie R (2014a) Dynamics of the bacterial community structure in the rhizosphere of a maize cultivar. Soil Biol Biochem 68:392–401

Li B, Li Q, Xu Z, Zhang N, Shen Q, Zhang R (2014b) Responses of beneficial *Bacillus amyloliquefaciens* SQR9 to different soilborne fungal pathogens through the alteration of antifungal compounds production. Front Microbiol 5:636

Li J, Liu W, Luo L, Dong D, Liu T, Zhang T, Lu C, Liu D, Zhang D, Wu H (2015) Expression of *Paenibacillus polymyxa* β-1,3-1,4-glucanase in *Streptomyces lydicus* A01 improves its biocontrol effect against *Botrytis cinerea*. Biol Control 90:141–147

Liu Y, Chen L, Zhang N, Li Z, Zhang G, Xu Y, Shen Q, Zhang R (2016) Plant-microbe communication enhances auxin biosynthesis by a root-associated bacterium, *Bacillus amyloliquefaciens* SQR9. Mol Plant-Microbe Interact 29:324–330

Llamas MA, Imperi F, Visca P, Lamont IL (2014) Cell-surface signalling in *Pseudomonas*: stress responses, iron transport, and pathogenicity. FEMS Microbiol Rev 38:569–597

Loper JE, Kobayashi DY, Paulsen IT (2007) The genomic sequence of *Pseudomonas fluorescens* Pf-5: insights into biological control. Phytopathology 97:233–238

Lugtenberg BJJ, Bloemberg GV (2004) Life in the rhizosphere. In: Ramos JL (ed) *Pseudomonas*: genomics, life style and molecular architecture, vol 1. Kluwer Acad/Plenum, New York, pp 403–430

Lugtenberg B, Kamilova F (2009) Plant-growth-promoting rhizobacteria. Annu Rev Microbiol 63:541–556

Ma JF (2005) Plant root responses to three abundant soil minerals: silicon, aluminum and iron. Crit Rev Plant Sci 24:267–281

Marilley L, Hartwig UA, Aragno M (1999) Influence of an elevated atmospheric CO_2 content on soil and rhizosphere bacterial communities beneath *Lolium perenne* and *Trifolium repens* under field conditions. Microb Ecol 38:39–49

Mark GL, Dow JM, Kiely PD, Higgins H, Haynes J, Baysse C, Abbas A, Foley T, Franks A, Morrissey J et al (2005) Transcriptome profiling of bacterial responses to root exudates identifies genes involved in microbe-plant interactions. Proc Natl Acad Sci U S A 102:17454–17459

Maróti G, Kondorosi E (2014) Nitrogen-fixing *Rhizobium*-legume symbiosis: are polyploidy and host peptide-governed symbiont differentiation general principles of endosymbiosis? Front Microbiol 5:326
Matilla MA, Espinosa-Urgel M, Rodríguez-Herva JJ, Ramos JL, Ramos-González MI (2007a) Genomic analysis reveals the major driving forces of bacterial life in the rhizosphere. Genome Biol 8:R179
Matilla MA, Ramos JL, Duque E, de Dios Alché J, Espinosa-Urgel M, Ramos-González MI (2007b) Temperature and pyoverdine-mediated iron acquisition control surface motility of *Pseudomonas putida*. Environ Microbiol 9:1842–1850
Matilla MA, Ramos JL, Bakker PAHM, Doornbos RD, Badri DV, Vivanco JM, Ramos-González MI (2010) *Pseudomonas putida* KT2440 causes induced systemic resistance and changes in *Arabidopsis* root exudation. Environ Microb Rep 2:381–388
Matilla MA, Pizarro-Tobías P, Roca A, Fernández M, Duque E, Molina L, Wu X, Van der Lelie D, Gómez MJ, Segura A, Ramos JL (2011a) Complete genome of the plant growth promoting rhizobacterium *Pseudomonas putida* BIRD-1. J Bacteriol 193:1290
Matilla MA, Travieso ML, Ramos JL, Ramos-González MI (2011b) Cyclic diguanylate turnover mediated by the sole GGDEF/EAL response regulator in *Pseudomonas putida*: its role in the rhizosphere and an analysis of its target processes. Environ Microbiol 13:1745–1766
Matilla MA, Stöckmann H, Leeper FJ, Salmond GPC (2012) Bacterial biosynthetic gene clusters encoding the anti-cancer haterumalide class of molecules: biogenesis of the broad spectrum antifungal and antioomycete compound, oocydin A. J Biol Chem 287:39125–39138
Matilla MA, Leeper FJ, Salmond GP (2015) Biosynthesis of the antifungal haterumalide, oocydin A, in *Serratia*, and its regulation by quorum sensing, RpoS and Hfq. Environ Microbiol 17:2993–3008
Matilla MA, Nogellova V, Morel B, Krell T, Salmond GP (2016a) Biosynthesis of the acetyl-CoA carboxylase-inhibiting antibiotic, andrimid in *Serratia* is regulated by Hfq and the LysR-type transcriptional regulator, AdmX. Environ Microbiol 18:3635–3650
Matilla MA, Drew A, Udaondo Z, Krell T, Salmond GPC (2016b) Genome sequence of *Serratia plymuthica* A153, a model rhizobacterium for the investigation of the synthesis and regulation of haterumalides, zeamine, and andrimid. Genome Announc 4:e00373–e00316
Matilla-Vázquez MA, Matilla AJ (2014) Ethylene: role in plants under environmental stress. In: Ahmad P, Wani MR (eds) Physiological mechanisms and adaptation strategies in plants under changing environment, vol 2. Springer, New York, pp 189–222
Matilla MA, Udaondo Z, Krell T, Salmond GPC (2017) Genome sequence of MSU97, a plant-associated bacterium that makes multiple antibiotics. Genome Announc 5(9):e01752–16
Mendes R, Kruijt M, de Bruijn I, Dekkers E, van der Voort M, Schneider JH, Piceno YM, DeSantis TZ, Andersen GL, Bakker PA, Raaijmakers JM (2011) Deciphering the rhizosphere microbiome for disease-suppressive bacteria. Science 332:1097–1100
Mendes R, Garbeva P, Raaijmakers JM (2013) The rhizosphere microbiome: significance of plant beneficial, plant pathogenic, and human pathogenic microorganisms. FEMS Microbiol Rev 37:634–663
Meyer SL, Halbrendt JM, Carta LK, Skantar AM, Liu T, Abdelnabby HM, Vinyard BT (2009) Toxicity of 2,4-diacetylphloroglucinol (DAPG) to plant-parasitic and bacterial-feeding nematodes. J Nematol 41:274–280
Meziane H, Van der Sluis I, Van Loon LC, Höfte M, Bakker PAHM (2005) Determinants of *Pseudomonas putida* WCS358 involved in inducing systemic resistance in plants. Mol Plant Pathol 6:177–185
Micallef SA, Shiaris MP, Colón-Carmona A (2009) Influence of *Arabidopsis thaliana* accessions on rhizobacterial communities and natural variation in root exudates. J Exp Bot 60:1729–1742
Miller BR, Gulick AM (2016) Structural biology of nonribosomal peptide synthetases. In: Bradley, Evans (eds) Nonribosomal peptide and polyketide biosynthesis: methods and protocols. Methods in molecular biology, vol 1401. Springer, New York, pp 3–29

Miller SA, Beed FD, Harmon CL (2009) Plant disease diagnostic capabilities and networks. Annu Rev Phytopathol 47:15–38

Milling A, Smalla K, Xaver F, Maidl K, Schloter M, Munch JC (2004) Effects of transgenic potatoes with an altered starch composition on the diversity of soil and rhizosphere bacteria and fungi. Plant Soil 266:23–39

Mohammadi K (2012) Phosphorus solubilizing bacteria: occurrence, mechanisms and their role in crop production. Resour Environ 2:80–85

Mohr PG, Cahill DM (2003) Abscisic acid influences the susceptibility of *Arabidopsis thaliana* to *Pseudomonas syringae* pv. *tomato* and *Peronospora parasitica*. Funct Plant Biol 30:461–469

Mougel C, Offre P, Ranjard L, Corberand T, Gamalero E, Robin C, Lemanceau P (2006) Dynamic of the genetic structure of bacterial and fungal communities at different developmental stages of *Medicago truncatula* Gaertn. cv. Jemalong line J5. New Phytol 170:165–175

Mousa WK, Raizada MN (2015) Biodiversity of genes encoding anti-microbial traits within plants associated microbes. Front Plant Sci 6:231

Moynihan JA, Morrissey JP, Coppoolse ER, Stiekema WJ, O'Gara F, Boyd EF (2009) Evolutionary history of the phl gene cluster in the plant-associated bacterium *Pseudomonas fluorescens*. Appl Environ Microbiol 75:2122–2131

Nagpure A, Choudhary B, Gupta RK (2014) Chitinases: in agriculture and human healthcare. Crit Rev Biotechnol 34:215–232

Nandi M, Selin C, Brassinga AKC, Belmonte MF, Fernando WGD, Loewen PC, Kievit TR (2015) Pyrrolnitrin and hydrogen cyanide production by *Pseudomonas chlororaphis* strain PA23 exhibits nematicidal and repellent activity against *Caenorhabditis elegans*. PLoS One 10:e0123184

Nascimento FX, Vicente CSL, Barbosa P, Espada M, Glick BR, Oliveira S, Mota M (2013) The use of the ACC deaminase producing bacterium Pseudomonas putida UW4 as a biocontrol agent for pine wilt disease. BioControl 58:427–433

Naseem M, Kaltdorf M, Dandekar T (2015) The nexus between growth and defence signalling: auxin and cytokinin modulate plant immune response pathways. J Exp Bot 66:4885–4896

Nett M, Ikeda H, Moore BS (2009) Genomic basis for natural product biosynthetic diversity in the actinomycetes. Nat Prod Rep 26:1362–1384

Neumann G, Römheld V (2007) The release of root exudates as affected by the plant physiological status. In: Pinton R, Varanini Z, Nannipieri P (eds) The rhizosphere: biochemistry and organic substances at the soil-plant interface. CRC Press, New York, pp 23–72

Newmann G, Römheld V (2007) The release of root exudates as affected by the plant physiological status. In: Pinton R, Varanini Z, Nannipieri P (eds) The rhizosphere: biochemistry and organic substances at the soil-plant interface. CRC Press, New York, pp 23–72

Nihorimbere V, Cawoy H, Seyer A, Brunelle A, Thonart P, Ongena M (2012) Impact of rhizosphere factors on cyclic lipopeptide signature from the plant beneficial strain *Bacillus amyloliquefaciens* S499. FEMS Microbiol Ecol 79:176–191

Oerke EC, Dehne HW (2004) Safeguarding production – losses in major crops and the role of crop protection. Crop Prot 23:275–285

Oliver KL, Hamelin RC, Hintz WE (2008) Effects of transgenic hybrid aspen over-expressing polyphenol oxidase on rhizosphere diversity. Appl Environ Microb 74:5340–5348

Ongena M, Jourdan E, Adam A, Paquot M, Brans A, Joris B et al (2007a) Surfactin and fengycin lipopeptides of *Bacillus subtilis* as elicitors of induced systemic resistance in plants. Environ Microbiol 9:1084–1090

Ongena M, Jourdan E, Adam A, Paquot M, Brans A, Joris B, Arpigny JL, Thonart P (2007b) Surfactin and fengycin lipopeptides of *Bacillus subtilis* as elicitors of induced systemic resistance in plants. Environ Microbiol 9:1084–1090

Ovadis M, Liu X, Ismailov Z, Chet I, Chernin L (2004) The global regulator genes from biocontrol strain *Serratia plymuthica* IC270: cloning, sequencing, and functional studies. J Bacteriol 186:4983–4993

Parray JA, Jan S, Kamili AN, Qadri RA, Egamberdieva D, Ahmad P (2016) Current perspectives on plant growth-promoting rhizobacteria. J Plant Growth Regul 35:877–902

Partida-Martinez LP, Hertweck C (2007) A gene cluster encoding rhizoxin biosynthesis in *Burkholderia rhizoxina*, the bacterial endosymbiont of the fungus *Rhizopus microsporus*. Chembiochem 8:41–45

Patten CL, Blakney AJC, Coulson TJD (2013) Activity, distribution and function of indole-3-acetic acid biosynthetic pathways in bacteria. Crit Rev Microbiol 39:395–415

Peberdy JF (1990) Fungal cell wall – a review. In: Kuhn PJ, Trinci APJ, Jung MJ, Goosey MM, Cooping IG (eds) Biochemistry of cell walls and membranes in fungi. Springer, Heidelberg, pp 5–24

Perrig D, Boiero ML, Masciarelli OA, Penna C, Ruiz OA, Cassan FD, Luna MV (2007) Plant-growth-promoting compounds produced by two agronomically important strains of *Azospirillum brasilense*, and implications for inoculants formulation. Appl Microbiol Biotech 75:1143–1150

Perry LG, Alford ER, Horiuchi J, Paschke MW, Vivanco JM (2007) Chemical signals in the rhizosphere: root-root and root-microbe communication. In: Pinton R, Varanini Z, Nannipieri P (eds) The rhizosphere: biochemistry and organic substances at the soil-plant interface, 2nd edn. CRC Press, New York, pp 297–330

Pidot SJ, Coyne S, Kloss F, Hertweck C (2014) Antibiotics from neglected bacterial sources. Int J Med Microbiol 304:14–22

Pieterse CMJ, Van der Does D, Zamioudis C, Leon-Reyes A, Van Wees SCM (2012) Hormonal modulation of plant immunity. Annu Rev Cell Develop Biol 28:489–521

Pieterse CM, Zamioudis C, Berendsen RL, Weller DM, Van Wees SC, Bakker PA (2014) Induced systemic resistance by beneficial microbes. Annu Rev Phytopathol 52:347–375

Pimentel D (2005) Environmental and economic costs of the application of pesticides primarily in the United States. Environ Dev Sustain 7:229–252

Prathap M, Ranjitha-Kumari BD (2015) A critical review on plant growth promoting rhizobacteria. J Plant Pathol Micro 6:1000266

Quecine MC, Araujo WL, Marcon J, Gai CS, Azevedo JL, Pizzirani-Kleiner AA (2008) Chitinolytic activity of endophytic *Streptomyces* and potential for biocontrol. Lett Appl Microbiol 47:486–491

Quintana-Rodriguez E, Morales-Vargas AT, Molina-Torres J, Ádame-Alvarez RM, Acosta-Gallegos JA, Heil M (2015) Plant volatiles cause direct, induced and associational resistance in common bean to the fungal pathogen *Colletotrichum lindemuthianum*. J Ecol 103:250–260

Raaijmakers JM, Leeman M, MMP VO, Van der Sluis I, Schippers B, PAHM B (1995) Dose–response relationships in biological control of fusarium wilt of radish by *Pseudomonas* spp. Phytopathology 85:1075–1081

Raaijmakers JM, Paulitz TC, Steinberg CH, Alabouvette C, Moënne-Loccoz Y (2009) The rhizosphere: a playground and battlefield for soilborne pathogens and beneficial microorganisms. Plant Soil 321:341–361

Raaijmakers JM, De Bruijn I, Nybroe O, Ongena M (2010) Natural functions of lipopeptides from *Bacillus* and *Pseudomonas*: more than surfactants and antibiotics. FEMS Microbiol Rev 34:1037–1062

Ramaekers L, Remans R, Rao IM, Blair MW, Vanderleyden J (2010) Strategies for improving phosphorus acquisition efficiency of crop plants. Field Crops Res 117:169–176

Reed MLE, Glick BR (2013) Applications of plant growth-promoting bacteria for plant and soil systems. Gupta, V.K, Schmoll, M., Maki, M., Tuohy, M., Mazutti, M.A., Applications of microbial engineering. Taylor and Francis, Enfield, 181–229

Reinhold-Hurek B, Bünger W, Burbano CS, Sabale M, Hurek T (2015) Roots shaping their microbiome: global hotspots for microbial activity. Annu Rev Phytopathol 53:403–424

Reyes-Darias JA, García V, Rico-Jiménez M, Corral-Lugo A, Lesouhaitier O, Juárez-Hernández D, Yang Y, Bi S, Feuilloley M, Muñoz-Rojas J, Sourjik V, Krell T (2015) Specific gamma-aminobutyrate chemotaxis in pseudomonads with different lifestyle. Mol Microbiol 97:488–501

Rezzonico F, Zala M, Keel C, Duffy B, Moënne-Loccoz Y, Défago G (2007) Is the ability of biocontrol fluorescent pseudomonads to produce the antifungal metabolite 2,4-diacetylphloroglucinol really synonymous with higher plant protection? New Phytol 173:861–872
Richardson AE, Barea JM, McNeil AM, Prigent-Combaret C (2009) Acquisition of phosphorus and nitrogen in the rhizosphere and plant growth promotion by microorganisms. Plant Soil 321:305–339
Roca A, Pizarro-Tobías P, Udaondo Z, Fernández M, Matilla MA, Molina-Henares MA, Molina L, Segura A, Duque E, Ramos JL (2013) Analysis of the plant growth-promoting properties encoded by the genome of the rhizobacterium Pseudomonas putida BIRD-1. Environ Microbiol 15:780–794
Rodríguez H, Fraga R (1999) Phosphate solubilizing bacteria and their role in plant growth promotion. Biotech Adv 17:319–339
Rodríguez H, Fraga R, González T, Bashan Y (2006) Genetics of phosphate solubilization and its potential applications for improving plant growth-promoting bacteria. Plant Soil 287:15–21
Roesch LF, Fulthorpe RR, Riva A, Casella G, Hadwin AK, Kent AD, Daroub SH, Camargo FA, Farmerie WG, Triplett EW (2007) Pyrosequencing enumerates and contrasts soil microbial diversity. ISME J 1:283–290
Rosas SB, Andres JA, Rovera M, Correa NS (2006) Phosphate-solubilizing Pseudomonas putida can influence the rhizobia-legume symbiosis. Soil Biol Biochem 38:3502–3505
Roselló-Mora R, Amann R (2001) The species concept of prokaryotes. FEMS Microbiol Rev 25:39–67
Rossolini GM, Shippa S, Riccio ML, Berlutti F, Macaskie LE, Thaller MC (1998) Bacterial nonspecific acid phosphatases: physiology, evolution and use as tools in microbial biotechnology. Cell Mol Life Sci 54:833–850
Rudrappa T, Splaine RE, Biedrzycki ML, Bais HP (2008) Cyanogenic pseudomonads influence multitrophic interactions in the rhizosphere. PLoS One 3:e2073
Rudrappa T, Biedrzycki ML, Kunjeti SG, Donofrio NM, Czymmek KJ, Paré PW et al (2010) The rhizobacterial elicitor acetoin induces systemic resistance in *Arabidopsis thaliana*. Commun Integr Biol 3:130–138
Ryu CM, Farag MA, Hu CH, Reddy MS, Wei HX, Paré PW, Kloepper JW (2003) Bacterial volatiles promote growth in Arabidopsis. Proc Natl Acad Sci U S A 100:4927–4932
Ryu CM, Farag MA, Hu CH, Reddy MS, Kloepper JW, Paré PW (2004) Bacterial volatiles induce systemic resistance in Arabidopsis. Plant Physiol 134:1017–1026
Saha R, Saha N, Donofrio RS, Bestervelt LL (2013) Microbial siderophores: a mini review. J Basic Microbiol 53:303–317
Saha M, Sarkar S, Sarkar B, Kumar B, Sharma BK, Bhattacharjee S, Tribedi P (2016) Microbial siderophores and their potential applications: a review. Environ Sci Pollut Res 23:3984–3999
Saharan BS, Nehra V (2011) Plant growth promoting rhizobacteria: a critical review. Life Sci Med Res 21:1–30
Salomon MV, Bottini R, de Souza Filho GA, Cohen AC, Moreno D, Gil M, Piccoli P (2014) Bacteria isolated from roots and rhizosphere of Vitis vinifera retard water losses, induce abscisic acid accumulation and synthesis of defense-related terpenes in in vitro cultured grapevine. Physiol Plant 151:359–374
Sashidhar B, Podile AR (2010) Mineral phosphate solubilization by rhizosphere bacteria and scope for manipulation of the direct oxidation pathway involving glucose dehydrogenase. J App Microb 109:1–2
Sattely ES, Fischbach MA, Walsh CT (2008) Total biosynthesis: in vitro reconstitution of polyketide and nonribosomal peptide pathways. Nat Prod Rep 25:757–793
Schalk IJ, Hannauer M, Braud A (2011) New roles of bacterial siderophores in metal transport and tolerance. Environ Microbiol 13:2844–2854
Schenk PM, Carvalhais LC, Kazan K (2012) Unraveling plant–microbe interactions: can multispecies transcriptomics help? Trends Biotechnol 30:177–184

Scherlach K, Graupner K, Hertweck C (2013) Molecular bacteria-fungi interactions: effects on environment, food, and medicine. Annu Rev Microbiol 67:375–397
Schöller CEG, Gürtler H, Petersen R, Molin S, Wilkins K (2002) Volatile metabolites from actinomycetes. J Agric Food Chem 50:2615–2621
Scott JC, Greenhut IV, Leveau JHJ (2013) Functional characterization of the bacterial iac genes for degradation of the plant hormone indole-3-acetic acid. J Chem Ecol 39:942–951
Sgroy V, Cassan F, Masciarelli O, Del Papa MF, Lagares A, Luna V (2009) Isolation and characterization of endophytic plant growth-promoting (PGPB) or stress homeostasis-regulating (PSHB) bacteria associated to the halophyte *Prosopisstrom bulifera*. Appl Microbiol Biotech 85:371–381
Shigenaga AM, Arqueso CT (2016) No hormone to rule them all: interactions of plant hormones during the responses of plants to pathogens. Semin Cell Dev Biol 56:174–189
Singh RP, Shelke GM, Kumar A, Jha PN (2015) Biochemistry and genetics of ACC deaminase: a weapon to "stress ethylene" produced in plants. Front Microbiol 6:937
Soares-Gomes E, Schuch V, Macedo-Lemos EG (2013) Biotechnology of polyketides: new breath of life for the novel antibiotic genetic pathways discovery through metagenomics. Braz J Microbiol 44:1007–1034
Somers E, Vanderleyden J, Srinivasan M (2004) Rhizosphere bacterial signalling: a love parade beneath our feet. Crit Rev Microbiol 30:205–240
Spaepen S, Vanderleyden J (2011) Auxin and plant-microbe interaction. Cold Spring Harb Perspect Biol 3:a001438
Spaepen S, Vanderleyden J, Remans R (2007) Indole-3-acetic acid in microbial and microorganism-plant signaling. FEMS Microbiol Rev 31:425–448
Spoel SH, Dong X (2012) How do plants achieve immunity? Defence without specialized immune cells. Nat Rev Immunol 12:89–100
Srivastava S, Chaudhry V, Mishra A, Chauhan PS, Rehman A, Yadav A, Tuteja N, Nautiyal CS (2012) Gene expression profiling through microarray analysis in *Arabidopsis thaliana* colonized by *Pseudomonas putida* MTCC5279, a plant growth promoting rhizobacterium. Plant Signal Behav 7:235–245
Steenhoudt O, Vanderleyden J (2000) *Azospirillum*, a free-living nitrogen-fixing bacterium closely associated with grasses: genetic, biochemical and ecological aspects. FEMS Microbiol Rev 24:487–506
Strieker M, Tanović A, Marahiel MA (2010) Nonribosomal peptide synthetases: structures and dynamics. Curr Opin Struct Biol 20:234–240
Thakore Y (2006) The biopesticide market for global agricultural use. Ind Biotechnol 2:194–208
Till M, Race PR (2014) Progress challenges and opportunities for the re-engineering of *trans*-AT polyketide synthases. Biotechnol Lett 36:877–888
Trapet P, Avoscan L, Klinguer A, Pateyron S, Citerne S, Chervin C, Mazurier S, Lemanceau P, Wendehenne D, Besson-Bard A (2016) The *Pseudomonas fluorescens* siderophore pyoverdine weakens *Arabidopsis thaliana* defense in favor of growth in iron-deficient conditions. Plant Physiol 171:675–693
Tripathi AK, Khanuja SPS, Kumar S (2002) Chitin synthesis inhibitors as insect-pest control agents. J Med Arom Plant Sci 24:104–122
Udwary DW, Zeigler L, Asolkar RN, Singan V, Lapidus A, Fenical W et al (2007) Genome sequencing reveals complex secondary metabolome in the marine actinomycete *Salinispora tropica*. Proc Natl Acad Sci U S A 104:10376–10381
Uren NC (2007) Types, amounts and possible functions of compounds released into rhizosphere by soil-grown plants. In: Pinton R, Varanini Z, Nannipieri P (eds) The rhizosphere: biochemistry and organic substances at the soil-plant interface. CRC Press, New York, pp 1–21
Vacheron J, Desbrosses G, Bouffaud ML, Touraine B, Moënne-Loccoz Y, Muller D, Legendre L, Wisniewski-Dyé F, Prigent-Combaret C (2013) Plant growth-promoting rhizobacteria and root system functioning. Front Plant Sci 4:356

van de Mortel JE, de Vos RC, Dekker E, Pineda A, Guillod L, Bouwmeester K et al (2012) Metabolic and transcriptomic changes induced in *Arabidopsis* by the rhizobacterium *Pseudomonas fluorescens* SS101. Plant Physiol 160:2173–2188

Van Der Voort M, Meijer HJ, Schmidt Y, Watrous J, Dekkers E, Mendes R, Dorrestein PC, Gross H, Raaijmaker JM (2015) Genome mining and metabolic profiling of the rhizosphere bacterium *Pseudomonas* sp. SH-C52 for antimicrobial compounds. Front Microbiol 6:693

Van Loon LC (2007) Plant responses to plant growth-promoting bacteria. Eur J Plant Pathol 119:243–254

van Loon LC, Bakker PA, Pieterse CM (1998) Systemic resistance induced by rhizosphere bacteria. Annu Rev Phytopathol 36:453–483

Van Peer R, Niemann GJ, Schippers B (1991) Induced resistance and phytoalexin accumulation in biological control of *Fusarium* wilt of carnation by *Pseudomonas* sp. strain WCS417r. Phytopathology 81:728–734

Vansuyt G, Robi A, Briat JF, Curie C, Lemanceau P (2007) Iron acquisition from Fe-pyoverdine by *Arabidopsis thaliana*. Mol Plant-Microbe Interact 20:441–447

Vassilev N, Vassileva M, Nicolaeva I (2006) Simultaneous P-solubilizing and biocontrol activity of microorganisms: potentials and future trends. Appl Microb Biotech 71:137–144

Vejan P, Abdullah R, Khadiran T, Ismail S, Boyce AN (2016) Role of plant growth promoting rhizobacteria in agricultural sustainability – a review. Molecules 21:3–17

Velivelli SL, De Vos P, Kromann P, Declerck S, Prestwich BD (2014) Biological control agents: from field to market, problems, and challenges. Trends Biotechnol 32:493–496

Venugopalan A, Srivastava S (2015) Endophytes as in vitro production platforms of high value plant secondary metabolites. Biotechnol Adv 33:873–887

Verhagen BW, Glazebrook J, Zhu T, Chang HS, van Loon LC, Pieterse CM (2004) The transcriptome of rhizobacteria-induced systemic resistance in Arabidopsis. Mol Plant-Microbe Interact 17:895–908

Vespermann A, Kai M, Piechulla B (2007) Rhizobacterial volatiles affect the growth of fungi and *Arabidopsis thaliana*. Appl Environ Microbiol 73:5639–5641

Vincent MN, Harrison LA, Brackin JM, Kovacevich PA, Mukerji P, Weller DM, Pierson EA (1991) Genetic analysis of the antifungal activity of a soilborne *Pseudomonas aureofaciens* strain. Appl Environ Microbiol 57:2928–2934

Visca P, Imperi F, Lamont IL (2007) Pyoverdine siderophores: from biogenesis to biosignificance. Trends Microbiol 15:22–30

Vlot AC, Dempsey DA, Klessig DF (2009) Salicylic acid, a multifaceted hormone to combat disease. Annu Rev Phytopathol 47:177–206

Voisard C, Keel C, Haas D, Défago G (1989) Cyanide production by *Pseudomonas fluorescens* helps suppress black root rot of tobacco under gnotobiotic conditions. EMBO J 8:351–358

Walker TS, Bais HP, Grotewold E, Vivanco JM (2003) Root exudation and rhizosphere biology. Plant Physiol 132:44–51

Walters DR, Ratsep J, Havis ND (2013) Controlling crop diseases using induced resistance: challenges for the future. J Exp Bot 64:1263–1280

Wang H, Fewer DP, Holm L, Rouhiainen L, Sivonen K (2014) Atlas of nonribosomal peptide and polyketide biosynthetic pathways reveals common occurrence of nonmodular enzymes. Proc Natl Acad Sci U S A 111:9259–9264

Webb BA, Helm RF, Scharf BE (2016) Contribution of individual chemoreceptors to *Sinorhizobium meliloti* chemotaxis towards amino acids of host and non-host seed exudates. Mol Plant-Microbe Interact 29:231–239

Weber T, Blin K, Duddela S, Krug D, Kim HU, Bruccoleri R, Lee SY, Fischbach MA, Müller R, Wohlleben W, Breitling R, Takano E, Medema MH (2015) antiSMASH 3.0 – acomprehensive resource for the genome mining of biosynthetic gene clusters. Nucleic Acids Res 43:W237–W243

Wei G, Kloepper JW, Tuzun S (1991) Induction of systemic resistance of cucumber to *Colletotrichum orbiculare* by select strains of plant-growth promoting rhizobacteria. Phytopathology 81:1508–1512

Weingart H, Ullrich H, Geider K, Völksch B (2001) The Role of Ethylene Production in Virulence of *Pseudomonas syringae* pvs. *glycinea* and *phaseolicola*. Phytopathology 91:511–518

Weller DM (2007) *Pseudomonas* biocontrol agents of soilborne pathogens: looking back over 30 years. Phytopathology 97:250–256

Weller DM, Mavrodi DV, van Pelt JA, Pieterse CMJ, van Loon LC, Bakker PAHM (2012) Induced systemic resistance in *Arabidopsis thaliana* against *Pseudomonas syringae* pv. *tomato* by 2,4-diacetylphloroglucinol-producing *Pseudomonas fluorescens*. Phytopathology 102:403–412

Wen F, Van Etten HD, Tsaprailis G, Hawes MC (2007) Extracellular proteins in pea root tip and border cell exudates. Plant Physiol 143:773–783

Wenke K, Weise T, Warnke R, Valverde C, Wanke D, Kai M, Piechulla B (2012) Bacterial volatiles mediating information between bacteria and plants. In: Witzany J, Baluška F (eds) Biocommunication of plants, vol 14. Springer, Berlin, pp 327–347

Wheatley RE (2002) The consequences of volatile organic compound mediated bacterial and fungal interactions. Antonie Van Leeuwenhoek 81:357–336

Workentine ML, Chang L, Ceri H, Turner RJ (2009) The GacS-GacA two-component regulatory system of *Pseudomonas fluorescens*: a bacterial two-hybrid analysis. FEMS Microbiol Lett 292:50–56

Xie X, Zhang H, Pare PW (2009) Sustained growth promotion in *Arabidopsis* with long-term exposure to the beneficial soil bacterium *Bacillus subtilis* (GB03). Plant Signal Behav 4:948–953

Xu J, Audenaert K, Hofte M, De Vleesschauwer D (2013) Abscisic acid promotes susceptibility to the rice leaf blight pathogen *Xanthomonas oryzae* pv. *oryzae* by suppressing salicylic acid-mediated defenses. PLoS One 8:e67413

Yang J, Kloepper JW, Ryu CM (2009) Rhizosphere bacteria help plants tolerate abiotic stress. Trends Plant Sci 14:1–4

Yasuda M, Ishikawa A, Jikumaru Y, Seki M, Umezawa T, Asami T, Maruyama-Nakashita A, Kudo T, Shinozaki K, Yoshida S, Nakashita H (2008) Antagonistic interaction between systemic acquired resistance and the abscisic acid-mediated abiotic stress response in Arabidopsis. Plant Cell 20:1678–1692

Yu X, Ai C, Xin L, Zhou G (2011) The siderophore-producing bacterium, *Bacillus subtilis* CAS15, has a biocontrol effect on *Fusarium wilt* and promotes the growth of pepper. Eur J Soil Biol 47:138–145

Yuan J, Zhang N, Huang Q, Raza W, Li R, Vivanco JM, Shen Q (2015) Organic acids from root exudates of banana help root colonization of PGPR strain *Bacillus amyloliquefaciens* NJN-6. Sci Rep 25:13438

Zdor RE (2015) Bacterial cyanogenesis: impact on biotic interactions. J Appl Microbiol 118:267–274

Zhang H, Kim MS, Krishnamachari V et al (2007) Rhizobacterial volatile emissions regulate auxin homeostasis and cell expansion in Arabidopsis. Planta 226:839–851

Zhang H, Xie X, Kim MS, Kornyeyev DA, Holaday S, Paré PW (2008) Soil bacteria augment Arabidopsis photosynthesis by decreasing glucose sensing and abscisic acid levels in planta. Plant J 56:264–273

Zúñiga A, Poupin MJ, Donoso R, Ledger T, Guiliani N, Gutierrez RA, González B (2013) Quorum sensing and indole-3-acetic acid degradation play a role in colonization and plant growth promotion of *Arabidopsis thaliana* by *Burkholderia phyto* Firmans PsJN. Mol Plant-Microbe Interact 26:546–553

Chapter 4
Arbuscular Mycorrhizal Fungi and Plant Stress Tolerance

Abeer Hashem, Elsayed Fathi Abd_Allah, Abdulaziz A. Alqarawi, and Dilfuza Egamberdieva

Abstract Arbuscular mycorrhizal fungi (AMF) that inhabit the rhizosphere and colonize plant roots are considered to be beneficial to plant growth. AMF improves the rhizosphere soil characteristics and assists host plants by supplying essential mineral nutrients, especially phosphorus, while inhibiting the translocation of toxic ions such as Na and other metals. Plants have several tolerance mechanisms for averting the negative effects of different environmental stresses they encounter. Among these mechanisms, the antioxidant system is the key tolerance tool and is supported by the accumulation of osmolytes and the selective absorption of ions. Many reports have been published on the potential of AMF in plant growth regulation. Most of the researchers have adhered to studying morphological changes in host plants and have rarely described the physiological, biochemical, and molecular mechanisms of AMF-induced growth promotion and stress tolerance. The present review explores the existing literature to report the current status of AMF research, with a special focus on the AMF-triggered changes in the antioxidant and osmolyte metabolisms of plants that ameliorate the negative effects of stresses. In addition, we identify some key potential future targets to enhance the understanding of the beneficial effects of AMF in plant growth improvement under normal and stress conditions.

A. Hashem (✉)
Botany and Microbiology Department, College of Science, King Saud University, Riyadh, Saudi Arabia

Mycology and Plant Disease Survey Department, Plant Pathology Research Institute, ARC, Giza, Egypt
e-mail: habeer@ksu.edu.sa

E.F. Abd_Allah • A.A. Alqarawi
Plant Production Department, College of Food and Agricultural Sciences, King Saud University, Riyadh 11451, Saudi Arabia

D. Egamberdieva
Faculty of Biology, National University of Uzbekistan, Tashkent, Uzbekistan

D. Egamberdieva, P. Ahmad (eds.), *Plant Microbiome: Stress Response*, Microorganisms for Sustainability 5, https://doi.org/10.1007/978-981-10-5514-0_4

Keywords AMF • Abiotic stresses • Antioxidants • Osmolytes

4.1 Introduction

During developmental events, plants are exposed to a variety of factors that may be either abiotic or biotic; therefore, there is a great chance that plants may exhibit altered growth patterns. The main determinants of altered growth in plants are environmental conditions, for example, the soil health including its salinity status and water level, environmental temperature, concentration of mineral nutrients, and presence or absence of metals and their metalloids. The net effect of these factors is seen on the growth and the productivity of the crops (Hashem et al. 2015). Modulations of the physiology and biochemistry of the plant species that is imparted by these environmental factors result in altered growth and consequent yield, thereby posing a major threat to the global food security (Abd_Allah et al. 2015). In addition to soil characteristics impacting the growth of existing flora, several environmental stress factors, either abiotic or biotic and individually or conjunctively, affect the growth of plants by directly or indirectly changing the soil characteristics (Hashem et al. 2014; Alqarawi et al. 2014a; Ahanger et al. 2014).

Environmental constraints impose stern effects on plant growth and development all over the world and particularly in arid and semiarid regions (Barnawal et al. 2014). Climate change has further aggravated the situation. The use of salt- and metal-rich water for irrigation purposes continuously adds to the problem, resulting in conversion of agriculturally productive land into unproductive land. Stresses induce toxic effects and hence affect the important physiological and biochemical processes including photosynthesis and ion homeostasis, ultimately leading to a reduction in growth (Tejera et al. 2004; Porcel et al. 2012; Khan et al. 2015). Stress-caused alterations in photosynthesis are associated with impeded carbon and nitrogen metabolism, and in legumes, salinity alters the nitrogen fixation and hence the growth as well as the yield (Tejera et al. 2004). More importantly, plants that are subjected to stresses exhibit greater production and accumulation of toxic reactive oxygen species (ROS), which have the potential to damage membrane lipids, oxidize proteins, and damage nucleic acids (Mittler 2002; Hashem et al. 2016). The key ROS that are generated under stress include superoxide, hydroxyl, and peroxide radicals (Ahmad et al. 2010). The ROS often accumulate in sensitive tissues such as leaves to cause oxidation of the abovementioned macromolecular structures, hence affecting the plant's cellular physiology. ROS are mainly generated in chloroplasts, mitochondria, and peroxisomes (Mittler 2002; Ahmad et al. 2010).

4.2 Stress Tolerance Mechanisms

To counteract the oxidative damage imparted by excess ROS, plants have developed several defense mechanisms to keep ROS at the optimal concentration. These systems include an antioxidant defense system that is composed of both enzymatic and nonenzymatic components. Among the enzymatic components are superoxide dismutase (SOD), catalase (CAT), guaiacol peroxidase (POD), ascorbate peroxidase (APX), monodehydroascorbate reductase (MDAR), dehydroascorbate reductase (DAR), and glutathione reductase (GR), while nonenzymatic components include certain vitamins such as ascorbic acid and tocopherol, polyphenols, and glutathione (Ahmad et al. 2010; Hashem et al. 2015, 2016). The antioxidant system neutralizes excess ROS to protect cells against stress-triggered oxidative damage. Antioxidant systems, compared to other tolerance mechanisms, are considered among the main defense systems for mitigating the deleterious effects of stress. Other major contributors to stress mitigation include the accumulation of osmolytes such as free proline, sugar alcohols, and amino acids, which leads to the protection of cellular structures and function (Ahmad and Sharma 2008). When antioxidant systems lose efficiency, osmolytes strengthen the antioxidant system by helping in the scavenging of ROS. Under salt-stressed conditions, osmoregulation is achieved by the efficient uptake and accumulation of ions such as sodium, chloride, and potassium, while in drought-stressed conditions, accumulation of compatible solutes and ions with the simultaneous sequestration and compartmentation of deleterious ions into vacuoles or apoplasts is considered a key trait that determines the stress tolerance (Ahmad and Sharma 2008; Azooz et al. 2011; Hashem et al. 2016).

For combating stress, at the initial stage of stress exposure, plants try to avoid stresses by minimizing the potential negative impacts of stresses; for example, morphological adaptations include leaf rolling, development of heavy pubescence on leaves, leaf falling, and modification of plant structures (Ruiz-Lozano et al. 2006; Hameed et al. 2014; Ahmad et al. 2016). In addition to the indigenous tolerance mechanism, plants adopt several alternative mechanisms for protecting metabolism at physiological and biochemical levels (Hameed et al. 2014). There have occasionally been several attempts made by various workers to design new management techniques to improve plant protection and productivity (Ahmad et al. 2011, 2015; Khan et al. 2014; Alqarawi et al. 2014a, b; Hashem et al. 2014, 2015, 2016; Abd_Allah et al. 2015, 2017). According to this research, one of the efficient efforts for achieving improved plant growth and sustaining the food security is the exploitation of naturally occurring soil microflora. There is considerable evidence available within the literature that demonstrates the beneficial role of microorganisms in the rhizospheric soil, of which arbuscular mycorrhizal fungi (AMF) and plant growth-promoting rhizobacteria (PGPR) constitute the key components (Lopez-Raez et al. 2010; Abdel Latef and Chaoxing 2014; Hameed et al. 2014; Ahanger et al. 2014; Hashem et al. 2015, 2016; Alqarawi et al. 2014a, b). AMF, microscopic filamentous fungi, have the

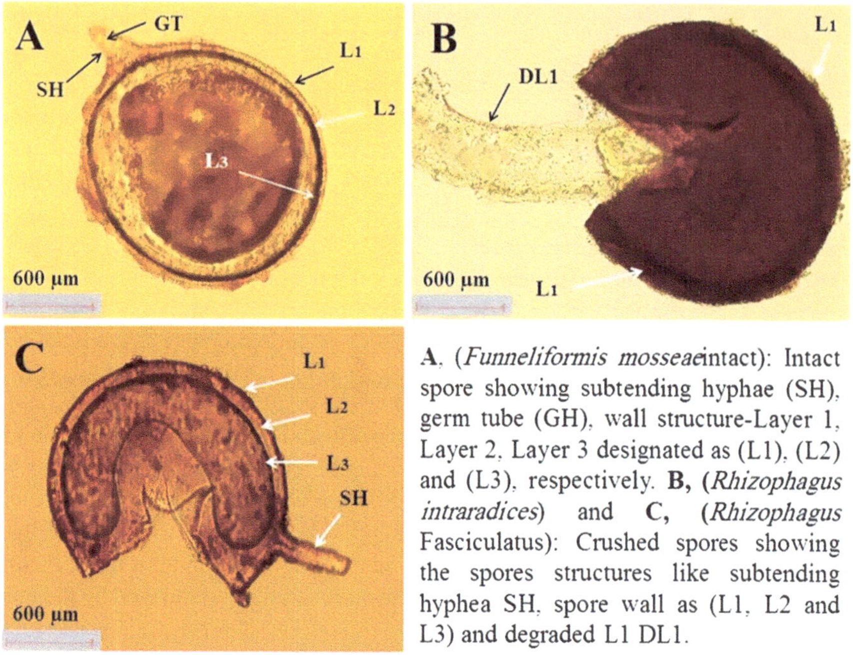

Fig. 4.1 Spore morphology of some arbuscular mycorrhizal fungi was isolated from soil samples of tomato. (**a**) (*Funneliformis mosseae* intact): intact spore showing subtending hyphae (*SH*), germ tube (*GH*), wall structure – Layer 1, Layer 2, and Layer 3 designated as (*L1*), (*L2*), and (*L3*), respectively. (**b**) (*Rhizophagus intraradices*) and (**c**) (*Rhizophagus* fasciculatus): crushed spores showing the spore structures like subtending hyphae (*SH*), spore wall (*L1*, *L2*, and *L3*), and degraded L1 DL1

capability to colonize most plants either directly in the root tissues or in the rhizosphere. AMF usually forms ramified filaments a few centimeters long to benefit the host plant (Abd_Allah et al. 2015). The spore morphology of some arbuscular mycorrhizal fungi that were isolated from soil samples of tomato was illustrated in Fig. 4.1a–c.

AMF form symbiotic associations with several plant species and have proved to possess the potential to improve soil structure and plant growth in normal as well as stressful environments (Tang et al. 2009; Navarro et al. 2013; Hashem et al. 2016). For example, under salt- and heavy metal stress conditions, AMF are inoculated and believed to act as essential bio-ameliorators, therefore helping plants in alleviating stress-induced damage (Rabie and Almadini 2005; Abd_Allah et al. 2015). AMF colonization has been reported to enhance plant growth and vigor (Evelin et al. 2009; Ahanger et al. 2014; Alqarawi et al. 2014a, b; Hashem et al. 2016). AMF colonization brings morphological, nutritional, and physiological changes and has also been reported to enhance resistance of plants to abiotic and biotic stresses. AMF modify root architecture so that roots access more water and nutrients (Aroca et al. 2008; Wu et al. 2010a, b; Ahmad et al. 2015). Mycorrhizal inoculation not

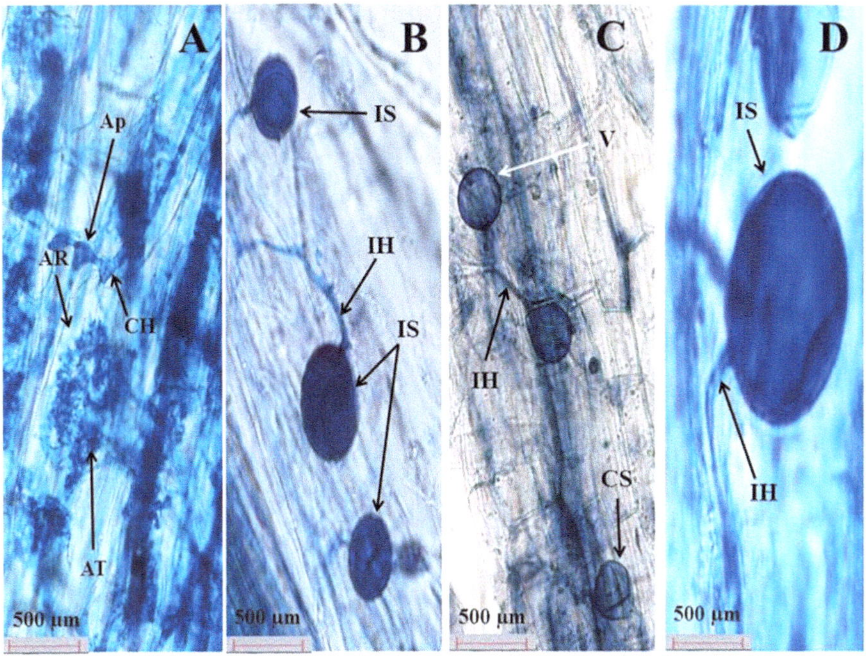

Fig. 4.2 **(a–d)** Photomicrographs of structural colonization of AMF in the root of tomato plant showed a complete abundance of AMF colonization. The AMF colonization was shown by the presence of appressorium which then spread and formed different morphological structures. *AT* arbuscular trunk hyphae, *CH* arbusculate hyphal coils, *AP* appressorium, *AR* arbuscules, *IS* interadical spore, *IH* interadical hyphae, *V* vesicles, *CS* crushed spores

only affects root morphology but also physiology in host plants (Badri et al. 2009). For example, citrus plants that are colonized with AMF have large leaf areas, higher phosphorus content, and increased photosynthesis compared to uncolonized plants (Sherstha et al. 1995). The intensity of structural colonization of AMF in the root of tomato plant showed complete abundance with different morphological structures initiated by appressorium which then spread and formed different morphological structures within the root tissues such as vesicles, arbuscules, interadical spores, arbuscular trunk hyphae, arbusculate hyphal coils, and interadical hyphae (Fig. 4.2a–d).

It has been proven that AMF symbionts are associated with more than 90% of plants and contribute markedly to growth regulation by providing multiple benefits to plants (Sherstha et al. 1995; Bonfante and Genre 2010; Hajiboland et al. 2010; Hajiboland 2012). Xie et al. (2014) have described AMF association and its ecological significance in plants that flourish in terrestrial and wetland environments. This review addresses the beneficial role of AMF in plant growth and the involvement of AMF in the protection of plants from oxidative damage that is induced by the environmental stresses, with special reference to the antioxidant system.

4.3 Arbuscular Mycorrhizal Fungi: Occurrence, Diversity, and Benefits

The literal meaning of the term mycorrhiza—"myco" means "fungus" and "rhiza" means "root"—reflects the common meaning, i.e., a fungus infecting or inhabiting the roots (Hameed et al. 2014). Studies have confirmed that AMF hyphae markedly improve soil structure by making the nutrients available and increasing the water-holding capacity (Candido et al. 2015; Hashem et al. 2014). In certain plants where AMF colonization is not inherent, characteristic inoculation of AMF makes hyphal networks that affect plant growth via regulating soil health and the composition of species. There are nearly 250 AMF species that have been identified to date (Kruger et al. 2012). This result suggests that AMF are a widely distributed group, but still, with such a low number of identified species, extensive research efforts are needed for further isolation and identification of soil-borne beneficial fungal species. This research will further develop ideas about the functional differences between different isolates. An effort has been made in this work to gather the existing knowledge about AMF research with an aim to identify key future research directions.

Until now, different AMF isolates have fit within six different genera and an order, i.e., *Glomales*, also referred as *Zygomycota* (Ahanger et al. 2014; Hameed et al. 2014). Molecular studies carried out by Schuβler et al. (2001) have confirmed through ribosomal RNA sequencing that all AMF species fall in the same monophyletic clade and that they are clearly segregated from other fungi. Advancement in molecular techniques has enabled direct identification of the AMF isolates from infecting roots or rhizospheres, and the genetic diversity in the different species has also been confirmed (Kruger et al. 2012). It has been earlier demonstrated that a single AMF spore can contain thousands of nuclei, reflecting a multigenomic nature (Kuhn et al. 2001). It shall be noted here that the efficiency of an AMF genotype to affect a host plant varies significantly with the attributes of the AMF or the host. For example, AMF-related attributes include the number of the extra-radical mycelia, mycorrhizal-specific expression of genes, and improvement in the nutrient absorption and usage efficiency resulting in markedly variable responses from or effects on host plants (Hashem et al. 2014, 2015; Abd_Allah et al. 2015). The exact mechanism underlying this behavior is largely unknown; however, many workers have put forward their suggestion, and experimental evidence is accumulating day by day. For example, the extra-radical mycelia of AMF are responsible for the absorption of mineral nutrients from the soil solution and their subsequent transport into the host plant. It has been reported that the constitutive expression of transport proteins that are involved in assimilation of essential elements such as nitrogen and phosphorus also affects the plant's potential to assimilate minerals, even if AMF are present (Lee et al. 2012). Therefore, it should be accepted that variations in the efficiency of different AMF isolates or species to improve the growth and protect the metabolism of plants exist and depend on the particular AMF isolate and host species.

4.4 Effect of Stresses on Plant Growth and the Role of AMF

As discussed in the previous sections, AMF promote plant growth and productivity under stressful conditions. AMF have been accepted as potential future biofertilizers and biocontrol agents, thereby proving to be considerably advantageous for plants (Hameed et al. 2014). The key mechanisms that are induced by AMF for ameliorating the oxidative damage to protect the plant cell function have been attributed to the AMF-triggered strengthening of the tolerance mechanism, improved water and nutrient uptake, and modulation in the expression of genes that are involved in signaling (and hence the stress response) (Hashem et al. 2015). The following sections will highlight the role of AMF in modulation of growth with special reference to their influence on the antioxidant system.

4.4.1 Water Stress and AMF

Water stress is an important abiotic stress factor that affects the growth of plants by hampering key metabolic attributes including nutrient uptake and assimilation, enzyme activity, photosynthesis, protein synthesis, and antioxidant metabolism (Hameed et al. 2014; Nazar et al. 2015). It has been observed that water stress is often accompanied by temperature stress, leading to extreme drying of soil and thereby resulting in drought stress. The unavailability of water for absorption by roots, by itself, is called water stress, and increased temperature accelerates the transpiration rate, which is concomitant with photosynthetic arrest due to excess generation and accumulation of ROS, a form of oxidative stress (Ahanger and Agarwal 2016; Nazar et al. 2015). However, plants that are in symbiosis with AMF or inoculated with AMF exhibit apparent increases in growth and metabolism that has been mostly concluded to be due to the greater uptake of essential minerals (Al-Ezerjawi and Kadhim 2014; Al-karaki et al. 2004; Hameed et al. 2014). Al-Ezerjawi and Kadhim (2014) have reported that *Trichoderma harzianum* protects *Triticum aestivum* from drought stress by improving the soil characteristics that result in greater chlorophyll production, yield, and nitrogen uptake. Increases in water and nutrient uptake and subsequent promotion of growth due to AMF inoculation may be a result of an intrusion of the hyphal networks and the production of glomalin; these factors improve the soil structure and thus the water relations (Wu and Zou 2017). As already noted, the increased plant growth that is mediated by AMF can be due to soil stabilization, in addition to improved health. Many workers have clearly ascribed the AMF-induced increased growth and drought stress mitigation to enhancement in the uptake and transfer of nutrients and water, which leads to better osmoregulation and water-use efficiency, in addition to reduced production of ROS. AMF inoculation protects photosynthesis and associated attributes such as stomatal conductance and gas exchange, transpiration, and water-use efficiency; the improvements in these attributes may be due to increased water uptake in

AMF-inoculated plants that is due to the presence of extra-radical hyphae (Gholamhoseini et al. 2013). Moreover, AMF inoculation stimulates growth under drought stress by increasing the hydraulic conductivity and osmotica accumulation (Hameed et al. 2014). The improved water absorption in AMF-inoculated plants is reported by the study of Li et al. (2013), who have demonstrated overexpression of root aquaporin gene, which is involved in water uptake, in *Zea mays*; this expression was specific in regions that were infected by hyphae. Cucumber plants inoculated with *Trichoderma harzianum* showed greater uptake of essential elements such as N, P, K, and Cu compared to uninoculated plants, suggesting the possibility of active involvement of the soil-borne microorganisms in directly modulating nutrient uptake in addition to improving the soil health (Yedidia et al. 2001).

Regarding the AMF-induced accumulation of osmolytes, the existing research reports seem to be contradictory. Stimulation of the biosynthesis of osmolytes such as proline, glycine betaine, sugars, and polyols can improve the growth-promoting effect of AMF in plants by maintaining the tissue water potential (Yooyongwech et al. 2013). In *Erythrina variegata*, inoculation with AMF brings down the synthesis of sugars compared to control plants. Similarly, plants inoculated with AMF showed differential synthesis of amino acids when exposed to drought stress (Ogawa and Yamauchi 2006). AMF-induced accrued production of amino acids can protect the functioning of the cell by maintaining the precursor pools of many important molecules such as redox components, e.g., glutathione. Reports advocating the ubiquitous role of osmolytes in water stress tolerance are extensive; however, the role of AMF and other soil-borne microflora is not fully known yet and requires due attention in the future. Proline synthesis is downregulated when samples are inoculated with AMF (Ruiz-Sanchez et al. 2010; Fan and Liu 2011; Asrar et al. 2012), and such experimental evidence can raise questions about the credibility of the use of AMF in water stress amelioration due to the key role of proline in stress tolerance. A reduction in the synthesis of biosynthetic precursors of proline has been suggested as a probable reason for reduced proline accumulation. Zou et al. (2013), while working on *Poncirus trifoliata* subjected to drought and inoculated with *Funneliformis mosseae*, observed significant growth improvement that resulted in increased biomass accumulation; however, the proline content had declined considerably.

It happens to be accepted that a common response of plants to stresses, i.e., the production of ROS such as superoxide anion radicals, singlet oxygen, hydrogen peroxide, and hydroxyl radicals, increases considerably because of water stress impeding cellular function, and under extreme conditions, it can cause death (Nazar et al. 2015). However, it shall be noted that an efficient antioxidant system averts the ROS-triggered oxidative stress and maintains the cellular redox balance at adequate levels (Ahmad et al. 2010; Hashem et al. 2015). Based on certain key experiments, it is being believed that inoculation of AMF upregulates the antioxidant system, resulting in quick elimination of ROS and thereby leading to maintenance of cellular redox levels (Hameed et al. 2014; Wu and Zou 2017). Wu et al. (2007), while detecting the efficiency of five different species of *Glomus*, observed a significant amelioration of drought stress-induced photosynthetic inhibition by upregulation of the activities of antioxidant enzymes such as SOD, POD, and CAT and the accumulation of osmolytes including soluble sugars and protein. They observed *Glomus mosseae* and *Glomus versiforme* to be more beneficial among the five tested species. Recently, Huang et al. (2017) have also reported that under drought stress, inoculating trifoliate orange with AMF significantly reduced the production of ROS including H_2O_2 and superoxide, ultimately resulting in greater membrane protection via improved efflux of H_2O_2 through roots. In addition, many other reports are available that describe the efficiency of AMF in mitigation of drought stress via the modulation of antioxidant systems (Ruiz-Lozano et al. 1996; Ruíz-Lozano and Aroca 2010; Wu and Zou 2010; Baslam and Goicoechea 2012; Armada et al. 2017). Ruiz-Sanchez et al. (2010) have significantly enhanced the synthesis of glutathione, an antioxidant molecule, thereby bringing stability to photosynthetic pathways. In *Zea mays*, inoculation with AMF improved the photosynthetic efficiency under drought stress by enhancing the uptake of essential nutrients and redox components such as glutathione and ascorbate, resulting in reduced oxidative damage via a reduction in the generation of ROS. AMF was found to boost the growth-promoting effect of PGPR when they were applied in combination (Armada et al. 2017). Inoculation with AMF upregulated the gene expression of SOD, CAT, APX, GR, and monodehydroascorbate reductase in watermelon leaves under drought stress, resulting in greater rubisco activity and the photosynthetic performance (Mo et al. 2016). In addition to improving the enzymatic and active components of the antioxidant machinery, AMF inoculation improves the antioxidant defense system by increasing the synthesis of flavonoids, which have been considered to be key ROS-neutralizing molecules (Abbaspour et al. 2012). Plant polyphenols have been ascribed antioxidant function, and improved secondary metabolite accumulation has been attributed to AMF inoculation (Walter and Strack 2011). Hazzoumi et al. (2015) have reported that AMF does not affect phenol content in *Ocimum gratissimum*.

4.4.2 Salinity Stress and AMF

Salinity is one of the major abiotic stress factors that affect growth and development by reducing photosynthetic efficiency, mineral assimilation, and antioxidant metabolism (Khan et al. 2014; Hashem et al. 2015: Abd_Allah et al. 2015). High salinity concentrations in the soil solution result in restricted water uptake, and entry of excessive amounts of toxic salts triggers the rapid generation of ROS, resulting in oxidative damage to important structures including proteins and nucleic acids and ultimately hampering metabolic homeostasis, thereby causing cessation (Khan et al. 2009; Ahmad et al. 2012; Hashem et al. 2014; Abd_Allah et al. 2015). From time to time, several mitigating strategies have been introduced to ameliorate the salinity-triggered deleterious effects on growth (Khan et al. 2014). High salinity alters ionic homeostasis, leading to altered nutrient acquisition and affecting the enzyme activity by affecting the redox balance (Khan et al. 2014; Iqbal et al. 2015). Apart from these strategies, the use of soil-borne AMF in improving plant growth, biomass accumulation, and productivity under saline conditions has been reported by several workers (Evelin et al. 2009; Aroca et al. 2008, 2013; Alqarawi et al. 2014a, b; Abdel Latef and Chaoxing 2014; Hashem et al. 2014, 2015; Abd_Allah et al. 2016; Elhindi et al. 2017). While considering these reports, it shall be noted here that inoculation of AMF improves water absorption, resulting in greater photosynthetic efficiency, maintenance of nutrient balance, enzyme activity, and, more importantly, upregulation of the antioxidant defense system. Hameed et al. (2014) have reviewed work that reports that under salt stress, AMF colonization protects plants by improving water absorption via the hyphal network, in addition to improving selective ion absorption and maintaining gas exchange capacity. AMF colonization improves growth of host plants by increasing the root growth, morphology and the hydraulic conductivity, stomatal conductance, net photosynthesis, transpiration rate, and water-use efficiency, in addition to improving the uptake of nitrogen and potassium (Lin et al. 2017; Elhindi et al. 2017). It has been reported that under salinity stress, AMF inoculation mitigates the negative impact of the stress on chlorophyll content and the uptake of essential elements such as N, P, K, Ca, and Mg (Borde et al. 2010; Cekic et al. 2012; Alqarawi et al. 2014a; Hashem et al. 2015; Scagel and Bryla 2017). AMF inoculation maintains the activity of photosynthetase under salinity conditions in addition to increasing the activity of the photosystems, the chlorophyll content, and the activity of carbonic anhydrase (Talaat and Shawky 2004). More interestingly, plants that have been inoculated with AMF have been observed to reduce the salt-triggered oxidative damage by protecting the membrane lipids from the oxidation by ROS (Alqarawi et al. 2014a; Yang et al. 2014). AMF protects membranes by maintaining greater contents of polyunsaturated fatty acid components (Alqarawi et al. 2014a), and this mechanism has been attributed to the upregulation of the antioxidant defense system (Hashem et al. 2014, 2015; Abd_Allah et al. 2015). Such an impact of AMF on the membrane fatty acids under salinity conditions results in improved plasma membrane integrity, thereby decreasing the leakage of cellular components (Garg and Manchanda 2009). Greater membrane stability, with a higher ratio of unsaturated fatty acids in AMF-inoculated plants, is also attributed to reduced production of ROS such as H_2O_2 and OH (Talaat and Shawky 2014; Hashem et al. 2015). Yang et al. (2014) have demonstrated amelioration of salinity stress-induced oxidative damage to membranes by AMF colonization. In addition, inoculation of AMF has been reported to modulate the biosynthesis of important osmoprotectants such as proline, glycine betaine, and soluble sugars, thereby protecting photosynthesis by improving chloroplastic membrane integrity (Yang et al. 2008). Plant species accumulating greater osmoprotectants under mycorrhizal-inoculated conditions protect and improve photosynthetic efficiency (Hashem et al. 2015; Hameed et al. 2014; Scagel and Bryla 2017). It shall be noted that the actual mechanisms underlying the protective pathways are not clearly known. The greater accumulation of osmolytes under salinity conditions that is due to inoculation with AMF makes AMF inoculation a suitable candidate for enhancing the salt stress tolerance of sensitive plant species (Ahanger et al. 2014; Hashem et al. 2016). AMF inoculation improves the activity of key antioxidant components, developing a strong protective mechanism against the harmful effects of ROS. Upregulation of the activities of SOD, CAT, POD, APX, and GR has been observed to be due to inoculation of plants with AMF (Alqarawi et al. 2014b; Abdel Latef and Chaoxing 2014; Evelin and Kapoor 2014; Hashem et al. 2015, 2016). It has been suggested that high salinity reduces the biosynthesis of strigolactone, and increased synthesis of strigolactone in AMF-inoculated plants has been reported to be responsible for improved development, symbiosis, and stress tolerance (Aroca et al. 2013).

More importantly, inoculation with AMF reduces the Na/K ratio via significant enhancement of the uptake of K^+ ions, which leads to maintenance of enzyme activity and other important metabolic pathways (Hashem et al. 2015). Uptake of essential elements including N, P, K, Mg, Ca, etc. is accompanied by reduced uptake of toxic ions such as Na^+ in AMF-inoculated plants (Hashem et al. 2016). Such effects of AMF inoculation on the host plant's mineral assimilation prevent ion toxicity and alleviate nutrient deficiency.

4.4.3 Metal Stress and AMF

It has raised many serious concerns that day-by-day increasing industrialization and inefficient damping systems continuously pollute soil with different toxic metals. The increasing accumulation of these toxic heavy metals such as lead, cadmium, and mercury hinders the growth of important crops, resulting in yield reductions (Ahmad et al. 2010, 2015). It is important to mention here that these heavy metals exhibit great mobility from soils into plants, resulting in their accumulation in the edible parts and thus causing several health risks in humans. Elevated toxic concentrations of different metals in the soils inhibit germination and shoot and root growth and promote photosynthetic and transpiration arrest and senescence (Asgher et al. 2014; Abd_Allah et al. 2015). Reduced leaf area, chlorophyll synthesis, and the rubisco activity under metal stress are the prime reasons for reduced growth, in addition to altered enzyme activity (Asgher et al. 2014). Heavy metals have a great affinity for active sites of the enzymes (Verma and Dubey 2003; Ahmad et al. 2010). However, there are some plant species that accumulate considerable concentrations of heavy metals and have developed mechanisms such as volatilization to expel them. These plants have the capability to support the profuse growth of symbionts (AMF or PGPR), thereby conferring their greater tolerance to toxic metals and justifying them as suitable candidates for phytoremediation. Soil-borne microflora change the oxidation state of the toxic metals, making them either less toxic or mediating their elimination as complexes (van Hullebusch et al. 2005). Similarly, AMF improves plant growth by changing the pH of rhizospheric soil and by chelation or precipitating metals in polyphosphate granules or other compounds that are secreted by the associated fungal species (Ahanger et al. 2014). Gonzalez-Chavez et al. (2004) have suggested that the glomalin that is secreted by AMF stops the movement of toxic metals such as cadmium and copper. AMF are prevalent in almost every type of soil, even under extreme saline conditions (Ahanger et al.

Table 4.1 Effect of *Fusarium oxysporum* f.sp. *lycopersici* (FOL) triggered wilt disease on relative water content (%) in *Solanum lycopersicum* with and without arbuscular mycorrhizal fungi (AMF) inoculation

Treatments	Relative water content (%)
Control	86.09b
FOL only	48.82d
FOL + AMF	64.14c
AMF only	92.40a
LSD at 0.05	2.57

Data presented is mean of three replicates

2014; Hameed et al. 2014). Behavior of AMF has been observed to display considerable variability, e.g., in some cases, it improves transfer of heavy metals to host plants while restricting their mobilization in others, giving way to two different mechanisms, called phyto-extraction and phyto-stabilization (Abdel Latef et al. 2016). However, it shall be noted here that the mechanisms mediating the specificity that is governed by the AMF are largely unknown. Lanfranco et al. (2002) have demonstrated that the gene encoding the metallothionein from *Gigaspora margarita* reduces metal toxicity.

However, reports about the chelation of toxic metals in the rhizosphere are available, e.g., broom sedge exposed to aluminum was inoculated with three AMF species including *Rhizophagus clarus*, *Acaulospora morrowiae*, and *Funneliformis heterogama*. Kelly et al. (2005) demonstrated that *Rhizophagus clarus* enabled the highest resistance to increased Al concentrations, imparting improved growth to sedge plants. The probable reason for this behavior is the variation in the production of glomalin (Wu and Zou 2017). In aluminum-contaminated soils, glomalin binds with most aluminum species and has been suggested to be the best determinant of AMF-induced growth enhancement under such conditions (Toljander et al. 2007).

Contradictory research reports are also available regarding the AMF-mediated metal stress amelioration, with some reports advocating that an increasing intensity of metals triggered deleterious effects due to AMF, while some report no change with AMF (Liao et al. 2003; Gamalero et al. 2009; Garg and Singla 2012; Yang et al. 2015). AMF-colonized plants show relatively less accumulation of toxic metal ions in shoot tissue. Garg and Singla (2012) have reported greater leaf relative water content and chlorophyll concentrations in AMF-inoculated pea seedlings that were subjected to arsenic stress; these effects were due to the cumulative effect of AMF on the accumulation of proline, glycine betaine, and total free amino acids. In another context, AMF-inoculated tomato plants depicted an improvement of 6.829% in the relative water content (RWC) as compared to control plants (Table 4.1) and triggered the imparted amelioration of electrolyte leakage (Fig. 4.3) under adverse impact of biotic stress (fusarium wilt disease). There are a considerable number of reports available that describe the role of AMF inoculation in the amelioration of metal toxicity via upregulation of the antioxidant system. Yang et al. (2015), while studying the responses of a leguminous tree (*Robinia pseudoacacia* L.) to two different AMF species (*Funneliformis mosseae* and *Rhizophagus intraradices*) for improving phyto-stabilization efficiency and tolerance to lead-contaminated soils,

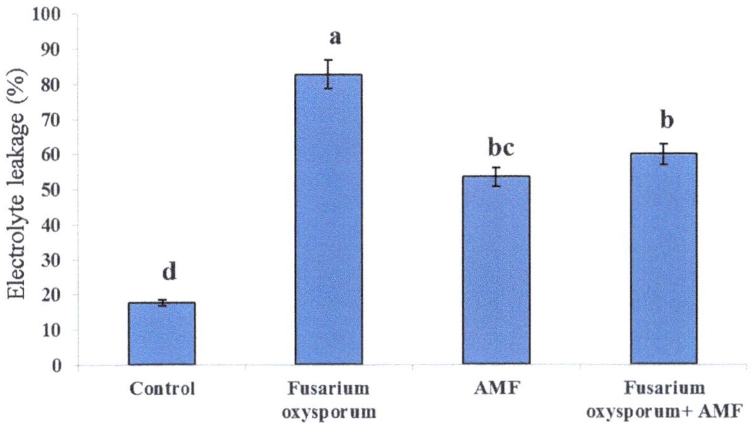

Fig. 4.3 Effect of *Fusarium oxysporum* f.sp. *lycopersici* (FOL) triggered wilt disease on electrolyte leakage with and without arbuscular mycorrhizal fungi (AMF) in *Solanum lycopersicum* L. Data presented are the means ± SE ($n = 5$)

observed a strong correlation between the AMF-induced enhancement in the activity of antioxidant enzymes including SOD, APX, and GPX that led to a reduction in the production of ROS and lipid peroxidation. Such AMF-triggered increases in antioxidant system were responsible for the protection and improvement of photosynthesis in *Robinia pseudoacacia*. FOL infection in tomato plants triggered the antioxidant metabolism with an increase in antioxidant enzymes as SOD, APX, DHAR, and GR activity. Plants treated with FOL + AMF showed 51.749%, 12.752%, 18.828%, and 26.385% increase in activity of SOD, APX, DHAR, and GR, while inoculation with AMF only resulted in 23.763%, 4.433%, 13.097%, and 8.266% increase in activity of SOD, APX, DHAR, and GR, respectively (Fig. 4.4a–d).

4.4.4 Temperature Stress and AMF

High and low (cold) temperatures, as with other stresses, are the prime environmental factors affecting plant growth maintenance. Various developmental processes and the key metabolic processes that contribute to the regulation of plant growth are affected by exposure to low temperatures. Average global agricultural productivity is severely affected by temperature stress, and more importantly, it increases the deleterious impact of other stresses such as drought, salinity, and mineral stresses (Machado and Paulsen 2001). *Hordeum spontaneum* that was subjected to combined temperature and drought stress exhibited a significant decline in biomass accumulation and grain yield by inhibiting PSII function, indicating an alteration in the functional integrity of the oxygen-evolving complex and the connectivity of PSII units (Jedmowski et al. 2015). Briefly, the exposure of plants to

Fig. 4.4 Effect of *Fusarium oxysporum* f.sp. *lycopersici* (FOL) triggered wilt disease on activity of (**a**) superoxide dismutase, (**b**) ascorbate peroxidase, (**c**) dehydroascorbate reductase, (**d**) glutathione reductase with and without arbuscular mycorrhizal fungi (AMF) in *Solanum lycopersicum* L. Data presented are the means ± SE (n = 5)

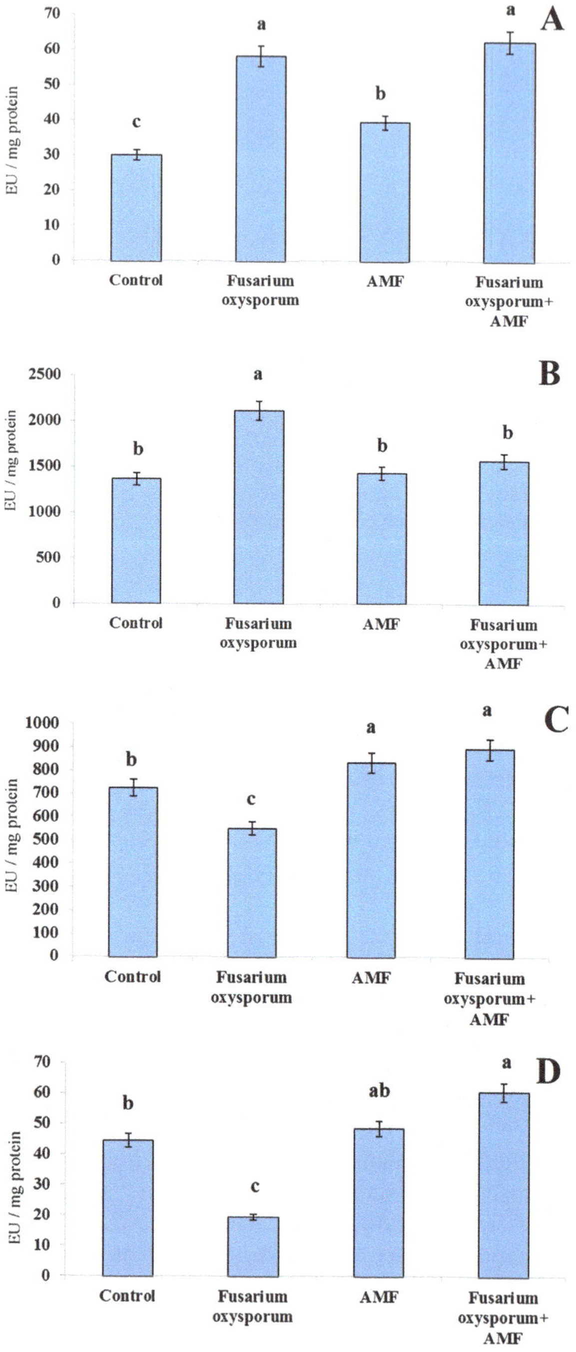

high-temperature regimes delays germination, reduces growth transitions and biomass accumulation, and promotes leaf drying and fall, photosynthetic arrest, and fruit discoloration and damage via greater accumulation of toxic ROS (Jedmowski et al. 2015).

The structure as well as activity of several important macromolecules is affected by low temperature, mainly by its deleterious effect on the cellular osmotic potential, plasma membrane integrity, and, more importantly, the restriction of antioxidant metabolism, leading to loss of the structural stability of major proteins (Yadav 2010; Liu et al. 2013; Paredes and Quiles 2015). Among the key visible effects of low temperature are reduced leaf area, growth retardation, chlorosis, and wilting; the key physiological impacts are a reduction in hydraulic conductance, a loss of stomatal movements, and decreased photosynthesis (Faroor et al. 2009; Adam and Murthy 2014; Liu et al. 2013; Paredes and Quiles 2015). In *Avena nuda*, Liu et al. (2013) have observed significant increases in the generation of ROS, leading to lipid peroxidation and a reduction in chlorophyll synthesis and growth due to cold stress (−10 °C). Like other stresses, to avert the low-temperature-induced ravages, the exploitation of AMF has been worked out, and it has been reported that low temperature not only affects the plant species but also considerably modulates the growth of the existing microflora by affecting spore germination and hyphal development, resulting in reduced root colonization (Wu and Zou 2010). Bunn et al. (2009), while exploring the effect of ambient and elevated temperatures on growth and AMF colonization in temperature-sensitive (*Agrostis scabra* and *Erythranthe guttata*) and temperature-tolerant (*Dichanthelium lanuginosum*) plant species, observed that the AMF colonization potential including spore germination and extra-radical hyphal growth was reduced considerably in the sensitive species compared with the thermal-tolerant plant. Contrarily to this finding, Heinemeyer and Fitter (2004) demonstrated that exposure of *Glomus mosseae* to 10–12 °C has no effect on the colonization and hyphal growth; however, when plants that were colonized with *Glomus mosseae* were exposed to low temperature, AMF colonization potential was significantly affected, indicating that AMF could have contributed to the temperature tolerance. To counteract the ill effects of low- and high-temperature stress-mediated ROS-generated oxidative damage, plants inoculated with AMF have been observed to show greater antioxidant enzyme activity (Liu et al. 2013; Chen et al. 2014). In another study by Zhu et al. (2010), the root colonization potential of *Glomus etunicatum* remained unaffected at 5 °C for at least 7 days of stress exposure.

Inoculation with AMF protected the net photosynthetic rate and associated characteristics including stomatal conductance, transpiration rate, and quantum efficiency of PSII, thereby protecting the metabolism by significantly increasing CO_2 assimilation (Zhu et al. 2012; Xu et al. 2016) and by improving the accumulation of proline and sugars (Xu et al. 2016). Moreover, in AMF-colonized plants, chloroplast thylakoid membranes and the photosynthetic apparatus exhibited much greater stability compared to that of uninoculated plants (Zhu et al. 2012). Though high temperatures affect the water relations by reducing the leaf water potential and the relative water content (Machado and Paulsen 2001), the hyphae of AMF increase root hydraulic conductivity and thereby contribute to improved water flow and osmotic adjustment (Evelin et al. 2009; Xu et al. 2016). Moreover, it has been reported that AMF inoculation improves water-use efficiency to improve metabolic capacity and yield potential of plants (Elhindi et al. 2017).

Reports describing the beneficial role of AMF colonization in the mitigation of high- and low-temperature stress are rare, particularly in relation to antioxidant systems, and extensive research efforts are needed to unravel the exact mechanisms underlying the tolerance mechanisms that are initiated by AMF inoculation. Chen et al. (2013) have observed that cucumber inoculated with AMF and exposed to low temperatures showed a greater accumulation of phenols, flavonoids, and lignins, resulting in reduction of H_2O_2 production and therefore providing a better correlation between AMF-triggered accumulation of polyphenols and the amelioration of oxidative damage. Maize plants inoculated with *Glomus etunicatum* and *Glomus intraradices* maintained greater activities of antioxidant enzymes such as CAT and POD, protecting the membranes from the low-temperature-induced metabolic hindrances (Chen et al. 2014). In addition, AMF-colonized plants maintained a greater accumulation of osmolytes. Inoculation of trifoliate orange (*Poncirus trifoliata*) with *Glomus mosseae* mitigated the negative effects of low (15 °C), optimum (25 °C), and high (35 °C) temperatures on growth by improving the root morphology and stimulating the activity of ROS-scavenging enzymes including SOD and CAT in addition to increasing soluble protein content that resulted in increased biomass accumulation (Wu 2011).

4.4.5 Nutrient Stress and AMF

It is well accepted that optimal growth under normal and stressful conditions is most often regulated by the optimal supplementation of mineral elements (Ahmad et al. 2015, 2016). Application of different nutrients protects plant metabolism by enhancing the enzyme activity, photosynthesis, and antioxidant metabolism and reducing the accumulation of toxic salts and metals (Asgher et al. 2014; Ahmad et al. 2015, 2016; Abd_Allah et al. 2017). The deficiency of mineral nutrients can influence the plant growth by altering its normal growth patterns, enzyme activity, photosynthetic efficiency, and antioxidant defense system, leading to net increase in stress susceptibility. It shall be noted here that AMF has been much more recognized for their role in improving the uptake of several mineral nutrients and, more importantly, the nutrients that exhibit low mobility such as Zn and Cu (Alqarawi et al. 2014a, b; Hashem et al. 2015). AMF-inoculated plants have greater amounts of mineral nutrients such as N, P, K, Ca, and Mg and significant declines in the amounts of toxic ions such as Na (under salinity conditions) (Hashem et al. 2015) and Cd (under metal stress) (Abd_Allah et al. 2015). It has been reported that AMF colonization improves the availability and uptake of nutrients as well as their assimilation, e.g., Abd_Allah et al. (2015) have demonstrated significant increases in the activity of phosphate-solubilizing enzymes due to AMF inoculation. This effect has been mostly attributed to the development of hyphal networks in the soil, resulting in protection from low pH conditions (Muthukumar et al. 2014). Rohyadi (2008) reported that maize that was inoculated with *Glomus margarita* increased its phosphorus uptake under acidic conditions. Smith et al. (2000) have estimated that AMF contributes nearly 80% of the total phosphorus to *Medicago truncatula* via the extra-radical mycelium; however, variations in the phosphorus uptake in different AMF species may be present (da Silva et al. 1994). In *Brachiaria decumbens*, Siqueira et al. (1990) demonstrated increased Ca levels due to AMF colonization under acidic soil conditions. In another study, Yano and Takaki (2005) have reported increased uptake of essential elements such as K, P, Ca, and Mg due to AMF inoculation in potato grown in acidic soils (pH 4.2–5.2). Several workers have attributed the amelioration of nutrient-deficit stress by AMF to the development of an extensive extra-radical hyphal network.

4.5 Conclusions and Future Prospects

AMF improves growth by enhancing mineral nutrition, particularly phosphorus metabolism, and significantly affecting the soil health status. The beneficial role of AMF in the mitigation of stress via modulations in the key defense mechanisms is obvious. However, understanding the exact mechanism of how amelioration is caused by AMF is in its infancy and needs a lot improvement; therefore, extensive research efforts are needed. Combining physiological, biochemical, and molecular approaches will prove helpful in investigating the regulatory pathways that are induced by AMF and maintain the structural and functional integrity of the cell under normal and stress conditions. AMF has been approved for its beneficial growth-promoting effects; however, the initiation and the actual implementation of the various defense mechanisms, whether directly or indirectly controlled by AMF at the proteomic or genetic levels, require due attention. In this direction, identification of genes and proteins that are expressed by or in response to AMF inoculation will prove to be an important platform.

References

Abbaspour H, Saeid-Sar S, Afshari H, Abdel-Wahhab MA (2012) Tolerance of mycorrhiza infected Pistachio (*Pistacia vera* L) seedlings to drought stress under glasshouse conditions. J Plant Physiol 169:704–709

Abd_Allah EF, Hashem A, Alqarawi AA, Alwathnani HA (2015) Alleviation of adverse impact of cadmium stress in sunflower (*Helianthus annuus* L.) by arbuscular mycorrhizal fungi. Pak J Bot 47(2):785–795

Abd_Allah EF, Hashem A, Alqarawi AA, Wirth S, Egamberdieva D (2017) Calcium application enhances growth and alleviates the damaging effects induced by Cd stress in sesame (Sesamum indicum L.) J Plant Int 12(1):237–243. https://doi.org/10.1080/17429145.2017.1319500

Abdel Latef AA, Chaoxing H (2014) Does the inoculation with *Glomus mosseae* improve salt tolerance in pepper plants? J Plant Growth Regul 33:644–653

Abdel Latef AAH, Hashem A, Rasool S, Abd_Allah EF, Alqarawi AA, Egamberdieva D, Jan S, Anjum NA, Ahmad P (2016) Arbuscular mycorrhizal symbiosis and abiotic stress in plants: a review. J Plant Biol 59:407–426. https://doi.org/10.1007/s12374-016-0237-7

Adam S, Murthy SDS (2014) Effect of cold Stress on photosynthesis of plants and possible protection mechanisms. In: Gaur RK, Sharma P (eds) Approaches to plant stress and their management. Springer, New Delhi. https://doi.org/10.1007/978-81-322-1620-9_12

Ahanger MA, Agarwal RM (2016) Potassium improves antioxidant metabolism and alleviates growth inhibition under water and osmotic stress in wheat (*Triticum aestivum* L). Protoplasma 1–11. https://doi.org/10.1007/s00709-016-1037-0

Ahanger MA, Hashem A, Ab_Adllah EF, Ahmad P (2014) Arbuscular mycorrhiza in crop improvement under environmental stress. In: Ahmad P, Rasool S (eds) Emerging technologies and management of crop stress tolerance, vol 2. Elsevier, New York, pp 69–95

Ahmad P, Sharma S (2008) Salt stress and phyto-biochemical responses of plants – a review. Plant Soil Environ 54(3):89–99

Ahmad P, Jaleel CA, Salem MA, Nabi G, Sharma S (2010) Roles of enzymatic and non-enzymatic antioxidants in plants during abiotic stress. Crit Rev Biotechnol 30:161–175. https://doi.org/10.3109/07388550903524243

Ahmad P, Nabi G, Ashraf M (2011) Cadmium-induced oxidative damage in mustard [*Brassica juncea* L. Czern.& Coss.] plants can be alleviated by salicylic acid. South Afri J Bot 77:36–44

Ahmad P, Hakeem KR, Kumar A, Ashraf M, Akram NA (2012) Salt induced changes in photosynthetic activity and oxidative defense system of three cultivars of mustard (*Brassica juncea* L.) Afr J Biotechnol 11(11):2694–2703

Ahmad P, Sarwat M, Bhat NA, Wani MR, Kazi AG, Tran LS (2015) Alleviation of cadmium toxicity in *Brassica juncea* L. (Czern. & Coss.) by calcium application involves various physiological and biochemical strategies. PLoS One 10(1):e0114571. https://doi.org/10.1371/journal.pone.0114571

Ahmad P, Ab_Adllah EF, Hashem A, Sarwat M, Gucel S (2016) Exogenous application of selenium mitigates cadmium toxicity in *Brassica juncea* L. (Czern & Cross) by up-regulating antioxidative system and secondary metabolites. J Plant Growth Regul 35(4):936–950. https://doi.org/10.1007/s00344-016-9592-3

Al-Ezerjawi NH, Kadhim JH (2014) Effect of two isolates of *Trichoderma harzianum* on total nitrogen, chlorophyll a & b contents and yield of wheat (*Triticum aestivum* L) Class Eba'a-95. Int J Sci Res 3(8):1078–1083

Al-Karaki G, McMichael B, Zak J (2004) Field response of wheat to arbuscular mycorrhizal fungi and drought stress. Mycorrhiza 14:263–269. DOI:10.1007/s00572-003-0265-2

Alqarawi AA, Abd_Allah EF, Hashem A (2014a) Alleviation of salt-induced adverse impact via mycorrhizal fungi in *Ephedra aphylla* Forssk. J Plant Interact 9(1):802–810. https://doi.org/10.1080/17429145.2014.949886

Alqarawi AA, Hashem A, Abd_Allah EF, Alshahrani TS, Huqail AA (2014b) Effect of salinity on moisture content, pigment system, and lipid composition in *Ephedra alata* Decne. Acta Biol Hung 65(1):61–71. https://doi.org/10.1556/ABiol.65.2014.1.6

Armada E, Azcon R, Lopez-Castillo OM, Calvo-Polanco M, Ruiz-Lozano JM (2017) Autochthonous arbuscular mycorrhizal fungi and *Bacillus thuringiensis* from a degraded Mediterranean area can be used to improve physiological traits and performance of a plant of agronomic interest under drought conditions. Plant Physiol Biochem 90:64–74

Aroca R, Vernieri P, Ruiz-Lozano JM (2008) Mycorrhizal and non-mycorrhizal *Lactuca sativa* plants exhibit contrasting responses to exogenous ABA during drought stress and recovery. J Exp Bot 59(8):2029–2041. https://doi.org/10.1093/jxb/ern057

Aroca R, Ruiz-Lozano JM, Zamarreno AM, Paz JA, Garcia-Mina JM, Pozo MJ, Lopez-Raez JA (2013) Arbuscular mycorrhizal symbiosis influences strigolactone production under salinity and alleviates salt stress in lettuce plants. J Plant Physiol 170:47–55

Asgher M, Khan NA, Khan MIR, Fatma M, Masood A (2014) Ethylene production is associated with alleviation of cadmium-induced oxidative stress by sulfur in mustard types differing in ethylene sensitivity. Ecotoxicol Environ Saf 106:54–61

Asrar AA, Abdel-Fattah GM, Elhindi KM (2012) Improving growth, flower yield, and water relations of snapdragon (*Antirrhinum majus* L.) plants grown under well-watered and water stress conditions using arbuscular mycorrhizal fungi. Photosynthetica 50:305–316

Azooz MM, Youssef AM, Ahmad P (2011) Evaluation of salicylic acid (SA) application on growth, osmotic solutes and antioxidant enzyme activities on broad bean seedlings grown under diluted seawater. Int J Plant Physiol Biochem 3(14):253–264

Badri DV, Weir TL, van der Lelie D, Vivanco JM (2009) Rhizosphere chemical dialogues: plant–microbe interactions. Curr Opin Biotechnol 20:642–650

Barnawal D, Bharti N, Maji D, Chanotiya CS, Kalra A (2014) ACC deaminase-containing Arthrobacter protophormiae induces NaCl stress tolerance through reduced ACC oxidase activity and ethylene production resulting in improved nodulation and mycorrhization in *Pisum sativum*. J Plant Physiol 171:884–894

Baslam M, Goicoechea N (2012) Water deficit improved the capacity of arbuscular mycorrhizal fungi (AMF) for inducing the accumulation of antioxidant compounds in lettuce leaves. Mycorrhiza 22:347–359

Bonfante P, Genre A (2010) Mechanisms underlying beneficial plant fungus interactions in mycorrhizal symbiosis. Nat Commun 1:48

Borde M, Dudhane M, Jite PK (2010) AM fungi influences the photosynthetic activity, growth and antioxidant enzymes in *Allium sativum* L. under salinity condition. Not Sci Biol 2:64–71

Bunn R, Lekberg Y, Zabinski C (2009) Arbuscular mycorrhizal fungi ameliorate temperature stress in thermophilic plants. Ecology 90(5):1378–1388

Candido CV, Campanelli G, D'Addabbo T, Castronuovo D, Perniola M, Camele I (2015) Growth and yield promoting effect of artificial mycorrhization on field tomato at different irrigation regimes. Sci Hortic 187:35–43

Cekic FO, Unyayar S, Ortas I (2012) Effect of arbuscular mycorrhizal inoculation on biochemical parameters in *Capsicum annum* grown under long term salt stress. Turk J Bot 36:63–72

Chen S, Jin W, Liu A, Zhang S, Liu D, Wang F, Lin X, He C (2013) Arbuscular mycorrhizal fungi (AMF) increase growth and secondary metabolism in cucumber subjected to low temperature stress. Sci Hortic 160:222–229

Chen X, Song F, Liu F, Tian C, Liu S, Xu H, Zhu X (2014) Effect of different arbuscular mycorrhizal fungi on growth and physiology of maize at ambient and low temperature regimes. Sci World J 2014:956141. https://doi.org/10.1155/2014/956141

da Silva LHB, de Miranda JCC, de Miranda LN (1994) Effect of vesicular-arbuscular mycorrhiza in the growth of wheat varieties with differing aluminum tolerance, in cerrado soil. R Bras Cienc Solo 18:407–414

Elhindi KM, El-Din AS, Elgorban AM (2017) The impact of arbuscular mycorrhizal fungi in mitigating salt-induced adverse effects in sweet basil (*Ocimum basilicum* L.) Saudi J Biol Sci 24(1):170–179. https://doi.org/10.1016/j.sjbs.2016.02.010

Evelin H, Kapoor R (2014) Arbuscular mycorrhizal symbiosis modulates antioxidant response in salt-stressed Trigonella foenum-graecum plants. Mycorrhiza 24:197–208. https://doi.org/10.1007/s00572-013-0529-4

Evelin H, Kapoor R, Giri B (2009) Arbuscular mycorrhizal fungi in alleviation of salt stress: a review. Ann Bot 104:1263–1280

Fan QJ, Liu JH (2011) Colonization with arbuscular mycorrhizal fungus affects growth, drought tolerance and expression of stress-responsive genes in *Poncirus trifoliata*. Acta Physiol Plant 33:1533–1542

Faroor M, Aziz T, Wahid A, Lee DJ, Siddique KHM (2009) Chilling tolerance in maize: agronomic and physiological approaches. Crop Pasture Sci 60:501–516

Gamalero E, Lingua G, Berta G, Glick BR (2009) Beneficial role of plant growth promoting bacteria and arbuscular mycorrhizal fungi on plant responses to heavy metal stress. Can J Microbiol 55:501–514

Garg N, Manchanda G (2009) Role of arbuscular mycorrhizae in the alleviation of ionic, osmotic and oxidative stresses induced by salinity in *Cajanus cajan* (L.) Millsp. (pigeon pea). J Agron Crop Sci 195:110–123

Garg N, Singla P (2012) The role of Glomus mosseae on key physiological and biochemical parameters of pea plants grown in arsenic contaminated soil. Sci Hortic 14:92–101

Gholamhoseini M, Ghalavand A, Dolatabadian A, Jamshidi E, Khodaei-Joghan A (2013) Effects of arbuscular mycorrhizal inoculation on growth, yield, nutrient uptake and irrigation water productivity of sunflowers grown under drought stress. Agric Water Manag 117:106–114

Gonzalez-Chavez MC, Carrillo-Gonzalez R, Wright SF, Nichols KA (2004) The role of glomalin, a protein produced by arbuscular mycorrhizal fungi, in sequestering potentially toxic elements. Environ Pollut 130:317–323

Hajiboland R (2012) Effect of micronutrient deficiencies on plants stress responses. In: Ahmad P, Prasad MNV (eds) Abiotic stress responses in plants: metabolism, productivity and sustainability. Springer Science+Business Media, LLC, New York, pp 283–329

Hajiboland R, Aliasgharzadeh A, Laiegh SF, Poschenrieder C (2010) Colonization with arbuscular mycorrhizal fungi improves salinity tolerance of tomato (*Solanum lycopersicum* L.) plants. Plant Soil 331:313–327

Hameed A, Wu QS, Abd_Allah EF, Hashem A, Kumar A, Lone HA, Ahmad P (2014) Role of AM fungi in alleviating drought stress in plants. In: Miransari M (ed) Use of microbes for the alleviation of soil stresses. Springer Science+Business Media, New York. https://doi.org/10.1007/978-1-4939-0721-2_4

Hashem A, Abd_Allah EF, Alqarawi AA, El-Didamony G, Alwhibi Mona S, Egamberdieva D, Ahmad P (2014) Alleviation of adverse impact of salinity on faba bean (*Vicia faba* L.) by arbuscular mycorrhizal fungi. Pak J Bot 46:2003–2020. https://www.pakbs.org/pjbot/PDFs/46(6)/10.pdf

Hashem A, Abd_Allah EF, Egamberdieva D, Ahmad P (2015) Effect of AM fungi on growth, physio-biochemical attributes, lipid peroxidation, antioxidant enzymes and plant growth regulators in *Lycopersicon esculentum* mill. Subjected to different concentration of NaCl. Pak J Bot 47:327–340

Hashem A, Abd_Allah EF, Alqarawi AA, Egamberdieva D (2016) Bioremediation of adverse impact of cadmium toxicity on *Cassia italica* Mill by arbuscular mycorrhizal fungi. Saudi J Bio Sci 23:39–47. https://doi.org/10.1016/j.sjbs.2015.11.007

Hazzoumi Z, Moustakime Y, Elharchli E, Joutei KA (2015) Effect of arbuscular mycorrhizal fungi (AMF) and water stress on growth, phenolic compounds, glandular hairs, and yield of essential oil in basil (*Ocimum gratissimum* L). Chem Biol Technol Agric 2:10. https://doi.org/10.1186/s40538-015-0035-3

Heinemeyer A, Fitter AH (2004) Impact of temperature on the arbuscular mycorrhizal (AM) symbiosis: growth responses of the host plant and its AM fungal partner. J Exp Botany 55(396):525–534. https://doi.org/10.1093/jxb/erh049

Huang YM, Zou YN, Wu QS (2017) Alleviation of drought stress by mycorrhizas is related to increased root H2O2 efflux in trifoliate orange. Sci Rep 7:42335. https://doi.org/10.1038/srep42335

Hullebusch V, Lens P, Tabak H (2005) Developments in bioremediation of soils and sediments polluted with metals and radionuclides. 3. Influence of chemical speciation and bioavailability on contaminants remobilization bio-processes. Rev Environ Sci Biotechnol 4:185–212

Iqbal N, Shahid Umar S, Khan NA (2015) Nitrogen availability regulates proline and ethylene production and alleviates salinity stress in mustard (*Brassica juncea*). J Plant Physiol 178:84–91

Jedmowski C, Ashoub A, Momtaz O, Brüggemann W (2015) Impact of drought, heat, and their combination on chlorophyll fluorescence and yield of wild barley (*Hordeum spontaneum*). J Bot 2015:120868. https://doi.org/10.1155/2015/120868

Kelly CN, Morton JB, Cumming JR (2005) Variation in aluminium resistance among arbuscular mycorrhizal fungi. Mycorrhiza 15:193–201

Khan NA, Nazar R, Anjum NA (2009) Growth, photosynthesis and antioxidant metabolism in mustard (*Brassica juncea* L.) cultivars differing in ATP-sulfurylase activity under salinity stress. Sci Hortic 122:455–460

Khan MIR, Asgher M, Khan NA (2014) Alleviation of salt-induced photosynthesis and growth inhibition by salicylic acid involves glycine betaine and ethylene in mungbean (*Vigna radiata* L.) Plant Physiol Biochem 80:67–74

Khan MIR, Nazir F, Asgher M, Per TS, Khan NA (2015) Selenium and sulfur influence ethylene formation and alleviate cadmium-induced oxidative stress by improving proline and glutathione production in wheat. J. Plant Physiol 173:9–18

Kruger M, Kruger C, Walker C, Stockinger H, Schubler A (2012) Phylogenetic reference data for systematic and phytlotaxonomy of arbuscular mycorrhizal fungi from phylum to species level. New Phytol 193:970–984

Kuhn G, Hijri M, Sanders IR (2001) Evidence for the evolution of multiple genomes in arbuscular mycorrhizal fungi. Nature 414:745–748. https://www.nature.com/articles/414745a.pdf

Lanfranco L, Bolchi A, Ros EC, Ottonello S, Bonfante P (2002) Differential expression of a metallothionein gene during the presymbiotic versus the symbiotic phase of an arbuscular mycorrhizal fungus. Plant Physiol 130:58–67. www.plantphysiol.org/cgi/doi/10.1104/pp.003525

Lee BR, Muneer S, Avice JC, Jin Jung W, Kim TH (2012) Mycorrhizal colonisation and P-supplementation effects on N uptake and N assimilation in perennial ryegrass under well watered and drought stressed conditions. Mycorrhiza 22:525–534

Li T, Hu Y, Hao Z, Li H, Wang Y, Chen B (2013) First cloning and characterisation of two functional aquaporin genes from am arbuscular mycorrhizal fungus *Glomus intraradices*. New Phytol 197:617–630

Liao JP, Lin XG, Cao Z, Shi YQ, Wong MH (2003) Interactions between arbuscular mycorrhizae and heavy metals under sand culture experiments. Chemosphere 50:847–853

Lin J, Wang Y, Sun S, Mu C, Yan X (2017) Effects of arbuscular mycorrhizal fungi on the growth, photosynthesis and photosynthetic pigments of *Leymus chinensis* seedlings under salt-alkali stress and nitrogen deposition. Sci Total Environ 576:234–241

Liu W, Yu K, He T, Li F, Zhang D, Liu J (2013) The low temperature induced physiological responses of *Avena nuda* L., a cold-tolerant plant species. Sci World J 2013:658793. https://doi.org/10.1155/2013/658793

Lopez-Raez JA, Verhage A, Fenandez I, Garcia JM, Azcon-Aguilar C, Flors V, Pozo MJ (2010) Hormonal and transcriptional profiles highlight common and differential host responses to arbuscular mycorrhizal fungi and the regulation of the oxylipin pathway. J Exp Bot 61:2589–2601

Machado S, Paulsen GM (2001) Combined effects of drought and high temperature on water relations of wheat and sorghum. Plant Soil 233(2):179–187

Mittler R (2002) Oxidative stress, antioxidants and stress tolerance. Trends Plant Sci 7:405–410

Mo Y, Wang Y, Yang R, Zheng J, Liu C, Li H, Ma J, Zhang Y, Wei C, Zhang X (2016) Regulation of plant growth, photosynthesis, antioxidation and osmosis by an arbuscular mycorrhizal fungus in watermelon seedlings under well-watered and drought conditions. Front Plant Sci 7:644. https://doi.org/10.3389/fpls.2016.00644

Muthukumar T, Priyadharsini P, Uma E, Jaison S, Pandey RR (2014) Role of arbuscular mycorrhizal fungi in alleviation of acidity stress on plant growth. In: Miransari M (ed) Use of microbes for the alleviation of soil stresses, vol 1. Springer Science+Business Media, New York, pp 43–71

Navarro JM, Perez-Tornero O, Morte A (2013) Alleviation of salt stress in citrus seedlings inoculated with arbuscular mycorrhizal fungi depends on the rootstock salt tolerance. J Plant Physiol 171:76. https://doi.org/10.1016/j.jplph.2013.06.006

Nazar R, Umar S, Khan NA, Sareer O (2015) Salicylic acid supplementation improves photosynthesis and growth in mustard through changes in proline accumulation and ethylene formation under drought stress. South Afr J Bot 98:84–94

Ogawa A, Yamauchi A (2006) Root osmotic adjustment under stress in maize seedlings. Mode of accumulation of several solutes for osmotic adjustment in root. Plant Prod Sci 9:39–46

Paredes M, Quiles MJ (2015) The effects of cold stress on photosynthesis in hibiscus plants. PLoS One 10(9):e0137472. https://doi.org/10.1371/journal.pone.0137472

Porcel R, Aroca R, Ruiz-Lozano JM (2012) Salinity stress alleviation using arbuscular mycorrhizal fungi. A review. Agron Sust Dev 32:181–200

Rabie GH, Almadini AM (2005) Role of bioinoculants in development of salt-tolerance of Vicia faba plants under salinity stress. Afr J Biotechnol 4:210–222

Rohyadi A (2008) Growth responses of external hyphae of arbuscular mycorrhizal fungi to acidic conditions and their effects on cowpea growth. Microbiology 2:22–26

Ruíz-Lozano JM, Aroca R (2010) Modulation of aquaporin genes by the arbuscular mycorrhizal symbiosis in relation to osmotic stress tolerance. In: Sechback J, Grube M (eds) Symbiosis and stress. Springer, Berlin, pp 359–374. https://doi.org/10.1007/978-90-481-9449-0

Ruiz-Lozano JM, Azcon R, Palma JM (1996) Superoxide dismutase activity in arbuscular-mycorrhizal Lactuca sativa L. plants subjected to drought stress. New Phytol 134:327–333

Ruiz-Lozano JM, Porcel R, Aroca R (2006) Does the enhanced tolerance of arbuscular mycorrhizal plants to water deficit involve modulation of drought-induced plant genes? New Phytol 171:693–698

Ruiz-Sanchez M, Aroca R, Munoz Y, Armada E, Polon R, Ruiz-Lozano JM (2010) The arbuscular mycorrhizal symbiosis enhances the photosynthetic efficiency and the antioxidative response of rice plants subjected to drought stress. J Plant Physiol 167:862–869

Scagel CF Bryla DR (2017) Salt exclusion and mycorrhizal symbiosis increase tolerance to nacl and $CaCl_2$ salinity in 'Siam Queen' basil. Hortic Sci 52(2):278–287. https://doi.org/10.21273/HORTSCI11256-16

Schuβler A, Schwarzott D, Walker C (2001) A new fungal phylum, the Glomeromycota: phylogeny and evolution. Mycol Res 105:1413–1421

Sherstha YH, Ishii T, Kadoya K (1995) Effect of vesicular–arbuscular mycorrhizal fungi on the growth, photosynthesis, transpiration and the distribution of photosynthates of bearing Satsuma mandarin trees. J Jpn Soc Hortic Sci 64:517–525

Siqueira JO, Rocha WF Jr, Oliveira E, Colozzi-Filho A (1990) The relationship between vesicular-arbuscular and lime: associated effects on the growth and nutrition of Brachiaria grass (Brachiaria decumbens). Biol Fertil Soil 10:65–71

Smith FW, Rae AL, Hawkesford MJ (2000) Molecular mechanisms of phosphate and sulphate transport in plants. Biochem Biophys Acta 1465:236–245

Talaat NB, Shawky BT (2004) Protective effects of arbuscular mycorrhizal fungi on wheat (Triticum aestivum L) plants exposed to salinity. Environ Exp Bot 98:20–31. DOI:10.1002/jpln.201200618

Tang M, Chen H, Huang JC, Tian ZQ (2009) AM fungi effects on the growth and physiology of *Zea mays* seedlings under diesel stress. Soil Bio Biochim 41:936–940

Tejera NA, Campos R, Sanjuan J, Lluch C (2004) Nitrogenase and antioxidant enzyme activities in *Phaseolus vulgaris* nodules formed by *Rhizobium tropici* isogenic strains with varying tolerance to salt stress. J Plant Physiol 161:329–338

Toljander JF, Lindahl BD, Paul LR, Elfstrand M, Finlay RD (2007) Influence of arbuscular mycorrhizal mycelial exudates on soil bacterial growth and community structure. FEMS Microbiol Ecol 61:295–304

Verma S, Dubey R (2003) Lead toxicity induces lipid peroxidation and alters the activities of antioxidant enzymes in growing rice plants. Plant Sci 164:645–655

Walter MH, Strack D (2011) Carotenoids and their cleavage products: biosynthesis and functions. Nat Prod Rep 28:663–692

Wu QS (2011) Mycorrhizal efficacy of trifoliate orange seedlings on alleviating temperature stress. Plant Soil Environ 57(10):459–464
Wu QS, Zou YN (2010) Beneficial roles of arbuscular mycorrhizal in citrus seedlings at temperature stress. Sci Hortic 125:289–293
Wu QS, Zou YN (2017) Arbuscular mycorrhizal fungi and tolerance of drought stress in plants. In: Wu QS (ed) Arbuscular mycorrhizas and stress tolerance of plants. Springer, Singapore
Wu QS, Zou YN, Xia RX, Wang MY (2007) Five *Glomus* species affect water relations of citrus tangerine during drought stress. Bot Stud 48:147–154
Wu QS, Zou YN, He XH (2010a) Contributions of arbuscular mycorrhizal fungi to growth, photosynthesis, root morphology and ionic balance of citrus seedlings under salt stress. Acta Physiol Plant 32:297–304
Wu QS, Zou YN, Liu W, Ye XF, Zai HF, Zhao LJ (2010b) Alleviation of salt stress in citrus seedlings inoculated with mycorrhiza: changes in leaf antioxidant defense systems. Plant Soil Environ 56:470–475
Xie X, Weng B, Cai B, Dong Y, Yan C (2014) Effects of arbuscular mycorrhizal inoculation and phosphorus supply on the growth and nutrient uptake of *Kandelia obovata* (Sheue, Liu and Yong) seedlings in autoclaved soil. Appl Soil Ecol 75:162–171
Xu H, Lu Y, Zhu X (2016) Effects of arbuscular mycorrhiza on osmotic adjustment and photosynthetic physiology of maize seedlings in black soils region of northeast China. Braz Arch Bio Tech 59. https://doi.org/10.1590/1678-4324-2016160392
Yadav SK (2010) Cold stress tolerance mechanisms in plants. A review. Agron Sustain Dev 30:515–527. https://doi.org/10.1051/agro/2009050
Yang X, Liang Z, Wen X, Lu C (2008) Genetic engineering of the biosynthesis of glycine betaine leads to increased tolerance of photosynthesis to salt stress in transgenic tobacco plants. Plant Mol Biol 66:73–86
Yang Y, Tang M, Sulpice R, Chen H, Tian S, Ban Y (2014) Arbuscular mycorrhizal fungi alter fractal dimension characteristics of *Robinia pseudoacacia* L seedlings through regulating plant growth, leaf water status, photosynthesis and nutrient concentration under drought stress. J Plant Growth Regul 3:612–625
Yang Y, Han X, Liang Y, Ghosh A, Chen J, Tang M (2015) The combined effects of arbuscular mycorrhizal fungi (AMF) and lead (Pb) stress on Pb accumulation, plant growth parameters, photosynthesis, and antioxidant enzymes in *Robinia pseudoacacia* L. PLoS One 10(12):e0145726. https://doi.org/10.1371/journal.pone.0145726
Yano K, Takaki M (2005) Mycorrhizal alleviation of acid soil stress in the sweet potato (Ipomoea batatas). Soil Biol Biochem 37:1569–1572. https://doi.org/10.1016/j.soilbio.2005.01.010
Yedidia I, Srivastva AK, Kapulnik Y, Chet I (2001) Effect of *Trichoderma harzianum* on microelement concentrations and increased growth of cucumber plants. Plant Soil 235:235–242
Yooyongwech S, Phaukinsang N, Cha-Um S, Supaibulwatana K (2013) Arbuscular mycorrhiza improved growth performance in *Macadamia tetraphylla* L. grown under water deficit stress involves soluble sugar and proline accumulation. Plant Growth Regul 69:285–293
Zhu XC, Song FB, Xu HM (2010) Effect of arbuscular mycorrhizal fungi on photosynthetic characteristics of maize under low temperature stress. Acta Ecol Sin 21:470–475. https://www.ncbi.nlm.nih.gov/pubmed/20462022
Zhu XC, Song FB, Liu SQ, Liu TD, Zhou X (2012) Arbuscular mycorrhizae improves photosynthesis and water status of *Zea mays* L. under drought stress. Plant Soil Environ 58:186–191
Zou YN, Wu QS, Huang YM, Ni QD, He XH (2013) Mycorrhizal-mediated lower proline accumulation in Poncirus trifoliata under water deficit derives from the integration of inhibition of proline synthesis with increase of proline degradation. PLoS One 8:1–8. https://doi.org/10.1371/journal.pone.0080568

Chapter 5
Trichoderma: Beneficial Role in Sustainable Agriculture by Plant Disease Management

Laith Khalil Tawfeeq Al-Ani

Abstract *Trichoderma* as biological control agents have been widely used against many plant pathogens, such as viruses, bacteria, fungi, nematodes, and higher parasitic plants. This species of fungi has been considered to be very beneficial for different levels of life. It features remote sensing and is fast in attacking and suppressing the growth of plant pathogens, and it improves plant growth. It can produce different secondary compounds and readily activates others fungi, producing very significant enzymes, such as chitinase, proteases, and β-1,3-glucanase, inducing plant defense, systemic resistance, and strong and active competition against plant pathogens. It is party to an important detoxification process to reduce the toxicity secreted by plant pathogens. It is therefore necessary to clarify the significance of *Trichoderma* in the control of plant diseases that results in improvements in sustainable agriculture. This should include coverage of the different aspects of the interaction between *Trichoderma* and the various kingdoms of organisms. Here is provided an excellent guide to the importance of *Trichoderma* as biological control agents (BCAs) in sustainable agriculture through reducing plant diseases and increasing field production. *Trichoderma* can combine several advantages in one product – the control of different of plant diseases, enhancement of plant growth, and the provision of a clean environment for the benefit of sustainable agriculture.

Keywords Biological control • Induced systemic resistance • Plant disease • Trichoderma

---

L.K.T. Al-Ani (✉)
Department of Plant Protection, College of Agriculture, University of Baghdad, Baghdad, Iraq

School of Biology Science, Universiti Sains Malaysia, Minden, Pulau Pinang, Malaysia
e-mail: cmv_virus2002@yahoo.com; laith.kt77@gmail.com

D. Egamberdieva, P. Ahmad (eds.), *Plant Microbiome: Stress Response*, Microorganisms for Sustainability 5, https://doi.org/10.1007/978-981-10-5514-0_5

5.1 Introduction

A recent challenge in the agricultural sector is to increase yields and decrease plant disease to a minimum level. Traditional methods such as the use of fungicides, nematicides, herbicides, and fertilizer are among general methods in plant disease management and crop yield improvement. Although these mechanisms have the ability to control plant disease and suppress plant pathogens, they are not eco-friendly. Continuous usage of chemical-based methods has also caused the pathogens develop more resistance toward pesticides. The use of agrochemical pesticides containing various hazardous chemicals such as ethylated, methylated, and aromatic substances also adversely affect and pollute the atmosphere and water, thereby harming fish, beneficial insects such as the honeybee, non-target organisms such as plant growth promoting rhizobacteria, and plant growth promoting fungi (PGPF). Chemical residues can also burn or cause yellowing effects to plant leaves.

Today, many researchers are searching for alternative methods with significant eco-friendly activity. Biological control agents (BCAs), natural enemies of plant pathogens, are a very strong candidate to replace conventional methods. BCAs are commonly isolated from the rhizosphere, phyllosphere, and soil. They comprise several agents such as PGPF, non-pathogenic fungi, mycorrhizal, entomopathogenic fungi, mycoparasitic fungi, and endophytic fungi (Steyaert et al. 2003; Hermosa et al. 2012; Murali et al. 2012; Sylla 2013; Doni et al. 2013). Several fungi agents reported as biological controls are *Coniothyrium minitans* (anamorphs of *Paraphaeosphaeria minitans*) (Verkley et al. 2004; Chitrampalam et al. 2011), *Clonostachys rosea* (Sutton et al. 2002), *Trichoderma* spp. (Al-Ani et al. 2013a; Al-Ani 2017), *Fusarium oxysporum* f.sp. cubense, *F. oxysporum*, *F. solani*, and *F. fujikuroi* (Al-Ani et al. 2013b; Al-Ani 2010, 2017), *Piptocephalis virginiana* (Berry and Barnett 1957), *Gonatobotrys simplex* (Hoch 1977), *Pythium paroecandrum* (Abdelghani et al. 2004), *Chaetomium* (Hung et al. 2015), *Sphaerodes quadrangularis* (Goh and Vujanovic 2010), *Cryphonectria parasitica* (Kunova et al. 2016), and *Rhizoctonia solani* and *Rhizoctonia* (BNR) spp. (Hwang and Benson 2003).

Of these, *Trichoderma* is the most versatile genus of fungi worldwide that have been used to control plant pathogenic fungi and manage plant diseases and plant growth. Historically, *Trichoderma* was introduced as antagonistic fungi and has been known as a biocontrol agent of several plant pathogens since the 1920s (Weindling 1934; Samuels 1996). Several advantages, such as ubiquitous distribution, ease of isolation and culture, and rapid growth on many substrates attract the researchers to use it for sustainable agriculture. *Trichoderma* spp. have been reported to control important plant pathogenic fungi such as *Fusarium*, *Phytophthora*, *Pythium*, *Colletotrichum*, *Fulvia fulva*, *Rhizoctonia*, *Plasmopara viticola*, *Pseudoperonospora cubensis*, *Monilia laxa*, *Rhizopus*, *Botrytis*, *Alternaria*, *Cladosporium*, *Gaeumannomyces*, *Verticillium*, and *Sclerotinia* (Table 5.1). *Trichoderma* is also widely used to control plant diseases such as Fusarium wilt (Al-Ani 2017), bacterial wilt (Yuana et al. 2016), sheath blight (de França et al. 2015), mosaic virus (Luo et al. 2010), southern stem rot (Sennoi et al. 2013), downy

Table 5.1 *Trichoderma* spp. used as biocontrol agents (Al-Ani 2017)

Trichoderma species	Pathogen	Crop	References
T. harzianum	*F. oxysporum* f.sp. *lycopersici*	Tomato	Sivan (1987)
	Macrophomina phaseolina	Cowpea	Adekunle et al. (2001)
	F. oxysporum f.sp. *cubense*	Banana	Saravanan et al. (2003) and Nan et al. (2014)
	F. oxysporum f.sp. *lycopersici*	Tomato	Marzano et al. (2013)
	Fusarium oxysporum f. sp. cucumerinum	Cucumber	Zhang et al. (2013)
	Sclerotinia sclerotiorum, Rhizoctonia solani, Verticillium dahliae, Phytophthora nicotianae and P. cinnamomi	Soil-borne disease	Aleandri et al. (2015)
T. koningii	*Fusarium oxysporum* f.sp. *cucumerinum*, and *Pythium* spp	Pea seeds	Lifshitz et al. (1986)
	Rhizoctonia solani and *Pythium ultimum*	*Capsicum annuum*	Harris (1999)
	Sclerotium rolfsii	Soil-borne disease	Tsahouridou and Thanassoulopoulos (2001)
	Macrophomina phaseolina	Cowpea	Adekunle et al. (2001)
	Rhizoctonia solani	Cotton	Hanson and Howell (2002)
	Sclerotium rolfsii	Tomato seeds	Tsahouridou and Thanassoulopoulos (2002)
T. viride	*Sclerotium rolfsii*	Tomato	Wokocha (1990)
	F. oxysporum f.sp. *cubense*	Banana	Saravanan et al. (2003)
T. asperellum	*Fusarium oxysporum* f.sp. *dianthi*	Carnation	Sant et al. (2010)
	Phytophthora ramorum	Soil-borne disease	Widmer (2014)
	Sclerotinia sclerotiorum, Rhizoctonia solani, Verticillium dahliae, Phytophthora nicotianae and P. cinnamomi	Soil-borne disease	Aleandri et al. (2015)
T. parareesei	*Botrytis cinerea*	Tomato	Rubio et al. (2014)
T. hamatum	*Sclerotinia sclerotiorum, Rhizoctonia solani, Verticillium dahliae, Phytophthora nicotianae and P. cinnamomi*	Soil-borne disease	Aleandri et al. (2015)
T. atroviride	*Armillaria gallica*	Orchards	Pellegrini et al. (2014)
T. virens	*Rhizoctonia solani*	Cotton	Hanson and Howell (2002)

mildew (Perazzolli et al. 2012), seed-rotting fungi (Hadar et al. 1984), powdery mildew and grey mold (Elad et al. 1998), gummosis of citrus (Bicici et al. 1992), root-knot nematode (Al-Hazmi and Javeed 2016), root rot, damping-off, stem rot, Aspergillus crown rot, charcoal rot, red rot, Rhizoctonia black scurf, turfgrass diseases, decay in tree wounds, and internal decay of wood products (Gnanamanickam 2002).

Trichoderma has many mechanisms very useful for plants, such as to improve plant growth, to enhance the solubilization of mineral nutrients, to induce secondary metabolites production, to produce growth-regulating compounds, stimulation of plant defense, and production of siderophores. These mechanisms confirm the *Trichoderma* species to be a suitable biocontrol agent in plant disease management by developing biopesticides. There are several commercially available *Trichoderma*-based products such as Biobus 1.00WP (*Trichoderma viride*), Promot PlusWP Promot PlusDD (*Trichoderma* spp. *Trichoderma koningii*, and *Trichoderma harzianum*), RiB1, TRICÔ-ĐHCT, VI – ĐK, Bio – Humaxin Sen Vàng 6SC, and Fulhumaxin 5.15SC (*Trichoderma* spp.).

5.2 Mechanisms of *Trichoderma* spp. as BCAs Against Plant Pathogens

The mode of action involves several mechanisms such as mycoparasites, competition, antibiotics production, and the ability to induce plant defense and systemic resistance (Naher et al. 2014). The modes of action are as follows.

5.2.1 Mycoparasitism

The main mechanism in the antagonism of *Trichoderma* against fungal plant pathogens is mycoparasitic (Elad et al. 1982). Mycoparasitism is probably a factor and one of the most outstanding features of this fungal genus. The direct attack on another fungus is a compound process that involves successive events, including infection and penetration, subsequently killing the opponent fungus. *Trichoderma* spp. may exert direct biocontrol by parasitizing the broad range of fungi and growing toward them. The mycoparasite activity of *Trichoderma* starts with coils around the host hyphae followed by producing hooks with appressorium-like bodies, eventually penetrating the host cell wall (Elad et al. 1983; Inbar and Chet 1992; Ojha and Chatterjee 2011). The ability of *Trichoderma* spp. to act as mycoparasitic fungi is because of the production of cell wall degrading enzymes (CWDEs) such as chitinases, proteases, and ß-1,3-glucanases. These enzymes are involved in CWDEs of *R. solani*, *Sclerotium rolfsii*, and *Pythium aphanidermatum* (Elad et al. 1982; Sivan and Chet 1989; Harman et al. 2004). The adhesion to a host surface is just one step in a series of interaction events, see the interaction steps that described by Tunlid et al. (1992). *Trichoderma* can respond and recognize the host in different environmental conditions, and the successful colonization of rhizosphere, plants, and soil is relevant with the presence of a host (Mendoza-Mendoza et al. 2003). *Trichoderma* can respond to the host by the successive expression of pathogenesis-related proteins comprising chitinases, proteases, and glucanases (Harman et al. 2004). The method of induction varies from one *Trichoderma* strain or species to another, and *Trichoderma* secrete chitinases that degrade fungal cell-walls to liberate the oligomers, which induces exochitinases, and parasitism begins (Gajera et al. 2013). The induction differs between species and strains of *Trichoderma*, which may be unable to interact with the host, or affect the gene to gene reactions that apply to the ability of *Trichoderma* to become a parasite on the host.

5.2.2 Non-mycoparasitism

Some strains or species of *Trichoderma* are not mycoparasites of other fungi, or at least that phase of their life cycle is not involved in the biocontrol phenomenon. The mechanisms employed by these *Trichoderma* strains or species consist of antibiosis (production of antimicrobial and secondary metabolites), competition (on site, on nutrient, or in combination), involving antibiosis and competition, energizing plant defense before infection by the pathogen, improving the tolerance of infection with plant pathogens (Howell 2003), and leading to detoxification of phytotoxin for plant pathogens (Aggarwal et al. 2011).

5.3 *Trichoderma* as Biological Control Agents of Plant Pathogens

Biological controls are another alternative to chemical pesticides and have received mounting attention over the last 20 years (Paulitz and Belanger 2001). *Trichoderma* spp. have been reported to control many major plant pathogens, including viruses, fungi, bacteria, and nematodes. The classical category of biological control includes several mechanisms such as direct antagonism (mycoparasites), indirect antagonism (non-mycoparasites) such as antibiosis, competition, CWDEs, and induction of systemic resistance (Park 1960; Lo 1998).

5.3.1 Biocontrol of Plant Viruses

Plant viruses are among the important pathogens which are widespread and cause damage to plants. Damage caused by viruses simultaneously has harmful effects on sustainable agriculture. In general, some plant viruses depend on vectors such as nematodes, plant parasites, insects, seeds, and fungi. Chemical and biological controls measures have successfully decreased the rate of virus spread by applying Integrated Pest Management (IPM), improving host resistance, and enhancing plant growth, which increases plant tolerance to plant viruses' infection. *Trichoderma* showed the capacity to induce plant defense and stimulate resistance (as ISR (induced systemic resistance) and SAR (systemic acquired resistance)) by producing secondary metabolites, culture filters, and CWDEs (Luo et al. 2010).

For example, antimicrobial peptaibols from *T. pseudokoningii* SMF2, known as Trichokonin, have ISR and defense response against tobacco mosaic virus infection in tobacco (*Nicotiana tabacum* var. Samsun NN) (Luo et al. 2010). The production of a reactive oxygen species (ROS) and phenolic compounds in tobacco increased when treated with Trichokonin. Activities of pathogenesis-related enzymes PAL and POD significantly increased, and the expression of several plant defense genes was also upregulated.

Elsharkawy et al. (2013) found the strain of *T. asperellum* SKT-1 was able to induce resistance in the *Arabidopsis* plant against CMV (cucumber mosaic virus) infection by increasing expression levels of SA (salicylic acid)- and JA (jasmonic acid)/ET(ethylene)-inducible genes in leaves. Although the pre-treatment of *Arabidopsis* root with the culture filter of *T. asperellum* SKT-1 led to induction of defense mechanisms against CMV (Elsharkawy et al. 2013), Vitti et al. (2015) found the ability of *T. harzianum* T-22 strain (T22) to induce the defense responses in tomato (*Solanum lycopersicum* var. *cerasiforme*) against CMV. Histochemical analysis has revealed a different increase in the hydrogen peroxide and superoxide anion, suggesting the involvement of ROS in plant defense responses.

5.3.2 *Biocontrol of Plant Bacteria*

Plant bacteria are important to plants, mostly live either as endophytes or saprophytes, and coexist in the rhizosphere, soil, and phyllosphere. Some strains of bacteria are pathogenic to plants and can cause major plant diseases worldwide (Agrios 2005). Bacterial pathogens induce many types of symptoms on leaves, fruits, crown, roots, and vascular tissues. These symptoms appear as blights and spots on leaves, soft rots of fruits, wilts, scabs, and cankers (Agrios 2005).

However, bacterial diseases of plants are often not easy to control. There are many methods for control, such as protecting crop fields from infection using healthy seeds, preventing the spread of bacterial pathogens through insect-infected plants or any part of the plants, and decontaminating tools, machines, and hands after planting. Other control measures include using chemical controls (bactericides), physical controls by burning the plant infected with bacteria, introducing plant breeding programmes, and selecting high-resistance crop varieties. Biological controls by using BCAs such as plant growth promoting rhizobacteria, endophytic bacteria, mycorrhizal, endophytic fungi, and *Trichoderma* have also been used to suppress the pathogens. All of these methods can be used alone or in combination to control plant bacterial pathogens effectively. Of these, the best techniques to control the pathogens are chemical methods, followed by plant breeding and biological methods. However, plant breeding and biocontrol methods are safer for the environment.

For example, *T. asperellum* T203 conferred the protective effect by inducing systemic resistance against the cucumber leaf pathogen *Pseudomonas syringae* pv. *lachrymans* that is involved in the JA/ET signaling pathways of ISR (Lox1, Pal1, ETR1, and CTR1) in cucumber plants. Meanwhile, *T. pseudokoningii* SMF2 showed antimicrobial activity against a broad-spectrum of both Gram-positive bacteria (Shi et al. 2012) and Gram-negative bacteria (Li et al. 2014), where the strain was able to control *Pectobacterium carotovorum* sub sp. *carotovorum* (*Pcc*), which caused a soft rot disease of Chinese cabbage (Li et al. 2014). The strain produced Trichokonins which inhibited the growth of *Pcc* (Gram-negative) and induced resistance of cabbage plants. Trichokonins were able to increase the production of ROS, pathogenesis-related protein gene acidic PR-1a, and activation of SA (Li et al. 2014).

5.3.3 Biocontrol of Phytopathogenic Fungi

Phytopathogenic fungi are very harmful to plants, causing several major diseases and having more harmful effects on sustainable agriculture compared to phytopathogenic viruses or bacteria. Fungi can attack plants and incidence may be localized or systemic. Usually, fungi can infect all parts of plant-leaf, root, and seeds as well as stored seeds. In general, infection by fungi causes symptoms such as root rot, necrosis, wilts, spots, stunting, powdery mildew, downy mildew, blight, canker, dieback, damping-off, crown rot, smut, basal stem rot, anthracnose, rusts, scab, and general decline (Agrios 2005).

The most common methods used to control phytopathogenic fungi include: (1) protective methods, such as resistant plant varieties and use of pathogen-free seed; (2) culture methods such as crop rotation; (3) chemical methods; (4) biocontrol methods by using antagonistic microorganisms such as bacteria and fungi. These methods can reduce disease caused by fungal pathogens. Chemical methods are very effective on fungal pathogens but prolonged use of chemical-based methods has a harmful effect on the environment. Biocontrol agents are also effective in controlling fungal pathogens and are safer for the environment. The application of fungicides and consumer acceptance of resistant cultivars can be very complex, which makes biological control of phytopathogenic fungi an attractive alternative. *Trichoderma* spp. are well-known for their activity against many plant pathogens that cause major problems worldwide (Table 5.1) (Sharma et al. 2011). *Trichoderma* can be a potential alternative to control charcoal rot in soybean (Khalili et al. 2016).

The interactions between a fungus and another fungus are very attractive, involving: (1) a mutually beneficial relationship that may increase the infection in plants, such as avirulent fungi with virulent fungi, low virulent fungi with high virulent fungi; (2) interspecific interaction such as hyphal interaction and somatic; (3) an antagonistic relationship, including (a) avirulent fungi against virulent fungi, (b) highly virulent fungi against virulent fungi, (c) parasitic fungi against virulent fungi, and (d) parasitic fungi against benefit fungi. However, Boddy (2016) has explained the spectrum of fungus–fungus interactions in a scheme.

Trichoderma spp. use many mechanisms against phytopathogenic fungi, which can divide in two ways – direct and indirect. The direct mechanism is accomplished when it reduces the pathogen population by antagonistic effects which include competition, antibiosis, and parasitism. The indirect mechanism is achieved by reaction between *Trichoderma* and plant against phytopathogenic fungi. This interaction is also denoted as 'cross-protection' or 'induced resistance' and is based on the creation of the host's own defense system (Marois 1990). Therefore, many biocontrol agents employ more than one mechanism to protect plants (Fravel and Engelkes 1994).

There are seven different mechanisms which have been suggested for suppression of phytopathogenic fungi. These modes of actions probably include all antagonistic effects against plant pathogens. The mechanisms are as follows:

1. Mycoparasitism refers to parasitism on mycelium or spores of fungal hosts caused by the production of cell wall degrading enzymes (CWDEs) or lysis enzymes that degrade the cell wall, such as β-1, 3-gluconase, chitinases, cellulases, lipases, and proteases (Van den Boogert 1996; Viterbo et al. 2002a, b; Gajera et al. 2012; Saravanakumar et al. 2016a).
2. Competition for space, nutrients including carbon and iron, and infection sites such as by modifying the rhizosphere by acidifying the soil (Benítez et al. 2004; Arst and Penalva 2003) so the pathogens cannot grow.
3. Antibiosis – production of antifungal compounds, volatile and non-volatile metabolite compounds, and stopping the growth after spore germination as fungistatic.
4. Induction of systemic resistance such as ISR and SAR.
5. Induction of plant defense such as rhizosphere modification and colonization of the plant root. This mechanism causes the change in physiological responses.
6. Detoxification that produced by plant fungal pathogens (Vázquez et al. 2015).
7. Biofertilizers and PGPF, through enhancement of plant growth, enhance the solubilization of mineral nutrients, improve the media of rhizosphere and soil, and colonize root intercellular spaces (Hermosa et al. 2012). These multiple mechanisms are used either in combination or individually to control phytopathogenic fungi (Elad 2000; Bae et al. 2016).

Mycoparasitism has been demonstrated by many *Trichoderma* species on different fungal pathogens. *T. koningii* MTCC 796 and *T. harzianum* T12 were able to parasitize the mycelia of *Macrophomina phaseolina* as well as induce the enzyme activities of CWDEs (Gajera et al. 2012; Khalili et al. 2016). Meanwhile, *T. harzianum* Tveg1 and *Trichoderma atroviride* TR10 could inhibit the mycelium growth of *F. oxysporum* f.sp. *cubense* tropical race 4 (*Foc*TR4) in in vitro experiment by Al-Ani et al. (2013a). Saravanakumar et al. (2016a) also found the strain of *T. asperellum* CCTCC-RW0014 showed mycoparasitic activity on *F. oxysporum* f. sp. *cucumerinum* by producing various CWDEs such as chitinase, cellulase, protease, and β (1–3) glucanase. The growth of *F. solani* was inhibited when *T. hamatum* URM 6656 was applied, which can be attributed to the production of lysis enzymes called chitinases (da Silva et al. 2016). *T. harzianum* species (THSC) attacked the plant fungal pathogens *Ceratocystis radicicola* of date palm and showed lysis of the hyphal pathogen and phialoconidia along with scattered aleurioconidia in vitro (Al-Naemi et al. 2016). de Lima et al. (2016) found that *T. atroviride* T17 showed high antagonistic activity against *Guignardia citricarpa* of citriculture that was associated with the secretion of proteins, including chitinase, mutanase, a-1,3-glucanase, a-1,2-mannosidase, carboxylic hydrolase ester, carbohydrate-binding module family 13, glucan 1,3-β-glucosidase, α-galactosidase, and neutral protease. Al-Ani (2017) screened 31 isolates of *Trichoderma* against *Foc*TR4 in vitro and found 12 isolates of *Trichoderma* (*T. harzianum, T. parareesei, T. reesei, T. capillare, T. atroviride*, and *T. koningii*) overgrow the *Foc*TR4 after the ninth days of inoculation.

Examining the competition for nutrients, Sivan and Chet (1989) found that *T. harzianum* was able to compete with *F. oxysporum* f. sp. *vasinfectum* for carbon in vitro by inhibiting chlamydospores germination and simultaneously suppressed *Fusarium* wilt of cotton in vivo. Sarrocco et al. (2009) found that *T. virens* I10 can compete with *R. solani* for carbon in soil by producing a cellulose enzyme. Three isolates of *Trichoderma* (*T. atroviride* P1, *T. harzianum* T22, and *T. viride*) showed strong competitiveness with *Phytophthora cinnanerium, Botrytis cinaria*, and *R. solani* (Olabiyi and Ruocco 2013). However, several *Trichoderma* can stop the growth of other fungi by producing siderophores (iron-chelating compounds) (Chet and Inbar 1994). *Trichoderma asperellum* strain T34 has the potential to control Fusarium wilt of tomato caused by *F. oxysporum* f.sp. *lycopersici* (Fol) through competition for iron (Segarra et al. 2010). Lehner et al. (2013) suggested that *Trichoderma* spp. can produce siderophores through screening using LC-HRMS/MS. Al-Ani (2017) found four isolates of *Trichoderma* (two isolates of *T. harzianum* (TL5, Tveg1), *T. parareesei* T26, and *T. koningii* TR102) were capable of competing with *Foc*TR4 for iron by producing siderophores.

In antibiosis production, *T. parareesei* inhibited the growth of *Foc*TR4 by up to 96% in vitro by producing secondary metabolites (Al-Ani et al. 2013a). Bae et al. (2016) found *T. atroviride/petersenii* (KACC, 40557) showed the highest inhibition of *Phytophthora* growth. *T. harzianum* T12 produced many volatile compounds to control charcoal rot in soybean caused by *M. phaseolina* (Khalili et al. 2016). *T. harzianum* Th-Sks showed high efficacy against *F. oxysporum* and *Pythium aphanidermatum*, which caused damping off and wilt diseases of brinjal and okra by producing volatile and non-volatile compounds (Sain and Pandey 2016). The culture filtrates containing volatile compounds of THSC were able to decrease the size of necroses caused by *C. radicicola* of date palm in vivo (Al-Naemi et al. 2016). Seven volatile compounds, possibly with antifungal activity, produced by *T. parareesei* T26 inhibited *Foc*TR4 in vitro and managed to reduce the disease severity of Fusarium wilt by up to 100% in vivo (Al-Ani 2017).

The ISR of *T. harzianum* in the roots of cucumber was observed through the change in structural compounds, the deposition of newly formed barriers, and strengthening of the epidermal and cortical cell walls (Yedidia et al. 1999). *T. harzianum* was able to induce systemic resistance in carrot against *Alternaria radicina* and *Botrytis cinerea* by using chitinase and CHIT36 expressed in the plant

(Baranski et al. 2008). *Trichoderma virens* and *T. atroviride* induced plant defense and activated the signaling pathway including SA and/or JA, as well as camalexin, conferring resistance in *Arabidopsis thaliana* against necrotrophic fungus *Botrytis cinerea* (Contreras-Cornejo et al. 2011). *Trichoderma asperellum* was able to induce acquired resistance in cucumber by activating peroxidase and a boost in SA (Hermosa et al. 2012). The ISR in plants was triggered by increasing the ET or JA pathways by *Trichoderma* cellulase complexes (Hermosa et al. 2013). *T. harzianum* triggered the transient production of ROS by *Thph1* and *Thph2* proteins, which required enhancing ISR in maize leaf (Saravanakumar et al. 2016b).

In induced plant defense mechanisms, *T. virens* has induced plant defense through seed treatment using terpenoid synthesis in cotton root against *R. solani* (Howell et al. 2000). *T. viride* JAU60 stimulated the specific defense enzymes of polyphenol oxidase, β-1,3 glucanase, phenylalanine ammonia lyase, and chitinase against collar rot disease caused by *Aspergillus niger* Van *Tieghem* (Gajera et al. 2015). *T. asperellum* induced the plant defense-related genes in the banana plant against *Fusarium oxysporum* f.sp. *cubense* (Foc) (Raman et al. 2016). *Trichoderma virens* (KACC 40929) stimulated defense-related genes against *Phytophthora* infection and changes of plant hormonal (Bae et al. 2016). *T. aureoviride* URM 5158 was capable of reducing disease severity to 60% in the shoot and 84% in the root of cassava plants by inducing plant defense. The strain changed the physiological response and maximized the enzyme activity of ROS groups (da Silva et al. 2016). Seed treatment with *T. viride* JAU60 has increased the activity of ROS enzymes and reduced 51–58% collar rot disease incidence by rot pathogen *Aspergillus niger* (Gajera et al. 2016). *T. asperellum* BHUT8 induced plant defense in tomato seedlings which include phenylalanine ammonia-lyase (PAL), peroxidase (PO), polyphenol oxidase (PPO), lignifications, and the accumulation of some secondary metabolites such as shikimic acid and gallic acid (Singh et al. 2016).

In detoxification mechanism, several isolates of *T. viride* were able to detoxify *R. solani* toxin (Sriram et al. 2000). Aggarwal et al. (2011) also found that *T. viride* (TV5-2) detoxified the *Bipolaris sorokiniana* toxin and reduced the disease severity of spot blotch in wheat. Tian et al. (2016) found eight strains of *Trichoderma* (*T. harzianum* GIM3.442, *T. harzianum JF309, T. koningii* GIM3.137, *T. longibrachiatum* GIM3.534, *T. harzianum* Q710613, *T. atroviride* Q710251, *T. asperellum* Q710682, and *T. virens* Q710925) that showed antagonistic activity against *F. graminearum*, the causal agent of *Fusarium* head blight (FHB), by inhibiting the mycelium growth and detoxifying deoxynivalenol (DON) to deoxynivalenol-3-glucoside (D3G).

In biofertilizers and PGPF mechanisms, *Trichoderma* spp. can colonize the roots of plants which can improve the plant growth, increase the crop productivity, form a strong resistance to plant pathogens, and improve nutrient uptake (Arora et al. 1992). PGPF *T. harzianum* T-22 has solubilized and chelated various plant nutrient compounds and further enhanced plant growth (Altomare et al. 1999). *Trichoderma hamatum* and *T. koningii* could increase crop productivity up to 300% (Benítez et al. 2004). Qi and Zhao (2012) revealed that *T. asperellum* strain Q1 acted as PGPF and increased the content of osmosis molecules and chlorophyll, plant biomass, and enhanced the activity of osmosis molecules and antioxidant enzymes under saline environments. *T. asperellum* T34 plays a role in increasing the accumulation of Cu and Fe in the aerial parts of cucumber plants, as well as Zn and Mn according to the availability in the soil (de Santiago et al. 2013). *T. atroviride* LU132 was capable of colonizing oilseed rape (*Brassica napus*) and increased the biomass of root and shoot (Maag et al. 2013). *T. harzianum* (ANR-1) increased the plant height and dry weight of tomato plants (Sundaramoorthy and Balabaskar 2013). Li et al. (2015) suggested that *T. harzianum* strain SQR-T037 can promote plant growth by development of the root and the increased nutrient uptake, as well as dissolution (i.e., most likely chelating for Cu, chelating for Fe, acidification, and redox). *T. harzianum* T12 is able to enhance the plant growth of soybean (Khalili et al. 2016). *Trichoderma* can enhance plant vigor of *Miscanthus x giganteus* (Mxg), including growth, chlorophyll concentration, plant height, and shoot dry weight (Chirino-Valle et al. 2016). Singh et al. (2016) found that seeds of some plants, such as tomato, ridge gourd, chilli, and guar, treated with *T. asperellum* BHUT8 can improve seed germination and radicle length. Al-Ani (2017) found that *T. harzianum* Tveg1 can improve the plant vigor of banana, such as plant height, the content of chlorophyll, plant biomass, and number of the leaves. Sain and Pandey (2016) showed that the plant height and fruit yield of brinjal and okra increased when *T. harzianum* Th-Sks was used to treat the seeds.

5.3.4 Biocontrol of Plant-Parasitic Nematodes

Nematodes are worm-like but quite distinct taxonomically from the true worms. There are several hundred species, and they obtain their food by feeding on living plants using spears or stylets. This feeding method has caused major plant disease worldwide. These nematodes have an effect on sustainable agriculture amounting to 11–14%, involving such crops as legumes, cereals, banana, vegetable, cassava, fruits, and nonedible field crops (Agrios 2005). Control measures are often difficult, particularly involving systemic nematicides and insecticide treatments to decrease the nematodes and vectors (Agrios 2005). Biological control, cultural, and physical methods are other general measures for controlling nematodes. Oil-cakes, residues from leguminous crops, other materials with a low C/N ratio, and animal manures can also be added to soil (Stirling 2011).

Biological control is an alternative management system to control plant-parasitic nematodes which are suppressed by pathogen-specific agents comprising many enemies, such as viruses, bacteria, fungi, nematodes, microarthropods, and protozoa. Fungi are registered to be biocontrol agents for nematodes. Soil contains a large range of fungi species which are able to suppress plant-parasite nematodes and are called nematophagous fungi. More than 200 species of nematophagous fungi have been described (Tunlid and Ahrén 2011). Some isolates or strains of *Trichoderma* have been considered to be in the nematophagous fungi group. Biocontrol activity of *Trichoderma* spp. against plant-parasitic nematodes is exploited by difference mechanisms. These mechanisms include parasitism, competition, antibiosis, induction of plant defense, systemic resistance, enhancement of plant growth, tolerance of the infection, and impact on the life cycle of the plant-parasitic nematode.

For parasitism, *Trichoderma* attacks eggs, juveniles (larvae), and adults. In the parasitism the mode of action is attached to the nematode and then the parasite. The special chemical is the gelatinous matrixes (gm) that envelop the eggs and play a very essential role to attach the mycelium of *Trichoderma* to the eggs, egg masses, and the second stage juveniles (J2s) (Sharon et al. 2007, 2011). The *Trichoderma* secretion enhances the ability of the plant-parasitic nematodes. These enzymes include proteases, cellulases, hemicellulases, chitinases, and glucanases (Viterbo et al. 2002a, b; Keswani et al. 2013). *T. harzianum* is able to attack and colonize the eggs, and egg masses of the root-knot nematode *Meloidogyne javanica* by instigating proteolytic activity (Sharon et al. 2001). The two isolates of *T. asperellum*-203 and *T. atroviride*-IMI 206040 attack the eggs and J2s of *Meloidogyne javanica* by the formation of coiling and appressorium-like structures and then parasitism (Sharon et al. 2007).

For competition, *Trichoderma* maybe competes with plant parasitic nematodes. The competition for space and feeding sites may occur with plant parasitic nematodes (Hussey and Roncadori 1978). For the antibiosis, nematicidal, anti-nematode and secondary metabolites as volatile and non-volatile compounds are included. For example, *T. viride* metabolites impacted on reproduction and development of *Meloidogyne javanica* through implementing root-dip treatments with the fungal culture filtrate (Khan and Saxena 1997). Strains of *T. harzianum* have the capability to produce the anti-nematode and nematicidal against the root-knot nematode *Meloidogyne javanica* that appears as immobilized J2 s and reduces the penetration of the root by the nematode (Sharon et al. 2001). The culture filtrate of *T. harzianum* was then able to inhibit egg hatching of the nematode *M. javanica* at the standard concentration in vitro that may release toxic metabolites/enzymes into the medium (Khattak et al. 2008). Yang et al. (2012) found three metabolites of *Trichoderma* sp. YMF 1.00416 comprising a new compound, 1β-vinylcyclopentane-1α,3α-diol (1), and two known metabolites, 6-pentyl-2*H*-pyran-2-one (2) and 4-(2-hydroxyethyl) phenol (3); compound 2 was nematicidal and killed 85% of *Bursaphelenchus xylophilus*, *Panagrellus redivivus*, and *Caenorhabditis elegans*. *T. harzianum* strains were able to produce antibiosis as a mechanism antagonistic against the nematode *M. cionopagum* (Szabó et al. 2012).

To induce resistance, for example, *Trichoderma* can stimulate resistance as a localized or systemic response against nematodes, which may occur on the surface of the roots, inside the roots, and in the soil (Sharon et al. 2011). *Trichoderma* primes JA- and SA-dependent defenses in tomato roots against the root knot nematode *M. incognita* (Martínez-Medina et al. 2017). *T. harzianum* isolate T10 was able to change the chemical and physical reactions in tomato against the invasion of *M. javanica* nematode because of two different kinds of systemic resistances, ISR and SAR (Selim et al. 2014). *T. atroviride* could induce systemic resistance against *Meloidogyne javanica* caused by root-knot nematodes of tomato in vivo by a split-root occurrence that is a trigger of SA-, JA- and ET-dependent defense pathways (de Medeiros et al. 2017). For the stimulation of plant defense, for example, *T. harzianum* isolate ITEM908 stimulated the plant defense by the expression of patterns of the genes *PR1* in tomato against the invasion of *M. incognita* (Leonetti et al. 2014).

For the enhancement of plant growth, for example, *T. harzianum* and *T. lignorum* isolates improved growth as fresh weight against the root-knot nematode *Meloidogyne javanica* (Spiegel and Chet 1998). *T. harzianum* is efficient in controlling *Meloidogyne javanica* on tomato at the highest used density (10^{10} spore/g soil). *T. atroviride* enhanced plant growth against *Meloidogyne javanica* (de Medeiros et al. 2017).

Tolerance of the infection is reflected in a decreased number and size of root galling, cessation of the growth and reproduction of plant-parasitic nematodes inside the plant, increase of crop yield, and a halt to the plant-parasitic nematode from completing its life cycle. *T. asperellum* T-16 led to a reduction of the number and size of galls and enhanced the tomato yields that were infected by *M. incognita* and caused root-knot nematodes in vegetables (Affokpon et al. 2011). *T. harzianum* T22 enhanced plant height, number of branches, and yield growth against *Meloidogyne incognita* of soybeans (Izuogu and Abiri 2015). Javeed et al. (2016) mention that three isolates of *Trichoderma* reduced root galling in tomatoes that resulted from infection by *M. javanica*. Al-Hazmi and Javeed (2016) stated that *T. harzianum* are able to inhibit root galling of *Meloidogyne javanica* on tomato. *T. atroviride* impacted on *Meloidogyne javanica* by reducing the number of galls (de Medeiros et al. 2017).

Impact on the life cycle of a plant-parasitic nematode can be reflected in nematode reproduction, the type and number of gender (the number of males compared with females), larval mortality, egg hatching, the number of eggs, egg masses, size and movement of the nematode or juveniles, and the general physiological functions of the plant-parasitic nematode. *T. asperellum* T-16 inhibited the densities of J2s of *Meloidogyne incognita* in the roots, but *T. asperellum* T-12 inhibited the densities of J2s in the soils, and *T. brevicompactum* T-3 suppressed egg production (Affokpon et al. 2011). Four strains of *Trichoderma* – *T. harzianum*, *T. virens*, *T. atroviride*, and *T. rossicum* – triggered higher and faster mortality of plant-parasitic nematode *Xiphinema index*, which is capable of transmitting several plant viruses (Daragó et al. 2013).

T. harzianum T22 inhibited the development and parasitic effects of *Meloidogyne incognita* and led to a reduction of the soil nematode population that caused root knot nematode of soybeans (Izuogu and Abiri 2015). Javeed et al. (2016) found three isolates of *Trichoderma* – *T. harzianum* (isolate No.27), *T. hamatum* (isolate No.5), and *T. viride* (isolate No.8) – which decreased the number of juveniles (larvae) and inhibited the reproduction of *M. javanica*. *T. harzianum* can suppress nematode reproduction via egg masses, eggs, and J2s of *Meloidogyne javanica* on tomato (Al-Hazmi and Javeed 2016). *T. atroviride* decreased the number of eggs and adult nematodes that impact on *Meloidogyne javanica* (de Medeiros et al. 2017).

5.3.5 Biocontrol of Parasitic Higher Plants by Trichoderma

The parasitic higher plants are vascular plants that penetrate the tissue of hosts (vascular plants) and absorb nutrients from the host by unique connections. Some of the parasitic plants are major factors that impacts on production of many crops. The parasitic plants have no or little chlorophyll, false roots, but can produce flower and seeds. There are more than 2,500 known species of higher plants (Sharma 2006) that live on other plants parasitically. For control of parasitic plants comprise many methods such as chemical herbicides, biocontrol, breeding and selection of resistant crop, soil fumigation, solarization, parasitic-plant-free seed, cultural practices, and the use of trap and catch crops (Abdel-Kader and El-Mougy 2009). Biological control is an alternative method to control parasitic plants instead of chemical herbicides. *Trichoderma* is a biological control or bioherbicide agent that uses many mechanisms against plant diseases as parasitic plants. *T. harzianum* (T1 and T3) and *T. viride* (T2) applied of tomato plants by the foliar spray and the soil drench method that reduced both infection and intensity of attack by the plant pathogens of broomrapes *Orobanche ramose* (Abdel-Kader and El-Mougy 2007, 2009). *T. harzianum* attacked living tissues of broomrapes *Orobanche ramose* that could cause soft rot, black lesion, a reduction of number of *Orobanche* shoots, and complete deterioration within 7 days (100%) (Nawar and Sahab 2011). *T. harzianum* reduced the number root-parasitic weeds such as *Phelipanche* and *Orobanche* spp. that impact on the plant hosts by producing some effective secondary metabolites as 5-deoxystrigol and 4-deoxyorobanchol (Boari et al. 2016).

5.3.6 Biocontrol of Plant Viroids by Trichoderma

Plant viroids is a plant pathogen since 1971, and it is causing slight damage to the agricultural economy but sometime be more catastrophic. Plant viroids can cause several important diseases such as citrus exocortis, apple scar skin, potato spindle tuber, the cadang-cadang disease of coconut, and avocado sunblotch (Agrios 2005). Viroids are very important plant pathogens because they are composed of a (1) short stretch of circular, (2) nonprotein-coding, (3) single-stranded RNA with autonomous replication. The control of plant viroids is a complex method. Some methods to control this plant pathogen include eradication, cultural controls, elimination of insect vectors, and inducing resistance. Trichoderma has not yet been registered as a biocontrol agent against plant viroids, but I assume they can be used. Why is Trichoderma considered a biocontrol agent against plant viroids? The genus Trichoderma has many potential mechanisms to control different plant diseases, such as the indirect method that may be effective on this pathogen. This indirect method includes inducing resistance and defense and control of the vectors, as well as enhanced growth and tolerance of the plant against viroid infection. Trichoderma is able to induce resistance against plant virus, as described in Sect. 5.3.1. As the life cycle of plant virus is the same as that of plant viroids, use of Trichoderma as a future biocontrol agent against plant viroids is recommended, which may lead to obtaining greater insight about the relationship between Trichoderma and plant viroids.

5.3.7 *Biocontrol of Phytoplasma by Trichoderma*

Phytoplasma is a plant pathogen that was detected in 1967 and causes around 200 plant diseases worldwide. However, Phytoplasmas lack cell walls and have ribosomes, cytoplasm, and strands of nuclear material which are bounded by a "unit" membrane (Agrios 2005). Phytoplasmas cause very significant diseases such as European stone fruit yellows, apple proliferation, lethal coconut yellowing, peach X disease, grapevine yellows, pear decline, and aster yellows (Agrios 2005). Controlling of phytoplasma is very difficult. Phytoplasma can be controlled by various common methods, as described in Sect. 5.3.6. Trichoderma is not a registered biocontrol agent against phytoplasma. Also, my assumption here is the same as in Sect. 5.3.6. Why is Trichoderma considered a biocontrol agent against phytoplasma? Trichoderma has many indirect mechanisms that can be effective against phytoplasma, which are mentioned in Sect. 5.3.6. Therefore, we need to know the nature of relationship between Trichoderma and phytoplasma, which may be beneficial in the future, to obtain a cleaner method for the ecosystem.

5.4 Conclusion

All *Trichoderma* species are of interest to researchers because of their use in the biocontrol and reduction of plant pathogens (viruses, bacteria, fungi, plant parasitic nematode, and parasitic higher plants) as well as the enhancement of tolerance in the plant against plant pathogens and plant growth promotion. This is achieved through the supply of plant nutrient by the secretion of some very interesting items such as chelating compounds, for example, siderophores, chelating for Cu, and acidification. *Trichoderma* has many mechanisms, such as direct parasitic, via appressorium-like structures, secreting enzymes or volatile compounds inhibiting the host growth. Indirect impact by detoxification against diseases of the plants and induction of plant defense by changing the activity of plant physiology, such as by activating very important enzymes, are also involved. This confers protection to plants and also induces systemic resistance, changing the physical and chemical barriers to prevent plant pathogens from entering the plant. *Trichoderma* can therefore be used as biobactericides, biofungicides, and bioherbicides to control the major plant pathogens, and as biofertilizers to enhance plant growth for the major plants. The combination of biocontrol with biofertilization in same product may also contribute to increasing sustainable agriculture, making the environment cleaner, free from residues from chemical pesticides and fertilizers. The final result is the production of cleaner food.

References

Abdelghani E-Y, Bala K, Paul B (2004) Characterisation of *Pythium paroecandrum* and its antagonism towards *Botrytis cinerea*, the causative agent of grey mould disease of grape. FEMS Microbiol Lett 230:177–183

Abdel-Kader MM, El-Mougy NS (2007) Applicable control measure against *Orobanche ramose* in tomato plants. Aust Plant Pathol 36:160–164

Abdel-Kader MM, El-Mougy NS (2009) Prospects of mycoherbicides for control of broomrapes (*Orobanche* spp.) in Egypt. J Plant Prot Res 49(1):63–75

Adekunle AT, Cardwell KE, Florini DA, Ikotun T (2001) Seed treatment with *Trichoderma* species for control of damping-off of cowpea caused by *Macrophomina phaseolina*. Biochem Sci Technol 11:449–457

Affokpon A, Coyne DL, Htay CC, Agbèdè RD, Lawouin L, Coosemans J (2011) Biocontrol potential of native *Trichoderma* isolates against root-knot nematodes in West African vegetable production systems. Soil Biol Biochem 43:600–608

Aggarwal R, Gupta S, Singh VB, Sharma S (2011) Microbial detoxification of pathotoxin produced by spot blotch pathogen Bipolaris sorokiniana infecting wheat. J Plant Biochem Biotechnol 20(1):66–73

Agrios GN (2005) Plant pathology. Harcourt Academic Press, New York, pp 397–398, 403, 616, 625

Al-Ani LKT (2010) Biological control of Fusarium wilt of banana by non pathogenic Fusarium oxysporum. PPSKH colloquium, Pust Pengajian Sains Kajihayat/School of Biological Sciences, USM, June, p 10

Al-Ani LKT (2017) Potential of utilizing biological and chemical agents in the control of Fusarium wilt of banana. PhD, School of Biology Science, Universiti Sains Malaysia, Pulau Pinang, Malaysia, p 259

Al-Ani LKT, Salleh B, Ghazali AHA (2013a) Biocontrol of Fusarium wilt of banana by Trichoderma spp. 8th PPSKH colloquium, Pust Pengajian Sains Kajihayat/School of Biological Sciences, USM, June 5–6

Al-Ani LKT, Salleh B, Mohammed AM, Ghazali AHA, Al-Shahwany AW, Azuddin NF (2013b) Biocontrol of Fusarium wilt of banana by non-pathogenic Fusarium spp. International symposium on tropical fungi, ISTF, IPB International Convention Center, Bogor, Indonesia; 09/2013, pp 50–51

Aleandri MP, Chilosi G, Bruni N, Tomassini A, Vettraino AM, Vannini A (2015) Use of nursery potting mixes amended with local *Trichoderma* strains with multiple complementary mechanisms to control soil-borne diseases. Crop Prot 67:269–278

Al-Hazmi AS, Javeed MT (2016) Effects of different inoculum densities of *Trichoderma harzianum* and *Trichoderma viride* against *Meloidogyne javanica* on tomato. Saudi J Biol Sci 23:288–292

Al-Naemi FA, Ahmed TA, Nishad R, Radwan O (2016) Antagonistic effects of *Trichoderma harzianum* isolates against *Ceratocystis radicicola*: pioneering a biocontrol strategy against black scorch disease in date palm trees. J Phytopathol 164(7–8):464–475

Altomare C, Norvell WA, BjöRkman T, Harman GE (1999) Solubilization of phosphates and micronutrients by the plant-growth-promoting and biocontrol fungus *Trichoderma harzianum* Rifai 1295-22. Appl Environ Microbiol 65(7):2926–2933

Arora DK, Elander RP, Mukerji KG (1992) Handbook of applied mycology, Fungal Biotechnology, vol 4. Marcel Dekker, New York

Arst JH, Penalva M (2003) PH regulation in *Aspergillus* and parallels with higher eukaryotic regulatory systems. Trends Genet 19:224–231

Bae S-J, Mohanta TK, Chung JY, Ryua M, Park G, Shim S, Hong S-B, Seo H, Bae D-W, Bae I, Kima J-J, Bae H (2016) *Trichoderma* metabolites as biological control agents against *Phytophthora* pathogens. Biol Control 92:128–138

Baranski R, Klocke E, Nothnagel T (2008) Chitinase CHIT36 from *Trichoderma harzianum* enhances resistance of transgenic carrot to fungal pathogens. J Phytopathol 156:513–521

Benítez T, Rincón AM, Limón MC, Codón AC (2004) Biocontrol mechanisms of *Trichoderma* strains. Int Microbiol 7:249–260

Berry CR, Barnett HL (1957) Mode of parasitism and host range of *Piptocephalis virginiana*. Mycologia 49:374–386

Bicici M, Dede Y, Çinar A (1992) *Trichoderma* species against gummosis disease in lemon trees. In: Tjamos EC, Papavizas GC, Cook RJ (eds) Biological control of plant diseases: progress and challenges for the future, NATO ASI Series. Springer, New York, pp 193–196

Boari A, Ciasca B, Pineda-Martos R, Lattanzio VMT, Yoneyama K and Vurro M (2016) Biological control of parasitic weeds by using strigolactone-degrading fungi (579). In: 7th international weed science congress June 19–25 Prague, Czech Republic, p 138

Boddy L (2016) Interactions between fungi and other microbes. In: Watkinson SC, Boddy L, Money NP (eds) The fungi. Elsevier, London, p 340

Chet I, Inbar J (1994) Biological control of fungal pathogens. Appl Biochem Biotechnol 48:37–43

Chirino-Valle I, Kandula D, Littlejohn C, Hill R, Walker M, Shields M, Cummings N, Hettiarachchi D, Wratten S (2016) Potential of the beneficial fungus *Trichoderma* to enhance ecosystem-service provision in the biofuel grass *Miscanthus x giganteus* in agriculture. Sci Rep 6:25109

Chitrampalam P, Wu BM, Koike ST, Subbarao KV (2011) Interactions between *Coniothyrium minitans* and *Sclerotinia minor* affect biocontrol efficacy of *C. minitans*. Phytopathology 101(3):358–366

Contreras-Cornejo HA, Macías-Rodríguez L, Beltrán-Peña E, Alfredo Herrera-Estrella A, López-Bucio J (2011) *Trichoderma*-induced plant immunity likely involves both hormonal- and camalexin-dependent mechanisms in *Arabidopsis thaliana* and confers resistance against necrotrophic fungus *Botrytis cinerea*. Plant Signal Behav 6(10):1554–1563

da Silva JAT, de Medeiros EV, da Silva JM, Tenório DdA, Moreira KA, Nascimento TCEdS, Souza-Motta C (2016) *Trichoderma aureoviride* URM 5158 and *Trichoderma hamatum* URM 6656 are biocontrol agents that act against Cassava root rot through different mechanisms. J Phytopathol 164(11–12):1003–1011

Daragó A, Szabó M, Hrács K, Takács A, Nagy PI (2013) In vitro investigations on the biological control of *Xiphinema index* with *Trichoderma* species. Helminthologia 50(2):132–137

de França SKS, Cardoso AF, Lustosa DC, Ramos EMLS, de Filippi MCC, da Silva GB (2015) Biocontrol of sheath blight by *Trichoderma asperellum* in tropical lowland rice. Agron Sustain Dev 35(1):317–324

de Lima FB, Félix C, Osório N, Alves A, Vitorino R, Domingues P, Correia A, Ribeiro RTdS, Esteves AC (2016) Secretome analysis of *Trichoderma atroviride* T17 biocontrol of *Guignardia citricarpa*. Biol Control 99:38–46

de Medeiros HA, Filho JVdA, de Freitas LG, Castillo P, Rubio MB, Hermosa R, Monte E (2017) Tomato progeny inherit resistance to the nematode *Meloidogyne javanica* linked to plant growth induced by the biocontrol fungus *Trichoderma atroviride*. Sci Rep 7:40216

de Santiago A, García-López AM, Quintero JM, Avilés M, Delgado A (2013) Effect of *Trichoderma asperellum* strain T34 and glucose addition on iron nutrition in cucumber grown on calcareous soils. Soil Biol Biochem 57:598–605

Doni F, Al-Shorgani NKN, Tibin EMM, Abuelhassan NN, Anizan I, Che Radziah CMZ, Wan Mohtar WY (2013) Microbial involvement in growth of paddy. Curr Res J Biol Sci 4(6):285–290

Elad Y (2000) Biological control of foliar pathogens by means of *Trichoderma harzianum* and potential modes of action. Crop Prot 19:709–714

Elad Y, Chet I, Henis Y (1982) Degradation of plant pathogenic fungi by *Trichoderma harzianum*. Can J Microbiol 28:719–725

Elad Y, Sadovski Z, Chet I (1983) Detection of mycoparasitism by infrared photomicrography. Microb Ecol 9:185–187

Elad Y, Kirshner B, Yehuda N, Sztejnberg A (1998) Management of powdery mildew and gray mold of cucumber by *Trichoderma harzianum* T39 and *Ampelomyces quisqualis* AQ10. BioControl 43:241–251

Elsharkawy MM, Shimizu M, Takahashi H, Ozaki K, Hyakumachi M (2013) Induction of systemic resistance against cucumber mosaic virus in *Arabidopsis thaliana* by *Trichoderma asperellum* SKT-1. Plant Pathol J 29(2):193–200

Fravel DR, Engelkes CA (1994) Biological management. In: Campbell CL, Benson DM (eds) Epidemiology and management of root diseases. Springer verslag, Berlin, pp 293–308

Gajera HP, Bambharolia RP, Patel SV, Khatrani TJ, Goalkiya BA (2012) Antagonism of Trichoderma spp. against Macrophomina phaseolina: evaluation of coiling and cell wall degrading enzymatic activities. J Plant Pathol Microbiol 3:7

Gajera H, Domadiya R, Patel S, Kapopara M, Golakiya B (2013) Molecular mechanism of *Trichoderma* as bio-control agents against phytopathogen system – a review. Curr Res Microbiol Biotechnol 1(4):133–142

Gajera HP, Savaliya DD, Patel SV, Golakiya BA (2015) *Trichoderma viride* induces pathogenesis related defense response against rot pathogen infection in groundnut (*Arachis hypogaea* L.) Infect Genet Evol 34:314–325

Gajera HP, Katakpara ZA, Patel SV, Golakiya BA (2016) Antioxidant defense response induced by *Trichoderma viride* against *Aspergillus niger* Van Tieghem causing collar rot in groundnut (*Arachis hypogaea* L.) Microb Pathog 91:26–34

Gnanamanickam SS (2002) Biological control of crop diseases. Marcel Dekker, New York, p 449

Goh YK, Vujanovic V (2010) *Sphaerodes quadrangularis* biotrophic mycoparasitism on *Fusarium avenaceum*. Mycologia 102(4):757–762

Hadar Y, Harman GE, Taylor AG (1984) Evaluation of *Trichoderma koningii* and *T. harzianum* from New York soils for biological control of seed rot caused by *Pythium* spp. Phytopathology 74:106–110

Hanson LE, Howell CR (2002) Biocontrol efficacy and other characteristics of protoplast fusants between *Trichoderma koningii* and *T. virens*. Mycol Res 106(3):321–328

Harman GE, Howell CR, Viterbo A, Chet I, Lorito M (2004) *Trichoderma* species – opportunistic, avirulent plant symbionts. Nat Rev Microbiol 2:43–56

Harris AR (1999) Biocontrol of *Rhizoctonia solani* and *Pythium ultimum* on Capsicum by *Trichoderma koningii* in potting medium. Microbiol Res 154(2):131–135

Hermosa R, Viterbo A, Chet I, Monte E (2012) Plant-beneficial effects of *Trichoderma* and of its genes. Microbiology 158:17–25

Hermosa R, Rubio MR, Cardoza RE, Nicolás C, Monte E, Gutiérrez S (2013) The contribution of *Trichoderma* to balancing the costs of plant growth and defense. Int Microbiol 16:69–80

Hoch HC (1977) Mycoparasitic relationships: *Gonatobotrys simplex* parasitic on *Alternaria tenuis*. Phytopathology 67:309–314

Howell CR (2003) Mechanisms employed by *Trichoderma* species in the biological control of plant diseases: the history and evolution of current concepts. Plant Dis 87:4–10

Howell CR, Hanson LE, Stipanovic RD, Puckhaber LS (2000) Induction of terpenoid synthesis in cotton roots and control of *Rhizoctonia solani* by seed treatment with *Trichoderma virens*. Phytopathology 90:248–252

Hung PM, Wattanachai P, Kasem S, Poaim S (2015) Biological control of *Phytophthora palmivora* causing root rot of pomelo using *Chaetomium* spp. Mycobiology 43(1):63–70

Hussey RS, Roncadori RW (1978) Interaction of *Pratylenchus brachyurus* and *Gigaspora margarita* on cotton. J Nematol 10:16–20

Hwang J, Benson DM (2003) Expression of induced systemic resistance in poinsettia cuttings against *Rhizoctonia* stem rot by treatment of stock plants with binucleate *Rhizoctonia*. Biol Control 27:73–80

Inbar J, Chet I (1992) Biomimics of fungal cell-cell recognition by use of lectin-coated nylon fibers. J Bacteriol 174:1055–1059

Izuogu NB, Abiri TO (2015) Efficacy of *Trichoderma harzianum*T22 as a biocontrol agent against root-knot nematode (*Meloidogyne incognita*) on some soybean varieties. Croat J Food Sci Technol 7(2):47–51

Javeed MT, Al-Hazmi AS, Molan YY (2016) Antagonistic effects of some indigenous isolates of *Trichoderma* spp. against *Meloidogyne javanica*. Pak J Nematol 34(2):183–191

Keswani C, Singh SP, Singh HB (2013) A superstar in biocontrol Enterprise: *Trichoderma* spp. Biotechnol Today 3(2):27–30

Khalili E, Javed MA, Huyop F, Rayatpanah S, Jamshidi S, Wahab RA (2016) Evaluation of *Trichoderma* isolates as potential biological control agent against soybean charcoal rot disease caused by *Macrophomina phaseolina*. Agric Environ Biotechnol 30(3):479–488

Khan TA, Saxena SK (1997) Effect of root-dip treatment with fungal filtrates on root penetration, development and reproduction of *Meloidogyne javanica* on tomato. Int J Nematol 7:85–88

Khattak B, Saifullah, Stephen M (2008) Effect of some indigenous isolates of *Trichoderma harzianum* on root knot nematode, *Meloidogyne javanica* (TREUB) chitwood. Sarhad J Agric 24:285–288

Kunova A, Pizzatti C, Cerea M, Gazzaniga A, Cortesi P (2016) New formulation and delivery method of Cryphonectria parasitica for biological control of chestnut blight. J Appl Microbiol 122(1):180–187

Lehner SM, Atanasova L, Neumann NKN, Krska R, Lemmens M, Druzhinina IS, Schuhmacher R (2013) Isotope-assisted screening for iron containing metabolites reveals a high degree of diversity among known and unknown siderophores produced by *Trichoderma* spp. Appl Environ Microbiol 79(1):18–31

Leonetti P, Costanza A, Zonno MC, Molinari S, Altomare C, C. (2014) How fungi interact with nematode to activate the plant defence response to tomato plants. Commun Agric Appl Biol Sci 79(3):357–363

Li Y-H, Luo Y, Zhang X-S, Shi W-L, Gong Z-T, Shi M, Chen L-L, Chen X-L, Zhang Y-Z, Song X-Y (2014) Trichokonins from *Trichoderma pseudokoningii* SMF2 induce resistance against Gram-negative *Pectobacterium carotovorum* subsp. carotovorum in Chinese cabbage. FEMS Microbiol Lett 354:75–82

Li R-X, Cai F, Pang G, Shen Q-R, Li R, Chen W (2015) Solubilisation of phosphate and micronutrients by *Trichoderma harzianum* and its relationship with the promotion of tomato plant growth. PLoS One 10(6):e0130081

Lifshitz R, Windham MT, Baker R (1986) Mechanism of biological control of pre-emergence damping-off of pea by seed treatment with *Trichoderma* spp. Phytopathology 76:720–725

Lo C-T (1998) General mechanisms of action of microbial biocontrol agents. Plant Pathol Bull 7:155–166

Luo Y, Zhang D-D, Dong X-W, Zhao P-B, Chen L-L, Song X-Y, Wang X-J, Chen X-L, Shi M, Zhang Y-Z (2010) Antimicrobial peptaibols induce defense responses and systemic resistance in tobacco against tobaccomosaic virus. FEMS Microbiol Lett 313:120–126

Maag D, Kandula DRW, Müller C, Mendoza-Mendoza A, Wratten SD, Stewart A, Rostás M (2013) *Trichoderma atroviride* LU132 promotes plant growth but not induced systemic resistance to *Plutella xylostella* in oilseed rape. BioControl 59(2):241–252

Marois JJ (1990) Biological control of diseases caused by *Fusarium oxysporum*. In: Ploetz RC (ed) Fusarium wilt of banana. APS Press, St Paul, pp 77–81

Martínez-Medina A, Fernandez I, Lok GB, Pozo MJ, Pieterse CMJ, Van Wees SCM (2017) Shifting from priming of salicylic acid- to jasmonic acid-regulated defences by *Trichoderma* protects tomato against the root knot nematode *Meloidogyne incognita*. New Phytol 213(3):1363–1377

Marzano M, Gallo A, Altomare C (2013) Improvement of biocontrol efficacy of Trichoderma harzianum vs. Fusarium oxysporum f. sp. lycopersici through UV-induced tolerance to fusaric acid. Biol Control 67(3):397–408

Mendoza-Mendoza A, Pozo MJ, Grzegorski D, Martínez P, García JM, Olmedo-Monfil V, Cortés C, Kenerley C, Herrera-Estrella A (2003) Enhanced biocontrol activity of *Trichoderma* through inactivation of a mitogen-activated protein kinase. Proc Natl Acad Sci 100(26):15965–15970

Murali M, Amruthesh KN, Sudisha J, Niranjana SR, Shetty HS (2012) Screening for plant growth promoting fungi and their ability for growth promotion and induction of resistance in pearl millet against downy mildew disease. J Phytol 4(5):30–36

Naher L, Yusuf UK, Ismail A, Hossain K (2014) *Trichoderma* spp.: a biocontrol agent for sustainable management of plant diseases. Pak J Bot 46(4):1489–1493

Nan Z, Xin H, Juan Z, Raza W, Xing-Ming Y, Yun-Ze R, Qi-Rong S, Qi-Wei H (2014) Suppression of fusarium wilt of banana with application of bio-organic fertilizers. Pedosphere 24(5):613–624

Nawar LS, Sahab AF (2011) Evaluation of compost fortified with *Trichoderma* Spp. isolates as biological agents against broomrape of chamomile herbs. Nat Sci 9(8):229–236

Ojha S, Chatterjee NC (2011) Mycoparasitism of *Trichoderma* spp. in biocontrol of Fusarial wilt of tomato. Arch Phytopathol Plant Protect 44(8):771–782

Olabiyi TI, Ruocco M (2013) *In-vitro* competition bio-assay experiment on the effect of *Trichoderma* species and some crop pathogenic fungi. J Biol Agric Healthc 3(12):2224–3208

Park D (1960) Antagonism – the background of soil fungi. In: Parkinson D, Waid JS (eds) Ecology of soil-borne plant pathogens. Liverpool Univ. Press, Liverpool, pp 148–159

Paulitz TC, Belanger RR (2001) Biological control in greenhouse systems. Annu Rev Phytopathol 39:103–133

Pellegrini A, Prodorutti D, Pertot I (2014) Use of bark mulch pre-inoculated with *Trichoderma atroviride* to control Armillaria root rot. Crop Prot 64:104–109

Perazzolli M, Moretto M, Fontana P, Ferrarini A, Velasco R, Moser C, Delledonne M, Pertot I (2012) Downy mildew resistance induced by *Trichoderma harzianum* T39 in susceptible grapevines partially mimics transcriptional changes of resistant genotypes. BMC Genomics 13:660

Qi W, Zhao L (2012) Study of the siderophore-producing *Trichoderma asperellum* Q1 on cucumber growth promotion under salt stress. J Basic Microbiol 53:355–364

Raman T, Gopalakrishnan V, Perumal GD (2016) Identification of differentially expressed genes from *Fusarium oxysporum* f. sp *cubense* and *Trichoderma asperellum* (prr2) interaction in the susceptible banana cultivar Grand Naine. Turk J Bot 40:480–487

Rubio MB, Quijada NM, Pérez E, Domínguez S, Monte E, Hermosa R (2014) Identifying *Trichoderma parareesei* beneficial qualities for plants. Appl Environ Microbiol 80(6):1864–1873

Sain SK, Pandey AK (2016) Evaluation of some Trichoderma harzianum isolates for the management of soil borne diseases of brinjal and okra. Proc Natl Acad Sci India Sect B: Biol Sci, 1–10

Samuels GJ (1996) Trichoderma: a review of biology and systematic of the genus. Mycol Res 100:923–935

Sant D, Casanova E, Segarra G, Avilés M, Reis M, Trillas MI (2010) Effect of *Trichoderma asperellum* strain T34 on Fusarium wilt and water usage in carnation grown on compost-based growth medium. Biol Control 53(3):291–296

Saravanakumar K, Yu C, Dou K, Wang M, Li Y, Chen J (2016a) Synergistic effect of *Trichoderma*-derived antifungal metabolites and cell wall degrading enzymes on enhanced biocontrol of *Fusarium oxysporum* f. sp. *cucumerinum*. Biol Control 94:37–46

Saravanakumar K, Fan L, Fu K, Yu C, Wang M, Xia H, Sun J, Li Y, Chen J (2016b) Cellulase from *Trichoderma harzianum* interacts with roots and triggers induced systemic resistance to foliar disease in maize. Sci Rep 6:35543

Saravanan T, Muthusamy M, Marimuthu T (2003) Development of integrated approach to manage the fusarial wilt of banana. Crop Prot 22:1117–1123

Sarrocco S, Guidi L, Fambrini S, Degl'Innocenti E, Vannacci G (2009) Competition for cellulose exploitation between *Rhizoctonia solani* and two *Trichoderma* isolates in the decomposition of wheat straw. J Plant Pathol 91(2):331–338

Segarra G, Casanova E, Avilés M, Trillas I (2010) *Trichoderma asperellum* strain T34 controls Fusarium wilt disease in tomato plants in soilless culture through competition for iron. Fungal Microbiol 59:141–149

Selim ME, Mahdy ME, Sorial ME, Dababat AA, Sikora RA (2014) Biological and chemical dependent systemic resistance and their significance for the control of root-knot nematodes. Nematology 16(8):917–927

Sennoi R, Singkham N, Jogloy S, Boonlue S, Saksirirat W, Kesmala T, Patanothai A (2013) Biological control of southern stem rot caused by *Sclerotium rolfsii* using *Trichoderma harzianum* and arbuscular mycorrhizal fungi on Jerusalem artichoke (*Helianthus tuberosus* L.) Crop Prot 54:148–153

Sharma PD (2006) Plant pathology. Alpha Science International, New Delhi, p 21.1

Sharma P, Vignesh Kumar P, Ramesh R, Saravanan K, Deep S, Sharma M, Mahesh S, Dinesh S (2011) Biocontrol genes from Trichoderma species: a review. Afr J Biotechnol 10(86):19898–19907

Sharon E, Bar-Eyal M, Chet I, Herrera-Estrella A, Kleifeld O, Spiegel Y (2001) Biological control of the root-knot nematode *Meloidogyne javanica* by *Trichoderma harzianum*. Phytopathology 91:687–693

Sharon E, Chet I, Viterbo A, Bar-Eyal M, Nagan H, Samuels GJ, Spiegel Y (2007) Parasitism of *Trichoderma* on *Meloidogyne javanica* and role of the gelatinous matrix. Eur J Plant Pathol 118:247–258

Sharon E, Chet I, Spiegel Y (2011) *Trichoderma* as a biological control agent. In: Davies K, Spiegel Y (eds) Biological control of plant-parasitic nematodes: building coherence between microbial ecology and molecular mechanisms. Springer, Berlin, pp 183–201

Shi M, Chen L, Wang X-W, Zhang T, Zhao P-B, Song X-Y, Sun C-Y, Chen X-L, Zhou B-C, Zhang Y-Z (2012) Antimicrobial peptaibols from *Trichoderma pseudokoningii* induce programmed cell death in plant fungal pathogens. Microbiology 158:166–175

Singh V, Upadhyay RS, Sarma BK, Singh HB (2016) *Trichoderma asperellum* spore dose depended modulation of plant growth in vegetable crops. Microbiol Res 193:74–86

Sivan A (1987) Biological control of Fusarium crown rot of tomato by *Trichoderma harzianum* under field conditions. Plant Dis 71:587–592

Sivan A, Chet I (1989) Degradation of fungal cell walls by lytic enzymes of *Trichoderma harzianum*. J Gen Microbiol 135:675–682

Spiegel Y, Chet I (1998) Evaluation of *Trichoderma* spp. as a biocontrol agent against soil borne fungi and plant-parasitic nematodes in Israel. Integr Pest Manag Rev 3:169–175

Sriram S, Raguchander T, Babul S, Nanadakumar R, Shanmugam V, Balasubramaniam P, Muthukrishnan S, Samiyappan R (2000) Inactivation of phytotoxins produced by Rhizoctonia solani the rice sheath blight pathogens. Can J Microbiol 46:520–524

Steyaert JM, Ridgway HJ, Elad Y, Stewart A (2003) Genetic basis of mycoparasitism: a mechanism of biological control by species of *Trichoderma*. N Z J Crop Hortic Sci 31:281–291

Stirling GR (2011) Biological control of plant-parasitic nematodes: an ecological perspective, a review of progress and opportunities for further research. In: Davies K, Spiegel Y (eds) Biological control of plant-parasitic nematodes: building coherence between microbial ecology and molecular mechanisms. Springer, Dordrecht, pp 1–38

Sundaramoorthy S, Balabaskar P (2013) Biocontrol efficacy of *Trichoderma* spp. against wilt of tomato caused by *Fusarium oxysporum* f.sp. *lycopersici*. J Appl Biol Biotechnol 1(03):036–040

Sutton JC, Liu W, Huang R, Owen-Going N (2002) Ability of *Clonostachys rosea* to establish and suppress sporulation potential of *Botrytis cinerea* in deleafed stems of hydroponic greenhouse tomatoes. Biocontrol Sci Tech 12:413–425

Sylla J (2013) Phyllosphere of organically grown strawberries. PhD. Department of Biosystems and Technology, Swedish University of Agricultural Sciences, Alnarp, p 68

Szabó M, Csepregi K, Gálber M, Virányi F, Fekete C (2012) Control plant-parasitic nematodes with *Trichoderma* species and nematode-trapping fungi: the role of Chi18-5 and chi18-12 genes in nematode egg-parasitism. Biol Control 63(2):121–128

Tian Y, Tan Y, Liu N, Yan, Z, Liao Y, Chen J, de Saeger S, Yang H, Zhang Q, Wu A (2016) Detoxification of deoxynivalenol via glycosylation represents novel insights on antagonistic activities of *Trichoderma* when confronted with *Fusarium graminearum*. Toxins, 8(11):335

Tsahouridou PC, Thanassoulopoulos CC (2001) *Trichoderma koningii* as a potential parasite of sclerotia of *Sclerotium rolfsii*. Cryptogam Mycol 22(4):289–295

Tsahouridou PC, Thanassoulopoulos CC (2002) Proliferation of *Trichoderma koningii* in the tomato rhizosphere and the suppression of damping-off by *Sclerotium rolfsii*. Soil Biol Biochem 34(6):767–776

Tunlid A, Ahrén D (2011) Molecular mechanisms of the interaction between nematode-trapping fungi and nematodes: lessons from genomics. In: Davies K, Spiegel Y (eds) Biological control of plant-parasitic nematodes: building coherence between microbial ecology and molecular mechanisms. Springer, Dordrecht, pp 145–169

Tunlid A, Jansson HB, Nordbring-Hertz B (1992) Fungal attachment to nematode. Mycol Res 96:401–412

Van den Boogert PHJF (1996) Mycoparasitism and biocontrol. In: Sneh B, Jabaji-Hare S, Neate S, Dijst G (eds) Rhizoctonia species_ taxonomy, molecular biology, ecology, pathology and disease control. Springer, Netherlands, pp 485–493

Vázquez MB, Barrera V, Bianchinottiac V (2015) Molecular identification of three isolates of *Trichoderma harzianum* isolated from agricultural soils in Argentina, and their abilities to detoxify in-vitro metsulfuron methyl. Botany 93:793–800

Verkley GJM, da Silva M, Wicklow DT, Crous PW (2004) Paraconiothyrium, a new genus to accommodate the mycoparasite Coniothyrium minitans, anamorphs of Paraphaeosphaeria, and four new species. Stud Mycol 50(2):323–336

Viterbo A, Montero M, Ramot O, Friesem D, Monte E, Llobell A, Chet I (2002a) Expression regulation of the endochitinase chit36 from *Trichoderma asperellum* (*T. harzianum* T-203). Curr Genet 42:114–122

Viterbo A, Ramot O, Chernin L, Chet I (2002b) Significance of lytic enzymes from *Trichoderma* spp. in the biocontrol of fungal plant pathogens. Antonie Leeuwenhoek 81:549–556

Vitti A, Monaca EL, Sofo A, Scopa A, Cuypers A, Nuzzaci M (2015) Beneficial effects of *Trichoderma harzianum* T-22 in tomato seedlings infected by cucumber mosaic virus (CMV). BioControl 60:135–147

Weindling R (1934) Studies on lethal principle effective in the parasitic action of *Trichoderma lignorum* on *Rhizoctonia solani* and other soil fungi. Phytopathology 24:1153–1179

Widmer TL (2014) Screening *Trichoderma* species for biological control activity against *Phytophthora ramorum* in soil. Biol Control 79:43–48

Wokocha RC (1990) Integrated control of *Sclerotium rolfsii* infection of tomato in the Nigerian Savanna: effect of *Trichoderma viride* and some fungicides. Crop Prot 9:231–234

Yang Z, Yu Z, Lei L, Xia Z, Shao L, Zhang K, Li G (2012) Nematicidal effect of volatiles produced by *Trichoderma* sp. J Asia Pac Entomol 15:647–650

Yedidia I, Benhamou N, Chet I (1999) Induction of defense responses in cucumber plants (*Cucumis sativus* L.) by the biocontrol agent *Trichoderma harzianum*. Appl Environ Microbiol 65(3):1061–1070

Yuana S, Lia M, Fanga Z, Liua Y, Shia W, Pana B, Wuc K, Shib J, Shena B, Shena Q (2016) Biological control of tobacco bacterial wilt using *Trichoderma harzianum* amended bioorganic fertilizer and the arbuscular mycorrhizal fungi *Glomus mosseae*. Biol Control 92:164–171

Zhang F, Zhua Z, Yanga X, Rana W, Shena Q (2013) *Trichoderma harzianum* T-E5 significantly affects cucumber root exudates and fungal community in the cucumber rhizosphere. Appl Soil Ecol 72:41–48

Chapter 6
Role of Secondary Metabolites from Plant Growth-Promoting Rhizobacteria in Combating Salinity Stress

Jitendra Mishra, Tahmish Fatima, and Naveen Kumar Arora

Abstract Increasing salinity and decreasing crop productivity are the two parallel problems, currently faced in agroecosystems across the globe. In the absence of proper remediation of salt stress in soil, the loss of productivity and quality has raised manifold. The saline-tolerant plant growth-promoting rhizobacteria (PGPR) are being realized as alleviators of salt stress. The capability of salt-tolerant PGPR in harsh environmental conditions could be beneficial for plant survival in saline soils. Secondary metabolites produced by salt-tolerant PGPR have also shown clear-cut role in improving plant physiological conditions under osmotic stress. Salt-tolerant PGPR and their metabolites can be the solution for increasing the productivity and remediation of saline soils in an eco-friendly manner. However, research on salt-tolerant PGPR and their metabolites is still in its primary phase and requires more global attention and application.

Keywords Salinity stress • Plant growth promoting rhizobacteria • Secondary metabolites • Biocontrol

6.1 Introduction

At all stages of growth, crops can face different abiotic and biotic stresses (Atkinson and Urwin 2012; Rejeb et al. 2014). At the global level, these stresses have been found to cause substantial loss in crop productivity and quality (Suzuki et al. 2014; Rana et al. 2016). Although, individually, several vital activities of plants are affected either by single or multiple stresses, impact of salinity is much more serious on overall physiology of the plant. Soil salinization is a global problem, and level of salinity is increasing in several parts of the world including the Mediterranean

---

J. Mishra • T. Fatima • N.K. Arora (✉)
Department of Environmental Microbiology, BBA University, Lucknow, Uttar Pradesh, India
e-mail: nkarora.bbau@gmail.com

D. Egamberdieva, P. Ahmad (eds.), *Plant Microbiome: Stress Response*,
Microorganisms for Sustainability 5, https://doi.org/10.1007/978-981-10-5514-0_6

Basin, Australia, Central and Southern Asia, the Middle East, Europe, and Northern Africa (Nedjimi et al. 2006; Yensen 2006; FAO 2008; Ladeiro 2012). FAO Land and Plant Nutrition Management Service (2008) stated that over 6%, which accounts for more than 800 million ha of land, is affected either by salinity or sodicity. Qadir et al. (2014) reported that about 20% of irrigated land is affected by salinity and is prone to crop loss and yield; this problem can increase further depending on the agricultural practices and irrigation methods by up to 50% (Egamberdieva and Lugtenberg 2014). According to Zhu et al. (2005), salinization of agricultural land causes revenue losses in terms of $11.4 billion and $1.2 billion in irrigated and non-irrigated areas, respectively. This clearly indicates the improper ways of irrigation and need for corrections.

The electrical conductivity (EC) of the soil solution is considered as the primary criterion for measuring salinity. The US Salinity Laboratory Staff marked that EC above 4 dS m^{-1} is saline (Rengasamy 2006). The accumulation of different salts in excessive amounts causes ion toxicity or ion imbalance in plants (Yang et al. 2009; Gupta and Huang 2014; Abbasi et al. 2016). Among the different ions, Na^+ shows maximum toxicity effect, and majority of salt-affected soils contain a high concentration of sodium salts (Tavakkoli et al. 2011; Khorasgani et al. 2017). However, defining salinity stress in terms of soil chemistry is somehow complicated that involves many soil chemical processes and also beyond the scope of this review (for this, one can see the review by Rengasamy 2016). Increasing salinity level in soil has a negative impact on plant health (Colla et al. 2006; Paul and Lade 2014). Some of the drastic aspects of increased salinity are osmotic imbalance, increased ethylene production, low moisture retention, generation of reactive oxygen species (ROS), plasmolysis and nutrient imbalance, reduced photosynthetic capability, impaired nitrogen fixation, stomatal closure, inhibition of seed germination, early desiccation of flowers and fruits, reduction of root and shoot length, etc. (Zahran 1991; Stepien and Klobus 2005; Egamberdieva 2009, 2011; Maheshwari et al. 2012; Arora et al. 2012; Tewari and Arora 2013; Kaya et al. 2013).

Reclamation of saline soil by physical and chemical methods has been carried out for decades (Oster 1993). The most commonly used physical methods involve plowing, subsoiling, sanding, and profile inversion (Raychev et al. 2001). Chemicals such as gypsum, calcium chloride, limestone, sulfuric acid, sulfur, and iron sulfate are also used to neutralize the effect of salt in soil (da Silveira et al. 2008). Amendments of saline soils by these methods are very expensive and unaffordable to farmers. Apart from this, chemical amenders are required in large quantities, and this causes an ecological imbalance in nature. Over the past 50 years, green revolution resulted in indiscriminate usage of chemical pesticides and fertilizers, with an aim to feed increasing population both in developing and developed countries. Since 1945, about 17% agricultural lands have lost their fertility, where the use of pesticides and fertilizers served as the important causative factor (Tilman et al. 2002). Pesticides and fertilizers lead to degradation of soil due to their bioaccumulation and increasing salinity of the soil (Abolfazl et al. 2009). Excessive use of pesticides has now become an important anthropogenic factor in increasing soil salinization. Pesticides also attain further stability under saline conditions and become even more non-degradable. For example, parathion is reported to be more stable under saline conditions as compared to non-saline soils (Reddy and Sethunathan 1985).

In the last few years, attempts have been made to combat salinity stress in plants by eco-friendly means. In this regard, using salt-tolerant crop cultivars and co-inoculation of seed or plant with salt-tolerant PGPR has shown success. There are evidences showing that plants forming an association with PGP microbes show better adaptation against salinity stress (Rodriguez and Redman 2008). PGPR particularly those isolated from saline soils show intimate association with the plant and play a significant role in stress alleviation. All over the world (under varying salinity levels), PGPR have been found to enhance the growth of diverse crops (Table 6.1). The rhizosphere region is a microhabitat for microorganisms, and in saline soils, the mutualistic relationship between plant and PGPR can be very helpful for both the partners helping each other in survival and providing nutrients, hormones, and moisture and combating opportunistic and obligate phytopathogens (Bais et al. 2006). However, in salt-affected soils, abundance, diversity, composition, and functions of microbes are highly affected, and only those that survive can overrule the negative impact of salt stress (Oren 2008). According to Omar et al. (1994), with an increase in the salinity level to above 5%, the total count of bacteria and *Actinobacteria* may reduce drastically. Hence, it is important to select salt-tolerant

Table 6.1 Examples of some salt-tolerant PGPR along with their salinity tolerance as reported from different countries

S. no.	Country	Salt-tolerant PGPR	Salt tolerance level (mM)	Isolation source	References
1.	Algeria	*B. licheniformis* RBA32	400	*Solanum tuberosum*	Nabti et al. (2013)
2.	Argentina	*Ochrobactrum* sp. MEP_33b	1000	*Zea mays*	Principe et al. (2007)
3.	Australia	*Rhizobium* sp.	400	*Acacia* sp.	Thrall et al. (2009)
4.	China	*P. protegens* KY4410	1199	Saline soil	Wang et al. (2015)
		Klebsiella sp.	1530	*Festuca arundinacea*	Liu et al. (2014)
5.	Colombia	*Azotobacter* sp.	200	Saline soil	Rojas-Tapias et al. (2012)
6.	Denmark	*Enterobacter* sp. MN17	400	*Chenopodium quinoa*	Yang et al. (2016)
		Bacillus sp.	2911	Saline soil	Nielsen et al. (1995)
7.	Egypt	*B. subtilis*	514	*Solanum lycopersicum* L.	Abeer et al. (2015)
8.	France	*Rhizobium meliloti*	500	*Hedysarum coronarium*	Talibart et al. (1994)
		Paenibacillus sp.	1000	Saline soil	Ali et al. (2012)

(continued)

Table 6.1 (continued)

S. no.	Country	Salt-tolerant PGPR	Salt tolerance level (mM)	Isolation source	References
9.	India	*P. fluorescens* PF17	600	*Helianthus annuus*	Tewari and Arora (2016)
		Mesorhizobium loti	700	*Acacia catechu*	Kumar et al. (1999)
		B. licheniformis	1000	*Suaeda fruticosa* rhizosphere	Goswami et al. (2014)
		Serratia marcescens	1027	*Capparis decidua*	Singh and Jha (2016)
		Azotobacter chroococum	1369	Saline soil	Chaudhary et al. (2013)
		Pseudomonas sp.	1541	*Arthrocnemum indicum*	Sharma et al. (2016)
		P. aeruginosa PF07	1600	*Helianthus annuus*	Tewari and Arora (2014b)
		B. pumilus	2000	*Mangifera indica* L.	Kannan et al. (2014)
		P. aeruginosa PF23	2000	Saline soil	Tewari and Arora (2014a)
		Zhihengliuella sp.	2568	*Salicornia brachiate*	Jha et al. (2011)
		Halobacillus sp.	3425	Salt lake	Ramadoss et al. (2013)
10.	Iran	*Pseudomonas fluorescens*	400	*Triticum aestivum* L.	Safari et al. (2016)
11.	Italy	*R. leguminosarum*	343	*Lathyrus annuus*	Moschetti et al. (2005)
12.	Japan	*Rhizobium fredii*	400	*Glycine max*	Fujihara and Yoneyama (1993)
13.	Morocco	*Burkholderia phymatum* GR01N	400	*Phaseolus vulgaris*	Talbi et al. (2013)
		Sinorhizobium medicae	513	*Medicago sativa* L.	Elboutahiri et al. (2010)
		Sinorhizobium sp. S3G	2100	*Trigonella foenum* graecum	Abdelmoumen and El Idrissi (2009)
14.	Thailand	*S. marcescens*	1712	*Gynura pseudochina*	Nakbanpote et al. (2014)
15.	Tunisia	*S. meliloti*	684	*M. truncatula*	Mrabet et al. (2011)
		Rhizobium sp.	1000	Saline soil	Trabelsi et al. (2009)
		Halomonas sp.	3425	*Salicornia* sp.	Mapelli et al. (2013)

(continued)

Table 6.1 (continued)

S. no.	Country	Salt-tolerant PGPR	Salt tolerance level (mM)	Isolation source	References
16.	Turkey	*Bacillus gibsonii*	200	Saline soil	Orhan (2016)
		B. gibsonii	200	Saline soil	Orhan (2016)
		Arthrobacter sp.	750	*Salsola grandis*	Kataoka et al. (2017)
		Kocuria erythromyxa	1712	Saline soil	Karlidag et al. (2013)
17.	USA	*R. leguminosarum bv phaseoli* USDA 2671	685	*Phaseolus vulgaris* (USDA)	Abdelmoumen et al. (1999)
		B. thuringiensis AZP2	2000	*Pinus ponderosa*	Timmusk et al. (2014)
18.	Uzbekistan	*P. chlororaphis*	856	Culture collection	Egamberdieva et al. (2015)
19.	Vietnam	*Pseudomonas* sp.	1712	*Ipomoea aquatica* L.	Trung et al. (2016)

PGPR for such soils and use them in combating the harmful effects of salinity on plants. Saline-tolerant PGPR have developed complex physiological and biochemical mechanisms which maintain their survival and multiplication in saline conditions (Omar et al. 2009; Vaishnav et al. 2016; Habib et al. 2016). Studies on salt-tolerant PGPR indicate that under saline conditions, these microbes accumulate various metabolites to protect themselves and even their mutualistic partners (Bharti et al. 2016). However, under high soil salinity, the functional mechanisms of salt-tolerant PGPR are not fully understood and need to be explored further. The present review discusses and throws light on how metabolites produced by salt-tolerant PGPR play an important role in helping the plant and bacteria to survive under stressful conditions and also enhancing the growth of salt-stressed crops.

6.2 Metabolites Produced by Salt-Tolerant PGPR Under Salinity Stress

PGPR have been classified as biofertilizers, phytostimulators, rhizomediators, and biopesticides, depending upon their adapted functional strategies under various physiological conditions (Somers et al. 2004; Antoun and Prévost 2005; Ahemad and Kibret 2014). To perform each of the above roles, PGPR produce a variety of primary or low molecular weight secondary metabolites (Table 6.2). In addition, PGPR also are involved in suppressing pathogenic fungi which gain opportunities to infect plants immuno-compromised under salinity stress (Haas and Defago 2005; Saravanakumar and Samiyappan 2007; Barriuso et al. 2008; Dimkpa et al. 2009). The secondary metabolites although have little or no importance in the primary metabolism are important for survival and protection of the cell, particularly in

Table 6.2 Secondary compounds or enzymes produced by salt-tolerant PGPR, involved in salinity tolerance

S. no.	Secondary compounds or enzymes	Salt-tolerant PGPR	References
1.	Osmolytes		
	Glycine betaine	*P. alcaligenes*	Jha et al. (2011)
		Azospirillum brasilense	Chowdhury et al. (2007)
		B. subtilis	Bremer and Kramer (2000)
	Proline	*Burkholderia*	Barka et al. (2006)
		Bacillus sp.	Sziderics et al. (2007)
		P. fluorescens	Metwali et al. (2015)
		Pseudomonas strains	Naz and Bano (2015)
		P. pseudoalcaligenes	Hanson and Nelson (1978)
		Azospirillum sp.	Zarea et al. (2012)
		Azospirillum sp.	Bashan (1999)
		Azospirillum sp.	Casanovas et al. (2003)
		Azospirillum sp.	Bashan and Holguin (1997)
		Oceanobacillus profundus Pmt2	Qurashi and Sabri (2011)
		Exiguobacterium oxidotolerans STR36	Bharti et al. (2014)
		Dietzia natronolimnaea STR1	Bharti et al. (2016)
		P. koreensis AK-1	Kasotia et al. (2016)
		Pseudomonas sp.	Bano and Fatima (2009)
	Soluble sugars	*P. mendocina*	Kohler et al. (2009)
		B. amyloliquefaciens SQR9	Chen et al. (2016)
		A. brasilense	Bacilio-Jimenez et al. (2001)
		Serratia marcescens CDP-13	Singh and Jha (2016)
		Bacillus strains	Nayer and Reza (2008)
		Rhizobium tropici	Figueiredo et al. (2008)
	Alanine, serine, threonine aspartic acid, and other amino acids	*P. fluorescens* MSP-393	Paul and Lade (2014)
		A. braziliense	Hamdia et al. (2004)
		A. lipoferum	Qudsaia et al. (2013)
	Polyamine and amide		
	Cadaverine	*A. brasilense*	Cassan et al. (2009)
	N-acetylglutaminylglutamine amide (NAGGA)	*P. putida*	Kets et al. (1996)
		P. syringae B728a	Kurz et al. (2010)

(continued)

Table 6.2 (continued)

S. no.	Secondary compounds or enzymes	Salt-tolerant PGPR	References
2.	Exopolysaccharides	*Rhizobium* sp.	Ahemad and Khan (2012b)
		Rhizobium sp. strain YAS34	Alami et al. (2000)
		Enterobacter cloacae P6	Mahmood et al. (2016)
		R. meliloti	Lloret et al. (1998)
		B. megaterium	Nadeem et al. (2016)
		Bacillus sp.	Ashraf et al. (2004, 2006)
		B. tequilensis S4	Rolli et al. (2014)
		B. circulans	Khodair et al. (2008)
		P. putida GAP-P45	Sandhya et al. (2009)
		Zhihengliuella alba	Siddikee et al. (2011)
		B. subtilis	Han and Lee (2005)
		Proteus penneri Pp1	Naseem and Bano (2014)
		Pseudomonas sp.	Singh et al. (1992)
		P. aeruginosa PF23	Tewari and Arora (2014a, b)
	Stress alleviating enzymes		
1.	ACC deaminase	*Pseudomonas* sp.	Nadeem et al. (2007)
		Achromobacter piechaudii	Mayak et al. (2004)
		P. fluorescens	Saravanakumar and Samiyappan (2007)
		Burkholderia phytofirmans	Akhtar et al. (2015)
		B. subtilis	Abeer et al. (2015)
		Enterobacter sp.	Habib et al. (2016)
		B. halodenitrificans	Ramadoss et al. (2013)
		B. licheniformis K11	Lim and Kim (2013)
		Acidovorax sp.	Esquivel-Cote et al. (2010)
		Rhizobium strain Mk3	Ahmad et al. (2013)
		Pseudomonas stutzeri C4	Tank and Saraf (2010)
		P. fluorescens	Zahir et al. (2009)
		Enterobacter sp. EJ01	Kim et al. (2014)
		S. quinivirans	Belimov et al. (2005)
		S. marcescens KiSII	George et al. (2013)
		Arthrobacter protophormiae	Barnawal et al. (2014)
		P. koreensis AK-1	Kasotia et al.(2016)
		P. fluorescens	Nadeem et al. (2016)
		Bradyrhizobium japonicum	Shaharoona et al. (2006)
	ROS scavenging enzymes		
1.	Superoxide dismutase	*Achromobacter xylosoxidans*	Karthikeyan et al. (2012)
		S. marcescens CDP-13	Singh and Jha (2016)
		P. putida H-2-3	Kang et al. (2014a)

(continued)

Table 6.2 (continued)

S. no.	Secondary compounds or enzymes	Salt-tolerant PGPR	References
2.	Ascorbate peroxidase	*B. lentus*	Golpayegani and Tilebeni (2011)
		B. megaterium	Habib et al. (2016)
		B. lentimorbus NRRL B-30488	Nautiyal et al. (2008)
		B. safensis	Chakraborty et al. (2013)
		Burkholderia cepacia	Kang et al. (2014b)
		Dietzia natronolimnaea STR1	Bharti et al. (2016)
		P. stutzeri MBE04	Sharma et al. (2016)
		Pseudomonas sp. MBE05	
3.	Catalase	*Enterobacter* sp.	Habib et al. (2016)
		Halomonas desiderata STR8	Bharti et al. (2014)
		P. mendocina	Kohler et al. (2010)
		P. putida AKMP7	Shaik et al. (2011)

adverse conditions. Perhaps the most important types of secondary metabolites reported by salt-tolerant PGPR are osmoprotectants (Han et al. 2014; Egamberdieva and Lugtenberg 2014). In the following sections, the role and applications of secondary metabolites involved in combating salt stress are discussed.

6.2.1 Osmoprotectants/Osmolytes/Compatible Solutes

Under saline conditions to defend the osmotic upshift and efflux of K^+ ions, salt-tolerant bacteria accumulate intracytoplasmic soluble sugars such as sucrose, trehalose (Breedveld et al. 1993), maltose, cellobiose, turanose, gentiobiose, and palatinose (Gouffi et al. 1999) or solutes like amino acids, e.g., glutamate, proline, alanine, serine, threonine, and aspartic acid; quaternary amines, e.g., glycine betaine and carnitine; imino acids (pipecolate); K^+; and tetrahydropyrimidines (ectoines). All these are known as compatible solutes (Wood et al. 2001; Alloing et al. 2006; Fernandez-Aunı́on et al. 2010; Qurashi and Sabri 2013; Tewari and Arora 2013; Orhan 2016). These help in maintaining the equilibrium across the membranes, reducing cell osmotic potential and maintaining high turgor pressure (Prado et al. 2000; Ashraf and Harris 2004; Chaum et al. 2004; De Lacerda et al. 2005), stabilizing the proteins and ensuring correct folding of polypeptides under denaturing conditions (Street et al. 2006; Paul and Lade 2014) at high salt concentration, and alleviating toxicity of NaCl (Kavi Kishor et al. 2005). Bacteria either synthesize these solutes de novo, like trehalose and sucrose (Miller and Wood 1996), or uptake it from the surrounding environment (Vriezen et al. 2007). Oren (1999) reported that uptake is preferable than synthesizing because the latter consumes more energy, and under stress environment, energy conservation is better for survival. This adaptive mechanism has been reported as stress mediated, i.e., the metabolism of accumulation slows down when the osmotic balance shifts towards the equilibrium. Talbi et al. (2013) demonstrated that *Burkholderia phymatum* GR01N under 200 mM NaCl accumulated trehalose, sugar alcohol mannitol, and alanine which were osmotically driven. Karunya and Reetha (2014) found that strains of *Azospirillum brasilense*, *Bacillus subtilis*, and *Pseudomonas fluorescens* showed a significant production of proline at 4% NaCl. Other workers have also discussed osmolyte production by salt-tolerant PGPR under stress conditions increasing their survival chances (Paul and Lade 2014; Chen et al. 2016; Nadeeem et al. 2016). Lópéz-Leal et al. (2014) suggested that *Rhizobium etli* uses *tre*YZ pathway to synthesize sugars such as trehalose and sucrose during osmotic stress. Talibart et al. (1994) concluded that glutamate, N-acetylglutaminylglutamine amide (NAGGN), and trehalose were major accumulates in *Rhizobium meliloti* under 0.5 M NaCl concentration maintaining osmotic balance.

Hyperionic and hyperosmotic environment under saline conditions also challenges the survival and productivity of plants impacting nodulation, photosynthesis, nitrogen fixation, seed germination, alteration of protein, lipid and energy metabolism (Parida and Das 2005; Dantas et al. 2005; Rabie et al. 2005; Golpayegani and Tilebeni 2011; Paul and Lade 2014), membrane destabilization, nutrient paucity (Hasegawa et al. 2000), increased desiccation, and defoliation (Shannon and Grieve 1999). Inoculation of salt-tolerant PGPR helps in mitigating the growth adversity due to stress by alleviating the toxicity of Na^+ and ion imbalance through enhanced level of osmoprotectant synthesis in plants (Azooz et al. 2004; Ramezani et al. 2011). Singh and Jha (2016) discussed that inoculation of wheat plant with *Serratia marcescens* CDP-13 under saline conditions modulated the plant concentration of proline and soluble sugars in comparison to uninoculated control plants. Rojas-Tapias et al. (2012) observed that upon inoculation with *Achromobacter chroococcum* C5 and C9, there was a significant decrease and increase of Na^+ and K^+ concentration, respectively, in *Zea mays* under stress conditions, which improved the K^+/Na^+ ratio, thereby reducing NaCl toxicity. Qurashi and Sabri (2011), in their experiments, found that there was 98% increase in accumulation of endogenous proline in *Lens esculenta* upon inoculation with *Oceanobacillus profundus* Ptm2 under saline conditions, which thereby oppressed NaCl toxicity and improved plant growth. Ghorai et al. (2015) remarked increased accumulation of free amino acids in groundnut when inoculated with *Pseudomonas aeruginosa* AMAAS57 under salinity. Fan et al. (2016) detected PGP-related gene *cod*A in *Arthrobacter* strain TF4 and TF7, which was responsible for synthesis of glycine betaine under saline conditions, resisting the abiotic stress in tomato. Aydin et al. (2012) summarized two- and four-fold higher accumulations of soluble sugars in *Rhizobium gallicum*—common bean symbionts under low and high osmotic stress, respectively. Arora et al. (2006) reported the synthesis of an active microbial polyester, PHB (poly-β-hydroxybutyrate) by *Sinorhizobium* JB 1 at 700 mM concentration of NaCl, which helped in maintaining the osmotic balance in rhizobia and also served as an energy and carbon source for the bacteria under salt stress. Maintaining the osmotic balance is the primary defense mechanism for microbes and plants surviving under saline conditions. Therefore, osmolytes act as first-line immunity in sheathing the toxicity of NaCl.

6.2.2 Polysaccharides

Under saline conditions, bacteria secrete polysaccharides to promote adherence to environmental surfaces and formation of organo-mineral sheath (biofilm), providing physical and functional protection against desiccating situation (Geddie and Sutherland 1993; Fernandez-Aunión et al. 2010; Awad et al. 2012) and escaping the constraints of high salinity (Chen et al. 2008; Vyrides and Stuckey 2009; Poli et al. 2010). Bacterial polysaccharides have been characterized as capsular polysaccharides (CPS), exopolysaccharides (EPS), lipopolysaccharides (LPS), and β-1,2-glycans (Ormeño-Orrillo et al. 2012). Among these, extracellular polysaccharides have been identified as important components in biofilm formation (Rudrappa et al. 2008; Koo et al. 2013) and most promising in alleviating salinity stress (Lloret et al. 1998; D'Haeze and Holsters 2004; Upadhyay et al. 2011). EPS-producing PGPR help in forming an association of roots and bacteria called rhizosheath, which acts as protective and active site for nutrient recycling, ion balancing, availing water to plants, monitoring cationic intake, and helping in maintaining the symbiotic relation between the bacteria and the plant, and in legumes this can even help in the process of nodulation (Czarnes et al. 2000; Ashraf et al. 2004; Sandhya et al. 2009; Awad et al. 2012; Bhargava et al. 2016). EPS have also been reported in forming soil aggregates improving vegetation, reducing evapotranspiration, and forming water micropores in silt soil. EPS prevent the contact between rhizobia and saline environment (Elsheikh and Wood 1990; Poli et al. 2010), protecting the nodules from oxygen toxicity and thereby maintaining proper activity of nitrogenase in nodules (Bhargava et al. 2016), increasing phosphate solubilization by legumes from organic and inorganic sources (Alikhani et al. 2006), protecting bacteria against plant antimicrobial compounds (Skorupska et al. 2006; Singh et al. 2015), and also neutralizing the harmful effects of ROS (reactive oxygen species) produced under salinity stress (Lehman and Long 2013; Karmakar et al. 2015). Qurashi and Sabri (2012) highlighted that under saline conditions, to overcome the competition of nutrients, bacteria retrogress from planktonic stage to sessile stage adhering to the solid surface. Under such circumstances, EPS provide them a watery layer mantling against the toxicity of salts. Immobilization of Na^+ ions is also facilitated by EPS to reduce the toxicity and osmotic unbalancing of salts (Grover et al. 2011; Dodd and

Perez-Alfocea 2012; Kumari and Khanna 2015; Ribeiro and Burkert 2016). EPS like biopolymers have also been regarded beneficial in ameliorating and improvising marginal and degraded wastelands (Letey 1994). Mahmood et al. (2016) confirmed that under saline conditions, inoculation of mung bean with EPS-producing *Enterobacter cloacae* P6 and *Bacillus drentensis* P16 increased nutrient and water availability to the crop due to its altered rhizosphere, correlating to the formation of biofilm on the root surface. Successful plant-associated biofilms are highly capable of resisting abiotic stress responses in plants and microbes supporting elevated growth, yield, and better crop quality as reported by many workers (Ramey et al. 2004; Saleh-Lakha and Glick 2006; Lugtenberg and Kamilova 2009). Ashraf et al. (2004) explained alleviation of salinity in plants by EPS-producing PGPR inoculation rendering reduced Na^+ uptake in roots and therewith its restricted transfer to leaves. Also their work suggested that due to the formation of plant microbe biofilm, higher root proportion of inoculated seedlings was insulated from NaCl toxicity; hence, there was a lesser apoplastic flow of sodium ions into stele (Kasotia et al. 2016). Study of Awad et al. (2012) implied that EPS-producing *Azotobacter chroococcum* inoculation resulted in better root system in maize, subsequently improving shoot growth under salt stress. Bezzate et al. (2000) confirmed that increased biofilm is correlated to increased EPS production by salt-tolerant PGPR attributing to better bacterial survival. Kasotia et al. (2016) demonstrated that EPS increased with elevation of NaCl concentration (up to 500 mM) in salt-tolerant *Pseudomonas* spp. Qurashi and Sabri (2013) also reported increment in EPS production up to 2.5 M NaCl in strains of *Halomonas variabilis* (HT1) and *Planococcus rifietoensis* (RT4). Vriezen et al. (2007) delineated that under NaCl toxicity, high molecular weight (HMW) succinoglycan production is elevated which increased survival of *S. meliloti* under desiccation. Recently, Kumari and Khanna (2015) stated that EPS production increases in saline conditions providing more desiccation resistance to both plant and microbe. In *S. meliloti*, two categories of polysaccharides have been reported, i.e., succinoglycans and galactoglucans (Reuber and Walker 1993). Similarly, *R. leguminosarum* showed production of surface polysaccharides, neutral glucomannan, and gel-forming polysaccharides (Laus et al. 2006). Tewari and Arora (2014a, b) concluded that EPS produced by salt-tolerant *P. aeruginosa* PF23 showed antifungal activity and was effective in biocontrol of *M. phaseolina* up to 500 mM NaCl in sunflower. Thus, EPS not only serve as important osmoprotectants but also help in biocontrol activity. Inoculation of plants with efficient EPS-producing PGPR can be a strategy to improvise the quality and yield of crops under saline conditions, and it can also be used to enhance the fertility of salt-affected marginal lands. EPS-amended bioformulations can also be a beneficial strategy in not only protecting the bacterial cells but also in maintaining shelf life of bacteria in inoculants (Arora and Mishra 2016).

6.2.3 ACC Deaminase

Ethylene stress is another stumbling block which alleviates plant growth by inhibiting root elongation (Glick 2005), nodulation (Ma et al. 2002; Middleton et al. 2007), defoliation, premature senescence (Lie et al. 2005; Kumari and Khanna 2015), and root growth (Madhaiyan et al. 2007) under salinity (Abeles et al. 1992; Bari and Jones 2009). Ahmad et al. (2011) described classical triple response as reliable marker or bioassay to illustrate the adverse effects of ethylene on plant growth under saline conditions. It includes stress responses of plant seedlings to accelerated level of salinity, i.e., stunted seedling length, increased shoot diameter, and directional alternation of growth (Neljubow 1901; Shaharoona et al. 2006; Ahmad et al. 2011). There are many reports showing that with increase in the level of ACC (1-aminocyclopropanecarboxylic acid: precursor of ethylene), classical triple response relatively increases in seedlings grown under salt stress (Shaharoona et al. 2006, 2007; Nadeem et al. 2009). Thus, alleviation of salinity stress demands lowering of ethylene level. To cope up with this, salt-tolerant PGPR initiate the production of ACC deaminase which acts as a sink to accumulating ethylene by taking up ACC and converting it into ammonia and α-ketoglutarate and also supplying nitrogen and energy as a result of degradation (Mayak et al. 1999, 2004; Tahir et al. 2006; Selvakumar et al. 2012; Egamberdieva and Lugtenberg 2014). This degradation is responsible for diluting the classical triple response of ethylene stress, resultantly promoting the growth of plants by increasing the seedling vigor index and yield. Furthermore, ACC deaminase producing PGPR enhance uptake of important nutrients like N, P, and K which correlatively increase K+ /Na^+ ratio in stress-mediated plants (Nadeem et al. 2009). Ethylene production by plants involves two peaks: first, small peak which is important for plant defense and second which is produced at high levels in deleterious conditions. ACC deaminase action involves reducing the second peak synthesis of ethylene, without affecting the former. ACC deaminase is already present in the plant system but in small quantity, and its alleviation demands the presence of an inducer, i.e., ACC (ethylene). Under salt stress, the concentration of ACC increases, correspondingly increasing the synthesis of ACC deaminase. This enhanced the level of ACC deaminase which significantly reduces the second peak of ethylene by 50–90% (Glick 2014). When ACC deaminase producing PGPR are inoculated to seeds, they get bound to the seed coat and lower the ethylene level at the site of initial root formation, thereby, ensuring the survival of seedlings and formation of longer roots (Penrose and Glick 2003). The effect of ACC deaminase activity of salt-tolerant PGPR on growth promotion and salt stress abatement has been studied for various crops like tomato, rice, pea, wheat, mung bean, groundnut, soybean, etc. (Mayak et al. 2004; Saravanakumar and Samiyappan 2007; Zahir et al. 2008; Bal et al. 2013; Rajput et al. 2013; Kim et al. 2014; Kasotia et al. 2016; Mahmood et al. 2016). Tank and Saraf (2010) concluded that PGPR showed ACC deaminase activity even at 6% NaCl and displayed increased root elongation of tomato plant as a result of decreased ethylene levels. Ahmad et al. (2013) opined that PGPR strains when grown on ACC showed varied cell density, depending on their efficacy of utilizing ACC as sole nitrogen source which was rather a parameter related to ACC deaminase production. Growth promotion of chickpea was also observed by Nascimento et al. (2012), through bacterial strain expressing exogenous ACC deaminase gene. Similar result was also reported by George et al. (2013) showing growth improvement of *Cocos nucifera* L. when inoculated with ACC deaminase producing *S. proteamaculans* and *S. quinivirans*. Cheng et al. (2007) showed that there was inhibition in root growth of plant inoculated with salt-tolerant bacteria which produced IAA without producing ACC deaminase, signifying the importance of ACC deaminase in retrogressing the concentration of elevated ACC under salinity stress. PGPR, viz., *P. syringae*, *Enterobacter aerogenes*, *P. fluorescens* (Mayak et al. 2004; Nadeem et al. 2007; Akhgar et al. 2014), and *B. mojavensis* (Pourbabaee et al. 2016), have been reported as efficient ACC deaminase producers under saline stress. Application of ACC deaminase producing bacteria to salt-stressed plants and their successful salinity alleviation have been extensively studied and reported. Still, work on ACC deaminase and other exogenous enzymes can be further extended. Loss of land due to salinity alarms to shift the paradigm in agriculture toward sustainability, and it can be best optimized with the use of such efficient PGPR.

6.2.4 Antioxidant System

Partial reduction of oxygen due to increased salinity leads to production of reactive oxygen species (ROS) such as superoxide radical (O_2^-), hydroxyl radical (OH^-), and hydrogen peroxide (H_2O_2) (Scandalios 2002). The main reason behind the ROS production involves over-reduction of photosynthetic electrons by reduced photosynthetic activity (Johnson et al. 2003; Hichem et al. 2009; Paul and Lade 2014). ROS increase oxidative damage in plant and microbial cells due to alteration in membranous protein lipids, nucleic acid disturbing metabolic enzyme activity, and cell homeostasis (Hong-bo et al. 2006; Mhadhbi et al. 2009; Sharma et al. 2012). In legumes, nodule is negatively affected by ROS, impairing its tissue integrity and function (Becana et al. 2000; Hernandez-Jimenez et al. 2002; Matamoros et al. 2003; Mhadhbi et al. 2011); there is slowing down of nitrogenase activity, decreasing nodule protein content and leghemoglobin (Mhadhbi et al. 2008, 2011). Functioning of nitrogenase is dependent on oxygen concentration (Puppo and Halliwell 1988; Aydi et al. 2004; Kratsch and Graves 2005). Oxygen transport in nodule is conducted by leghemoglobin contributing in the process of bacteroid and cell respiration. Due to impaired activity of leghemoglobin by ROS, oxygen conduction to nodules is adversely affected (Mhadhbi et al. 2009, 2015). To combat the detrimental effects of ROS in bacterial and plant cells, salt-tolerant PGPR adopt antioxidant systems to scavenge the oxidative radicals (Jebara et al. 2005, 2010; Wang et al. 2009; Amudha and Balasubramani 2011; Farrisi et al. 2013). PGPR enhance the production of antioxidant enzymes in plants to higher levels in comparison to untreated control plants (Nautiyal et al. 2008; Chakraborty et al. 2013).

Antioxidant enzymes include superoxide dismutase (SOD), guaiacol peroxidase, ascorbate peroxidase (APX), catalase (CAT), polyphenol oxidase, and glutathione reductase (GR) which are generated in secondary metabolic pathways (Ghoulam et al. 2002). CAT and APX enzymes detoxify the effects of hydrogen peroxidase by converting it into H_2O and O_2 and are crucial for ROS detoxification (Scandalios et al. 1997). Bharti et al. (2014) observed elevated level of catalase and APX activity in PGPR-inoculated plants at different salinity levels in comparison to uninoculated plants, thereby physiologically protecting the plant against oxidative damage and promoting plant growth. Kim et al. (2014) concluded that there was higher APX activity in *Enterobacter* sp. EJ01-treated *Arabidopsis* seedlings under salt shock in comparison to treatment-free seedlings, improving the salt tolerance level and physiological activity of plant. The work of Habib et al. (2016) showed higher production of APX and CAT by *B. megaterium* UPMR2 and *Enterobacter* sp. UPMR18 under saline conditions. Some other workers have also reported increased level of antioxidant enzymes under salt stress conditions (Vardharajula et al. 2011; Tewari and Arora 2013; Gururani et al. 2013; Bharti et al. 2014). Antioxidant enzymes need to be further studied particularly in relation to PGPR to note the difference in synthesis and molecular mechanism involved in their production under saline and non-saline conditions.

6.3 Plant Growth Promotion Under Salinity Stress

The plant growth-promoting effects of the salt-tolerant PGPR can be very crucial in enhancing productivity of saline soils. Salt-tolerant PGPR produce an array of secondary compounds including phytohormones, siderophores, and organic acids. These metabolites help in plant growth promotion by diverse ways.

6.3.1 Phytohormone Production

It has been noticed that salt toxicity reduces phytohormone levels in plants (Egamberdieva 2009, 2013; Alqarawi et al. 2014). Prevalence of salt-tolerant PGPR in rhizosphere of salt-affected plant assures proper growth via synthesizing and secreting plant growth regulators. According to Patten and Glick (1996), 80% of rhizosphere microbes isolated from various crops possess the ability to synthesize and release auxins as secondary metabolites. Indole-3-acetic acid (IAA) is a naturally occurring auxin to regulate several aspects of plant development including embial activity, abscission of leaves, and induction of flowering and fruiting (Zhao 2010: Ramalingam and In-Jung 2013). High salt stress causes modification in root structure by altering auxin accumulation and its redistribution (Wang et al. 2009). Isolation and exploration of salt-tolerant PGPR with the capability to synthesize phytohormones is a quest for various workers (Table 6.3). There are few studies which proved that VOCs emitted by *B. subtilis* GB03 can trigger phytohormone signaling including auxin, cytokinins, salicylic acid, and gibberellins in *Arabidopsis thaliana* (Ryu et al. 2004; Liu and Zhang 2015). Kang et al. (2014b) showed that saline-tolerant *Burkholderia cepacia* SE4, *Promicromonospora* sp. SE188, and *Acinetobacter calcoaceticus* SE370 strains reduced adverse effects of salinity and osmotic stress by secreting phytohormone gibberellin and antioxidants in *Cucumis sativus*. Production of exogenous phytohormones during abiotic stresses as a product of secondary metabolism has been reported by several workers (Hamayun et al. 2010; Iqbal and Ashraf 2013; Kang et al. 2014a, b). Recently, Orhan (2016) showed that under salt stress, IAA produced by halotolerant and halophilic PGPR increased the root and shoot length and total fresh weight of the wheat plants. Several workers have now reported halotolerant or halophilic PGPR which can protect the plant from the deleterious effects of high salinity (Goswami et al. 2014; Sharma et al. 2016; Hingole and Pathak 2016).

Table 6.3 Phytohormones produced by salt-tolerant PGPR and effect on plant growth under salinity stress

Phytohormones	Salt-tolerant PGPR	Crops	Plant growth parameters under salinity	References
Auxins	*P. fluorescens* Mk25	*Vigna radiata*	Enhanced total dry weight and salt tolerance index	Ahmad et al. (2013)
	Streptomyces sp.	*Triticum aestivum*	Increased plant growth	Sadeghi et al. (2012)
	P. putida Rs-198	*Gossypium hirsutum* L.	Increment in growth parameters and germination rate	Yao et al. (2010)
	P. aeruginosa T15	*Solanum lycopersicum*	Increased root and shoot length and number of leaves	Tank and Saraf (2010)
	Enterobacter sp. EJ01	*Arabidopsis*	Increased biomass	Kim et al. (2011, 2014)
	Klebsiella oxytoca Rs-5	*Gossypium hirsutum* L.	Increased growth promotion and salt tolerance	Liu et al. (2013a)
	Pseudomonas sp. KM3113	*Brassica napus*	Increase in root length	Wang et al. (2015)
	Planococcus rifietoensis SAL-15	*T. aestivum* L.	Increased plant height and biomass	Rajput et al. (2013)
	Halomonas sp.	*Salicornia*	Increased in plant growth parameters	Marasco et al. (2016)

(continued)

Table 6.3 (continued)

Phytohormones	Salt-tolerant PGPR	Crops	Plant growth parameters under salinity	References
	Zhihengliuella halotolerans	*T. aestivum*	Increased total weight and root and shoot length	Orhan (2016)
	Brachybacterium saurashtrense JG-06	*Arachis hypogea*	Increased root and shoot dry weight and length	Shukla et al. (2012)
	S. marcescens CDP-13	*Triticum aestivum* L.	Increased root length and chlorophyll b content	Singh and Jha (2016)
	B. drentensis P16	*Vigna radiata*	Increased shoot and root length and fresh weight	Mahmood et al. (2016)
	Achromobacter sp.	*Z. mays*	Increased root and shoot weight	Arruda et al. (2013)
	P. fluorescens biotype F	*Helianthus annuus*	Reduced Na^+ accumulation in leaves, root and shoot increase	Shilev et al. (2010)
	A. chroococcum C5	*Z. mays*	Increased shoot length and dry weight and increase in polyphenol content of leaves	Tapias et al. (2012)
	S. meliloti	*Medicago truncatula*	Increased plant growth parameters	Bianco and Defez (2009)
	Klebsiella sp. MBE02	*Arachis hypogea*	Increased shoot and root length and dry weight	Sharma et al. (2016)
	P. putida GR12–2	*Vigna radiata*	Enhanced root development	Mayak et al. (1999)
	P. fluorescens	*Cucumis sativus*	Increase in root and shoot length and total biomass of plant	Nadeem et al. (2016)
	E. hormaechei	*S. lycopersicum*	Increased stress tolerance and growth promotion	Egamberdieva et al. (2014)

(continued)

Table 6.3 (continued)

Phytohormones	Salt-tolerant PGPR	Crops	Plant growth parameters under salinity	References
Gibberellic acid	*Burkholderia cepacia* SE4	*Cucumis sativus* L.	Improved shoot and root growth	Kang et al. (2014a)
	P. putida H-2-3	*Glycine max*	Enhanced shoot length, plant fresh weight, and chlorophyll content	Kang et al. (2014b)
	Sphingomonas sp. *LK11*	*S. lycopersicum*	Increased plant growth parameters and biomass	Halo et al. (2015)
	P. fluorescens	*Raphanus sativus*	Increased fresh weight of root and leaves	Mohamed and Gomaa (2012)
	Azospirillum lipoferum	*Z. mays*	Increased plant growth parameters	Cohen et al. (2009)
Cytokinin	*B. subtilis*	*Lactuca sativa*	Increased shoot biomass under stressed condition	Arkhipova et al. (2007)
	B. subtilis	*Platycladus orientalis*	Increased root and shoot dry weight	Liu et al. (2013b)

6.3.2 Nitrogen (N_2) Fixation

The input of nitrogen by biological fixation is a very important activity in agricultural soils. Salt stress is found detrimental to the majority of N_2-fixing microbes. Salt stress affects legume-*Rhizobium* symbioses to a greater extent. High salt concentration inhibits bacterial colonization, root hair curling, and infection thread formation, reduction in respiration of the nodules, and production of cytosolic leghemoglobin protein (Zahran 1999). These symptoms also result in a reduction of dry weight and N content in the shoot (Zahran 1999). In saline soils, the association of salt-tolerant N_2-fixing bacteria with their host plants received attention and has been examined extensively in improving the fertility and productivity of low-N soils (Kumar et al. 1999). In this context, application of salt-tolerant PGPR in the form of bioinoculants has also proven very useful in legume crops. In a study, Garg and Singla (2005) showed that under salt stress, chickpea seeds inoculated with salt-tolerant *Mesorhizobium ciceri* increased nodule number and mass. Similarly, salt-tolerant strains of *R. leguminosarum* bv. *ciceri* isolated from wild chickpeas were found to increase dry weights of root and shoot, the root-to-shoot ratio (RSR), number and dry weights of nodules, chlorophyll, and N content of the chickpea plant at

50 and 100 mM NaCl concentrations (Hatice et al. 2010). In a study, Keneni et al. (2010) compared N_2 fixation potential of acid- and salt-tolerant native strains of *Rhizobium* with exotic strains, on faba beans. They found that native strains tolerated higher salt concentration (5% NaCl) and performed well, and apart from N_2 fixation, dry matter yield, nodulation, and nodule wet weight of faba bean were also enhanced. In a study, Arora et al. (2000) also tested potential of *Rhizobium meliloti* strain in salt tolerance and found that the strain was able to tolerate 850 mM NaCl concentration and enhanced the growth of *Mucuna pruriens*.

Free-living diazotrophs have also led to significant increase in total N input in agricultural soils. In a study, Zahran (1997) found that salt-tolerant *Azotobacter* can fix nitrogen at salinity level of 5–10% NaCl. Under salinity stress, Silini et al. (2016) determined the effect of salt-tolerant *A. chroococcum* AZ6 along with osmolytes on the growth parameters of durum wheat varieties, and they found that both in the presence or absence of osmolytes, *A. chroococcum* AZ6 reduced the effect of salt stress. Recently, Khalid et al. (2017) showed that under salinity conditions, *Azospirillum brasilense* enhanced growth attributes including shoot height, root length, fresh and dry weight, leaf area, and chlorophyll content in *Trifolium repens* in comparison to uninoculated controls. Salt-tolerant *Azotobacter* sp. and *Azospirillum* sp. can also be used as biofertilizers for agriculture in saline soils (Bapurao 2012; Akhter et al. 2012).

6.3.3 Mineral Uptake by Plants

Salinity impairs mineral uptake by plants. Salinity stress causes nutritional imbalance due to higher levels of Na^+/Ca^{2+}, Na^+/K^+, Na^+/Mg^{2+}, Cl^-/NO_3^-, and $Cl^-/H_2PO_4^-$, thus causing plant growth retardation (Grattan and Grieve 1999). It has been noticed that crop irrigation with saline water usually decreases infiltration that results in low uptake of plant available water and affects performance, survivability, germination, and emergence in plant (Chaichi et al. 2016). In the last few years, salt-tolerant strains of PGPR have been reported from extreme soils and tested for their capability to solubilize minerals (Srinivasan et al. 2012). In a study, Nautiyal et al. (2000) isolated salt-tolerant phosphate-solubilizing strains of PGPR showing phosphate solubilization in the presence of 10% salt, pH 12, and temperature up to 45 °C. Cherif-Silini et al. (2013) isolated phosphate-solubilizing salt-tolerant *Bacillus* strains from the rhizosphere of wheat. Salt stress is also known to decrease iron availability to plants (Turan et al. 2012). Pakroo and Kashirad (1981) showed that in saline conditions, Fe application increased uptake of other elements by roots and shoots. In microbial system, Fe uptake is driven by synthesis of low molecular weight compounds known as siderophores. Hence, it has been the emphasis of several workers to isolate salt-tolerant PGPR having siderophore production capability. In a study, Prabhavati and Mallaiah (2008) showed siderophore production by *Rhizobium* sp. up to 600 mM salt concentration. Tank and Saraf (2010) also reported PGPR strains showing siderophore production in saline conditions. However, further research is required to elucidate the role of PGPR in uptake of minerals, particularly under saline conditions.

6.4 Metabolites for Biocontrol

All over the world, phytopathogenic bacteria and fungi cause a significant loss in quantity and quality of economically important crops (Oerke et al. 1994; Vidhyasekaran 2002; Madden and Wheelis 2003). Apart from health hazards, the use of synthetic pesticides is not preferable due to ecological perspectives (Sheng et al. 2005). In salt-affected soils, sorption of pesticides decreases, and their uptake by plants increases (Kookana et al. 2014). It has also been found that increased soil salinity also favors survival of some phytopathogenic fungi. In a study, Daami-Remadi et al. (2009) observed that soil salinity (2–10 g of NaCl) increases the population of phytopathogen *Fusarium oxysporum* f. sp. *lycopersici* in the soil. Similarly, Goudarzi et al. (2011) also reported that by an increase of soil salinity levels by up to 1400 mg of NaCl kg^{-1}, shoot and root colonization by charcoal rot fungi, *Macrophomina phaseolina*, significantly increased. According to Besri (1993), in saline conditions, even pathogen-resistant varieties of plants may be invaded by more salt-resilient fungal pathogens such as *Fusarium* and *Verticillium*. Egamberdieva et al. (2011) and Goudarzi et al. (2011) also claimed that salt stress increases the susceptibility of plants toward various phytopathogens.

Under saline conditions, PGPR can improve plant health. There have been studies performed where salt-tolerant strains of PGPR are tested for their biocontrol activity against phytopathogenic fungi and bacteria under saline conditions (Kumar et al. 2005; Principe et al. 2007; Berg et al. 2013; Arora 2015). Some reports showed promising results and salt-tolerant strains were found to be very effective in controlling phytopathogens in saline soils (Yue et al. 2007; Arora et al. 2016). In a study, El-Sayed et al. (2014) isolated 66 rhizobacteria (*Bacillus*, *Enterobacter*, and *Pseudomonas*) associated with wild plants grown in arid soils and showed their biocontrol activity against *F. oxysporum* and *Sclerotinia sclerotiorum*. Wang et al. (2015) also observed that production of secondary metabolites with biocontrol activity was effective in salt-tolerant PGPR strains even under saline conditions.

Production of secondary metabolites has been the main mechanism of biocontrol by PGPR. Siderophores, organic acids, and antibiotics are main metabolites involved in biocontrol. However, only little attention has been paid to study the metabolite production by saline-tolerant PGPR (Mavrodi et al. 2012; Wang et al. 2015). Some reports claimed *Stenotrophomonas* as an efficient salt-tolerant PGPR (Messiha et al. 2007; Ryan et al. 2009; Carmody et al. 2011). *Stenotrophomonas* excretes volatile organic compounds (VOCs): β-phenylethanol and dodecanal which negatively influence the growth of phytopathogenic fungi (Kai et al. 2007). The members of fluorescent pseudomonad group are also known for the production of secondary metabolites valuable in biocontrol of bacteria, fungi, nematodes, and viruses (Maurhofer et al. 1992; Battu and Reddy 2009; Darabpour et al. 2010). Even in saline conditions, they may produce these metabolites in adequate quantity so as to inhibit the growth of phytopathogens (Mavrodi et al. 2012). In a study, Egamberdieva (2012) showed that under saline conditions, *P. chlororaphis* TSAU13 strain was able to control foot and root rot of cucumber and tomato caused by *F. solani*. The potential of soil bacilli to produce several antibiotics including bacteriocins and antibiotics has been recognized since more than 50 years now (Sansinenea and Oritz 2011). In a study, Sharma et al. (2015) screened salt-tolerant *Bacillus* from saline soils of eastern Indo-Gangetic plains of India and found that out of total, half showed 4% NaCl tolerance and also were able to produce volatile secondary metabolite HCN under salinity stress. Amaresan et al. (2016) isolated NaCl tolerating (up to 10%) *Bacillus* from Andaman and Nicobar Islands showing strong antagonistic activity against *Sclerotium rolfsii*. Tewari and Arora (2016) also showed that EPS-producing salt-tolerant PGPR may be very useful in biocontrol of phytopathogenic fungi in saline soils.

6.5 Future Direction

The role of secondary metabolites from salt-tolerant PGPR is generally overlooked. Salt-tolerant PGPR and their metabolites can have tremendous applications in the reclamation of salinity stress in salt-affected regions of the world. Conventional approaches of treating saline soils with physical and chemical methods are not up to mark. Bio-based approaches are gaining momentum for sustainable agriculture and preserving the soil and its useful flora and fauna. In this context, bioformulation or bioinoculants containing salt-tolerant PGPR can be very useful in remediation of saline and arid soils with very low productivities. Metabolites form PGPR which can play very important role in their survival can also be useful in protecting the symbiotic partner (plant) under stress conditions. PGPR can also be utilized for triggering the defense mechanisms of the plant under salinity stress or protecting it from stress-related metabolites such as ethylene and ROS. The use of EPS and osmoprotectants producing PGPR has been successful in lab and field trials. Metabolites such as EPS and osmoprotectants can also be used in the protection of PGPR in bioinoculants and even when introduced in the soils. Salt-tolerant PGPR, with multiple traits such as protection of plant from stress and phytopathogens, can be a reasonable and eco-friendly answer for improving the productivity of saline soils and bringing them back to normal. Metabolomic and molecular approaches can be used to further know the exact mechanisms and triggering of secondary metabolites in PGPR under the stress conditions so as to develop tailor-made formulations for saline soils. Till date we have only been able to explore and identify only a miniscule of secondary metabolites from bacteria and even lesser from PGPR. Secondary metabolites from salt-tolerant PGPR are needed to be explored further for utilizing them or their producers in overcoming the ever-increasing menace of salinity in the agricultural soils.

6.6 Conclusion

Currently, soil salinization is one of the major problems around the globe affecting agricultural productivity. Apart from primary salinization, secondary salinization caused by anthropogenic activities has a far greater effect on crop productivity and on food security too. Till date not many options have been successful for reclamation of saline soils. The use of saline-tolerant PGPR and their metabolites might play a very important role in increasing productivity of saline soils and eventually leading to their remediation. With the help of salt-tolerant PGPR, it is also possible to protect the crops (grown under salt stress) from phytopathogens. Research has shown that application of salt-tolerant PGPR in saline soils has proven to be an effective and sustainable approach. Hence, bioformulations can be developed using such strains or their metabolites for effective improvement of saline soils and making them green once again.

References

Abbasi H, Jamil M, Haq A, Ali S, Ahmad R, Malik Z (2016) Salt stress manifestation on plants, mechanism of salt tolerance and potassium role in alleviating it: a review. Žemdirbystė (Agriculture) 103(2):229–238

Abdelmoumen H, El Idrissi MM (2009) Germination, growth and nodulation of *Trigonella foenum graecum* (Fenu Greek) under salt stress. Afr J Biotechnol 8(11):2489–2496

Abdelmoumen HA, Filali-Maltout M, Neyra A, Belabed MM, Idrissi E (1999) Effect of high salts concentrations on the growth of rhizobia and responses to added osmotic. J Appl Microbiol 86(6):889–898

Abeer H, Abd Allah EF, Alqarawi AA, Al-Huqail AA, Alshalawi SRM, Wirth S, Dilfuza E (2015) Impact of plant growth promoting *Bacillus subtilis* on growth and physiological parameters of *Bassia indica* (Indian bassia) grown under salt stress. Pak J Bot 47(5):1735–1741

Abeles FB, Morgan PW, Saltveit ME Jr (1992) Ethylene in plant biology, 2nd edn. Academic, New York

Abolfazl A, Zabihi RH, Movafegh S, Hossein AAM (2009) The efficiency of plant growth promoting rhizobacteria (PGPR) on yield and yield components of two varieties of wheat in salinity conditions. Am Eurasian J Sustain Agric 3(4):824–828

Ahemad M, Khan MS (2012) Ecological assessment of biotoxicity of pesticides towards plant growth promoting activities of pea (*Pisum sativum*)-specific *Rhizobium* sp. strain MRP1. Emirates J Food Agric 24:334–343

Ahemad M, Kibret M (2014) Mechanisms and applications of plant growth promoting rhizobacteria: current perspective. J King Saud Univ Sci 26:1–20

Ahmad M, Zahir ZA, Asghar HN, Asghar M (2011) Inducing salt tolerance in mung bean through coinoculation with rhizobia and plant-growth-promoting rhizobacteria containing 1-aminocyclopropane-1-carboxylate deaminase. Can J Microbiol 57:578–589

Ahmad M, Zahir ZA, Nazli F, Akram F, Arshad M, Khalid M (2013) Effectiveness of halo-tolerant, auxin producing *Pseudomonas* and *Rhizobium* strains to improve osmotic stress tolerance in mung bean (*Vigna radiata* L.) Braz J Microbiol 44:1341–1348

Akhgar AR, Arzanlou M, Bakker PAHM, Hamidpour M (2014) Characterization of 1-aminocyclo propane-1-carboxylate (ACC) deaminase-containing *Pseudomonas* spp. in the rhizosphere of salt-stressed canola. Pedosphere 24:461–468

Akhtar SS, Andersen MN, Naveed M, Zahir Z, Liu F (2015) Interactive effect of biochar and plant growth-promoting bacterial endophytes on ameliorating salinity stress in maize. Funct Plant Biol 42:770–781

Akhter SM, Hossain SJ, Hossain SA, Datta RK (2012) Isolation and characterization of salinity tolerant *Azotobacter* sp. Green J Biol Sci 2(3):043–051

Alami Y, Achouak W, Marol C, Heulin T (2000) Rhizosphere soil aggregation and plant growth promotion of sunflowers by an exopolysaccharide-producing *Rhizobium* sp. strain isolated from sunflower roots. Appl Environ Microbiol 66:3393–3398

Ali S, Charles TC, Glick BR (2012) Delay of flower senescence by bacterial endophytes expressing 1-aminocyclopropane-1-carboxylate deaminase. J Appl Microbiol 113:1139–1144

Alikhani HA, Saleh-Rastin N, Antoun H (2006) Phosphate solubilization activity of rhizobia native to Iranian soils. Plant Soil 287:35–41

Alloing G, Travers I, Sagot B, Le Rudulier D, Dupont L (2006) Proline betaine uptake in *Sinorhizobium meliloti*: characterization of Prb, an Opp-Like ABC transporter regulated by both proline betaine and salinity stress. J Bacteriol 188:6308–6317

Alqarawi AA, Hashem A, Abd Allah EF, Alshahrani TS, Huqail AA (2014) Effect of salinity on moisture content, pigment system, and lipid composition in *Ephedra alata* Decne. Acta Biol Hung 65(1):61–71

Amaresan N, Kumar K, Madhuri K, Kannamma GU (2016) Isolation and characterization of salt tolerant plant growth promoting rhizobacteria from plants grown in tsunami affected regions of Andaman and Nicobar Islands. Geomicrobiol J 36(20):942–947

Amudha J, Balasubramani G (2011) Recent molecular advances to combat abiotic stress tolerance in crop plants. Biotechnol Mol Biol Rev 6(2):31–58

Antoun H, Prévost D (2005) Ecology of plant growth promoting rhizobacteria. In: Siddiqui ZA (ed) PGPR: biocontrol and biofertilization. Springer, Dordrecht, pp 1–38

Arkhipova TN, Prinsen E, Veselov SU, Martinenko EV, Melentiev AI, Kudoyarova GR (2007) Cytokinin producing bacteria enhance plant growth in drying soil. Plant Soil 292:305–315

Arora NK (ed) (2015) Plant microbes symbiosis: applied facets. Springer, India

Arora NK, Mishra J (2016) Prospecting the roles of metabolites and additives in future bioformulations for sustainable agriculture. Appl Soil Ecol 107:405–407

Arora N, Kumar V, Maheshwari D (2000) Isolation of both fast and slow growing rhizobia effectively nodulating a medicinal legume, *Mucuna pruriens*. Symbiosis 29(2):121–137

Arora NK, Singhal V, Maheshwari DK (2006) Salinity induced accumulation of poly-β- hydroxybutyrate in rhizobia indicating its role in cell protection. World J Microbiol Biotechnol 22:603–606

Arora NK, Tewari S, Singh S, Lal N, Maheshwari DK (2012) PGPR for protection of plant health under saline conditions. In: Maheshwari DK (ed) Bacteria in agrobiology: stress management. Springer, Berlin, pp 239–258

Arora NK, Verma M, Prakash J, Mishra J (2016) Regulation of biopesticides: global concerns and policies. In: Arora NK, Mehnaz S, Balestrini R (eds) Bioformulations: for sustainable agriculture. Springer, New Delhi. https://doi.org/10.1007/978-81-322-2779-3_16

Arruda L, Beneduzi A, Martins A, Lisboa B, Lopes C, Bertolo F, Passaglia LMP, Vargas LK (2013) Screening of rhizobacteria isolated from maize (*Zea mays* L.) in Rio Grande do Sul State (South Brazil) and analysis of their potential to improve plant growth. Agric Ecosyst Environ Appl Soil Ecol 63:15–22

Ashraf M, Harris PJC (2004) Potential biochemical indicators of salinity tolerance in plants. Plant Sci 166:3–16

Ashraf M, Hasnain S, Berge O, Mahmood T (2004) Inoculating wheat seedlings with exopolysaccharides producing bacteria restricts sodium uptake and stimulates plant growth under salt-stress. Biol Fertil Soils 40:157–162

Ashraf M, Hasnain S, Berge O (2006) Effect of exopolysaccharides producing bacterial inoculation on growth of roots of wheat (*Triticum aestivum* L.) plants grown in a salt-affected soil. Int J Environ Sci Technol 3:43–51

Atkinson NJ, Urwin PE (2012) The interaction of plant biotic and abiotic stresses: from genes to the field. J Exp Bot 63:3523–3543

Awad N, Turky AM, Attia M (2012) Ameliorate of environmental salt stress on the growth of *Zea mays* L. plants by exopolysaccharides producing bacteria. J Appl Sci Res 8:2033–2044

Aydi S, Drevon JJ, Abdelly C (2004) Effect of salinity on root–nodule conductance to the oxygen diffusion in the *Medicago truncatula–Sinorhizobium meliloti* symbiosis. Plant Physiol Biochem 42:833–840

Aydin A, Kant C, Turan M (2012) Humic acid application alleviate salinity stress of bean (*Phaseolus vulgaris* L.) plants decreasing membrane leakage. Afr J Agric Res 7:1073–1086

Azooz MM, Shaddad MA, Abdel-Latef AA (2004) Leaf growth and K^+ /Na^+ ratio as an indication of the salt tolerance of three sorghum cultivars grown under salinity stress and IAA treatment. Acta Agron Hung 52:287–296

Bacilio-Jimenez M, Aguilar-Flores S, del Valle MV, Prez A, Zepeda A, Zenteno E (2001) Endophytic bacteria in rice seeds inhibit early colonization of roots by *Azospirillum brasilense*. Soil Biol Biochem 33:167–172

Bais HP, Weir TL, Perry LG, Gilroy S, Vivanco JM (2006) The role of root exudates in rhizosphere interactions with plants and other organisms. Annu Rev Plant Biol 57(1):233–266

Bal HB, Nayak L, Das S, Adhya TK (2013) Isolation of ACC deaminase PGPR from rice rhizosphere and evaluating their plant growth promoting activity under salt stress. Plant Soil 366:93–105

Bano A, Fatima M (2009) Salt tolerance in *Zea mays* (L.) following inoculation with *Rhizobium* and *Pseudomonas*. Biol Fertil Soils 45:405–413

Bapurao ND (2012) Potential halotolerant *Azospirillum brasilense* N-30 as a biofertilizer for saline soils. Acta Biol Indica 1(2):198–206

Bari R, Jones JD (2009) Role of plant hormones in plant defence responses. Plant Mol Biol 69:473–488

Barka AE, Nowak J, Clement C (2006) Enhancement of chilling resistance of inoculated grapevine plantlets with a plant growth-promoting rhizobacterium, *Burkholderia phytofirmans* strain PsJN. Appl Environ Microbiol 72:7246–7252

Barnawal D, Bharti N, Maji D, Chanotiya CS, Kalra A (2014) ACC deaminase containing *Arthrobacter protophormiae* induces NaCl stress tolerance through reduced ACC oxidase activity and ethylene production resulting in improved nodulation and mycorrhization in *Pisum sativum*. J Plant Physiol 171:884–894

Barriuso J, Solano BR, Gutierrez Manero FJ (2008) Protection against pathogen and salt stress by four plant growth-promoting rhizobacteria isolated from *Pinus* sp. on *Arabidopsis thaliana*. Phytopathology 98:666–672

Bashan Y (1999) Interaction of *Azospirillum* spp. in soils: a review. Biol Fertil Soils 29:245–256

Bashan Y, Holguin G (1997) *Azospirillum*-plant relationships: environmental and physiological advances (1990–1996). Can J Microbiol 43:103–121

Battu PR, Reddy MS (2009) Isolation of secondary metabolites from *Pseudomonas fluorescens* and its characterization. Asian J Res Chem 2(10):26–29

Becana M, Dalton DA, Moran JF, Iturbe-Ormaetxe I, Matamoros MA, Rubio MC (2000) Reactive oxygen species and antioxidants in legume nodules. Physiol Plant 109:372–381

Belimov AA, Hontzeas N, Safronova VI, Demchinskaya SV, Piluzza G, Bullitta S, Glick BR (2005) Cadmium-tolerant plant growth-promoting bacteria associated with the roots of Indian mustard (*Brassica juncea* L. Czern.) Soil Biol Biochem 37:241–250

Berg G, Alavi M, Schmidt CS (2013) Biocontrol and osmoprotection for plants under salinated conditions. In: de Bruijn F (ed) Molecular microbial ecology of the rhizosphere, vol 2. Wiley, Hoboken, p 587

Besri M (1993) Effects of salinity on plant disease development. In: Lieth H, Al Masoom A (eds) Towards the rational use of high salinity tolerant plants, 2. Springer, Dordrecht, pp 67–74

Bezzate S, Aymerich S, Chambert R, Czarnes S, Berge O, Heulin T (2000) Disruption of the *Paenibacillus polymyxa* levansucrase gene impairs its ability to aggregate soil in the wheat rhizosphere. Environ Microbiol 2:333–342

Bhargava Y, Murthy JSR, Rajesh Kumar TV, Narayana Rao M (2016) Phenotypic, stress tolerance and plant growth promoting characteristics of rhizobial isolates from selected wild legumes of semiarid region, Tirupati, India. Adv Microbiol 6:1–12

Bharti N, Barnawal D, Awasthi A, Yadav A, Kalra A (2014) Plant growth promoting rhizobacteria alleviate salinity induced negative effects on growth, oil content and physiological status in *Mentha arvensis*. Acta Physiol Plant 36:45–60

Bharti N, Pandey SS, Barnawal D, Patel VK, Kalra A (2016) Plant growth promoting rhizobacteria *Dietzia natronolimnaea* modulates the expression of stress responsive genes providing protection of wheat from salinity stress. Sci Rep 6:34768

Bianco C, Defez R (2009) *Medicago truncatula* improves salt tolerance when nodulated by an indole-3-acetic acid over-producing *Sinorhizobium meliloti* strain. J Exp Bot 60(11):3097–3107

Breedveld MW, Dijkema C, Zevenhuizen L, Zehnder AJB (1993) Response of intracellular carbohydrates to a NaCl shock in *Rhizobium leguminosarum biovar trifolii* TA1 and *Rhizobium meliloti* SU47. J Gen Microbiol 139:3157–3163

Bremer E, Kramer R (2000) Coping with osmotic challenges: osmoregulation through accumulation and release of compatible solutes in bacteria. In: Storz G, Hengge-Aronis R (eds) Bacterial stress responses, vol 6. ASM Press, Washington, DC, pp 79–97. pp 501

Carmody LA, Spilker T, LiPuma JJ (2011) Reassessment of *Stenotrophomonas maltophilia* phenotype. J Clin Microbiol 49:1101–1103

Casanovas EM, Barassi CA, Andrade FH, Sueldo RJ (2003) *Azospirillum*- inoculated maize plant responses to irrigation restraints imposed during flowering. Cereal Res Commun 31:395–402

Cassán F, Perrig D, Sgroy V, Masciarelli O, Penna C, Luna V (2009) *Azospirillum brasilense* Az39 and *Bradyrhizobium japonicum* E109, inoculated singly or in combination, promote seed germination and early seedling growth in corn (*Zea mays* L.) and soybean (*Glycine max* L.) Eur J Soil Biol 45:28–35

Chaichi MR, Afshar RK, Saberi M, Falahtabar N (2016) Alleviation of salinity and drought stress in corn production using a non-ionic surfactant. J Anim Plant Sci 26(4):1042–1047

Chakraborty U, Chakraborty BN, Chakraborty AP, Dey PL (2013) Water stress amelioration and plant growth promotion in wheat plants by osmotic stress tolerant bacteria. World J Microbiol Biotechnol 29:789–803

Chaudhary D, Narula N, Sindhu SS, Behl RK (2013) Plant growth stimulation of wheat (*Triticum aestivum* L.) by inoculation of salinity tolerant *Azotobacter* strains. Physiol Mol Biol Plants 19(4):515–519

Chaum S, Kirdmanee C, Supaibulwatana K (2004) Biochemical and physiological responses of thai jasmine rice (*Oryza sativa* L. ssp. indica cv. KDML105) to salt stress. Sci Asia 30:247–253

Chen BD, Huang GB, Chen YL, Pei HD, Cui YL, Yue Y (2008) Effects of salt stress on root activity and seedling growth of oil-sunflower. Chin J Oil Crop Sci 30:327–330

Chen L, Liu Y, Wu G, Njeri KV, Shen Q, Zhang N, Zhang R (2016) Induced maize salt tolerance by rhizosphere inoculation of *Bacillus amyloliquefaciens* SQR9. Physiol Plant 158:34–44

Cheng Z, Park E, Glick BR (2007) 1-Aminocyclopropane-1-carboxylate deaminase from *Pseudomonas putida* UW4 facilitates the growth of canola in the presence of salt. Can J Microbiol 53(7):912–918

Cherif-Silini H, Silini A, Ghoul M, Yahiaoui B, Arif F (2013) Solubilization of phosphate by the *Bacillus* under salt stress and in the presence of osmoprotectant compounds. Afr J Microbiol Res 7:4562–4571

Chowdhury SP, Nagarajan T, Tripathi R, Mishra MN, Le Rudulier D, Tripathi AK (2007) Strain-specific salt tolerance and osmoregulatory mechanisms in *Azospirillum brasilense*. FEMS Microbiol Lett 267:72–79

Cohen AC, Travaglia CN, Bottini R, Piccoli PN (2009) Participation of abscisic acid and gibberellins produced by endophytic *Azospirillum* in the alleviation of drought effects in maize. Botany 87:455–462

Colla G, Rouphael Y, Cardarelli M, Rea E (2006) Effect of salinity on yield, fruit quality, leaf gas exchange, and mineral composition of grafted watermelon plants. Hortscience 41:622–627

Czarnes S, Hallett PD, Bengough AG, Young IM (2000) Root- and microbial-derived mucilages affect soil structure and water transport. Eur J Soil Sci 51:435–443

D'Haeze W, Holsters M (2004) Surface polysaccharides enable bacteria to evade plant immunity. Trends Microbiol 12:555–561

Da Silveira KR, Ribeiro MR, De Oliveira LB, Heck RJ, Da Silveira RR (2008) Gypsum-saturated water to reclaim alluvial saline sodic and sodic soils. Sci Agric 65(1):69–76

Daami-Remadi M, Ben Oun H, Souissi A, Mansour M, Jabnoun-Khiareddine H, Nasraoui B (2009) Effects of saline irrigation water on *Verticillium* wilt severity and tomato growth. Plant Stress 3:40–48

Dantas BF, Lucianade S, Carlos A (2005) Physiological response of cowpea seeds to salinity stress. J Am Soc Hortic Sci 27(1):144–148

Darabpour E, Motamedi H, Seyyed Nejad SM (2010) Antimicrobial properties of *Teucrium polium* against some clinical pathogens. Asian Pac J Trop Med 3:124–127

Dimkpa CO, Merten D, Svatos A, Büchel G, Kothe E (2009) Siderophores mediate reduced and increased uptake of cadmium by *Streptomyces tendae* F4 and sunflower (*Helianthus annuus*), respectively. J Appl Microbiol 107:1687–1696

Dodd IC, Pérez-Alfocea F (2012) Microbial amelioration of crop salinity stress. J Exp Bot 63:3415–3428

Egamberdieva D (2009) Alleviation of salt stress by plant growth regulators and IAA producing bacteria in wheat. Acta Physiol Plant 31:861–864

Egamberdieva D (2011) Survival of *Pseudomonas extremorientalis* TSAU20 and *P. Chlororaphis* TSAU13 in the rhizosphere of common bean (*Phaseolus vulgaris*) under saline conditions. Plant Soil Environ 57(3):122–127

Egamberdieva D (2012) *Pseudomonas chlororaphis*: a salt-tolerant bacterial inoculant for plant growth stimulation under saline soil conditions. Acta Physiol Plant 34(2):751–756

Egamberdieva D (2013) The role of phytohormone producing bacteria in alleviating salt stress in crop plants. In: Miransari M (ed) Biotechnological techniques of stress tolerance in plants. Stadium Press, Auburn, pp 21–39

Egamberdieva D, Lugtenberg B (2014) Use of plant growth-promoting rhizobacteria to alleviate salinity stress in plants. In: Miransari M (ed) Use of microbes for the alleviation of soil stresses, vol 1. Springer Science+Business Media, New York, pp 73–96

Egamberdieva D, Kucharova Z, Davranov K, Berg G, Makarova N, Azarova T, Chebotar V, Tikhonovich I, Kamilova F, Validov S, Lugtenberg B (2011) Bacteria able to control foot and root rot and to promote growth of cucumber in salinated soils. Biol Fertil Soils 47:197–205

Egamberdieva D, Botir H, Hashem A, Abd-Allah EF (2014) Characterization of salt tolerant *Enterobacter hormaechei* strain associated with tomato root grown in arid saline soil. J Pure Appl Microbiol 8(5):4231–4239

Egamberdieva D, Li L, Lindström K, Räsänen L (2015) A synergistic interaction between salt tolerant *Pseudomonas* and *Mesorhizobium* strains improves growth and symbiotic performance of liquorice (*Glycyrrhiza uralensis* Fish.) under salt stress. Appl Microbiol Biotechnol 100:2829–2841

Elboutahiri N, Thami-Alami I, Udupa SM (2010) Phenotypic and genetic diversity in *Sinorhizobium meliloti* and *S. medicae* from drought and salt affected regions of Morocco. BMC Microbiol 10:15

El-Sayed SW, Akhkha A, El-Naggar MY, Elbadry M (2014) In vitro antagonistic activity, plant growth promoting traits and phylogenetic affiliation of rhizobacteria associated with wild plants grown in arid soil. Front Microbiol 5:651. https://doi.org/10.3389/fmicb.2014.00651

Elsheikh EAE, Wood M (1990) Effect of salinity on growth, nodulation and nitrogen yield of chickpea (*Cicer arietinum* L.) J Exp Bot 41:1263–1269

Esquivel-Cote R, Ramirez-Gama RM, Tsuzuki-Reyes G, OrozcoSegovia A, Huante P (2010) *Azospirillum lipoferum* strain AZm5 containing 1-aminocyclopropane-1-carboxylic acid deaminase improves early growth of tomato seedlings under nitrogen deficiency. Plant Soil 337(1–2):65–75

Fan P, Chen D, He Y, Zhou Q, Tian Y, Gao L (2016) Alleviating salt stress in tomato seedlings using *Arthrobacter* and *Bacillus megaterium* isolated from the rhizosphere of wild plants grown on saline – alkaline lands. Int J Phytoremediation 18(11):1113–1121

FAO (2008) Food and agricluture organization. Land and plant nutrition management service. http://www.fao.org/ag/AGL/public.stm/

Farrisi M, Ghoulam C, Bouizgaren A (2013) Change in water deficit saturation and photosynthetic pigments of alfalfa populations under salinity and assessment of proline role in salt tolerance. Agric Sci Res J 3:29–35

Fernandez-Aunión C, Ben-Hamouda T, Iglesias-Guerra F, Argandona M, Reina-Bueno M, Nieto JJ, Aouani ME, Vargas C (2010) Biosynthesis of compatible solutes in rhizobial strains isolated from *Phaseolus vulgaris* nodules in Tunisian fields. BMC Microbiol 10(192):3–16

Figueiredo MVB, Burity HA, Martìnez CR, Chanway CP (2008) Alleviation of drought stress in the common bean (*Phaseolus vulgaris* L.) by co-inoculation with *Paenibacillus polymyxa* and *Rhizobium tropici*. Appl Soil Ecol 40:182–188

Fujihara S, Yoneyama T (1993) Effects of pH and osmotic stress on cellular polyamine contents in the soybean *Rhizobia Rhizobium fredii* P220 and *Bradyrhizobium japonicum* A1017. Appl Environ Microbiol 59(4):1104–1109

Garg N, Sinhgla R (2005) Influence of salinity on growth and yield attributes in chickpea cultivars. Turk J Agric For 29:231–235

Geddie JL, Sutherland IW (1993) Uptake of metals by bacterial polysaccharides. J Appl Bacteriol 74(4):467–472

George P, Gupta A, Gopal M, Thomas L, Thomas GV (2013) Multifarious beneficial traits and plant growth promoting potential of *Serratia marcescens* KiSII and *Enterobacter* sp. RNF 267 isolated from the rhizosphere of coconut palms (*Cocos nucifera* L.) World J Microbiol Biotechnol 29:109–117

Ghorai S, Pal KK, Dey R (2015) Alleviation of salinity stress in groundnut by application of PGPR. Int Res J Eng Technol 2:742–750

Ghoulam C, Foursy A, Fares K (2002) Effects of salt stress on growth, inorganic ions and proline accumulation in relation to osmotic adjustment in five sugar beet cultivars. Environ Exp Bot 47:39–50

Glick BR (2005) Modulation of plant ethylene levels by the bacterial enzyme ACC deaminase. FEMS Microbiol Lett 251:1–7

Glick BR (2014) Bacteria with ACC deaminase can promote plant growth and help to feed the world. Microbiol Res 169:30–39

Golpayegani A, Tilebeni HG (2011) Effect of biological fertilizers on biochemical and physiological parameters of Basil (*Ociumum basilicm* L.) medicine plant. Am–Eur J Agric Environ Sci 11(3):411–416

Goswami D, Dhandhukia P, Patel P, Thakker JN (2014) Screening of PGPR from saline desert of Kutch: Growth promotion in *Arachis hypogea* by *Bacillus licheniformis* A2. Microbiol Res 169:66–75

Goudarzi S, Banihashemi Z, Maftoun M (2011) Effect of salt and water stress on root infection by *Macrophomina phaseolina* and ion composition in shoot in sorghum. Iran J Plant Pathol 47(3):69–83

Gouffi K, Pica N, Pichereau V, Blanco B (1999) Disaccharides as a new class of non-accumulated osmoprotectants for *Sinorhizobium meliloti*. Appl Environ Microbiol 65:1491–1500

Grattan SR, Grieve CM (1999) Mineral nutrient acquisition and response by plants grown in saline environments. In: Pessarakli M (ed) Handbook of plant and crop stress. Marcel Dekker, New York, pp 203–229

Grover M, Ali SKZ, Sandhya V, Rasul A, Venkateswarlu B (2011) Role of microorganisms in adaptation of agriculture crops to abiotic stresses. World J Microbiol Biotechnol 27:1231–1240

Gupta B, Huang B (2014) Mechanism of salinity tolerance in plants: physiological, biochemical, and molecular characterization. Int J Genom 2014:18

Gururani MA, Upadhyaya CP, Baskar V, Venkatesh J, Nookaraju A, Park SW (2013) Plant growth-promoting rhizobacteria enhance abiotic stress tolerance in *Solanum tuberosum* through inducing changes in the expression of ROS-scavenging enzymes and improved photosynthetic performance. J Plant Growth Regul 32:245–258

Haas D, Défago G (2005) Biological control of soil-borne pathogens by fluorescent pseudomonads. Nat Rev Microbiol 3:307–319

Habib SH, Kausar H, Saud HM, Ismail MR, Othman R (2016) Molecular characterization of stress tolerant plant growth promoting rhizobacteria (PGPR) for growth enhancement of rice. Int J Agric Biol. https://doi.org/10.1155/2016/6284547

Halo BA, Khan AL, Waqas M, Al-Harrasi A, Hussain J, Ali L, Adnan M, Lee IJ (2015) Endophytic bacteria (*Sphingomonas* sp. LK11) and Gibberellin can improve *Solanum lycopersicum* growth and oxidative stress under salinity. J Plant Interact 10:117–125

Hamayun M, Khan SA, Khan AL, Rehman G, Kim Y, Iqbal I, Hussain J, Sohn E, Lee IJ (2010) Gibberellins production and plant growth promotion from pure cultures of *Cladosporium* sp. MH-6 isolated from cucumber (*Cucumis sativus*. L.) Mycologia 102:989–995

Hamdia MBE, Shaddad MAK, Doaa MM (2004) Mechanisms of salt tolerance and interactive effects of *Azospirillum brasilense* inoculation on maize cultivars grown under salt stress conditions. Plant Growth Regul 44(2):165–174

Han HS, Lee KD (2005) Physiological responses of soybean inoculation of *Bradyrhizobium japonicum* with PGPR in saline soil conditions. Res J Agric Biol Sci 1:216–221

Han QQ, Lü XP, Bai JP, Qiao Y, Paré PW, Wang SM, Zhang JL, Wu YN, Pang XP, Xu WB, Wang ZL (2014) Beneficial soil bacterium *Bacillus subtilis* (GB03) augments salt tolerance of white clover. Front Plant Sci 5:525

Hanson AD, Nelson CE (1978) Betaine accumulation and (^{14}C) formate metabolism in water stressed barley leaves. Plant Physiol 62:305–312

Hasegawa PM, Bressan RA, Zhu JK, Bohnert HJ (2000) Plant cellular and molecular responses to high salinity. Annu Rev Plant Physiol Plant Mol Biol 51:463–499

Hatice Ö, Caner K, Erdal E (2010) Effect of *Rhizobium* strains isolated from wild chickpea on the growth and symbiotic performance of chickpea (*Cicer arietinum* L.) under salt stress. Turk J Agric For 34:361–371

Hernández-Jiménez MJ, Lucas MM, de Felipe MR (2002) Antioxidant defense and damage in senescing lupin nodules. Plant Physiol Biochem 40:645–657

Hichem H, Mounir D, Naceur EA (2009) Differential responses of two maize (*Zea mays* L.) varieties to salt stress: Changes on polyphenols composition of foliage and oxidative damages. Ind Crop Prod 30:144–151

Hingole SS, Pathak AP (2016) Saline soil microbiome: A rich source of halotolerant PGPR. J Crop Sci Biotechnol 19:231–239

HongBo S, ZongSuo L, MingAn S (2006) Osmotic regulation of 10 wheat (*Triticum aestivum* L.) genotypes at soil water deficits. Colloids Surf B Biointerfaces 47:132–139

Iqbal M, Ashraf M (2013) Gibberellic acid mediated induction of salt tolerance in wheat plants: growth, ionic partitioning, photosynthesis, yield and hormonal homeostasis. Environ Exp Bot 86:76–85

Jebara M, Mohamed EA, Payre H, Jean-Jacques D (2005) Nodule conductance varied among common bean (*Phaseolus vulgaris* L.) genotypes under phosphorus deficiency. J Plant Physiol 162:309–315

Jebara S, Drevon JJ, Jebara M (2010) Modulation of symbiotic efficiency and nodular antioxidant enzyme activities in two *Phaseolus vulgaris* genotypes under salinity. Acta Physiol Plant 32:925

Jha Y, Subramanian RB, Patel S (2011) Combination of endophytic and rhizospheric plant growth promoting rhizobacteria in *Oryza sativa* shows higher accumulation of osmoprotectant against saline stress. Acta Physiol Plant 33:797–802

Johnson C, Boden E, Arias J (2003) Salicylic acid and NPR1 induce the recruitment of trans-activating TGA factors to a defense gene promoter in *Arabidopsis*. Plant Cell 15:1846–1858

Kai M, Effmert U, Berg G, Piechulla B (2007) Volatiles of bacterial antagonists inhibit mycelial growth of the plant pathogen *Rhizoctonia solani*. Arch Microbiol 187:351–360

Kannan R, Damodaran T, Pandey BK, Umamaheswari S, Rai RB, Jha SK, Mishra VK, Sharma DK, Sah V (2014) Isolation and characterization of endophytic plant growth-promoting bacteria (PGPB) associated to the sodicity tolerant polyembryonic mango (*Mangifera indica* L.) root stock and growth vigour in rice under saline sodic environment. Afr J Microbiol Res 8(7):628–636

Kang SM, Radhakrishnan R, Khan AL, Kim MJ, Park JM, Kim BR (2014a) Gibberellin secreting rhizobacterium, Pseudomonas putida H-2-3 modulates the hormonal and stress physiology of soybean to improve the plant growth under saline and drought conditions. Plant Physiol Biochem 84:115–124

Kang SM, Khan AL, Waqas M, You YH, Kim JH, Kim JG, Hamayun M, Lee IJ (2014b) Plant growth-promoting rhizobacteria reduce adverse effects of salinity and osmotic stress by regulating phytohormones and antioxidants in Cucumis sativus. J Plant Interact 9:673–682

Karlidag H, Yildirim E, Turan M, Pehluvan M, Donmez F (2013) Plant growth-promoting rhizobacteria mitigate deleterious effects of salt stress on strawberry plants (*Fragaria x ananassa*). Hortscience 48(5):563–567

Karmakar K, Rana A, Rajwar A, Sahgal M, Johri BN (2015) Legume-rhizobia symbiosis under stress. In: Arora NK (ed) Plant microbes Symbiosis: applied facets. Springer, New Delhi, pp 241–258

Karthikeyan B, Joe MM, Islam MR, Sa T (2012) ACC deaminase containing diazotrophic endophytic bacteria ameliorate salt stress in *Catharanthus roseus* through reduced ethylene levels and induction of antioxidative defense systems. Symbiosis 56:77–86

Karunya SK, Reetha D (2014) Screening of plant growth promoting rhizobacteria for ACC deaminase activity. Int J Curr Res Chem Pharma Sci 1(3):65–70

Kasotia A, Varma A, Tuteja N, Choudhary DK (2016) Amelioration of soybean plant from saline-induced condition by exopolysaccharide producing *Pseudomonas*-mediated expression of high affinity K^+ transporter (HKT1) gene. Curr Sci 111(12):25

Kataoka R, Güneri E, Turgay OC, Yaprak AE, Sevilir B, Başköse I (2017) Sodium-resistant plant growth-promoting rhizobacteria isolated from a halophyte, *Salsola grandis*, in saline-alkaline soils of Turkey. Eur J Soil Sci 6(3):216–225

Kavi Kishor PB, Sangam S, Amrutha RN, Sri Laxmi P, Naidu KR, Rao KR (2005) Regulation of proline biosynthesis, degradation, uptake and transport in higher plants: its implications in plant growth and abiotic stress tolerance. Curr Sci 88:424–438

Kaya C, Ashraf M, Dikilitas M, Atilla L (2013) Alleviation of salt stress-induced adverse effects on maize plants by exogenous application of indoleacetic acid (IAA) and inorganic nutrients – a field trial. Aust J Crop Sci 7(2):249–254

Keneni A, Assefa F, Prabu PC (2010) Characterization of acid and salt tolerant rhizobial strains isolated from faba bean fields of Wollo, northern Ethiopia. J Agric Sci Technol 12:365–376

Kets EPW, de Bont JAM, Heipieper HJ (1996) Physiological response of *Pseudomonas putida* S12 subjected to reduced water activity. FEMS Microbiol Lett 139:133–137

Khalid M, Bilal M, Hassani D, Iqbal HMN, Wang H, Huang D (2017) Mitigation of salt stress in white clover (*Trifolium repens*) by *Azospirillum brasilense* and its inoculation effect. Bot Stud 58:5

Khodair TA, Galal GF, El-Tayeb TS (2008) Effect of inoculating wheat seedlings with exopolysaccharide- producing bacteria in saline soil. J Appl Sci Res 4:2065–2070
Khorasgani OA, Mortazaeinezhad F, Rafiee P (2017) Variations in the growth, oil quantity and quality and mineral nutrients of chamomile genotypes under salinity stress. J Cent Eur Agric 18(1):150–169
Kim YC, Leveau J, McSpadden Gardener BB, Pierson EA, Pierson LS, Ryu CM (2011) The multifactorial basis for plant health promotion by plant-associated bacteria. Appl Environ Microbiol 77:1548–1555
Kim K, Jang YJ, Lee SM, Oh BT, Chae JC, Lee KJ (2014) Alleviation of salt stress by *Enterobacter* sp. EJ01 in tomato and *Arabidopsis* is accompanied by up-regulation of conserved salinity responsive factors in plants. Mol Cell 37:109–117
Kohler J, Hernández JA, Caravaca F, Roldána A (2009) Induction of antioxidant enzymes is involved in the greater effectiveness of a PGPR versus AM fungi with respect to increasing the tolerance of lettuce to severe salt stress. Environ Exp Bot 65:245–252
Kohler J, Caravaca F, Roldan A (2010) An AM fungus and a PGPR intensify the adverse effects of salinity on the stability of rhizosphere soil aggregates of *Lactuca sativa*. Soil Biol Biochem 42:429–434
Koo H, Falsetta ML, Klein MI (2013) The exopolysaccharide matrix: a virulence determinant of cariogenic biofilm. J Dent Res 92(12):1065–1073
Kookana RS, Boxall ABA, Reeves PT, Ashauer R, Beulke S, Chaudhry Q, Cornelis G, Fernandes TF, Gan J, Kah M, Lynch I, Ranville J, Sinclair C, Spurgeon G, Tiede K, Brink VDPJ (2014) Nanopesticides: Guiding principles for regulatory evaluation of environmental risks. J Agric Food Chem 62:4227–4240
Kratsch HA, Graves WR (2005) Oxygen concentration affects nodule anatomy and nitrogenase activity of *Alnus maritime*. Plant Cell Environ 28:688–696
Kumar H, Arora NK, Kumar V, Maheshwari DK (1999) Isolation, characterization and selection of salt tolerant rhizobia nodulating *Acacia catechu* and *A nilotica*. Symbiosis 26:279–288
Kumar R, Goyal V, Kuhad MS (2005) Influence of fertility–salinity interactions on growth, water status and yield of Indian mustard (*Brassica juncea*). Indian J Plant Physiol 10:139–144
Kumari P, Khanna V (2015) ACC-deaminase and EPS production by salt tolerant rhizobacteria augment growth in chickpea under salinity stress. Int J Bio-Resour Stress Manag 6(5):558–565
Kurz M, Burch AY, Seip B, Lindow SE, Gross H (2010) Genome-driven investigation of compatible solute biosynthesis pathways of *Pseudomonas syringae pv. syringae* and their contribution to water stress tolerance. Appl Environ Microbiol 76(16):5452–5462
Lacerda De CF, Cambraia J, Oliva MA, Ruiz HA (2005) Changes in growth and in solute concentrations in sorghum leaves and roots during salt stress recovery. Environ Exp Bot 54:69–76
Ladeiro B (2012) Saline agriculture in the 21st century: using salt contaminated resources to cope food requirements. J Bot 2012(310705):7. https://doi.org/10.1155/2012/310705
Laus MC, Logman TJ, Lamers GE, Van Brussel AAN, Carlson RW, Kijne JW (2006) A novel polar surface polysaccharide from *Rhizobium leguminosarum* binds host plant lectin. Mol Microbiol 59:1704–1713
Lehman AP, Long SR (2013) Exopolysaccharides from *Sinorhizobium meliloti* can protect against H_2O_2-dependent damage. J Bacteriol 195:5362–5369
Letey J (1994) Adsorption and desorption of polymers on soil. Soil Sci 158:244–248
Lie Q, Saleh-lakha S, Glick BR (2005) The effect of native and ACC deaminase-containing *Azospirillum brasilense* Cdl843 on the rooting of carnation cuttings. Can J Microbiol 51:511–514
Lim SJ, Kim SD (2013) Induction of drought stress resistance by multi-functional PGPR *Bacillus licheniformis* K11 in pepper. Plant Pathol J 29(2):201–208
Liu XM, Zhnag H (2015) The effects of bacterial volatile emissions on plant abiotic stress tolerance. Front Plant Sci 6:774
Liu Y, Shi ZQ, Yao LX, Yue HT, Li H, Li C (2013a) Effect of IAA produced by *Klebsiella oxytoca* Rs-5 on cotton growth under salt stress. J Gen Appl Microbiol 59:59–65

Liu F, Xing S, Ma H (2013b) Cytokinin-producing, plant growth-promoting rhizobacteria that confer resistance to drought stress in *Platycladus orientalis* container seedlings. Appl Microbiol Biotechnol 97:9155

Liu W, Hou J, Wang Q, Dinga L, Luo Y (2014) Isolation and characterization of plant growth-promoting rhizobacteria and their effects on phytoremediation of petroleum-contaminated saline-alkali soil. Chemosphere 117C(1):303–308

Lloret J, Wulff BBH, Rubio JM, Downie JA, Bonilla I, Rivilla R (1998) Exopolysaccharide II production is regulated by salt in the halotolerant strain *Rhizobium meliloti* EFB1. Appl Environ Microbiol 64:1024–1028

López-Leal G, Tabche ML, Castillo-Ramírez S, Mendoza-Vargas A, Ramírez-Romero MA, Dávila G (2014) RNA-seq analysis of the multipartite genome of *Rhizobium etli* CE3 shows different replicon contributions under heat and saline shock. BMC Genomics 15:770

Lugtenberg B, Kamilova F (2009) Plant-growth-promoting rhizobacteria. Annu Rev Microbiol 63:541–556

Ma W, Sebestianova SB, Sebestian J, Burd GI, Guinel FC, Glick BR (2002) Prevalence of 1-aminocyclopropane-1- carboxylate deaminase in rhizobia spp. Antonie van Leeuwenhoek, in press

Madden LV, Wheelis M (2003) The threat of plant pathogens as weapons against U.S. corps. Annu Rev Phytopathol 41:155–176

Madhaiyan M, Poonguzhali S, Sa T (2007) Characterization of 1-aminocyclopropane-1-carboxylate (ACC) deaminase containing *Methylobacterium oryzae* and interactions with auxins and ACC regulation of ethylene in canola (*Brassica campestris*). Planta 226:867–876

Maheshwari DK, Dubey RC, Aeron A, Kumar B, Kumar S, Tewari S, Arora NK (2012) Integrated approach for disease management and growth enhancement of *Sesamum indicum* L. utilizing *Azotobacter chroococcum* TRA2 and chemical fertilizer. World J Microbiol Biotechnol 28(10):3015–3024

Mahmood S, Daur I, Al-Solaimani SG, Ahmad S, Madkour MH, Yasir M, Hirt H, Ali S, Ali Z (2016) Plant growth promoting rhizobacteria and silicon synergistically enhance salinity tolerance of mung bean. Front Plant Sci 7:876

Mapelli F, Marasco R, Rolli E, Barbato M, Cherif H, Guesmi A, Ouzari I, Daffonchio D, Borin S (2013) Potential for plant growth promotion of rhizobacteria associated with Salicornia growing in Tunisian hypersaline soils. Biomed Res Int 2013(248078):13. https://doi.org/10.1155/2013/248078

Marasco R, Mapelli F, Rolli E, Mosqueira MJ, Fusi M, Bariselli P, Reddy M, Cherif A, Tsiamis G, Borin S, Daffonchio D (2016) *Salicornia strobilacea* (synonym of *Halocnemum strobilaceum*) grown under different tidal regimes selects rhizosphere bacteria capable of promoting plant growth. Front Microbiol 7:1286

Matamoros MA, Dalton DA, Ramos J, Clemente MR, Rubio MC, Becana M (2003) Biochemistry and molecular biology of antioxidants in the rhizobia–legume symbiosis. Plant Physiol 133:499–509

Maurhofer M, Keel C, Schnider U, Voisard C, Haas D, Défago G (1992) Influence of enhanced antibiotic production in *Pseudomonas fluorescens* strain CHA0 on its disease suppressive capacity. Phytopathology 82:190–195

Mavrodi DV, Mavrodi OV, Parejko JA, Bonsall RF, Kwak YS, Paulitz TC, Thomashow LS, Weller DM (2012) Accumulation of the antibiotic phenazine-1-carboxylic acid in the rhizosphere of dryland cereals. Appl Environ Microbiol 78:804–812

Mayak S, Tirosh T, Glick BR (1999) Effect of wild-type and mutant plant growth-promoting rhizobacteria on the rooting of mung bean cuttings. J Plant Growth Regul 18:49–53

Mayak S, Tirosh T, Glick BR (2004) Plant growth promoting bacteria confer resistance in tomato plants to salt stress. Plant Physiol Biochem 42:565–572

Messiha NAS, van Diepeningen AD, Farag NS, Abdallah SA, Janse JD, van Bruggen AHC (2007) *Stenotrophomonas maltophilia*: a new potential biocontrol agent of *Ralstonia solanacearum*, causal agent of potato brown rot. Eur J Plant Pathol 118:211–225

Metwali EM, Abdelmoneim TS, Bakheit MA, Kadasa NM (2015) Alleviation of salinity stress in faba bean (*Vicia faba* L.) plants by inoculation with plant growth promoting rhizobacteria (PGPR). Plant Omics 8(5):449

Mhadhbi H, Jebara M, Zitoun A, Limam F, Aouani ME (2008) Symbiotic effectiveness and response to mannitol mediated osmotic stress of various chickpea-rhizobia associations. World J Microbiol Biotechnol 24:1027–1035

Mhadhbi H, Mhamdi R, Jebara M, Limam F, Aouani ME (2009) Legume-rhizobia symbiotic interaction under salt and drought constraints: generation of reactive oxygen species and protective role of antioxidant enzymes within nodules. In: Hemantaranjan A (ed) Advances in plant physiology, vol 11. Scientific Publishers (India), Jodhpur, pp 1–21

Mhadhbi H, Fotopoulos V, Mylona PV, Jebara M, Aouani ME, Polidoros AN (2011) Antioxidant gene-enzyme responses in *Medicago truncatula* genotypes with different degree of sensitivity to salinity. Physiol Plant 141:201–214

Mhadhbi M, Chaouch M, Ajroud K, Darghouth MA, BenAbderrazak S (2015) Sequence polymorphism of cytochrome b gene in Theileria annulata Tunisian isolates and its association with buparvaquone treatment failure. PLoS One 10:e0129678

Middleton PH, Jakab J, Penmetsa RV, Starker CG, Doll J, Kalo P, Prabhu R, Marsh JF, Mitra RM, Kereszt A (2007) An ERF transcription factor in *Medicago truncatula* that is essential for Nod factor signal transduction. Plant Cell 19:1221–1234

Miller KJ, Wood JM (1996) Osmoadaptation by rhizosphere bacteria. Annu Rev Microbiol 50:101–136

Mohamed HI, Gomaa EZ (2012) Effect of plant growth promoting *Bacillus subtilis* and *Pseudomonas fluorescens* on growth and pigment composition of radish plants (*Raphanus sativus*) under NaCl stress. Photosynthesis 50:263–272

Moschetti G, Peluso A, Protopapa A, Anastasio M, Pepe O, Defez R (2005) Use of nodulation pattern, stress tolerance, nodC gene amplification, RAPD-PCR and RFLP-16S rDNA to discriminate genotypes of *Rhizobium leguminosarum biovar viciae*. Syst Appl Microbiol 28:619–631

Mrabet M, Abdellatif E, Zribi K, Mhamdi R, Djébali N (2011) *Sinorhizobium meliloti* can protect *Medicago truncatula* from infection by *Phoma medicaginis*. Phytopathol Mediterr 50:183–191

Nabti EH, Mokrane N, Ghoul M, Manyani H, Dary M, Megias MG (2013) Isolation and characterization of two halophilic *Bacillus* (*B. licheniformis* and *Bacillus* sp) with antifungal activity. J Ecol Heal Environ 1(1):13–17

Nadeem SM, Zahir ZA, Naveed M, Arshad M (2007) Preliminary investigations on inducing salt tolerance in maize through inoculation with rhizobacteria containing ACC-deaminase activity. Can J Microbiol 53:1141–1149

Nadeem SM, Zahir ZA, Naveed M, Arshad M (2009) Rhizobacteria containing ACC deaminase confer salt tolerance in maize grown on salt affected soils. Can J Microbiol 55:1302–1309

Nadeem SM, Ahmad M, Naveed M, Imran M, Zahir ZA, Crowley DE (2016) Relationship between in vitro characterization and comparative efficacy of plant growth-promoting rhizobacteria for improving cucumber salt tolerance. Arch Microbiol 198(4):379–387

Nakbanpote W, Panitlurtumpai N, Sangdee A, Sakulpone N, Sirisom P, Pimthong A (2014) Salt-tolerant and plant growth-promoting bacteria isolated from Zn/Cd contaminated soil: identification and effect on rice under saline conditions. J Plant Interact 9(1):379–387

Nascimento FX, Brígido C, Glick BR, Oliveira S, Alho L (2012) *Mesorhizobium ciceri* LMS-1 expressing an exogenous ACC deaminase increases its nodulation abilities and chickpea plant resistance to soil constraints. Lett Appl Microbiol 55:15–21

Naseem H, Bano A (2014) Role of plant growth-promoting rhizobacteria and their exopolysaccharide in drought tolerance of maize. J Plant Interact 9(1):689–701

Nautiyal CS, Bhadauria S, Kumar P, Lal H, Mondal R, Verma D (2000) Stress induced phosphate solubilization in bacteria isolated from alkaline soils. FEMS Microbiol Lett 182:291–296

Nautiyal C, Govindarajan R, Lavania M, Pushpangadan P (2008) Novel mechanism of modulating natural antioxidants in functional foods: involvement of plant growth promoting rhizobacteria NRRL B-30488. J Agric Food Chem 56:4474–4481

Nayer M, Reza H (2008) Drought-induced accumulation of soluble sugars and proline in two maize varieties. World Appl Sci J 3:448–453
Naz R, Bano A (2015) Molecular and physiological responses of sunflower (*Helianthus Annuus* L.) to PGPR and SA under salt stress. Pak J Bot 47(1):35–42
Nedjimi B, Daoud Y, Touati M (2006) Growth, water relations, proline and ion content of in vitro cultured *Atriplex halimus* subsp. *schweinfurthii* as affected by CaCl2. Commun Biometry Crop Sci 1(2):79–89
Neljubow D (1901) Uber die horizontale Nutation der Stengel von *Pisum sativum* und einiger anderer Pflanzen. Beih Bot Zbl 10:128–138
Nielsen P, Fritze D, Priest F (1995) Phenetic diversity of alkaliphilic *Bacillus* strains: proposal for nine new species. Microbiology 141:1745–1761
Oerke EC, Dehne HW, Nbeck S, Fweber A (1994) Crop production and crop protection – estimated losses in major food and cash crops. Elsevier Sci, Amsterdam
Omar SA, Abdel-Sater MA, Khallil AM, Abdalla MH (1994) Growth and enzyme activities of fungi and bacteria in soil salinized with sodium chloride. Folia Microbiol 39(1):23–28
Omar MNA, Osman MEH, Kasim WA, El-Daim A (2009) Improvement of salt tolerance mechanisms of barely cultivated under salt stress using *Azospirillum brasilense*. In: Ashraf M, Ozturk M, Athar HR (eds) Salinity and water stress. Springer, Netherland, pp 133–147
Oren A (1999) Bioenergetics aspects of halophilism. Microbiol Mol Biol Rev 63:334–348
Oren A (2008) Microbial life at high salt concentrations: phylogenetic and metabolic diversity. Saline Sys 4:2
Orhan F (2016) Alleviation of salt stress by halotolerant and halophilic plant growth-promoting bacteria in wheat (*Triticum aestivum*). Braz J Microbiol 47(3):621–627
Ormeño-Orrillo E, Menna P, Almeida LGP, Ollero FJ, Nicolás MF, Rodrigues EP, Nakatami AS, Batista JSS, Chueire LMO, Souza RC, Vasconcelos ATR, Megías M, Hungria M, Martínez-Romero E (2012) Genomic basis of broad host range and environmental adaptability of *Rhizobium tropici* CIAT 899 and *Rhizobium* sp. PRF 81 which are used in inoculants for common bean (*Phaseolus vulgaris* L.) BMC Genomics 13:735
Oster JD (1993) Sodic soil reclamation. In: Leith HA, Masoom AL (eds) Towards the rational use of high salinity tolerant plants, vol 1. Kluwer Academic, Amsterdam, pp 485–490
Pakroo N, Kashirad A (1981) The effect of salinity and iron application on growth and mineral uptake of sunflower (*Helianthus annuus* L.) J Plant Nutr 4(1):45–56
Parida AK, Das AB (2005) Salt tolerance and salinity effects on plants: a review. Ecotoxicol Environ Saf 60(3):324–349
Patten CL, Glick BR (1996) Bacterial biosynthesis of indole-3-acetic acid. Can J Microbiol 42:207–220
Paul D, Lade H (2014) Plant-growth-promoting rhizobacteria to improve crop growth in saline soils: a review. Agron Sustain Dev, Springer, Verlag 34:737–752
Penrose DM, Glick BR (2003) Methods for isolating and characterizing ACC deaminase-containing plant growth-promoting rhizobacteria. Physiol Plant 118:10–15
Poli A, Anzelmo G, Nicolaus B (2010) Bacterial exopolysaccharides from extreme marine habitats: production, characterization and biological activities. Mar Drugs 8:1779–1802
Pourbabaee AA, Bahmani E, Alikhani HA, Emami S (2016) Promotion of wheat growth under salt stress by halotolerant bacteria containing ACC deaminase. J Agric Sci Technol 18:855–864
Prabhavati E, Mallaiah KV (2008) Production of indole acetic acid by *Rhizobium* sp. nodulating *Macrotyloma uniflorum* (Lam.) Verdc Asian J Soil Sci 3(1):146–148
Prado FE, Boero C, Gallardo M, Gonzalez JA (2000) Effect of NaCl on germination, growth and soluble sugar. Bot Bull Acad Sin 41:27–34
Principe A, Alvarez F, Castro MG, Zachi L, Fischer SE, Mori GB, Jofre E (2007) Biocontrol and PGPR features in native strains isolated from saline soils of Argentina. Curr Microbiol 55:314–322
Puppo A, Halliwell B (1988) Formation of hydroxyl radicals from hydrogen peroxide in the presence of iron. Is haemoglobin a biological Fenton reagent? Biochem J 249:185–190

Qadir M, Quillerou E, Nangia V (2014) Economics of salt-induced land degradation and restoration. Nat Resour Forum 38:282–295

Qudsaia B, Noshinil Y, Asghari B, Nadia Z, Abida A, Fayazul H (2013) Effect of *Azospirillum* inoculation on maize (*Zea mays* L.) under drought stress. Pak J Bot 45:13–20

Qurashi AW, Sabri SN (2011) Osmoadaptation and plant growth promotion by salt tolerant bacteria under salt stress. Afr J Microbiol Res 5(2):3546–3554

Qurashi WA, Sabri NA (2012) Bacterial exopolysaccharide and biofilm formation stimulate chickpea growth and soil aggregation under salt stress. Braz J Microbiol 43:1183–1191

Qurashi AW, Sabri AN (2013) Osmolyte accumulation in moderately halophilic bacteria improves salt tolerance of chickpea. Pak J Bot 45(3):1011–1016

Rabie GH, Aboul-Nasr MB, Al-Humiany A (2005) Increased salinity tolerance of cowpea plants by dual inoculation of an arbuscular mycorrhizal fungus *Glomus clarum* and a nitrogen-fixer *Azospirillum brasilense*. Mycobiology 33:51–60

Rajput L, Imran A, Mubeen F, Hafeez FY (2013) Salt-tolerant pgpr strain *Planococcus rifietoensis* promotes the growth and yield of wheat (*Triticum aestivum* l.) cultivated in saline soil. Pak J Bot 45(6):1955–1962

Ramadoss D, Lakkineni VK, Bose P, Ali S, Annapurna K (2013) Mitigation of salt stress in wheat seedlings by halotolerant bacteria isolated from saline habitats. Springer Plus 2:1–7

Ramalingam R, In-Jung L (2013) Ameliorative effects of spermine against osmotic stress through antioxidants and abscisic acid changes in soybean pods and seeds. Acta Physiol Plant 35:263–269

Ramey BE, Koutsoudis M, von Bodman SB, Fuqua C (2004) Biofilm formation in plant-microbe associations. Curr Opin Microbiol 7:602–609

Ramezani E, Sepanlou MG, Badi HAN (2011) The effect of salinity on the growth, morphology and physiology of *Echium amoenum* Fisch & Mey. Afr J Biotechnol 10:8765–8773

Rana DS, Dass A, Rajanna GA, Kaur R (2016) Biotic and abiotic stress management in pulses. Indian J Agron 61:238–248

Raychev T, Popandova S, Jozefaciuk G, Hajnos M, Sokolowska Z (2001) Physicochemical reclamation of saline soils using coal powder. Int Agrophys 15:51–54

Reddy BR, Sethunathan N (1985) Salinity and the persistence of parathion in flooded soil. Soil Biol Biochem 17:235–239

Rejeb KB, Abdelly C, Savoure A (2014) How reactive oxygen species and proline face stress together. Plant Physiol Biochem 80:278–284

Rengasamy P (2006) World salinization with emphasis on Australia. J Exp Bot 57:1017–1023

Rengasamy P (2016) Soil chemistry factors confounding crop salinity tolerance—a review. Agronomy 6:53. https://doi.org/10.3390/agronomy6040053

Reuber TL, Walker GC (1993) Biosynthesis of succinoglycan, a symbiotically important exopolysaccharide of *Rhizobium meliloti*. Cell 74:269–280

Ribeiro VA, Burkert CAV (2016) Exopolysaccharides produced by *Rhizobium*: production, composition and rheological properties. J Polym Biopolym Phys Chem 4(1):1–6

Rodriguez R, Redman R (2008) More than 400 million years of evolution and some plants still can't make it on their own: plant stress tolerance via fungal symbiosis. J Exp Bot 59(5):1109–1114

Rojas-Tapias D, Moreno-Galván A, Pardo-Díaz S, Obando M, Rivera D, Bonilla R (2012) Effect of inoculation with plant growth-promoting bacteria (PGPB) on amelioration of saline stress in maize (*Zea mays*). Appl Soil Ecol 61:264–272

Rolli E, Marasco R, Vigani G, Ettoumi B, Mapelli F, Deangelis ML (2014) Improved plant resistance to drought is promoted by the root-associated microbiome as a water stress-dependent trait. Environ Microbiol 17:316–331

Rudrappa T, Biedrzycki ML, Bais HP (2008) Causes and consequences of plant-associated biofilms. FEMS Microbiol Ecol 64:153–166

Ryan RP, Monchy S, Cardinale M, Taghavi S, Crossman L, Avison MB, Berg G, van der Lelie D, Dow JM (2009) The versatility and adaptation of bacteria from the genus *Stenotrophomonas*. Nat Rev Microbiol 7:514–552

Ryu CM, Farag MA, Hu CH, Reddy MS, Wei HX, Kloepper JW, Pare PW (2004) Bacterial volatiles induce systemic resistance in *Arabidopsis*. Plant Physiol 134:1017–1026
Sadeghi A, Karimi E, Dahazi PA, Javid MG, Dalvand Y, Askari H (2012) Plant growth promoting activity of an auxin and siderophore producing isolate of *Streptomyces* under saline soil condition. World J Microbiol Biotechnol 28:1503–1509
Safari D, Jamali F, Nooryazdan HR, Bayat F (2016) Screening fluorescent pseudomonads isolated from wheat rhizosphere for plant growth- promoting and salt tolerance properties. Biol Forum – Int J 8(1):35–42
Saleh-Lakha S, Glick BR (2006) Plant growth-promoting bacteria. In: van Elsas JD, Jansson JK, Trevors JT (eds) Modern soil microbiology. CRC/Thomson Publishing, Boca Raton, pp 503–520
Sandhya V, Ali SKZ, Minakshi G, Reddy G, Venkateswarlu B (2009) Alleviation of drought stress effects in sunflower seedlings by the exopolysaccharides producing *Pseudomonas putida* strain GAP-P45. Biol Fertil Soils 46:17–26
Sansinenea E, Ortiz A (2011) Secondary metabolites of soil *Bacillus* spp. Biotechnol Lett 33(8):1523–1538
Saravanakumar D, Samiyappan R (2007) ACC deaminase from *Pseudomonas fluorescens* mediated saline resistance in groundnut (*Arachis hypogea*) plants. J Appl Microbiol 102:1283–1292
Scandalios JG (2002) The rise of ROS. Trends Biochem Sci 27:483–486
Scandalios JG, Guan L, Polidoros AN (1997) Catalases in plants, gene structure, properties, regulation and expression. In: Scandalios JG (ed) Oxidative stress and the molecular biology of antioxidant defences cold spring. Harbor Laboratory Press, New York, pp 343–406
Selvakumar G, Panneerselvam P, Ganeshamurthy AN (2012) Bacterial mediated alleviation of abiotic stress in crops. In: Maheshwari DK (ed) Bacteria in agrobiology: stress management. Springer-Verlag, Berlin, pp 205–224
Shaharoona B, Arshad M, Zahir ZA (2006) Effect of plant growth promoting rhizobacteria containing ACC-deaminase on maize (*Zea mays* L.) growth under axenic conditions and on nodulation in mung bean (*Vigna radiata* L.) Lett Appl Microbiol 42:155–159
Shaharoona B, Jamro GM, Zahir ZA, Arshad M, Memon KS (2007) Effectiveness of various *Pseudomonas* spp. and *Burkholderia caryophylli* containing ACC-deaminase for improving growth and yield of wheat (*Triticum aestivum* L.) J Microbiol Biotechnol 17:1300–1307
Shaik ZA, Vardharajula S, Minakshi G, Venkateswar RL, Bandi V (2011) Effect of inoculation with a thermotolerant plant growth promoting *Pseudomonas putida* strain AKMP7 on growth of wheat (*Triticum* spp.) under heat stress. J Plant Interact 6:239–246
Shannon MC, Grieve CM (1999) Tolerance of vegetable crops to salinity. Sci Hortic 78:5–38
Sharma A, Singh P, Kumar S, Kashyap PL, Srivastava AK, Chakdar H, Sharma AK (2015) Deciphering diversity of salt-tolerant bacilli from saline soils of eastern Indo-gangetic plains of India. Geomicrobiol J 32:170–180
Sharma P, Jha AB, Dubey RS, Pessarakli M (2012) Reactive oxygen species, oxidative damage, and anti-oxidative defense mechanism in plants under stressful conditions. J Bot 2012(217037):26. https://doi.org/10.1155/2012/217037
Sharma S, Kulkarni J, Jha B (2016) Halotolerant rhizobacteria promote growth and enhance salinity tolerance in peanut. Front Microbiol 7:1600
Sheng G, Yang Y, Huang M, Yang K (2005) Influence of pH on pesticide sorption by soil containing wheat residue-derived char. Environ Pollut 134:457–463
Shilev S, Sancho ED, Gonzalez MB (2010) Rhizospheric bacteria alleviate salt-produced stress in sunflower. J Environ Manag 95:S37–S41
Shukla PS, Agarwal PK, Jha B (2012) Improved salinity tolerance of *Arachis hypogaea* (L.) by the interaction of halotolerant plant-growth-promoting rhizobacteria. J Plant Growth Regul 31(2):195–206
Siddikee MA, Glick BR, Chauhan PS, Yim WJ, Sa T (2011) Enhancement of growth and salt tolerance of red pepper seedlings (*Capsicum annuum* L.) by regulating stress ethylene synthesis

with halotolerant bacteria containing 1-aminocyclopropane-1-carboxylic acid deaminase activity. Plant Physiol Biochem 49:427–443

Silini A, Cherif-Silini H, Yahiaoui B (2016) Growing varieties durum wheat (*Triticum durum*) in response to the effect of osmolytes and inoculation by *Azotobacter chroococcum* under salt stress. Afr J Microbiol Res 10:387–399

Singh RP, Jha PN (2016) The multifarious PGPR *Serratia marcescens* CDP-13 augments induced systemic resistance and enhanced salinity tolerance of wheat (*Triticum aestivum* L.) PLoS One 11(6):e0155026

Singh S, Koehler B, Fett WF (1992) Effect of osmolarity and dehydration on alginate production by fluorescent pseudomonads. Curr Microbiol 25:335–339

Singh RS, Kaur N, Kennedy JF (2015) Pullulan and pullulan derivatives as promising biomolecules for drug and gene targeting. Carbohydr Polym 123:190–207

Skorupska A, Janczarek M, Marczak M, Krol AMJ (2006) Rhizobial exopolysaccharides: genetic control and symbiotic functions. Microb Cell Factories 5:7

Somers E, Vanderleyden J, Srinivasan M (2004) Rhizosphere bacterial signalling: a love parade beneath our feet. Crit Rev Microbiol 30:205–240

Srinivasan R, Alagawadi AR, Mahesh S, Meena KK, Saxena AK (2012) Characterization of phosphate solubilizing microorganisms from salt-affected soils of India and their effect on growth of sorghum plants [*Sorghum bicolor* (L.) Moench]. Ann Microbiol 62:93–105

Stepien P, Klobus G (2005) Antioxidant defense in the leaves of C3 and C4 plants under salinity stress. Physiol Plant 125:31–40

Street TO, Bolen DW, Rose GD (2006) A molecular mechanism for osmolyte-induced protein stability. Proc Natl Acad Sci U S A 103:13997–14002

Suzuki N, Rivero RM, Shulaev V, Blumwald E, Mittler R (2014) Abiotic and biotic stress combinations. New Phytol 203:32–43

Sziderics AH, Rasche F, Trognitz F, Sessitsch A, Wilhelm E (2007) Bacterial endophytes contribute to abiotic stress adaptation in pepper plants (*Capsicum annuum* L.) Can J Microbiol 53:1195–1202

Tahir M, Arshad M, Naveed M, Zahir ZA, Shaharoona B, Ahmad R (2006) Enrichment of recycled organic waste with N fertilizer and PGPR containing ACC-deaminase for improving growth and yield of tomato. Soil Environ 25:105–112

Talbi C, Argandoña M, Salvador M, Alché JD, Vargas C, Bedmar EJ, Delgado MJ (2013) *Burkholderia phymatum* improves salt tolerance of symbiotic nitrogen fixation in *Phaseolus vulgaris*. Plant Soil 367:673–685

Talibart R, Jebbar M, Gouesbet G (1994) Osmoadaptation in rhizobia: ectoine-induced salt tolerance. J Bacteriol 176:5210–5217

Tank N, Saraf M (2010) Salinity-resistant plant growth promoting rhizobacteria ameliorates sodium chloride stress on tomato plants. J Plant Interact 5(1):51–58

Tavakkoli E, Fatehi F, Coventry S, Rengasamy P, McDonald GK (2011) Additive effects of $Na+$ and Cl^- ions on barley growth under salinity stress. J Exp Bot 62:2189–2203

Tewari S, Arora NK (2013) Plant growth promoting rhizobacteria for ameliorating abiotic stresses triggered due to climatic variability. Clim Chang Environ Sustain 1(2):95–103

Tewari S, Arora NK (2014a) Multifunctional exopolysacccharides from *Pseudomonas aeruginosa* PF23 involved in plant growth stimulation, biocontrol and stress amelioration in sunflower under stress conditions. Curr Microbiol 69:484–494

Tewari S, Arora NK (2014b) Talc based exopolysaccharides formulation enhancing growth and production of *Helianthus annuus* under saline conditions. Cell Mol Biol 60(5):73–81

Tewari S, Arora NK (2016) Fluorescent *Pseudomonas* sp. PF17 as an efficient plant growth regulator and biocontrol agent for sunflower crop under saline conditions. Symbiosis. https://doi.org/10.1007/s13199-016-0389-8

Thrall PH, Broadhurst LM, Hoque MS, Bagnall DJ (2009) Diversity and salt tolerance of native *Acacia* rhizobia isolated from saline and non-saline soils. Aust Ecol 34:950–963

Tilman D, Cassman KG, Matson PA, Naylor R, Polasky S (2002) Agricultural sustainability and intensive production practices. Nature 418:672–675

Timmusk S, Abd El-Daim I, Copolovici L, Tanilas T, Kännaste A, Behers L, Nevo E, Seisenbaeva G, Stenström E, Niinemets U (2014) Drought-tolerance of wheat improved by rhizosphere bacteria from harsh environments: enhanced biomass production and reduced emissions of stress volatiles. PLoS One 9:e96086. https://doi.org/10.1371/journal.pone.0096086

Trabelsi D, Mengoni A, Aouani ME, Bazzicalupo M (2009) Genetic diversity and salt tolerance of bacterial communities from two Tunisian soils. Ann Microbiol 59:1–8

Trung NT, Hieu HV, Thuan NH (2016) Screening of strong 1-aminocyclopropane-1-carboxylate deaminase producing bacteria for improving the salinity tolerance of cowpea. Appl Micro Open Access 2:1000111. https://doi.org/10.4172/2471-9315.1000111

Turan M, Gulluce M, Sahin F (2012) Effects of plant-growth-promoting rhizobacteria on yield, growth and some physiological characteristics of wheat and barley plants. Commun Soil Sci Plant Anal 43:1658–1673

Upadhyay SK, Singh JS, Singh DP (2011) Exopolysaccharide-producing plant growth-promoting rhizobacteria under salinity condition. Pedosphere 21(2):214–222

Vaishnav A, Kumari S, Jain S, Varma A, Tuteja N, Choudhary DK (2016) PGPR-mediated expression of salt tolerance gene in soybean through volatiles under sodium nitroprusside. J Basic Microbiol. https://doi.org/10.1002/jobm.201600188

Vardharajula S, Zulfikar SA, Grover M, Reddy G, Bandi V (2011) Drought-tolerant plant growth promoting *Bacillus* spp., effect on growth, osmolytes, and antioxidant status of maize under drought stress. J Plant Interact 6:1–14

Vidhyasekaran P (2002) Bacterial disease resistance in plants: molecular biology and biotechnological applications. Haworth Press, Binghamton, p 452

Vriezen JAC, De Bruijn FJ, Nüsslein K (2007) Responses of rhizobia to desiccation in relation to osmotic stress, oxygen and temperature. Appl Environ Microbiol 73:3451–3459

Vyrides I, Stuckey DC (2009) Effect of fluctuations in salinity on anaerobic biomass and production of soluble microbial products (SMPs). Biodegradation 20:165–175

Wang CM, Zhang JL, Liu XS, Li Z, Wu GQ, Cai JY (2009) *Puccinellia tenuiflora* maintains a low Na^+ level under salinity by limiting unidirectional Na^+ influx resulting in a high selectivity for K^+ over Na^+. Plant Cell Environ 32:486–496

Wang X, Mavrodi DV, Ke L, Mavrodi OV, Yang M, Thomashow LS (2015) Biocontrol and plant growth-promoting activity of rhizobacteria from Chinese fields with contaminated soils. Microb Biotechnol 8:404–418

Wood JM, Bremer E, Csonka LN, Kraemerd R, Poolmane B, HeideSmith TVD (2001) Osmosensing and osmoregulatory compatible solute accumulation by bacteria. Comp Biochem Physiol A Mol Integr Physiol 130:437–460

Yang CW, Xu HH, Wang LL, Liu J, Shi DC, Wang GD (2009) Comparative effects of salt-stress and alkali-stress on the growth, photosynthesis, solute accumulation, and ion balance of barley plants. Photosynthetica 47(1):79–86

Yang H, Hu J, Long X, Liu Z, Rengel Z (2016) Salinity altered root distribution and increased diversity of bacterial communities in the rhizosphere soil of Jerusalem artichoke. Sci Rep 6:2

Yao LX, Wu ZS, Zheng YY, Kaleem I, Li C (2010) Growth promotion and protection against salt stress by *Pseudomonas putida* Rs-198 on cotton. Eur J Soil Biol 46:49–54

Yensen NP (2006) Halophyte uses for the twenty-first century. In: Ajmal Khan M, Weber DJ (eds) Ecophysiology of high salinity tolerant plants. Springer, Berlin, pp 367–396

Yue HT, Mo WP, Li C, Zheng YY, Li H (2007) The salt stress relief and growth promotion effect of Rs-5 on cotton. Plant Soil 297:139–145

Zahir ZA, Munir A, Asghar HN, Shaharoona B, Arshad M (2008) Effectiveness of rhizobacteria containing ACC-deaminase for growth promotion of pea (*Pisum sativum*) under drought conditions. J Microbiol Biotechnol 18:958–963

Zahir AZ, Ghani U, Naveed M, Nadeem SM, Asghar HN (2009) Comparative effectiveness of *Pseudomonas* and *Serratia* sp. containing ACC deaminase for improving growth and yield of wheat (*Triticum aestivum* L.) under salt stressed conditions. Arch Microbiol 191:415–424
Zahran HH (1991) Conditions for successful *Rhizobium*-legume symbiosis in saline environments. Biol Fertil Soils 12:73–80
Zahran HH (1997) Diversity, adaptation and activity of microbial flora in saline environment
Zahran HH (1999) *Rhizobium*–legume symbiosis and nitrogen fixation under severe conditions and in an arid climate. Microbiol Mol Biol Rev 63:968–989
Zarea MJ, Hajinia K, Goltapeh EM, Rejali F, Varma A (2012) Effect of *Piriformospora indica* and *Azospirillum* strains from saline or non-saline soil on mitigation of the effects of NaCl. Soil Biol Biochem 45:139–146
Zhao Y (2010) Auxin biosynthesis and its role in plant development. Annu Rev Plant Biol 61:49–64
Zhu JK, Bressan RA, Hasegawa PM, Pardo JM, Bohnert HJ (2005) Salt and crops: salinity tolerance. News CAST: News Counc Agric Sci Technol 32:13–16

Chapter 7
Plant Hormones as Key Regulators in Plant-Microbe Interactions Under Salt Stress

Dilfuza Egamberdieva, Stephan Wirth, and Elsayed Fathi Abd_Allah

Abstract Salinity is a global problem that hinders the normal growth and development of most plants. The loss of arable land due to salinization directly affects the food requirements of the world's population. However, plants have their own tolerance mechanisms that can help to withstand a certain degree of salinity. Nonetheless, plants often fail to survive under high saline conditions. Many published studies have advocated the positive influence of phytohormones on the growth and stress tolerance of plants. In addition, the microbes associated with plants have the capability to synthesize plant growth hormones that play an important role in alleviating salt stress in plants. The biosynthesis of phytohormones such as auxins, gibberellic acid, salicylic acid, and abscisic acid by root-associated microbes is a compelling mechanism to alter plant physiology and the biochemical processes in plant tissue. This review summarizes the plant phytohormones and their metabolism and activity under abiotic stress. In addition, it addresses the microbes that produce phytohormones that are closely associated with plants, along with their roles and interactions with plants under various stress conditions.

Keywords Plant growth regulators • Salinity • Drought • Stress • Microbes

D. Egamberdieva (✉)
Faculty of Biology, National University of Uzbekistan, Tashkent, Uzbekistan
e-mail: egamberdievad@gmail.com

S. Wirth
Leibniz Centre for Agricultural Landscape Research (ZALF), Institute of Landscape Biogeochemistry, Müncheberg, Germany

E.F. Abd_Allah
Plant Production Department, College of Food and Agricultural Sciences, King Saud University, Riyadh 11451, Saudi Arabia

D. Egamberdieva, P. Ahmad (eds.), *Plant Microbiome: Stress Response*, Microorganisms for Sustainability 5, https://doi.org/10.1007/978-981-10-5514-0_7

7.1 Introduction

Several abiotic stress factors such as temperature, water stress, salinity stress, radiation stress, and heavy metals pose a great threat to a large number of food crops with a subsequent reduction in global food production (Ahmad 2010; Grayson 2013; Hashem et al. 2016). These increasing environmental threats have led to alarming projections that have resulted in the design and introduction of powerful additional strategies to provide food security to an ever increasing population. In addition, the productivity of irrigated arable areas does not benefit all of the global population, because increasing salinization is often followed by prolonged irrigation, making the problem almost unavoidable (Riadh et al. 2010). Primary salinity mainly results from factors such as excessive weathering of rocks, capillary rise from shallow brackish groundwater, intrusion of salt water from the sea along the coast, salt-laden sand blown by sea winds, and improper drainage. Secondary salinity is another drastic configuration that mainly results from the activities of humans, including increased urbanization and the excess use of saline water for irrigation purposes. Possible reasons for the secondary salinization involve excess irrigation without a proper drainage system, the disposal of industrial effluents in arable lands, excess fertilizer usage, deforestation, the flooding of salt rich waters, and poor quality groundwater used for irrigation.

Salt-affected soils can be saline or sodic soils. One of the main characteristics that distinguish these two types of soils is the presence or absence of specific anions that contribute to the pH of soils. It has been well documented that, due to the presence of large amounts of carbonate or bicarbonate ions in sodic soils, their pH exceeds 8.5 compared to saline soils that are dominated by chloride or sulphate ions that keep the pH below 8.5. There is currently a strong interest in studying current plant abiotic stress responses to help manage these problems. From time immemorial, salinity has adversely affected plant growth and therefore agricultural production. Such damaging effects of salt stress have remained a matter of great concern from ancient times onward especially since they have affected civilization (Qadir et al. 2014). It has been estimated that approximately 50% of the irrigated lands are affected by salt that inhibits plant growth. Saline soils with soluble salts affect plant growth at various stages leading to yield differences between crops and result in differences in their ion compositions at maturity.

Every crop has a threshold stress tolerance level specific for a specific crop variety, above which plant growth is drastically affected (Khan et al. 2006). Higher soil salinity hampers the growth of several crop plants mainly because of the reduction in the osmotic potential of the soil solution. This results in specific ion effects that lead to nutritional imbalances (Ahmad and Sharma 2008). Exposure of plants to salt stress causes alterations in several major plant processes such as photosynthesis, protein synthesis, respiration, water uptake, as well as energy and lipid metabolism. In addition, this salinity reduces membrane stability and mediates an increase in the production of toxic reactive oxygen species. Salinity-induced osmotic stress is mainly caused by the excess uptake and accumulation of Na^+ and Cl^- from the soil environment. A higher salt concentration in the soil leads to a reduction in the soil osmotic potential and therefore perturbs plant water uptake through the roots. Nevertheless, stresses such as higher salinity exert a more negative impact on plant growth. In response to this osmotic stress induced by higher salinity, the plants utilize different tolerance strategies such as the accumulation of low molecular mass compounds known as compatible solutes or osmolytes that involve glycine, betaine, proline, mannitol, sorbitol, and sugars. Salinity-induced osmotic stress results in a substantial increase in the production of toxic reactive oxygen species (ROS) resulting in oxidative stress. Toxic reactive oxygen species (ROS) include hydrogen peroxide (H_2O_2), singlet oxygen ($^1O_2$), superoxide ions (O_2^-), and peroxides. The most widely distributed of these reactive oxygen species is hydrogen peroxide (Ahmad et al. 2010). Levels of toxic ROS are maintained within limits under normal conditions. However, exposure to stressful environmental conditions increases their production and thus causes oxidative stress (Ahmad et al. 2010). Reactive oxygen species (ROS) induce damage to most of the sensitive and important cellular macromolecules including DNA (Tuteja et al. 2009). In addition, reactive oxygen species affect high-molecular mass molecules such as unsaturated membrane lipid components, resulting in the formation of lipid peroxides. These ROS primarily have such harmful effects because of their potential to trigger several autoxidative chain reactions that involve polyunsaturated fatty acids (Smirnoff 2000). ROS attacks on proteins induced by oxidative stress result in the site-specific amino acid modifications, the breakdown of peptide chains, the aggregation of cross-linked reaction products, and the increased susceptibility to proteolysis (Ahmad et al. 2010). These toxic ROS can also induce numerous lesions in the DNA by causing deletions and mutations as well as other associated lethal genetic effects (Srivalli et al. 2003; Tuteja et al. 2009).

Salt-affected soils can be reclaimed by actively implementing several physical and biological as well as chemical techniques. Physical methods involve scraping, leaching, and flushing. Chemical treatments can be another alternative to reclaim sodic and saline soils and make these waste soils arable. Chemical methods of reclamation involve treating soils with gypsum, sulfuric acid, and farmyard manure. Altering the biology of these soils affected by salt can be more effective at reclaiming and managing the land compared to the alternate techniques described above. The biological method involves the use of living systems such as salt-tolerant plants to reclaim salt-polluted wastelands. Reclaiming salinized unproductive waste soils by utilizing biological methods can be a more efficient and feasible alternative to convert salt-affected soils into productive arable lands. Most of the global agricultural lands are rain-fed. However, due to the lack of irrigation water, the irrigation of shallow and brackish groundwater in plantations of salt-tolerant crops reveals the urgent need to reclaim soils that are low to moderately saline. Using biotechnological and genomic approaches, salt-tolerant varieties of different crops such as wheat and rice have been produced (Ahmad et al. 2012; Singh et al. 2014). Plants employ several tolerance mechanisms to avert the effects of stresses, and these tolerance pathways are triggered whenever the plant senses alterations in metabolism. The induction of every type of tolerance mechanism depends on the synthesis of key regulatory molecules that elicit specific signaling events that result in regulation at the genetic and molecular levels. Phytohormones modulate growth, development, enzyme activity, nutrient allocation, source-sink regulation, and signaling. However, changes in the endogenous concentrations of the phytohormones in response to different stresses can have a negative impact on growth, and exogenously supplementing phytohormones has been adopted to improve growth and metabolism under such conditions. This review focuses on the past, current trends, and future needs regarding the regulation of phytohormone research to enhance plant performance under changing environments.

7.2 Plant Growth Regulators

Plant growth substances are involved in the regulation of growth and development of plants that grow under normal and stressed conditions. Indoleacetic acid (IAA) is a naturally occurring auxin that has a major role in plant growth regulation. It is involved in controlling vascular tissue development, cell elongation, and apical dominance (Wang et al. 2001). Little information is available on the relationship and impact of auxin on salt tolerance and the amelioration of salt stress. However, stress-induced alterations in the levels of IAA are similar to those of abscisic acid. Excess biosynthesis and accumulation of IAA are correlated with reduced growth, so reduced crop growth under stress could be due to alterations in the hormone balance (Ribaut and Pilet 1994). Therefore, exogenously applying growth hormone appears to be a promising strategy and a potent approach to counteract stress-induced changes. Most of the research work has reported that IAA levels decrease under salinity. For example, Prakash and Prathapasenan (1990) reported a reduction in IAA levels in NaCl-stressed rice leaves. In addition, they reported that applying GA3 under such conditions mitigates the effect of salinity on IAA levels demonstrating that salinity affects the hormonal balance by influencing plant growth and development. A significant reduction in the levels of IAA has been reported in salt-stressed rice (Nilsen and Orcutt 1996) and tomato (Dunlap and Binzel 1996). Nevertheless, several researchers have also reported that IAA mitigates salinity-induced damage in plants. For example, IAA has been reported to alleviate the inhibitory effects of salt stress in wheat. Akbari et al. (2007) also demonstrated that exogenous application of auxin enhanced the length of hypocotyls in fresh as well as dry biomass of wheat cultivars under saline conditions.

Several numbers of genes called primary auxin response genes are stimulated by auxin. To date, several auxin-responsive genes have been identified and well-characterized from different plant species such as *Arabidopsis* and rice (Hagen and Guilfoyle 2002). Auxin-responsive genes belong to three gene families including auxin/indoleacetic acid, *GH3*, and small auxin-up RNA (Guilfoyle et al. 1993).

Since abscisic acid (ABA) has such key metabolic functions as germination, maturity, and dormancy, it has been proposed that it mediates the plant's responses to a range of environmental stresses such as salinity and water stress (Baumann 2010). ABA mediates stomatal closure. In addition, it causes actin filament depolymerization that results in the dissociation of filaments and hence a reduction in the generation of new actin filaments, thus affecting cell polarity (Jae-Ung and Youngsook 2001). Actin filament depolymerization is believed to be an important basic step to initiate the signaling pathways necessary to bring about stomatal changes (opening and closing), and any modulations in actin interfere with the normal stomatal functioning. Shi and Zhu (2002) have demonstrated that ABA levels control the tissue distribution as well as the regulation of AtNHX1 expression under conditions of salt stress. Moreover, ABA has been reported to affect the expression of genes important for the functioning of vacuolar H^+-inorganic pyrophosphatase and its catalytic subunit (Fukuda and Tanaka 2006). In salt-stressed *Hordeum vulgare* L., Fukuda and Tanaka (2006) observed that the transcript levels of these genes were accompanied by an increase in the levels of the hormones responsible for the regulation of their expression. In addition, Keskin et al. (2010) reported the rapid induction and expression of the MAPK4 such as genes (TIP1 and GLP1) in response to exogenously applied ABA in wheat. These examples provide strong evidence in favour of ABA mediating the expression of stress-responsive genes and therefore growth promotion.

Cytokinins have the potential to break stress-induced seed dormancy of several plants including tomato, barley, and cotton (Bozcuk 1981). Zalewski et al. (2010) demonstrated that silencing the cytokinin oxidase gene leads to the improved productivity of barley under normal as well as stressed conditions. Such contradictory reports reveal that the actual mechanisms leading to cytokinin-induced tolerance in plants under stress are not fully known and that further elucidation at the molecular level could be illustrative. Nevertheless, studies using loss-of-function mutants of cytokinin revealed the role of cytokinins in the development of plant vasculature. Cytokinin-deficient plants exhibit a reduction in cell division and the meristematic activity of roots and shoots. Such reports that discuss the lower content of cytokinin in genetically engineered plants reflect its regulatory role in the hormonal control of meristematic activity and organ growth at the pre-embryonic development. It is believed that root-borne cytokinins behave as signals to control the long-term signal involved in the regulation of various processes including the nutrient status such as that of nitrogen at different sites (reviewed in Schmulling 2002). Therefore, it is clear that the cytokinin-induced changes in plant growth and metabolism are wide and diverse and require intensive investigations before one can conclude a positive or negative idea about them. Much of the dilemma about cytokinin can be ascribed to the lack of accurate knowledge, mutants, and biochemical tools for studying the consequences of cytokinin deficiency in plants. Transgenic *Arabidopsis* plants overexpressing the cytokinin oxidase/dehydrogenase gene exhibit increased cytokinin breakdown, concomitantly reduced cytokinin reporter gene, ARR5:GUS coding for β-glucuronidase reflecting in reduced cytokinin levels. Thus, these lines exhibit a reduced activity of vegetative and floral meristems and leaf primordia (Werner et al. 2003). Such reports support the role of cytokinin as the negative regulators of growth in terms of root and lateral root formation. They also detected AtCKX-green fluorescent proteins located in vacuoles, the endoplasmic reticulum, or the extracellular space implying the role of subcellular compartmentation in determining the role of cytokinin. Such activity was observed to remain predominantly confined to actively growing zones (Werner et al. 2003). Schafer et al. (2015) reviewed the role of *cis*-zeatin-type cytokinins in regulating plant development and metabolism under different environmental stresses including water, herbicide, herbivore attack, and pathogen stress. In light of the key role of cytokinins in plant growth regulation, it has been suggested that transgenic plants that have a reduced expression of cytokinin oxidase display enhanced growth by maintaining the concentration of cytokinins. For example, rice and barley that have been reported to show apparent yield improvements under normal conditions have been linked with a greater number of inflorescences (Ashikari et al. 2005; Zalewski et al. 2010).

Salicylic acid is another important endogenous plant growth hormone that is a phenolic compound. In plants, salicylic acid actively participates in mediating and regulating several essential physiological and biochemical processes such as growth, photosynthesis, nitrogen metabolism, ethylene production, and flowering (Hayat et al. 2010). In addition to its diverse physiological roles, salicylic acid provides protection against various environmental stresses including water stress (Senaratna et al. 2000), freezing (Tasgin et al. 2003), salinity (Azooz et al. 2011), and heavy metals (Ahmad et al. 2011). After exposure to stress, exogenously applied salicylic acid acts as a signal that is involved in the activation of specific response mechanisms in plants. Genes expressed due to salicylic acid treatment include the genes coding for several metabolically important constituents that are involved in several important processes (Jumali et al. 2011). Exogenous applications of salicylic acid have been reported to protect plants from the deleterious impact of stress factors by promoting several processes that contribute to enhanced stress tolerance (Jumali et al. 2011). The role of SA in the defence mechanism to alleviate salt stress in plants has been extensively studied (Hussein et al. 2007). Salicylic acid-induced amelioration of salinity stress has been observed in several crops such as faba bean (Azooz et al. 2011), maize (Gunes et al. 2007), *Vigna radiata* L. (Khan et al. 2014), and wheat (Shakirova et al. 2003). Azooz et al. (2011) reported that the application of salicylic acid to seawater-treated *Vicia faba* plants not only ameliorated the negative effect on the growth, biomass accumulation, and antioxidant system but also caused the efficient accumulation of organic osmolytes such as proline and free amino acids. Moreover, they also reported that the application of salicylic acid promoted the efficient sequestration and partitioning of deleterious ions such as Na. Increased synthesis and accumulation of proline and abscisic acid have been reported in salinity-stressed wheat seedlings contributing to better growth and yield (Shakirova et al. 2003). In *Vigna radiata* L., Khan et al. (2014) observed that the exogenous application of salicylic acid helped considerably in mitigating salt stress-induced changes. Salicylic acid-treated plants showed better growth in terms of biomass accumulation and displayed a higher photosynthetic rate and enhanced activity of antioxidant enzymes. Moreover, they also reported an enhancement in methionine and glycine betaine content due to the salicylic acid treatment. Salicylic acid promotes cell division at the apical meristem region of roots thereby leading to increased plant growth. In barley plants, salinity stress caused alterations in the rate of photosynthesis, membrane stability, and growth. However, these negative effects of salinity were ameliorated by the application of salicylic acid (El-Tayeb 2005). The direct addition of salicylic acid can also be an effective strategy to avoid salinity stress-induced damage to crop plants. In salinity-stressed maize, Gunes et al. (2007) reported that the addition of salicylic acid to the soil mitigates the salinity-induced negative impact by reducing the uptake of toxic ions such as Na and therefore reduced their accumulation within the sensitive plant parts. Salt stress enhances lipid peroxidation and reduces membrane stability, and the application of salicylic acid reduces these effects (Horvath et al. 2007). Wahid et al. (2007) reported a slight increase in hydrogen peroxide levels and explained that the pretreatment of wheat seeds with hydrogen peroxide resulted in enhanced salt tolerance. Salicylic acid alleviates salinity stress-induced oxidative damage due to the accumulation of sufficient levels of hydrogen peroxide. Applying salicylic acid to salt-stressed tomato directly in the root zone mitigates the salt-stress damage by maintaining the transpiration rate and stomatal conductance while reducing electrolyte leakage (Stevens et al. 2006). Rice seedlings subjected to salt stress resulted in the accumulation and increase in endogenous concentrations of salicylic acid due to upregulation in the activity of the salicylic acid biosynthetic pathway (Sawada et al. 2006). In salt-stressed *Vigna radiata* L., Khan et al. (2014) reported a reduction in the endogenous levels of ethylene due to salicylic acid application.

Accumulation of gibberellic acid occurs at higher rates when plants are exposed to environmental extremes. Plant scientists are experimenting widely with the use of exogenous application of phytohormones to improve the growth and yield of important crop plants. For example, Ahmad (2010) observed an increased growth in salinity-stressed *Brassica* due to the exogenous application of gibberellic acid. Moreover, an increase in the content of osmotic constituents was reported in salinity-stressed plants that was further increased by the application of gibberellic acid leading to more effective osmotic adjustment in gibberellic acid-treated plants even under salinity stress conditions (Ahmad 2010). The maintenance of tissue water content and gibberellic acid-induced mitigation of salinity effects on water content has been reported in *Brassica*, wheat (Manjili et al. 2012) and maize (Tuna et al. 2008). The exogenous application of gibberellic acid enhanced the activity of antioxidant enzyme activity promoting better and quick removal of toxic free radicals under salt stress. Lower levels of reactive oxygen species contribute to better growth and increased yield (Manjili et al. 2012). In addition, the exogenous application of gibberellic acid has been reported to avert the salinity-induced effects on germination and growth in *Arabidopsis thaliana* by mediating enhanced synthesis of salicylic acid that causes an increase in the activity of isochorismate synthase 1 (Alonso-Ramirez et al. 2009). They also demonstrated that overexpression of a gibberellin-responsive gene from *Fagus sylvatica* enhanced the salt tolerance of *Arabidopsis*. Several researchers have reported the efficiency of gibberellic acid at ameliorating salinity-induced deleterious changes that resulted in the prolonged growth of salt-stressed wheat and rice (Prakash and Prathapasenan 1990). Enhanced plant water uptake and reduced stomatal resistance were identified in gibberellic acid-treated tomato plants grown under saline conditions (Maggio et al. 2010). Gibberellic acid induces the efficient uptake as well as the partitioning of ions within the plant system leading to enhanced growth and maintained metabolism of the plant under normal as well as stress conditions (Iqbal and Ashraf 2013). Under salt stress conditions, improved germination and growth due to gibberellic acid has been reported by several researchers (Tuna et al. 2008; Manjili et al. 2012). Reduced peroxidation of lipids in salt-stressed plants and enhanced membrane stability as well as efficient free radical scavenging has also been reported (Manjili et al. 2012). The current literature clearly describes the ameliorative impact of gibberellic acid against salinity. In addition to this effect, gibberellins can show crosstalk with other phytohormones causing them to elicit certain important responses and mediate tolerance mechanisms to enhance stress tolerance. The synthesis of gibberellins can also be promoted through the application of another hormone such as auxin (Wolbang et al. 2004). The enhanced synthesis of gibberellic acid leads to enhanced ABA catabolism. Moreover, gibberellic acid directly affects growth, yield, and mineral nutrition as well as nitrogen metabolism. Khan et al. (2004) reported an increase in fruit yield, leaf area, nitrogen, phosphorous, and potassium uptake in tomato following the exogenous application of gibberellic acid. Moreover, enhanced lycopene content was also reported in gibberellic acid-treated plants resulting in increased nutritive value.

7.3 Microbial Phytohormones and Stress Response

Plant-associated microbes play a vital role in plant growth under abiotic stress by modifying the root system, enhancing mobilization and the uptake of several essential elements, and modulating physiological parameters. They are able to colonize root systems and plant tissues and facilitate the beneficial association with plants (Egamberdieva et al. 2011, 2013, 2015). The root-associated beneficial microbes, including root colonizing, endophytic, and symbiotic bacteria and fungi may stimulate root systems and facilitate greater absorption of water and nutrients from the subsoil, thus increasing plant growth and development (Cho et al. 2015; Parray et al. 2016). There are several mechanisms by which bacteria are able to stimulate root growth, protect plants from various soil-borne pathogens, stimulate stress tolerance, and induce systemic resistance, which include competition for nutrients and niches (Kamilova et al. 2006), solubilization of minerals (Sharma et al. 2013), and production of plant growth hormones (Egamberdieva and Kucharova 2009) and ACC deaminase enzyme (Glick 2014). The rhizosphere is rich in nutrients and secondary metabolites that attract microorganisms as nutrient sources (Shahab et al. 2009, Egamberdieva et al. 2017a, b). In response to root exudates, the microorganisms that colonize the rhizosphere produce various biologically active compounds, including phytohormones that are utilized by the plants as well.

Auxin, indole-3-acetic acid (IAA), is an important plant growth hormone and is synthesized by various microbial species (Amara et al. 2015). The additional supply of IAA in the root system stimulates its size, branching number, and the surface area and helps plants to absorb more nutrients from the soil and thus increase their growth and development. The microbes synthesize IAA through various routes, via L-tryptophan-dependent and independent pathways. The synthesis of indole 3-acetic acid (IAA) by microbes is a process that involves L-tryptophan metabolism, an independent pathway (Zhao 2012).

The abiotic stress strongly inhibits the synthesis of plant hormones (Debez et al. 2001) and the supplementation of plant growth regulators such as auxin (Khan et al. 2004), and gibberellins (Afzal et al. 2005) improves seed germination and plant growth (Afzal et al. 2005; Egamberdieva 2009). The microbial phytohormones have been reported to stimulate plant growth and development under various stress conditions, including salinity, heat, drought, and metal toxicity (Sgroy et al. 2009; Egamberdiyeva and Hoflich 2003, 2004; Liu et al. 2013; Hashem et al. 2016). In earlier studies, orchid-associated bacteria such as *Rhizobium*, *Microbacterium*, and *Mycobacterium* were able to produce IAA (Tsavkelova et al. 2007). The IAA synthesized by microbes is taken up by plant cells and can stimulate plant cell proliferation (Glick 2012). Bacterial phytohormones were reported to stimulate root growth and development. For example, the IAA-producing bacterial strains that were isolated from saline soil, including *Pseudomonas aureantiaca* TSAU22, *P. extremorientalis* TSAU20, and *P. putida* TSAU1, alleviated salt stress on seed germination by 79% (Egamberdieva 2009). Several salt-tolerant strains synthesizing IAA in culture medium such as *Serratia plymuthica* RR-2-5-10, *Stenotrophomonas rhizophila* e-p10, *P. fluorescens* SPB2145, *P. extremorientalis* TSAU20, and *P. fluorescens* PCL1751 significantly increased cucumber biomass and yield in greenhouse conditions (9–24%) (Egamberdieva et al. 2011). The strains *Pseudomonas* spp. that produce IAA in media containing 1.5% NaCl caused an enlargement of the root system and improved nutrient uptake, nodulation, and growth of goat's rue in salinized soil (Egamberdieva et al. 2013). Hashem et al. (2016) reported an improved shoot and root growth and nutrient uptake in *Acacia gerrardii* by IAA-producing *Bacillus subtilis* alone or combined with AMF under salt stress. A cytokinin-producing *Pseudomonas* strain showed a stimulatory effect on the plant growth of wheat and radish (de Salamone et al. 2001).

The cytokinin-producing root-associated bacteria strains *Arthrobacter*, *Bacillus*, *Azospirillum*, and *Pseudomonas* stimulated root and shoot growth of soybean and proline content under salt stress (Naz et al. 2009). In another study, *Bacillus megaterium* produced cytokinin and stimulated the growth of *A. thaliana* and *P. vulgaris* (Ortiz-Castro et al. 2008). The production of dihydrozeatin riboside and zeatin riboside was reported by *Pseudomonas* strains that showed a stimulatory effect on rice seedling growth (Karnwal and Kaushik 2011).

Aspergillus fumigatus-producing gibberellins (GAs) such as GA4 (24.8 ng/ml), GA9 (1.2 ng/ml), and GA12 (9.8 ng/ml) stimulated shoot biomass, leaf area, and chlorophyll contents of soybean under salt stress compared to non-inoculated plants (Khan et al. 2011). *Trichoderma asperellum* Q1 that produces indoleacetic acid (IAA), gibberellic acid (GA), and abscisic acid (ABA) stimulated the root length and fresh weight of cucumber seedlings under salt stress in comparison to the untreated control plants (Zhao and Zhang 2015). Many authors have reported that *Pseudomonas* species were the dominant producers of phytohormones among other root-associated microbes (Khakipour et al. 2008; Lawongsa et al. 2008). In another study, Khan and Doty (2009) found that the endophytic bacteria *Enterobacter*, *Pseudomonas*, *and Stenotrophomonas* associated with sweet potato plants produced IAA.

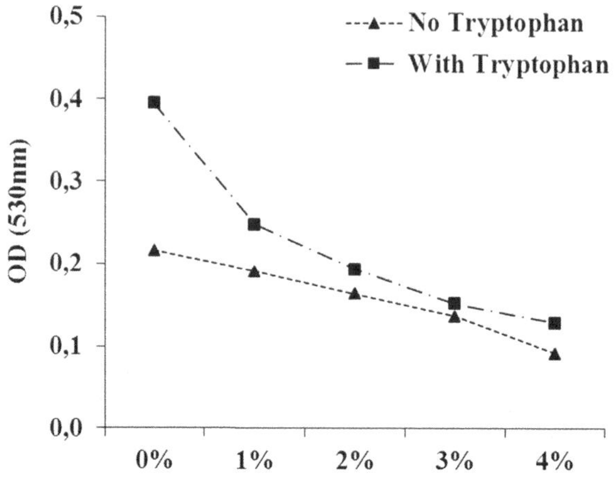

Fig. 7.1 The effect of NaCl concentrations of 1–4% (w/v) on indole-3-acetic acid production by *Pseudomonas putida* 1T1

The phytohormone-producing strains *Pseudomonas* sp. and *B. lentus* improved plant growth and physiological properties including the photosynthetic pigments of basil under salt stress conditions (Golpayegani and Tilebeni 2011). Cardinale et al. (2015) isolated *Curtobacterium flaccumfaciens* E108 and *Ensifer garamanticus* E110 from *Hordeum secalinum* and studied their effect on the plant growth of barley. The results showed that phytohormone-producing bacteria that were isolated stimulated root and shoot growth and stress tolerance of barley. The root colonizing halotolerant bacterium *B. licheniformis* HSW-16 was able to mitigate salt stress-induced damage and to stimulate growth of wheat through the production of IAA under saline soil condition (Singh and Jha 2016). Upadhyay et al. (2012) reported similar observations with the salt-tolerant bacterial strains *Bacillus subtilis* SU47 and *Arthrobacter* sp. SU18 that increased wheat biomass and total soluble sugars and reduced the sodium content in plant tissue. In an additional study, cucumber inoculated with *Trichoderma asperellum* Q1 contained higher concentrations of IAA, GA, and ABA in plant tissue under salt stress (Zhao and Zhang 2015). In another study, the salt-tolerant SA-producing bacterium *Serratia marcescens* NBRI1213 stimulated root and shoot growth and N, P, and K uptake by maize and increased the salt stress tolerance of the plants (Lavania and Nautiyal 2013). Gutierrez et al. (2009) reported that *Vibrio* spp. isolated from roots of the estuarine grasses *Spartina alterniflora* and *Juncus roemerianus* produced the phytohormone indole-3-acetic acid (IAA).

The root-associated microbes are able to produce phytohormones under saline condition as well. For example, the strains *Pseudomonas putida* 1T1 (A) and *Stenotrophomonas rhizophila* ep10 (B) were able to produce IAA in medium containing 1.5% NaCl (Fig. 7.1). Their growth was also not affected by salinity, while the bacterial strains were able to grow in up to 3% NaCl (Fig. 7.2). The root-

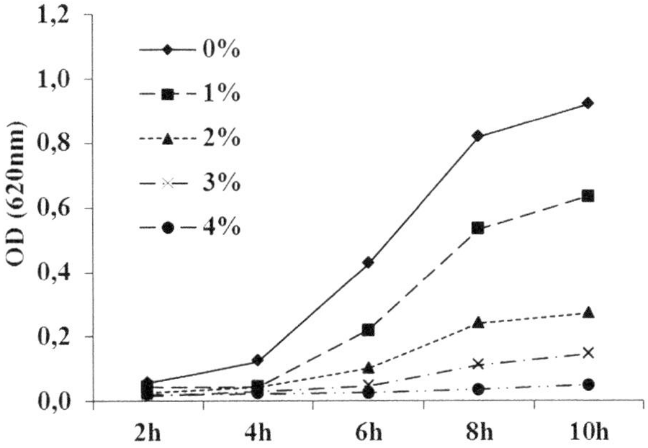

Fig. 7.2 The effect of NaCl concentration of 1–4% (w/v) on the growth of *Pseudomonas putida* 1T1

Table 7.1 Effect of selected plant growth-promoting bacteria on shoot and root length and dry weight of tomato growing in nonsaline (EC value 2.3 dSm^{-1}) and saline soil (EC value 7.1 dSm^{-1})

	Nonsaline			Saline			IAA[a]	
Bacterial strains	Shoot length[b]	Root length[c]	Dry weight[d]	Shoot length	Root length	Dry weight[d]	Tr −	Tr +
Control	8.19	5.75	0.151	7.65	5.64	0.14		
S. rhizophila ep-17	9.36*	7.69*	0.172*	6.39	5.52	0.117	7.4	10.6
*P. putida*1T1	9.06*	7.91*	0.166*	9.54*	7.54*	0.173*	8.3	10.9
P. trivialis 3Re2-7	8.74*	7.56*	0.155	8.37	6.79*	0.154	13.4	16.1
S. plymuthica RR2-5-10	10.24*	8.5*	0.189*	8.11	6.57	0.149	12.0	25.3
S. rhizophila e-p10	9.28*	7.33*	0.161	9.09*	7.82*	0.168*	14.6	14.8
P. chlororaphis RRj228	9.14*	7.47*	0.166*	7.25	6.83*	0.130	11.2	16.8

Tr Tryptophan
*Significantly different from the control at $P < 0.05$
[a]Expressed as gram per plant
[b]Expressed as cm per plant
[c]Expressed as cm per plant
[d]Auxin (IAA) level in μgml^{-1} after 5 days of incubation at 28 °C in medium supplemented with 1.5% NaCl

associated microbes that were able to produce IAA under saline conditions, namely, *S. rhizophila* ep-17, *P. putida* 1 T1, *P. trivialis* 3Re2–7, *S. plymuthica* RR2-5-10, *S. rhizophila* e-p10, and *P. chlororaphis* RRj228 stimulated the root and shoot growth of tomato under nonsaline and saline soil conditions (Table 7.1). Figure 7.3 shows the effect of IAA-producing *Pseudomonas extremorientalis* (TSAU20) stimulated the root system of tomato under saline soil condition (Fig. 7.3). In addition, these strains improved tomato biomass and yield under greenhouse conditions (Table 7.2). The microbial inoculants are more effective under nutrient-deficient soil conditions

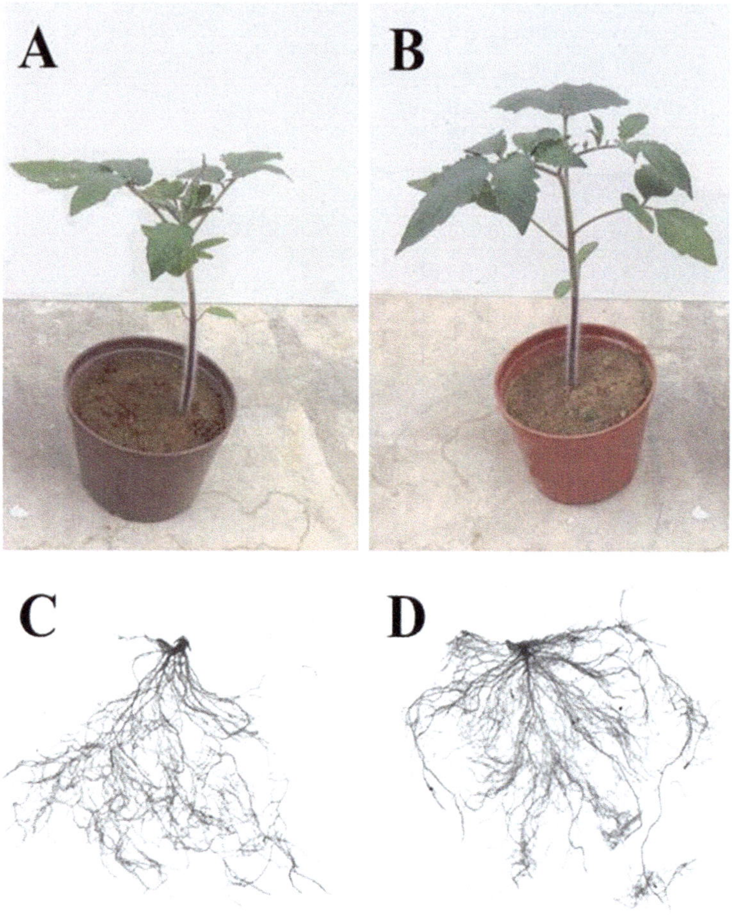

Fig. 7.3 The plant growth of tomato inoculated with IAA-producing *Pseudomonas extremorientalis* (TSAU20) under saline soil condition. (**a**) Control plant, (**b**) plant inoculated with TSAU20, (**c**) root of plant without inoculation, (**d**) root of plant inoculated with TSAU20

Table 7.2 Effect of selected bacterial strains on tomato (cv. *Bella*) seed germination, height, and fruit yield in greenhouse experiments

Treatment	Plant height (cm)	Seed germination (%)	Fruit yield (kg/m^2)	Fruit yield (%)
None (control)	125 ± 3.7	100	13.9 ± 0.9	100
P. putida 1T1	159[a] ± 3.4	127	16.6[a] ± 1.2	119
S. rhizophila e-p10	150[a] ± 4.7	120	15.9[a] ± 1.0	114
P. fluorescens PCL1751	155[a] ± 4.9	125	15.5[a] ± 1.3	112

[a]Tomato seeds were sown on 1.03.2009, and fruits were harvested on 20.06.2009; the temperature range was as follows: day 22–24 °C, night 12–14 °C

or abiotic stress. *Serratia* sp. isolated from chickpea nodules was found to produce IAA that leads an increased grain yield of chickpea in nutrient-deficient soil (Zaheer et al. 2016). The phytohormone-producing microbes were reported to stimulate the stress tolerance of plants to drought by modulating the antioxidant enzymes and physiological processes. The root and shoot biomass of clover was increased by *P. putida* and *B. megaterium* under drought stress, and this correlated with increased IAA concentration (Marulanda et al. 2009).

Yandigeri et al. (2012) isolated drought-tolerant endophytic actinobacteria *Streptomyces coelicolor* DE07, *S. olivaceus* DE10, and *Streptomyces geysiriensis* DE27 and found a significant increase in the seed germination of wheat. The strain *S. olivaceus* DE10 also increased yield (492.77 kg ha^{-1}). However, co-inoculation of *S. olivaceus* DE10 and *S. geysiriensis* DE27 revealed higher performance (550.09 kg ha^{-1}) under water stress. The strains that produced the highest concentration of IAA under water stress continued to do so until the end of the logarithmic phase of growth. A similar observation was reported by Salomon et al. (2014) who found that *Bacillus licheniformis* Rt4M10 and *Pseudomonas fluorescens* Rt6M10 isolated from the rhizosphere of *Vitis vinifera* stimulated the plant growth of grapevine under water stress through ABA production. Raza and Faisal (2013) also observed that the cytokinin-producing bacterium *Micrococcus luteus* chp37 isolated from the desert of Pakistan stimulated shoot and root biomass of maize under drought conditions. Liu et al. (2013) reported a similar observation and found that a cytokinin-producing *Bacillus subtilis* stimulated the shoot dry biomass by 19.2%, as well as the root biomass of *Platycladus orientalis* (oriental Thuja) by 13.9%, under drought stress. The SA-producing endophytic bacteria *Achromobacter xylosoxidans* and *Bacillus pumilus* enhanced the root and shoot growth of sunflower seedlings under water stress conditions (Forchetti et al. 2010). In another study, *Azospirillum lipoferum* that synthesized GA increased the stress tolerance of wheat to drought (Creus et al. 2004). According to Bianco and Defez (2009), the IAA is involved in enhanced cellular defence systems that protect the plants from external adverse conditions.

7.4 Conclusion and Future Prospects

Abiotic stress has been reported to perturb many physiological processes in plants and to modulate the metabolism and perception of their phytohormones. Plant-associated microbes modulate plant hormone levels and may affect the metabolism of endogenous phytohormones in the plant tissue. Such changes in metabolism can play an important role in plant development and protect plants from biotic and abiotic stresses, including drought, salt, nutrient deficiency, and heavy metals. Their beneficial effect on root and shoot growth, the physiological processes of plants, biomass, and yield of various plants have been reported in many studies. However, the underlying mechanisms and the interaction between phytohormones in the elicitation of the response are not fully understood. The biosynthesis of phytohormones such as IAA, GA, SA, ABA, and GB by root-associated microbes is a compelling mechanism to alter plant physiology and biochemical processes in plant tissues. Employing omics-based approaches, including proteomics, genomics, metagenomics, and metabolomics, on host-microbe-stress interaction studies is also important. Furthermore, studies on the performance of phytohormone-producing microbes in field sites are required to confirm their beneficial effect in natural environments where the competition for nutrient and niches are high.

References

Afzal I, Basra S, Iqbal A (2005) The effect of seed soaking with plant growth regulators on seedling vigor of wheat under salinity stress. J Stress Physiol Biochem 1:6–14

Ahmad P (2010) Growth and antioxidant responses in mustard (*Brassica juncea* L.) plants subjected to combined effect of gibberellic acid and salinity. Arch Agron Soil Sci 56(5):575–588

Ahmad P, Sharma S (2008) Salt stress and phyto-biochemical responses of plants. Plant Soil Environ 54(3):89–99

Ahmad P, Jaleel CA, Salem MA, Nabi G, Sharma S (2010) Roles of enzymatic and nonenzymatic antioxidants in plants during abiotic stress. Crit Rev Biotech 30(3):161–175

Ahmad P, Nabi G, Jeleel CA, Umar S (2011) Free radical production, oxidative damage and antioxidant defense mechanisms in plants under abiotic stress. In: Ahmad P, Umar S (eds) Oxidative stress, role of antioxidants in plants. Studium Press, New Delhi, pp 19–53

Ahmad P, Hakeem KR, Kumar A, Ashraf M, Akram NA (2012) Salt-induced changes in photosynthetic activity and oxidative defense system of three cultivars of mustard (*Brassica juncea* L.) Afr J Biotechnol 11(11):2694–2703

Akbari G, Sanavy SA, Yousefzadeh S (2007) Effect of auxin and salt stress (NaCl) on seed germination of wheat cultivars (*Triticum aestivum* L.) Pak J Biol Sci 10:2557–2561

Alonso-Ramirez A, Rodriguez D, Reyes D, Jimenez JA, Nicolas G, Lopez-Climent M, Gomez-Cadenas A, Nicolas C (2009) Evidence for a role of gibberellins in salicylic acid-modulated early plant responses to abiotic stress in Arabidopsis seeds. Plant Physiol 150:1335–1344

Amara U, Khalid R, Hayat R (2015) Soil bacteria and phytohormones for sustainable crop production. In: Maheshwari DK (ed) Bacterial metabolites in sustainable agroecosystem. Springer, Cham, pp 87–103. https://doi.org/10.1007/978-3-319-24654-3

Ashikari M, Sakakibara H, Lin S, Yamamoto T, Takashi T, Nishimura A, Angeles ER, Qian Q, Kitano H, Matsuoka M (2005) Cytokinin oxidase regulates rice grain production. Science 309:741–745

Azooz MM, Youssef AM, Ahmad P (2011) Evaluation of salicylic acid (SA) application on growth, osmotic solutes and antioxidant enzyme activities on broad bean seedlings grown under diluted seawater. Int J Plant Physiol Biochem 3:253–264

Baumann K (2010) Signalling: ABA's greatest hits. Nat Rev Mol Cell Biol 11(1):2–2

Bianco C, Defez R (2009) *Medicago truncatula* improves salt tolerance when nodulated by an indole-3-acetic acid-overproducing *Sinorhizobium meliloti* strain. J Exp Bot 60:3097–3107

Bozcuk S (1981) Effect of kinetin and salinity on germination of tomato, barley and cotton seeds. Ann Bot 48:81–84

Cardinale M, Ratering S, Suarez C, Montoya AMZ, Geissler-Plaum R, Schnell S (2015) Paradox of plant growth promotion potential of rhizobacteria and their actual promotion effect on growth of barley (*Hordeum vulgare* L.) under salt stress. Microb Res 181:22–32

Cho ST, Chang HH, Egamberdieva D, Kamilova F, Lugtenberg B, Kuo CH (2015) Genome analysis of *Pseudomonas fluorescens* PCL1751: a rhizobacterium that controls root diseases and alleviates salt stress for its plant host. PLoS One. https://doi.org/10.1371/journal.pone.0140231

Creus CM, Sueldo RJ, Barassi CA (2004) Water relations and yield in Azospirillum-inoculated wheat exposed to drought in the field. Can J Bot 82:273–281

de Salamone IEG, Hynes RK, Nelson LM (2001) Cytokinin production by plant growth promoting rhizobacteria and selected mutants. Can J Microbiol 47:404–411

Debez A, Chaibi W, Bouzid S (2001) Effet du NaCl et de regulateurs de croissance sur la germination d' *Atriplex halimus* L. Cah Agric 10:135–138

Dunlap JR, Binzel ML (1996) NaCl reduces indole-3-acetic acid levels in the roots of tomato plants independent of stress induced abscisic acid. Plant Physiol 112:379–384

Egamberdieva D (2009) Alleviation of salt stress by plant growth regulators and IAA producing bacteria in wheat. Acta Physiol Plant 31:861–864

Egamberdieva D, Kucharova Z (2009) Selection for root colonising bacteria stimulating wheat growth in saline soils. Biol Fertil Soils 45:561–573

Egamberdieva D, Kucharova Z, Davranov K, Berg G, Makarova N, Azarova T, Chebotar V, Tikhonovich I, Kamilova F, Validov S, Lugtenberg B (2011) Bacteria able to control foot and root rot and to promote growth of cucumber in salinated soils. Biol Fertil Soils 47:197–205

Egamberdieva D, Berg G, Lindström K, Räsänen LA (2013) Alleviation of salt stress of symbiotic *Galegaofficinalis* L. (goat's rue) by co-inoculation of rhizobium with root colonising *Pseudomonas*. Plant Soil 369(1):453–465

Egamberdieva D, Jabborova D, Hashem A (2015) *Pseudomonas* induces salinity tolerance in cotton (*Gossypium hirsutum*) and resistance to *Fusarium* root rot through the modulation of indole-3-acetic acid. Saudi J Biol 22(6):773–779

Egamberdieva D, Wirth S, Behrendt U, Parvaiz A, Berg G (2017a) Antimicrobial activity of medicinal plants correlates with the proportion of antagonistic endophytes. Front Microbiol 8:199

Egamberdieva D, Wirth S, Jabborova D, Räsänen LA, Berg G, Liao H (2017b) Coordination between *Bradyrhizobium* and root colonizing *Pseudomonas* alleviates salt stress in soybean (*Glycine max* L.) through altering root system architecture and improving nodulation. J Plant Interact 12(1):100–107

Egamberdiyeva D, Hoflich G (2003) Influence of growth promoting bacteria on the growth of wheat at different soils and temperatures. Soil Biol Biochem 35:973–978

Egamberdiyeva D, Hoflich G (2004) Importance of plant growth promoting bacteria on growth and nutrient uptake of cotton and pea in semi-arid region Uzbekistan. J Arid Environ 56:293–301

El-Tayeb MA (2005) Response of barley grains to the interactive effect of salinity and salicylic acid. Plant Growth Regul 45:215–224

Forchetti G, Masciarelli O, Izaguirre MJ, Alemano S, Alvarez D, Abdala G (2010) Endophytic bacteria improve seedling growth of sunflower under water stress, produce salicylic acid, and inhibit growth of pathogenic fungi. Curr Microbiol 61(6):485–493

Fukuda A, Tanaka Y (2006) Effects of ABA, auxin and gibberellin on the expression of genes for vacuolar H^+-inorganic pyrophosphatase, H^+ ATPase subunit A, and Na+/H+ antiporter in barley. Plant Physiol Biochem 44:351–358

Glick BR (2012) Plant growth promoting bacteria: mechanisms and applications. Scientifica 1–15

Glick BR (2014) Bacteria with ACC deaminase can promote plant growth and help to feed the world. Microbiol Res 169:30–39

Golpayegani A, Tilebeni HG (2011) Effect of biological fertilizers on biochemical and physiological parameters of Basil (*Ocimum basilicum* L.) medicine plant. Am Eurasian J Agric Environ Sci 11:411–416

Grayson M (2013) Agriculture and drought. Nature 501:S1. https://doi.org/10.1038/501S1a

Guilfoyle TJ, Hagen G, Li Y, Ulmasov T, Liu Z, Strabala T, Gee MA (1993) Auxin-regulated transcription. Aust J Plant Physiol 20:489–502

Gunes A, Inal A, Alpaslam M, Erslan F, Bagsi EG, Cicek N (2007) Salicylic acid induced changes on some physiological parameters symptomatic for oxidative stress and mineral nutrition in maize (*Zea mays* L.) grown under salinity. J Plant Physiol 164:728–736

Gutierrez CK, Matsui GY, Lincoln DE, Lovell CR (2009) Production of the phytohormoneindole-3-acetic acid by estuarine species of the genus Vibrio. Appl Environ Microbiol 75:2253–2258

Hagen G, Guilfoyle T (2002) Auxin-responsive gene expression: genes, promoters and regulatory factors. Plant Mol Biol 49:373–385

Hashem A, Abd_Allah EF, Alqarawi A, Al-Huqail AA, Wirth S, Egamberdieva D (2016) The interaction between arbuscular mycorrhizal fungi and endophytic bacteria enhances plant growth of *Acacia gerrardii* under salt stress. Front Plant Sci. https://doi.org/10.3389/fmicb.2016.01089

Hayat Q, Hayat S, Irfan M, Ahmad A (2010) Effect of exogenous salicylic acid under changing environment: a review. Environ Exp Bot 68:14–25

Horvath E, Szalai G, Janda T (2007) Induction of abiotic stress tolerance by salicylic acid signaling. J Plant Growth Regul 26:290–300

Hussein MM, Balbaa LK, Gaballah MS (2007) Salicylic acid and salinity effect on growth of maize plants. J Agric Biol Sci 3:321–328

Iqbal M, Ashraf M (2013) Gibberellic acid mediated induction of salt tolerance in wheat plants: growth, ionic partitioning, photosynthesis, yield and hormonal homeostasis. Environ Exp Bot 86:76–85

Jae-Ung H, Youngsook L (2001) Abscisic acid-induced actin reorganization in guard cells of day flower is mediated by cytosolic calcium levels and by protein kinase and protein phosphatase activities. Plant Physiol 125:2120–2128

Jumali SS, Said IM, Ismail I, Zainal Z (2011) Genes induced by high concentration of salicylic acid in *Mitragyna speciosa*. Aust J Crop Sci 5:296–303

Kamilova F, Kravchenko LV, Shaposhnikov AI, Azarova T, Makarova N, Lugtenberg B (2006) Organic acids, sugars, and L-tryptophane in exudates of vegetables growing on stonewool and their effects on activities of rhizosphere bacteria. Mol Plant-Microbe Interact 19:250–256

Karnwal A, Kaushik P (2011) Cytokinin production by fluorescent Pseudomonas in the presence of rice root exudates. Arch Phytopathol Plant Protect 44:1728–1735

Keskin BC, Sarikaya AT, Yuksel B, Memon AR (2010) Abscisic acid regulated gene expression in bread wheat. Aust J Crop Sci 4:617–625

Khakipour N, Khavazi K, Mojallali H, Pazira E, Asadirahmani H (2008) Production of auxin hormone by fluorescent *Pseudomonas*. Am Eurasian J Agric Environ Sci 4(6):687–669

Khan Z, Doty SL (2009) Characterization of bacterial endophytes of sweet potato plants. Plant Soil 322:197–207

Khan MA, Gul B, Weber DJ (2004) Action of plant growth regulators and salinity on seed germination of *Ceratoides lanata*. Can J Bot 82:37–42

Khan MA, Shirazi MU, Ali M, Mumtaz S, Sherin A, Ashraf MY (2006) Comparative performance of some wheat genotypes growing under saline water. Pak J Bot 38:1633–1639

Khan AL, Hamayun M, Kim Y-H, Kang SM, Lee JH, Lee IN (2011) Gibberellins producing endophytic *Aspergillus fumigatus* sp. LH02 influenced endogenous phytohormonal levels, isoflavonoids production and plant growth in salinity stress. Process Biochem 46:440–447

Khan MIR, Asgher M, Khan NA (2014) Alleviation of salt-induced photosynthesis and growth inhibition by salicylic acid involves glycine betaine and ethylene in mungbean (*Vigna radiata* L.) Plant Physiol Biochem 80:67–74

Lavania M, Nautiyal C (2013) Solubilization of tricalcium phosphate by temperature and salt tolerant *Serratia marcescens* NBRI1213 isolated from alkaline soils. Afr J Microbiol Res 7:4403–4413

Lawongsa P, Boonkerd N, Wongkaew S, O'Gara F, Teaumroong N (2008) Molecular and phenotypic characterization of potential plant growth-promoting *Pseudomonas* from rice and maize rhizospheres. World J Microbiol Biotechnol 24:1877–1884

Liu Y, Shi Z, Yao L, Yue H, Li H, Li C (2013) Effect of IAA produced by Klebsiellaoxytoca Rs-5 on cotton growth under salt stress. J Gen Appl Microbiol 59:59–65

Maggio A, Barbieri G, Raimondi G, De Pascale S (2010) Contrasting effects of ga3 treatments on tomato plants exposed to increasing salinity. J Plant Growth Regul 29:63–72

Manjili FA, Sedghi M, Pessarakli M (2012) Effects of phytohormones on proline content and antioxidant enzymes of various wheat cultivars under salinity stress. J Plant Nutr 35:1098–1111

Marulanda A, Barea JM, Azcón R (2009) Stimulation of plant growth and drought tolerance by native microorganisms (AM fungi and bacteria) from dry environments: mechanisms related to bacterial effectiveness. J Plant Growth Regul 28:115–124
Naz I, Bano A, Ul-Hassan T (2009) Isolation of phytohormones producing plant growth promoting rhizobacteria from weeds growing in Khewra salt range, Pakistan and their implication in providing salt tolerance to *Glycine max* L. Afr J Biotechnol 8(21):5762–5766
Nilsen E, Orcutt DM (1996) The physiology of plants under stress – abiotic factors. Wiley, New York, pp 118–130
Ortíz-Castro R, Valencia-Cantero E, López-Bucio J (2008) Plant growth promotion by *Bacillus megaterium* involves cytokinin signaling. Plant Sig Behav 3:263–265
Parray AP, Jan S, Kamili AN, Qadri RA, Egamberdieva D, Ahmad P (2016) Current perspectives on plant growth promoting rhizobacteria. Plant Growth Regul 35(3):877–902
Prakash L, Prathapasenan G (1990) NaCl and gibberellic acid induced changes in the content of auxin, the activity of cellulose and pectin lyase during leaf growth in rice (*Oryza sativa*). Ann Bot 365:251–257
Qadir M, Quillérou E, Nangia V, Murtaza G, Singh M, Thomas RJ, Drechsel P, Noble AD (2014) Economics of salt-induced land degradation and restoration. Nat Resour Forum 38:282–295
Raza A, Faisal M (2013) Growth promotion of maize by desiccation tolerant *Micrococcus luteus*-chp37 isolated from Cholistan desert, Pakistan. Aust J Crop Sci 7(11):1693–1698
Riadh K, Wided M, Hans-Werner K, Chedly A (2010) Responses of halophytes to environmental stresses with special emphasis to salinity. Adv Bot Res 53:117–145
Ribaut JM, Pilet PE (1994) Water stress and indole-3ylacetic acid content of maize roots. Planta 193:502–507
Salomon MV, Bottini R, de Souza Filho GA, Cohen AC, Moreno D, Gil M, Piccoli P (2014) Bacteria isolated from roots and rhizosphere of *Vitis vinifera* retard water losses, induce abscisic acid accumulation and synthesis of defense-related terpenes in in vitro cultured grapevine. Physiol Plant 51(4):359–374
Sawada H, Shim IS, Usui K (2006) Induction of benzoic acid 2-hydroxylase and salicylic acid biosynthesis: modulation by salt stress in rice seedlings. Plant Sci 171:263–270
Schafer M, Brütting C, Meza-Canales ID, Grokinsky DK, Vankova R, Baldwin IT, Meldau S (2015) The role of *cis*-zeatin-type cytokinins in plant growth regulation and mediating responses to environmental interactions. J Exp Bot 66(16):4873–4884
Schmulling T (2002) New insights into the functions of cytokinins in plant development. J Plant Growth Regul 21:40–49
Senaratna T, Touchell D, Bunns E, Dixon K (2000) Acetyl salicylic acid (aspirin) and salicylic acid induce multiple stress tolerance in bean and tomato plants. Plant Growth Regul 30:157–161
Sgroy V, Cassán F, Masciarelli O, Del Papa MF, Lagares A, Luna V (2009) Isolation and characterization of endophytic plant growth-promoting (PGPB) or stress homeostasis-regulating (PSHB) bacteria associated to the halophyte *Prosopis strombulifera*. Appl Microbiol Biotechnol 85:371–381
Shahab S, Nuzhat A, Nasreen SK (2009) Indole acetic acid production and enhanced plant growth promotion by indigenous PSBs. Afr J Agric Res 4:1312–1316
Shakirova FM, Sakhabutdinova AR, Bezrukova MV, Fatkhutdinova RA, Fatkhutdinova D (2003) Changes in the hormonal status of wheat seedlings induced by salicylic acid and salinity. Plant Sci 164:317–322
Sharma SB, Sayyed RZ, Trivedi MH, Gobi AT (2013) Phosphate solubilizing microbes: sustainable approach for managing phosphorus deficiency in agricultural soils. Springerplus 2:587
Shi HZ, Zhu JK (2002) Regulation of expression of the vacuolar $Na+/H+$ antiporter gene *AtNHX1* by salt stress and abscisic acid. Plant Mol Biol 50:543–550
Singh RP, Jha PN (2016) A halotolerant bacterium Bacillus licheniformis HSW-16 augments induced systemic tolerance to salt stress in wheat plant (*Triticum aestivum*). Front Plant Sci 7:1890

Singh D, Dadoria S, Agarwal RM, Shrivastava M, Ahanger MA (2014) A note on salt accumulation with ecophysiological comments. Bionature 30(4):9–14

Smirnoff N (2000) The role of active oxygen in the response of plants to water deficit and desiccation. New Phytol 125:27–58

Srivalli B, Chinnusamy V, Khanna-Chopra R (2003) Antioxidant defense in response to abiotic stresses in plants. J Plant Biol 30:121–139

Stevens J, Senaratna T, Sivasithamparam K (2006) Salicylic acid induces salinity tolerance in tomato (*Lycopersicon esculentum* cv. Roma): associated changes in gas exchange, water relations and membrane stabilisation. Plant Growth Regul 49:77–83

Tasgin E, Attici O, Nalbantogly B (2003) Effect of salicylic acid and cold on freezing tolerance in winter wheat leaves. Plant Growth Regul 41:231–236

Tsavkelova EA, Cherdyntseva TA, Klimova SY, Shestakov AI, Botina SG, Netrusov AI (2007) Orchid-associated bacteria produce indole-3-acetic acid, promote seed germination, and increase their microbial yield in response to exogenous auxin. Arch Microbiol 188:655–664

Tuna AL, Kaya C, Dikilitas M, Higgs D (2008) The combined effects of gibberellic acid and salinity on some antioxidant enzyme activities, plant growth parameters and nutritional status in maize plants. Environ Exp Bot 63:1–9

Tuteja N, Ahmad P, Panda BB, Tuteja R (2009) Genotoxic stress in plants, shedding light on DNA damage, repair and DNA repair helicases. Mutat Res 681:134–149

Upadhyay SK, Singh JS, Saxena AK, Singh DP (2012) Impact of PGPR inoculation on growth and antioxidants status of wheat plant under saline condition. Plant Biol 4:605–611

Wahid A, Perveen M, Gelani S, Basra SMA (2007) Pretreatment of seed with H_2O_2 improves salt tolerance of wheat seedlings by alleviation of oxidative damage and expression of stress proteins. J Plant Physiol 164:283–294

Wang Y, Mopper S, Hasentein KH (2001) Effects of salinity on endogenous ABA, IAA, JA, and SA in *Iris hexagona*. J Chem Ecol 27:327–342

Werner T, Motyka V, Laucou V, Smets R, Van Onckelen H, Schmülling T (2003) Cytokinin-deficient transgenic Arabidopsis plants show multiple developmental alterations indicating opposite functions of cytokinins in the regulation of shoot and root meristem activity. Plant Cell 15:2532–2550

Wolbang CM, Chandler PM, Smith JJ, Ross JJ (2004) Auxin from the developing inflorescence is required for the biosynthesis of active gibberellins in barley stems. Plant Physiol 134:769–776

Yandigeri MS, Meena KK, Singh D, Malviya N, Singh DP, Solanki MKA, Yadav K, Arora DK (2012) Drought-tolerant endophytic actinobacteria promote growth of wheat (*Triticum aestivum*) under water stress conditions. Plant Growth Regul 68(3):411–420

Zaheer A, Mirza BS, Mclean JE, Yasmin S, Shah SM, Malik KA, Mirza MS (2016) Association of plant growth-promoting *Serratia* spp. with the root nodules of chickpea. Res Microbiol 167:510–520

Zalewski W, Galuszka P, Gasparis S, Orczyk W, Nadolska-Orczyk A (2010) Silencing of the HvCKX1 gene decreases the cytokinin oxidase/dehydrogenase level in barley and leads to higher plant productivity. J Exp Bot 61:1839–1851

Zhao Y (2012) Auxin biosynthesis: a simple two-step pathway converts tryptophan to indole-3-acetic acid in plants. Mol Plant 5(2):334–338

Zhao L, Zhang Y (2015) Effects of phosphate solubilization and phytohormone production of *Trichoderma asperellum* Q1 on promoting cucumber growth under salt stress. J Integr Agric 14(8):1588–1597

Chapter 8
Nitric Oxide as a Signaling Molecule in Plant-Bacterial Interactions

Anukool Vaishnav, Sushil K. Sharma, Devendra Kumar Choudhary, Kanti Prakash Sharma, Ees Ahmad, Mahaveer P. Sharma, Aketi Ramesh, and Anil Kumar Saxena

Abstract Nitric oxide (NO), evolved during various biological processes occuring in soil, bacteria, and plants, is acting as signaling molecule to trigger different essential pathways involved in plant-microbe interactions. Reactive nitrogen species (RNS) is present at every developmental stage of plants and plays very important role in their life cycle. This valuable molecule also involved in signaling in response to biotic and abiotic stress in plants. Moreover, NO is very important or said to be a central molecule of nitrogen cycle. The NO is produced during different biological nitrogen transformation processes. Remarkably, the essential information of NO production and its efficient relations with plant and microbes are poorly characterized. This chapter covers the different processes of NO production in soil, bacteria, and plants and their role in different physiological processes. In particular, the role of NO is addressed as a signaling molecule in plant-microbe interactions including legume-rhizobium symbiosis.

Keywords Bacteria • Legumes • Nitric oxide • Nitrogen cycle • Plant

A. Vaishnav • S.K. Sharma (✉) • E. Ahmad • A.K. Saxena
ICAR-National Bureau of Agriculturally Important Microorganisms (ICAR-NBAIM), Maunath Bhanjan 275103, Uttar Pradesh, India
e-mail: sks_micro@rediffmail.com

D.K. Choudhary
Amity Institute of Microbial Technology (AIMT), Amity University Campus, Noida 201313, Uttar Pradesh, India

K.P. Sharma
Department of Biological Sciences, Mody University of Science & Technology (MUST), Sikar 332311, Rajasthan, India

M.P. Sharma • A. Ramesh
ICAR-Indian Institute of Soybean Research, Indore 452001, Madhya Pradesh, India

D. Egamberdieva, P. Ahmad (eds.), *Plant Microbiome: Stress Response*, Microorganisms for Sustainability 5, https://doi.org/10.1007/978-981-10-5514-0_8

8.1 Introduction

Nitric oxide (NO) is a reactive gaseous molecule, which has several regulatory functions in plant and microorganisms. Most of the mechanisms of NO production and catabolism are the same in both plants and microorganims. Nitric oxide is involved during initial stages of plant-microbe interactions. During the interactions, NO quickly diffuse in the form of reactive nitrogen species (RNS) across the plant membranes due to its lipophilic properties. NO, first identified as a vasodilatory messenger (Ignarro et al. 1987), is now recognized to be a signaling molecule in both eukaryotes and prokaryotes. Earlier NO was focused as atmospheric pollutant in plants and uptake by foliage caused phytotoxicity (Wellburn 1990; Hufton et al. 1996). Later, it was demonstrated that plants produce substantial amounts of NO, which is involved in growth, development, and defense responses (Noritake et al. 1996). The bacterial NO metabolism was spelled out in 1954 as a product of denitrification (Wijler and Delwiche 1954). The confirmation of NO as an intermediate of bacterial denitrification was reported in intact cell (Matsubara and Mort 1968) and cell-free extract of *Pseudomonas denitrificans* (Mlyata et al. 1969). Cueto et al. (1996) have reported nitric oxide synthase (NOS) activity in the root nodules of *Lupinus albus*. Later, NO was found as a regulating agent during plant defense (Delledonne et al. 1998). In addition, accumulation of NO and arginine in nodules suggested that NO is playing role in nodule development (Hérouart et al. 2002). Several other reports suggested that NO plays a role in root development by interaction with auxin (Guo et al. 2003; Pagnussat et al. 2004). Keeping in view of NO role in regulation of biological systems, in this chapter we are focusing on potential sources of NO production and its responses in plants and bacteria to understand further on how NO influences beneficial plant-bacterial interactions.

8.2 Production of NO in Soil

The NO is mainly produced from agricultural soils. In a survey report, global NO production from soil is around 8.9 $TgNa^{-1}$ (IPCC 2007). It depends on the several soil factors such as nitrogen content, moisture content, pH, temperature, etc. Figure 8.1 represents different processes in the soil leading to NO production.

Biological nitrogen transformation processes, such as nitrification and denitrification, are main sources of NO production in soils. Autotrophic nitrification process is performed by nitrifying bacteria in the presence of oxygen, in which NO and N_2O are produced as by-products during conversion of ammonium to nitrate. Nitrification process occurs through a two-step process: (1) ammonium-oxidizing bacteria such as *Nitrosomonas*, *Nitrosospira*, and *Nitrosococcus* convert ammonium into nitrite, and (2) nitrite-oxidizing bacteria like *Nitrobacter*, *Nitrospira*, *Nitrococcus*, and *Nitrospina* convert nitrite into nitrate (Fig. 8.2). Under anaerobic conditions, these reactions turn opposite and produce NO. During nitrification process, NO formation

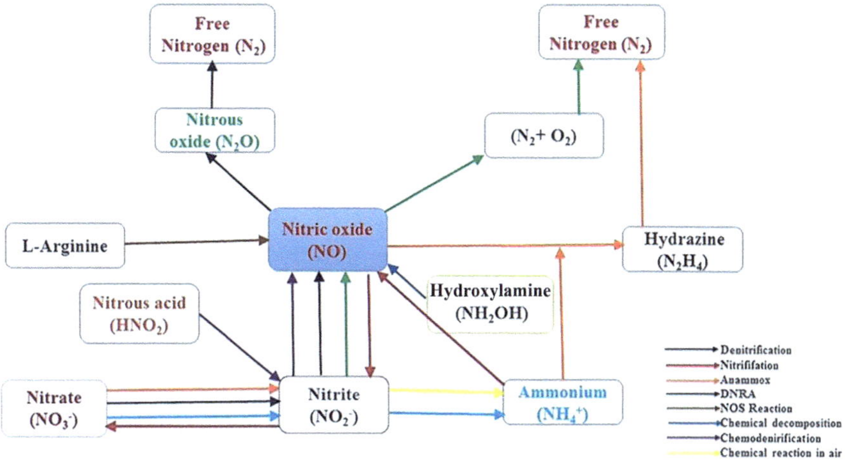

Fig. 8.1 Nitric oxide production in soil through different chemical and biological processes. [*DNRA* stands for dissimilatory nitrate reduction to ammonium, *anammox* stands for anaerobic ammonium oxidation, and *NOS* stands for nitric oxide synthase]

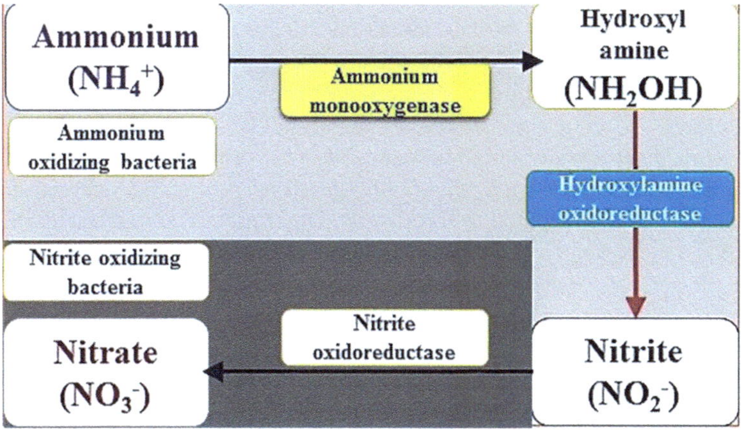

Fig. 8.2 Nitrification process performed by two different autotrophic bacteria *Nitrosomonas* and *Nitrobacteria* in soil. In this process, ammonium is finally converted into nitrate

rate has been determined up to 10% of gross ammonium oxidation (Garrido et al. 2002). In recent study, Daims et al. (2015) cultivated a bacterium namely *Nitrospira* as a complete nitrifying bacterium which is capable of directly oxidizing ammonia to convert nitrate via nitrite by ammonia monooxygenase and hydroxylamine dehydrogenase. This research have changed the fundamental aspect of nitrification and the nitrogen cycle.

Denitrification process is performed by denitrifiers, in which nitrate is reduced into gaseous product (N_2) with three intermediates, namely, nitrite, nitric oxide, and

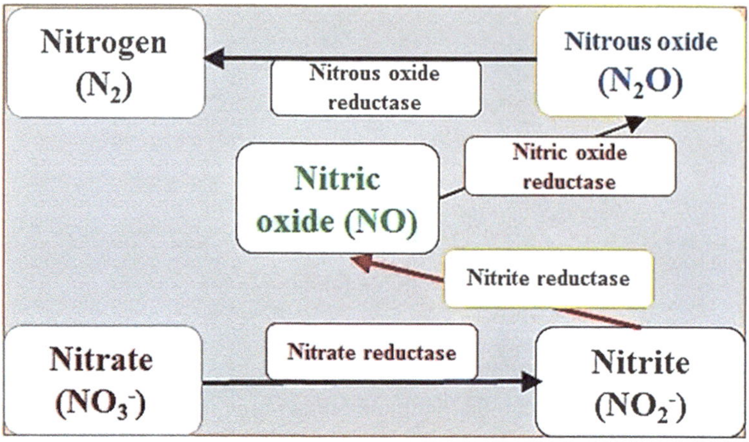

Fig. 8.3 Denitrification process catalyzed by four different enzymes, in which nitrate is finally converted into nitrogen gas

nitrous oxide. The complete reaction is catalyzed by four different enzymes including nitrate reductase (NR), nitrite reductase (NIR), nitric oxide reductase (NOR), and nitrous oxide reductase (N_2OR) (Fig. 8.3). Denitrification can occur under both aerobic and anaerobic conditions in which anaerobic condition is considered for NO production.

The denitrification rate is high in loam soil in the presence of nitrigenous fertilizers. In a survey study, Barton et al. (1999) have observed higher denitrification in agricultural soils as compared to forest soils. Other processes associated with NO production are codenitrification, DNRA, chemodenitrification, NOS, anammox, etc. (Medinets et al. 2015). Chemodenitrification is considered as an important process for NO production in soils. In this process, nitrite or nitrate is nonenzymatically converted into nitrogen gas at low pH. Chemodenitrification normally requires ammonium ions, reduced metals, high organic carbon, and soil water content. The DNRA is reported as a main pathway responsible for major consumption of nitrate in coastal, wetland, and terrestrial ecosystem (Rütting et al. 2011; Giblin et al. 2013).

8.3 Production and Response of NO in Plant and Bacteria

8.3.1 Nitric Oxide in Plant

There are many enzymatic and nonenzymatic reactions for NO production in plants (Fig. 8.4). In the enzymatic reaction, nitric oxide synthase (NOS) produces NO through oxidative pathways under normal oxygen status, wherein arginine works as

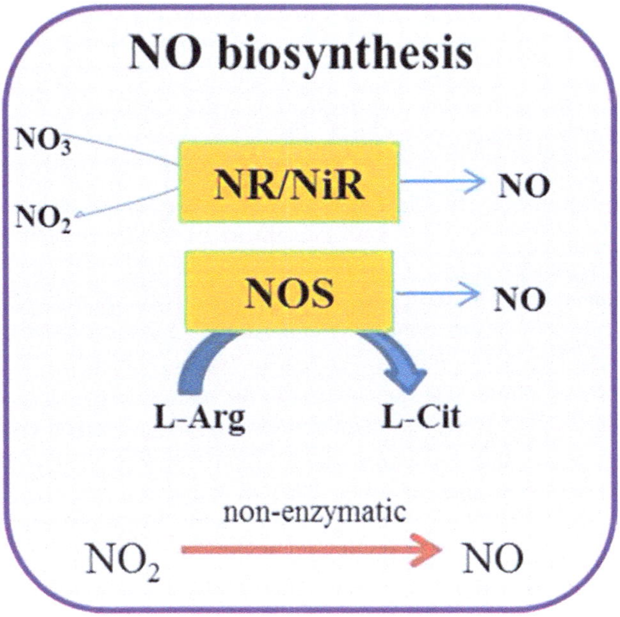

Fig. 8.4 Metabolic pathway involved in plant N transformation leading to NO production

a primary substrate (Gupta et al. 2011). The NOS activity in plants was first time detected indirectly through reaction of NOS inhibitor NG-monomethyl-L-arginine (NMMA) and its product L-citrulline (Cueto et al. 1996). A unique *AtNOS1* gene, encoding for NOS-like proteins, was isolated from *Arabidopsis*. NO contents were detected higher in wild type as compared to *Atnos1* mutant *Arabidopsis* plant (Guo et al. 2003). Other enzyme for NO production is nitrate reductase (NR), which catalyzes nitrite to NO (Yamasaki et al. 1999). This process has been reported in several plant species including maize, cucumber, spinach, and sunflower (Cohen et al. 2010). Yamasaki and Sakihama (2000) have confirmed NO production through NR activity by using sodium azide, a NR inhibitor that reduced NO production. Desikan et al. (2002) have used NR-mutated *Arabidopsis* plants, which showed lower NO production.

In nonenzymatic conversion, nitrite contents react with different plant metabolites under acidic conditions and produce NO (Wendehenne et al. 2001; Wojtaszek 2000). In a study, a nonenzymatic formation of NO was observed in the apoplast of barley aleurone cells (Bethke et al. 2004).

NO plays an important role in plant immunity. Several reports are available on NO role in plant stress tolerance mechanism (Tufan Oz et al. 2015; Arasimowicz-Jelonek and Floryszak-Wieczorek 2014). NO interacts with H_2O_2 and induces hypersensitive response during pathogen attack (Delledonne et al. 1998). During pathogen attack, NO regulates defense enzymatic activities and interacts with functional proteins by nitrosylating their cysteine residue (Tada et al. 2008). In a study, the soluble guanylate cyclase (sGC) produces NO and accumulates in plants during infection by pathogens, but plant enzyme was not responsible for production of GC like in animals, where NO exerts its action through cGMP (Meier et al. 2009; Isner and Maathuis 2011). However, NO production was recorded in *L. japonicus* upon infection of plant pathogens *Ralstonia solanacearum* and *Pseudomonas syringae* (Nagata et al. 2008).

Sodium nitroprusside (SNP)-elevated resistance to salt stress was found through regulation of alternative oxidase (AOX) pathway in *Medicago* plants. AOX could contribute to regulate the accumulation of reactive oxygen (ROS) and protect photosystem (Jian et al. 2015). Exogenous application of SNP has been found to induce antioxidant compounds and reduce phytochelatin level in response to arsenic toxicity (As) (Singh et al. 2015). Being a free radical, NO also played a role in alleviation of oxidative stress. Nitric oxide can neutralize the harmful effect of reactive oxygen species by directly interacting to them or inducing an antioxidant enzyme activity (Laspina et al. 2005). In a study, Fu et al. (2015) have reported that exogenous application of NO donor enhances chilling stress tolerance in plants through activating antioxidant system and plasma membrane transporter led to decreased ROS accumulation. NO also enhances iron uptake mechanism in plants. It activates iron transport and starvation pathway in root under limiting conditions (Graziano and Lamattina 2007; Singh et al. 2015). Several reports highlighted the role of NO in root morphogenesis that enhances uptake of nutrient content. In a study, NO treatment increased lateral root formation in rice plants that enhances N uptake and N use efficiency under low nitrate conditions (Sun et al. 2015). In addition NO is also reported to induce adventitious root formation through interaction with IAA hormone (Pagnussat et al. 2002).

8.3.2 Nitric Oxide in Bacteria

Mechanisms of bacterial NO production are more diverse than those of animals and plants. In contrast to eukaryotes, the formation of NO in prokaryotes has mainly been attributed to catabolic processes, NO being an intermediate in both denitrification (Zumft 1997) and nitrification pathways (Kuenen and Robertson 1994) (Fig. 8.5). Denitrification helps bacteria to respire under anaerobic environment in which nitrate is final electron acceptor instead of O_2. The end product of denitrification pathway is the startup of heterotrophic nitrification, in which NO produced through reduction of N_2 (Anderson et al. 1993). These processes are reported in plant growth-promoting rizobacteria (PGPR) including *Pseudomonas* spp., *Arthrobacter* spp., *Bacillus* spp., and *Azospirillum* spp. (Cutruzzolà 1999). Moreover, bacterial nitric oxide synthase (bNOS) activity is also reported for NO production in which oxidation of L-arginine to L-citrulline occurs. The first report on bacterial NOS was published by Chen and Rosazza (1994) in the genus *Nocardia*. The bNOS-dependent NO generation has been observed in many plant-associated and free-living bacteria including *Sinorhizobium*, *Mycobacterium*, *Nocardia*,

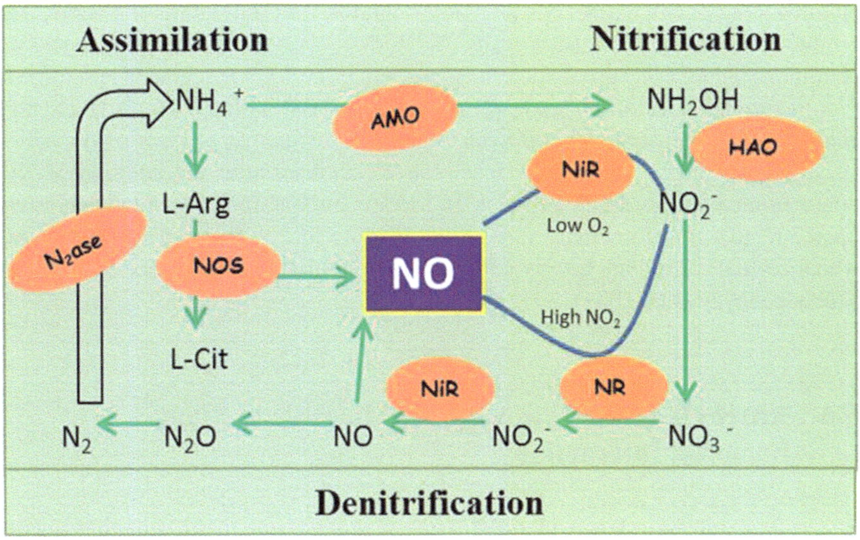

Fig. 8.5 Nitric oxide production pathways in bacteria. [*Orange ovals* indicate enzymes used by bacteria to synthesize NO. *HAO* hydroxylamine oxidoreductase, *NR* nitrate reductase, *NiR* nitrite reductase, *NOS* nitric oxide synthase, *N₂ase* nitrogenase, and *AMO* ammonia monooxygenase (Cohen et al. 2010)]

Rhodococcus, *Streptomyces*, *Bacillus*, *Geobacillus*, and *Paenibacillus* (Pii et al. 2007; Cohen et al. 2010).

Dissimilatory nitrate reduction to ammonium (DNRA) is often viewed as an important process for NO production in bacteria and conserves N within the ecosystem. In this process, NO_2 is converted into NH_4 in the presence of NO_3 as an electron acceptor and produces NO and N_2O as by-products. It is performed by heterotrophic organisms that uses organic carbon as the electron donor and by chemolithoautotrophic organisms that rely on inorganic substrates. Another source of NO is anaerobic ammonium oxidation (anammox) that occurs in anoxic, lithotrophic, and slow-growing bacteria. In this process, NO_2^- is converted into N_2 in the presence of NH_4^+, an electron donor. In the first step, NO_2 is reduced into NO, which reacts with NH_4 to form hydrazine (N_2H_4), and finally N_2H_4 is enzymatically dehydrogenized into N_2. Meanwhile some NO_2 is also oxidized in NO_3. Since NO is an important intermediates of anammox process, the anammox bacteria have the ability to tolerate high concentration of NO (Kartal et al. 2010). Anammox bacteria was first time discovered from bioreactors, and then these bacteria were reported from marine to agricultural ecosystems (Mulder et al. 1995; Medinets et al. 2015).

Bacterial nitric oxide synthase (bNOS) activity is also reported to protect host cells against antibiotics (Gusarov et al. 2009). In addition, bacterial heme-nitric oxide/oxygen-binding protein (H-NOX) is reported for biofilm formation through controlling c-di-GMP levels (Plate and Marletta 2012). *Nitrosomonas europaea* cultures treated with exogenous NO were found to turn into nonmotile forms and produce biofilms on the reactor walls (Schmidt et al. 2004). In a study, anaerobic conditions induced NO production in *Pseudomonas aeruginosa* cells that preformed biofilm formation and stimulated swarming motility (Barraud et al. 2006). Hochgrafe et al. (2008) demonstrated that *Bacillus subtilis* and *Staphylococcus aureus* synthesized more proteins involved in anaerobic metabolism after the addition of NO donor in medium. In an experiment, NO treatment was found to modulate ferric uptake regulator (*fur*) activity through its binding to Fe^{3+} in the Fe-*fur* complex, which regulates Fe homeostasis under stress conditions (D'Autréaux et al. 2002; Mukhopadhyay et al. 2004).

8.4 Nitric Oxide (NO) as a Signal Molecule in Plant-Bacterial Interactions

Recent research has shown that NO participates in early basal signaling during plant-bacterial interactions. NO modulates biological function in the form of RNS that react with proteins and alter their functions (Bellin et al. 2013). Bacteria and plants share several similar features for NO-producing pathways; due to this NO acts as signaling molecule during bacterial-root association. A list of important findings on NO role during plant-bacterial interactions is represented in Table 8.1.

The NO concentration inside the host and symbiont cells is tightly controlled by hemoglobins (Hbs) and thioredoxins/nitrosoglutathione (GSNO) reductase to limit toxic effects. The association of Hb and NO emerges as a key component between symbiotic and pathogenic interactions. Efforts has been undertaken to explore the role of NO in symbiotic interactions (Hichri et al. 2016). For example, at the basal level of infection, NO-dependent expression of plant defense-related genes is induced. After that the level of NO production depends on the type of interactions. In case of pathogen interaction, the Hb expression is maintained at low level, leading to a prolonged NO production and an elevated defense response. In case of symbiotic interaction, the Hb expression is rapidly induced, which decreases the NO level and establishment of the host process (Hichri et al. 2016). In the arbuscular mycorrhizae (AM) interaction, a NO scavenger gene VfLb29 was upregulated in arbuscule-containing cells that suppressed defense responses (Vieweg et al. 2004). The NR pathway is the best source for NO production in both pathogen and symbiotic interactions. Jian et al. (2015) demonstrated that NR-dependent NO increased the resistance against cucumber mosaic virus through regulating systemic acquired resistance (SAR) pathway. In another study, NO fumigation inhibited the disease symptoms of *Pseudomonas syringae* pv. tomato in *A. thaliana* NR double-deficient mutant plants (Vitor et al. 2013). Horchani et al. (2011) reported that NR activity in *M. truncatula* is involved in NO production during *S. meliloti* infection. The involvement of NR activity in NO production is also reported in *G. mosseae* arbuscular

Table 8.1 Some important findings on role of NO in plant-bacterial interactions

Bacteria	Plant	Role of NO	References
Sinorhizobium meliloti	*Medicago truncatula*	Plant NR produced NO present in root hair and nodule primodia	del Giudice et al. (2011)
			Horchani et al. (2011)
Bradyrhizobium japonicum	Soybean	NO activates bacterial responses to low O_2 tension in soybean	Leach et al. (2010)
			Sanchez et al. (2010)
			Mesa et al. (2003)
Mesorhizobium loti	*Lotus japonicus*	Class 1 plant hemoglobin genes enhance symbiotic nitrogen fixation	Shimoda et al. (2009)
Pseudomonas fluorescens	Tomato	Higher NO production inhibits *Ralstonia solanacearum*	Wang et al. (2005)
Azospirillum brasilense	Tomato	NO involved in lateral root formation	Creus et al. (2005)
			Molina-Favero et al. (2008)
Azospirillum brasilense	Wheat	Lectins of bacterium induced NO in seedlings	Alen'kina et al. (2014)
Cucumber mosaic virus	*Arabidopsis*	NO involved in resistance mechanism	Jian et al. (2015)
Botrytis cinerea	*Nicotiana benthamiana*	NO participates in disease resistance to necrotrophic pathogen	Asai and Yoshioka (2009)
Pseudomonas syringae	*Arabidopsis thaliana*	Nitrite as the major source of nitric oxide during infection	Modolo et al. (2005)
Sclerotinia sclerotiorum	*Arabidopsis thaliana*	NO participates in defense-related signaling pathways controlling disease resistance	Perchepied et al. (2010)
Botrytis cinerea	*Arabidopsis thaliana*	NO production mediates oligogalacturonides-triggered immunity and resistance	Rasul et al. (2012)
Pseudomonas fluorescens YT101	Maize	Respiratory nitrate reductase produced NO involved in bacterial colonization	Ghiglione et al. (2000)
Phyllobacterium sp.	*Arabidopsis thaliana*	NO involved in root architecture and N nutrition	Mantelin et al. (2006)
Pseudomonas simiae	Soybean	Exogenous NO enhanced root colonization and salt stress tolerance	Vaishnav et al. (2016)

mycorrhizal (AM) association with *M. truncatula* and *Nicotiana tabacum* (Moche et al. 2010; Calcagno et al. 2012).

The presence of NO in nodules was first time reported by Maskall et al. in 1977. After that, several studies observed NO production in different stages of symbiosis by using fluorescent probes. After 4 h of postinoculation with symbiont, NO production was shown in root of plants; after 4 days postinfection, NO was detected in root hair and nodule primordia, and at the mature stage of symbiosis,

NO production was observed in senescence zone (del Giudice et al. 2011; Baudouin et al. 2006; Cam et al. 2012; Nagata et al. 2008). Bloom et al. (2003) have reported the role of NO in root development. Nitric oxide regulates cell cycle regulatory genes that control lateral root formation (Lanteri et al. 2006. In addition, *Azospirillum*-mediated root development was also appeared to be NO dependent. *Azospirillum*-inoculated tomato roots displayed higher NO content as compared to non-inoculated roots (Creus et al. 2005). Similarly, NO has been reported in AM and lichen symbiosis. The production of NO is considered as a general response of lichen's rehydration (Hichri et al. 2016). NO accumulation was detected in the early stage of AM symbiosis with roots of *M. truncatula*. This accumulation stimulates lateral root formation, which was absent in nonsymbiotic plants (Calcagno et al. 2012).

Plant-derived products induce bNOS activity that improves chances of bacterial survival in plant environment. *Azolla pinnata*-derived sucrose was found to induce bNOS activity in *Rhodococcus* strain APG1 and exhibited potential significance in symbiotic association (Cohen et al. 2010). PGPR inoculation along with SNP was found to improve survival of soybean plant growth under salt stress. SNP-treated salt-stressed plants exhibited enhanced activities of different stress enzymes, [K^+]/[Na^+] ratio, and proline content as compared to non-treated salt-stressed plants (Vaishnav et al. 2013). *Streptomyces*-derived NO enhanced the uptake of iron in plants during iron deficiency. Nitric oxide induces the mobilization of iron in plants by forming iron nitrosyl complexes (Graziano and Lamattina 2005. Wang et al. (2005) examined the effect of NO as a biocontrol agent. They reported that a NO-overproducing mutant of *P. fluorescens* strain had showed higher antagonistic activity for *Ralstonia solanacearum* as compared to wild type. In a study, soil amended with glucosinolate increased systemic protection of *Brassica napus* plant affected with *Rhizoctonia solani.* This specific amendment changed the structure of the microbial community increasing nitrifiers, with the concomitant increase in NO production in the soil (Cohen et al. 2005).

Nitric oxide enhances biofilm formation in bacteria, which is useful for rhizoplane colonization of bacteria and protects roots against further aggression from the phytopathogen (Compant et al. 2010). Boddey et al. (1986) have found that a strain of *A. Brasilense* mutated for NO production was unable to colonize with plant root and produce plant growth stimulatory effects. Recently, Vaishnav et al. (2016) have reported that SNP treatment enhances biofilm formation in *P. simiae* that contributes in better colonization and plant growth promotion under saline stress (Figs. 8.6 and 8.7).

In continuation, SNP treatment altered root exudate pattern of soybean plants which were found to attract bacterial cells. In addition, SNP treatment induced two new bacterial volatile compounds including 4-nitroguaiacol and quinolone that showed potential for soybean seed germination under 100 mM NaCl stress. In gene expression analysis of soybean plant, the expression of plant genes was determined in response to NO under conditions of *P. simiae* bacterial inoculation. Interestingly, nitrite reductase and antioxidative enzyme genes were upregulated, whereas Na^+ transporter (HKT1) was downregulated under salt stress.

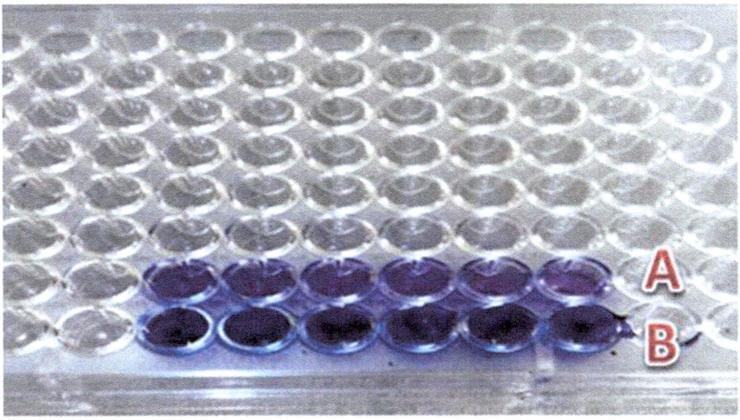

Fig. 8.6 Production of biofilm under SNP treatment. (*A*) *P. simiae* bacterial cells without treatment (*B*). *P. simiae* bacterial cells with SNP treatment. Higher bacterial OD_{570nm} was found in B cells (Vaishnav et al. 2016)

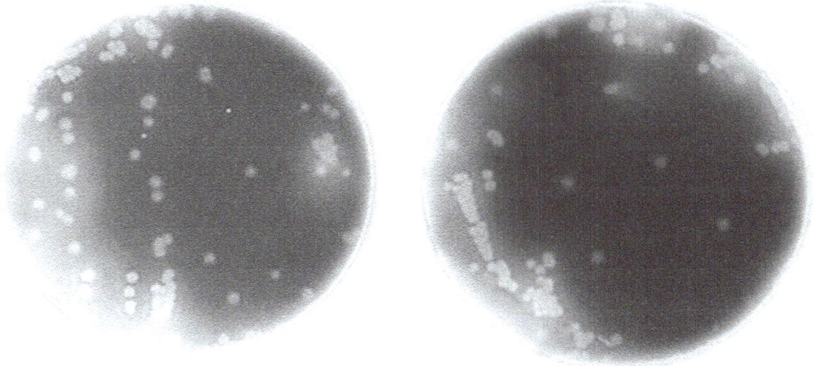

[A] [B]

Fig. 8.7 *Pseudomonas simiae* chemotaxis assay for SNP-treated (**A**) and non-treated (**B**) soybean root exudates. [SNP-treated root exudates attracted more bacterial colonies as compared to non-treated (Vaishnav et al. 2016)]

8.5 Role of NO in Symbiosis of *L. japonicus* and *M. loti*: A Case Study

The N_2-fixing symbiosis contains many steps including the cross identification between both symbiotic partners and the growth of nodules in the plants required for the entry of bacteria. Many researcher identified that NO acts as signaling molecule at different stages from initiation and interaction during nodule development. Recently, many research showed that nodulation process is controlled by NO through cytokinin, c-PTIO, flavonoid, terpene, and defense response proteins (Hichri et al. 2016). Nagata et al. (2008) found that NO was observed after few hours at the root surface of *L. japonicus* when inoculated with *Mesorhizobium loti* and after 10 and 24 h the production of NO was decreased. Whereas inoculation of nonsymbiont *Rhizobium* with *L. japonicus* does not induce the production of NO indicates that NO production mainly depends on the rhizobia and their host plant. The expression and repression of gene LjHB1 was checked by NO donor and NO scavenger, respectively (Shimoda et al. 2005). The LjHB1 gene encoding protein nsHb1was expressed after initial accumulation of NO in plant root of *L. japonicus* infected with *M. loti* (Nagata et al. 2008, 2009). These experiments indicate that the expression of Hb1 starts after formation of NO in plant roots and permits reception of symbiont by downregulating the NO in plants to reduce the response of plant defense. Murakami et al. (2011) validated that lipopolysaccharide induces the NO production in *M. loti* during symbiosis with *L. japonicus*. The release of NO was detected in *M. truncatula* and *S. meliloti* interaction by using diaminofluorescein 2-diacetate (DAF-2DA), NO-specific cell-permeable fluorescent probe, and NO biosensor bacterial strain (del Giudice et al. 2011). The dividing cortical cells in nodule produced NO during *M. Loti-L. japonicus* symbiosis. The transcriptomic analysis was performed with inoculated root of *M. truncatula* using RNA-Seq technology. The analysis showed that 2030 genes were affected by NO, out of them, many genes coded proteins which are involved in nodule development (del Giudice et al. 2011).

8.6 Conclusion and Future Prospects

A sustainable crop production requires a detailed knowledge of the interrelationships between plant and microbes present in the soil. Their interactions are responsible for important processes such as carbon sequestration, ecosystem functioning, and nutrient cycling. NO is recognized as a signaling molecule in diverse microbial symbioses with plants. Different NO sources have been identified, which are involved during symbiosis development. For example, NOS-like activities are observed in free-living *Rhizobium* cells during the early steps of interaction. Similarly, denitrification and NR system have been found in mature nodules. In contrast to NO role as an antimicrobial molecule during pathogen invasion, it also participates in beneficial associations to induce root hair formation and nodule development. NO changes the metabolic activities of plant, bacteria, and fungi leading to release of specific type of products which are helpful for beneficial interactions. Considering these findings, it is apparent that NO-producing/NO-tolerant microbes can be used for sustainable agriculture practices. We are at a very early stage in understanding the role of NO in both plants and microbes and their symbiosis establishment. Therefore, deciphering these mechanisms and relationship of NO appears to be another promising area of research for the future.

Acknowledgements This compilation was done under the project entitled "Development of Methods for Detection of Dissimilatory Nitrate Reduction to Ammonium (DNRA) Process in Bacteria from Different Agroecosystems" operating at ICAR-NBAIM Mau.

References

Alen'kina SA, Petrova LP, Sokolova MK, Chernyshova MP, Trutneva KA, Bogatyrev VA, Nikitina VE (2014) Comparative assessment of inductive effects of *Azospirillum lectins* with different antigenic properties on the signal systems of wheat seedling roots. Mikrobiologiia 83:336–345

Anderson IC, Poth M, Homstead J, Burdige D (1993) A comparison of NO and N_2O production by the autotrophic nitrifier *Nitrosomonas europaea* and the heterotrophic nitrifier *Alcaligenes faecalis*. Appl Environ Microbiol 59:3525–3533

Arasimowicz-Jelonek M, Floryszak-Wieczorek J (2014) Nitric oxide: an effective weapon of the plant or the pathogen? Mol Plant Pathol 15:406–416

Asai S, Yoshioka H (2009) Nitric oxide as a partner of reactive oxygen species participates in disease resistance to necrotrophic pathogen *Botrytis cinerea* in *Nicotiana benthamiana*. Mol Plant-Microbe Interact 22:619–629

Barraud N, Hassett DJ, Hwang SH, Rice SA, Kjelleberg S, Webb JS (2006) Involvement of nitric oxide in biofilm dispersal of *Pseudomonas aeruginosa*. J Bacteriol 188:7344–7353

Barton L, McLay CDA, Schipper LA, Smith CT (1999) Annual denitrification rates in agricultural and forest soils: a review. Aust J Soil Res 37:1073–1093

Baudouin E, Pieuchot L, Engler G, Pauly N, Puppo A (2006) Nitric oxide is formed in *Medicago truncatula-Sinorhizobium meliloti* functional nodules. Mol Plant-Microbe Interact 19:970–975

Bellin D, Asai S, Delledonne M, Yoshioka H (2013) Nitric oxide as a mediator for defense responses. Mol Plant-Microbe Interact 26:271–277

Bethke PC, Gubler F, Jacobsen JV, Jones RL (2004) Dormancy of *Arabidopsis* seeds and barley grains can be broken by nitric oxide. Planta 219:847–855

Bloom AJ, Meyerho VPA, Taylor AR, Rost TL (2003) Root development and absorption of ammonium and nitrate from the rhizosphere. J Plant Growth Regul 21:416–431

Boddey RM, Baldani VL, Baldani JI, Döbereiner J (1986) Effect of inoculation of *Azospirillum* spp. on nitrogen accumulation by field-grown wheat. Plant Soil 95:109–121

Calcagno C, Novero M, Genre A, Bonfante P, Lanfranco L (2012) The exudate from an arbuscular mycorrhizal fungus induces nitric oxide accumulation in *Medicago truncatula* roots. Mycorrhiza 22:259–269

Cam Y, Pierre O, Boncompagni E, Herouart D, Meilhoc E, Bruand C (2012) Nitric oxide (NO): a key player in the senescence of *Medicago truncatula* root nodules. New Phytol 196:548–560

Chen YJ, Rosazza JP (1994) A bacterial nitric oxide synthase from a *Nocardia* species. Biochem Biophys Res Commun 203:1251–1258

Cohen MF, Yamasaki H, Mazzola M (2005) *Brassica napus* seed meal soil amendment modifies microbial community structure, nitric oxide production and incidence of *Rhizoctonia* root rot. Soil Biol Biochem 37:1215–1227

Cohen MF, Lamattina L, Yamasaki H (2010) Nitric oxide signaling by plant-associated bacteria. In: Hayat S (ed) Nitric oxide in plant physiology. Wiley-VCH, Weinheim, pp 161–172

Compant S, Clément C, Sessitsch A (2010) Plant growth-promoting bacteria in the rhizo-and endosphere of plants: their role, colonization, mechanisms involved and prospects for utilization. Soil Biol Biochem 42:669–678

Creus CM, Graziano M, Casanovas EM, Pereyra MA, Simontacchi M, Puntarulo S, Barassi CA, Lamattina L (2005) Nitric oxide is involved in the *Azospirillum brasilense*-induced lateral root formation in tomato. Planta 221:297–303

Cueto M, Hernandez-Perera O, Martın R, Bentura ML, Rodrigo J, Lamas S, Golvano MP (1996) Presence of nitric oxide synthase activity in roots and nodules of *Lupinus albus*. FEBS Lett 398:159–164

Cutruzzolà F (1999) Bacterial nitric oxide synthesis. Biochim Biophys Acta 1411:231–249

Daims H, Lebedeva EV, Pjevac P, Han et al (2015) Complete nitrification by *Nitrospira* bacteria. Nature 528:504–509

del Giudice J, Cam Y, Damiani I, Fung-Chat F, Meilhoc E, Bruand C, Brouquisse R, Puppo A, Boscari A (2011) Nitric oxide is required for an optimal establishment of the *Medicago truncatula-Sinorhizobium meliloti* symbiosis. New Phytol 191:405–417

Delledonne M, Xia Y, Dixon RA, Lamb C (1998) Nitric oxide functions as a signal in plant disease resistance. Nature 394:585–588

Desikan R, Griffiths R, Hancock J, Neill S (2002) A new role for an old enzyme: nitrate reductase-mediated nitric oxide generation is required for abscisic acid-induced stomatal closure in *Arabidopsis thaliana*. PNAS 99:16314–16318

D'Autréaux B, Touati D, Bersch B, Latour JM, Michaud-Soret I (2002) Direct inhibition by nitric oxide of the transcriptional ferric uptake regulation protein via nitrosylation of the iron. Proc Natl Acad Sci 99:16619–16624

Fu J, Chu X, Sun Y, Miao Y, Xu Y, Hu T (2015) Nitric oxide mediates 5-aminolevulinic acid-induced antioxidant defense in leaves of Elymus nutans Griseb. exposed to chilling stress. PLoS One 10:e0130367

Garrido F, Henault C, Gaillard H, Perez S, Germon JC (2002) N_2O and NO emissions by agricultural soils with low hydraulic potentials. Soil Biol Biochem 34:559–575

Ghiglione J-F, Gourbiere F, Potier P, Philippot L, Lensi R (2000) Role of respiratory nitrate reductase in ability of *Pseudomonas fluorescens* YT101 to colonize the rhizosphere of maize. Appl Environ Microbiol 66:4012–4016

Giblin AE, Tobias CR, Song B, Weston N, Banta GT, Rivera-Monroy VH (2013) The importance of dissimilatory nitrate reduction to ammonium (DNRA) in the nitrogen cycle of coastal ecosystems. Oceanography 26:124–131

Graziano M, Lamattina L (2005) Nitric oxide and iron in plants: an emerging and converging story. Trends Plant Sci 10:4–8

Graziano M, Lamattina L (2007) Nitric oxide accumulation is required for molecular and physiological responses to iron deficiency in tomato roots. The Plant J 52:949–960

Guo FQ, Okamoto M, Crawford NM (2003) Identification of a plant nitric oxide synthase gene involved in hormonal signalling. Science 302:100–103

Gupta KJ, Fernie AR, Kaiser WM, van Dongen JT (2011) On the origins of nitric oxide. Trends Plant Sci 16:160–168

Gusarov I, Shatalin K, Starodubtseva M, Nudler E (2009) Endogenous nitric oxide protects bacteria against a wide spectrum of antibiotics. Science 325:1380–1384

Hérouart D, Baudouin E, Frendo P, Harrison J, Santos R, Jamet A, Van de Sype G, Touati D, Puppo. A (2002) Reactive oxygen species, nitric oxide, and glutathione: a key role in the establishment of the legume-Rhizobium symbiosis? Plant Physiol Biochem 40: 619–624

Hichri I, Boscari A, Meilhoc E, Catalá M, Barreno E, Bruand C, Lanfranco L, Brouquisse R (2016) Nitric oxide: a multitask player in plant–microorganism symbioses. In: Lamattina L, Garcıa-Mata C (eds) Gasotransmitters in plants. Springer, Cham, pp 239–268

Hochgrafe F, Wolf C, Fuchs S, Liebeke M, Lalk M, Engelmann S, Hecker M (2008) Nitric oxide stress induces different responses but mediates comparable protein thiol protectionin *Bacillus subtilis* and *Staphylococcus aureus*. J Bacteriol 190:4997–5008

Horchani F, Prevot M, Boscari A, Evangelisti E, Meilhoc E, Bruand C, Raymond P, Boncompagni E, Aschi-Smiti S, Puppo A, Brouquisse R (2011) Both plant and bacterial nitrate reductases contribute to nitric oxide production in *Medicago truncatula* nitrogen-fixing nodules. Plant Physiol 155:1023–1036

Hufton CA, Besford RT, Wellburn RA (1996) Effects of NO (+NO2) pollution on growth, nitrate reductase activities and associated protein contents in glasshouse lettuce grown hydroponically in winter with CO2 enrichment. New Phytol 133:495–501

Ignarro LJ, Buga GM, Wood KS, Byrns RE, Chaudhuri G (1987) Endothelium-derived relaxing factor produced and released from artery and vein is nitric oxide. Proc Natl Acad Sci U S A 84:9265–9269

IPCC (2007) Agriculture, forestry and other land use. In: Eggleston HS, Buendia L, Miwa K, Ngara T, Tanabe K (eds) IPCC guidelines for National Greenhouse gas inventories. IGES, Hayama

Isner JC, Maathuis FJ (2011) Measurement of cellular cGMP inplant cells and tissues using the endogenous fluorescent reporter FlincG. Plant J 65:329–334

Jian W, D-w Z, Zhu F, Wang S-x, Zhu T, X-j P, Zheng T, Feng H, Lin H-h (2015) Nitrate reductase-dependent nitric oxide production is required for regulation alternative oxidase pathway involved in the resistance to cucumber mosaic virus infection in Arabidopsis. Plant Growth Regul 77:99–107

Kartal B, Kuenen JV, Van Loosdrecht MCM (2010) Sewage treatment with anammox. Science 328:702–703

Kuenen JG, Robertson LA (1994) Combined nitrification-denitrification processes. FEMS Microbiol Rev 15:109–117

Lanteri ML, Pagnussat GC, Lamattina L (2006) Calcium and calcium-dependent protein kinases are involved in nitric oxide-and auxin-induced adventitious root formation in cucumber. J Exp Bot 57:1341–1351

Laspina NV, Groppa MD, Tomaro ML, Benavides MP (2005) Nitric oxide protects sunflower leaves against Cd-induced oxidative stress. Plant Sci 169:323–330

Leach J, Keyster M, Du Plessis M, Ludidi N (2010) Nitric oxide synthase activity is required for development of functional nodules in soybean. J Plant Physiol 167:1584–1591

Mantelin S, Desbrosses G, Larcher M et al (2006) Nitrate-dependent control of root architecture and N nutrition are altered by a plant growth-promoting *Phyllobacterium* sp. Planta 223:591–603

Maskall CS, Gibson JF, Dart PJ (1977) Electron-paramagnetic-resonance studies of leghaemoglobins from soya-bean and cowpea root nodules. Identification of nitrosyl-leghaemoglobin in crude leghaemoglobin preparations. Biochem J 167:435–445

Matsubara T, Mort T (1968) Studies on denitrification. IX. Nitrous oxide, its production and reduction to nitrogen. J Biochem 64:863–871

Medinets S, Skiba U, Rennenberg H, Butterbach-Bahl K (2015) A review of soil NO transformation: associated processes and possible physiological significance on organisms. Soil Biol Biochem 80:92–117

Meier S, Madeo L, Ederli L, Donaldson L, Pasqualini S, Gehring C (2009) Deciphering cGMP signatures and cGMP-dependent pathways in plant defence. Plant Signal Behav 4:307–309

Mesa S, Bedmar EJ, Chanfon A et al (2003) *Bradyrhizobium japonicum* NnrR, a denitrification regulator, expands the *FixLJ-FixK2* regulatory cascade. J Bacteriol 185:3978–3982

Mlyata M, Matsubara T, Mori T (1969) Studies on denitrification. XI Some properties of nitric oxide reductase. J Biochem 66:759–765

Moche M, Stremlau S, Hecht L, Gobel C, Feussner I, Stohr C (2010) Effect of nitrate supply and mycorrhizal inoculation on characteristics of tobacco root plasma membrane vesicles. Planta 231:425–436

Modolo LV, Augusto O, Almeida IMG, Magalhaes JR, Salgado I (2005) Nitrite as the major source of nitric oxide production by *Arabidopsis thaliana* in response to *Pseudomonas syringae*. Fed Eur Biochem Soc Lett 579:3814–3820

Molina-Favero C, Creus CM, Simontacchi M, Puntarulo S, Lamattina L (2008) Aerobic nitric oxide production by *Azospirillum brasilense* Sp245 and its influence on root architecture in tomato. Mol Plant-Microbe Interact 21:1001–1009

Mukhopadhyay P, Zheng M, Bedzyk LA, LaRossa RA, Storz G (2004) Prominent roles of the NorR and Fur regulators in the *Escherichia coli* transcriptional response to reactive nitrogen species. Proc Natl Acad Sci 101:745–750

Mulder A, van de Graaf AA, Robertson LA, Kuenen JG (1995) Anaerobic ammonium oxidation discovered in a denitrifying fluidized bed reactor. FEMS Microbiol Ecol 16:177–184

Murakami E, Nagata M, Shimoda Y, Kucho K, Higashi S, Abe M et al (2011) Nitric oxide production induced in roots of *Lotusjaponicus* by lipopolysaccharide from *Mesorhizobium loti*. Plant Cell Physiol 52:610–617

Nagata M, Murakami EI, Shimoda Y, Shimoda-Sasakura F, Kucho KI, Suzuki A, Abe M, Higashi S, Uchiumi T (2008) Expression of a class 1 hemoglobin gene and production of nitric oxide in response to symbiotic and pathogenic bacteria in Lotus japonicus. Mol Plant Microb Interact 21:1175–1183

Nagata M, Hashimoto M, Murakami EI, Shimoda Y, Shimoda-Sasakura F, Kucho KI, Suzuki A, Abe M, Higashi S, Uchiumi T (2009) A possible role of class 1 plant hemoglobin at the early stage of legume-rhizobium symbiosis. Plant Signal Behav 4:202–204

Noritake T, Kawakita K, Doke N (1996) Nitric oxide induces phytoalexin accumulation in potato tuber tissue. Plant Cell Physiol 37:113–116

Pagnussat GC, Simontacchi M, Puntarulo S, Lamattina L (2002) Nitric oxide is required for root organogenesis. Plant Physiol 129:954–956

Pagnussat GC, Lanteri ML, Lamattina L (2004) Nitric oxide mediates the indole acetic acid induction activation of a mitogen-activated protein kinase cascade involved in adventitious root development. Plant Physiol 1(35):279–286

Perchepied L, Balagué C, Riou C, Claudel-Renard C, Rivière N, Grezes-Besset B, Roby D (2010) Nitric oxide participates in the complex interplay of defense-related signaling pathways controlling disease resistance to *Sclerotinia sclerotiorum* in *Arabidopsis thaliana*. Mol Plant-Microbe Interact 23:846–860

Pii Y, Crimi M, Cremonese G, Spena A, Pandolfini T (2007) Auxin and nitric oxide control indeterminate nodule formation. BMC Plant Biol 7:21

Plate L, Marletta MA (2012) Nitric oxide modulates bacterial biofilm formation through a multicomponent cyclic-di-GMP signaling network. Mol Cell 46:449–460

Rasul S, Dubreuil-Maurizi C, Lamotte O, Koen E, Poinssot B, Alcaraz G, Wendehenne D, Jeandroz S (2012) Nitric oxide production mediates oligogalacturonides-triggered immunity and resistance to *Botrytis cinerea* in *Arabidopsis thaliana*. Plant Cell Environ 35:1483–1499

Rütting T, Boeckx P, Müller C, Klemedtsson L (2011) Assessment of the importance of dissimilatory nitrate reduction to ammonium for the terrestrial nitrogen cycle. Biogeosciences 8:1779–1791

Sanchez C, Gates AJ, Meakin GE, Uchiumi T, Girard L, Richardson DJ, Bedmar EJ, Delgado MJ (2010) Production of nitric oxide and nitrosylleghemoglobin complexes in soybean nodules in response to flooding. Mol Plant-Microbe Interact 23:702–711

Schmidt I, Steenbakkers PJM, opden Camp HJM, Schmidt K, Jetten MSM (2004) Physiologic and proteomic evidence for a role of nitric oxide in biofilm formation by *Nitrosomonas europaea* and other ammonia oxidizers. J Bacteriol 186:2781–2788

Shimoda Y, Nagata M, Suzuki A, Abe M, Sato S, Kato T, Tabata S, Higashi S, Uchiumi T (2005) Symbiotic rhizobium and nitric oxide induce gene expression of non-symbiotic hemoglobin in *Lotus japonicus*. Plant Cell Physiol 46:99–107

Shimoda Y, Shimoda-Sasakura F, Kucho K, Kanamori N, Nagata M, Suzuki A, Abe M, Higashi S, Uchiumi T (2009) Overexpression of class 1 plant hemoglobin genes enhances symbiotic nitrogen fixation activity between *Mesorhizobium loti* and *Lotus japonicus*. Plant J 57:254–263

Singh AP, Dixit G, Kumar A, Mishra S, Singh PK, Dwivedi S, Trivedi PK, Chakrabarty D, Mallick S, Pandey V, Dhankher OP (2015) Nitric oxide alleviated arsenic toxicity by modulation of antioxidants and thiol metabolism in rice (*Oryza sativa* L.). Front Plant Sci 6:1272

Sun H, Li J, Song W, Tao J, Huang S, Chen S, Zhang Y (2015) Nitric oxide generated by nitrate reductase increases nitrogen uptake capacity by inducing lateral root formation and inorganic nitrogen uptake under partial nitrate nutrition in rice. J Exp Bot 66:2449–2459

Tada Y, Spoel SH, Pajerowska-Mukhtar K, Mou Z, Song J, Wang C, Zuo J, Dong X (2008) Plant immunity requires conformational charges of NPR1 via S-nitrosylation and thioredoxins. Science 321:952–956

Tufan Oz M, Eyidogan F, Yucel M, Oktem HA (2015) Functional role of nitric oxide under abiotic stress conditions. In: Khan MN (ed) Nitric oxide action in abiotic stress responses in plants. Springer International Publishing, Switzerland, pp 21–41

Vaishnav A, Jain S, Kasotia A, Kumari S, Gaur RK, Choudhary DK (2013) Effect of nitric oxide signaling in bacterial-treated soybean plant under salt stress. Arch Microbiol 195:571–577

Vaishnav A, Kumari S, Jain S, Varma A, Tuteja N, Choudhary DK (2016) PGPR-mediated expression of salt tolerance gene in soybean through volatiles under sodium nitroprusside. J Basic Microbiol 56:1274–1288

Vieweg MF, Fruhling M, Quandt HJ, Heim U, Baumlein H, Puhler A, Kuster H, Andreas MP (2004) The promoter of the *Vicia faba* L. leghemoglobin gene VfLb29 is specifically activated in the infected cells of root nodules and in the arbuscule-containing cells of mycorrhizal roots from different legume and nonlegume plants. Mol Plant Microbe Interact 17:62–69

Vitor SC, Duarte GT, Saviani EE, Vincentz MG, Oliveira HC, Salgado I (2013) Nitrate reductase is required for the transcriptional modulation and bactericidal activity of nitric oxide during the defense response of *Arabidopsis thaliana* against *Pseudomonas syringae*. Planta 238:475–486

Wang Y, Yang Q, Tosa Y, Nakayashiki H, Mayama S (2005) Nitric oxide-overproducing transformants of *Pseudomonas fluorescens* with enhanced biocontrol of tomato bacterial wilt. J Gen Plant Pathol 71:33–38

Wellburn AR (1990) Why are atmospheric oxides of nitrogen usually phytotoxic and not alternative fertilizers? New Phytol 115:395–429

Wendehenne D, Pugin A, Klessig DF, Durner J (2001) Nitric oxide: comparative synthesis and signaling in animal and plant cells. Trends Plant Sci 6:177–183

Wijler J, Delwiche CC (1954) Investigation on the denitrifylng process in soil. Plant Soil 5(155):169

Wojtaszek P (2000) Nitric oxide in plants: to NO or not to NO. Phytochemistry 54:1–4

Yamasaki H, Sakihama Y (2000) Simultaneous production of nitric oxide and peroxynitrite by plant nitrate reductase: in vitro evidence for the NR-dependent formation of active nitrogen species. FEBS Lett 468:89–92

Yamasaki H, Sakihama Y, Takahashi S (1999) An alternative pathway for nitric oxide production in plants: new features of an old enzyme. Trends Plant Sci 4:128–129

Zumft WG (1997) Cell biology and molecular basis of denitrification. Microbiol Mol Biol Rev 61:533–616

Chapter 9
Quorum Sensing: Melody Beneath the Ground

Javid A. Parray, Sumira Jan, Mohammad Yaseen Mir, Nowsheen Shameem, and Azra N. Kamili

Abstract Quorum sensing (QS) systems allow bacterial organisms to coordinate their behavior depending on the local population density and are used in many artificial systems requiring cell-to-cell communication. The accumulation of optimum stimulatory levels of autoinducers is detected by bacteria that affect the gene expression of their behavior. QS bacteria defer the production of virulence factors until cell numbers reach optimum levels, resulting in infection leading to activation of the host immune system by secretion of virulence factors causing productive infection. It is a successful technique that coordinates the gene expression of groups of organisms. In this chapter, the mechanisms pertaining to varied bacterial QS systems are presented and discussed. The differences between two definitive bacterial signal transduction systems are also discussed. We contend that the bacterial QS systems are optimally intended to specifically translate extracellular autoinducer information into internal changes via gene expression. In addition, the different bacterial QS systems used in deciphering basic mechanisms underlying the advancement of bacterial communities are discussed. Here, we review the updated advancements of the genetic approaches in engineering QS circuits to utilize bacterial communication in environmental biotechnology.

Keywords Quorum sensing • Autoinducer • Bacteria • Biofilm

J.A. Parray (✉)
Department of Environmental Science, Govt SAM Degree College,
Budgam 191111, Jammu & Kashmir, India
e-mail: Javid06@gmail.com

S. Jan
CITH Sanatnagar, Srinagar, India

M.Y. Mir • N. Shameem • A.N. Kamili
Centre of Research for Development, University of Kashmir, Srinagar, India

D. Egamberdieva, P. Ahmad (eds.), *Plant Microbiome: Stress Response*,
Microorganisms for Sustainability 5, https://doi.org/10.1007/978-981-10-5514-0_9

9.1 Introduction

The quorum sensing (QS) bacterial species release chemical signals/substances known as autoinducers. The external concentration of these substances increases with cell population density. The accumulation of minimal stimulatory concentrations of autoinducers affects behavioral response via alteration of gene expression (Booth et al. 1997). Bacteria coordinate separate activities of a whole population using the signal response system, and thus function as multicellular organisms. In this review, specific QS systems and their similarities and dissimilarities are illustrated. The similarities between these systems may be due to the inherent ability of bacteria to communicate, while differences between the systems arise because each system has to be optimized for survival in a particular species niche. Basically, different types of signal receptors, their mechanism of signal transduction, and target outputs of each QS system imitate the unique ecology carried out by a specific bacterial species (Cheung et al. 1997; Cirioni et al. 2007; DiMango et al. 1995).

9.2 Discovery of Cell-to-Cell Communication

Various fascinating studies are currently being carried out with regard to cell-to-cell communication wherein both intra- and inter-species QS are analyzed to predict the particular bacterium's chance of survival or division of labor in an entire community. QS regulates virulence in both plant and human pathogens as they have a peculiar property of delaying virulence factor production by avoiding the activation of the host immune system (Steinmoen et al. 2003). This is illustrated by the agr QS system in *Staphylococus aureus*, which generally regulates the production of the virulence factors by enhancing attachment to host cells, as well as through other factors that encourage bacterial internalization and host cell apoptosis (Dufour et al. 2002). Moreover, autoinducers produced by various *S. aureus* strains vary from strain to strain, particularly with reference to agr-mediated QS (Schu et al. 2014). However, QS via HSL autoinducer signaling has been revealed to play a critical role in the proper development of bacterial biofilms (Mayville et al. 1999). Likewise, in biofilms, bacteria are organized into elaborate structures composed of either single or multiple species possessing aqueous channels that promote the transport of nutrients and hence prevent desiccation. Each bacterial strain has unique patterns for gene expression and differentiation (Grumbein et al. 2014). These characters of biofilms indicate that the bacterial community within them have augmented their chances of existence and propagation. Similarly, QS controls the production of antibiotics like phenazine in plant pathogens such as *Pseudomonas aureofaciens* as part of interspecies communication. It has been documented that antibiotic production is also controlled by bacterial species other than *P. aureofaciens* (Mulcahy et al. 2010). This may perhaps be due to the sensitivity of *P. aureofaciens* in intense nutrient competition. However, QS controls processes deleterious to the host by antagonist production, which interferes with autoinducer reception (Roux et al. 2009). For example, the seaweed *Delisea pulchra* produces various halogenated furanones and enones interfering with HSL-mediated processes such as swarming in *Serratia liquefaciens* (Roux et al. 2009). In another study, the furanones have been shown to directly bind to the HSL-binding site in LuxR and to displace the cognate HSL autoinducer in *D. pulchra*. It has also been observed that QS inhibition was correlated with the furanone ability to compete with HSL autoinducer binding (Gillaspy et al. 1995; Goerke et al. 2000; Heyer et al. 2002; Novick and Geisinger 2008).

9.2.1 Mechanism of Action in Gram-Positive Bacteria

Gram-positive bacteria communicate by utilizing signals of modified oligopeptides and sensor histidine kinase receptors, which are part of a "two-component" membrane system. A phosphorylation cascade mediates the cell signaling, which in turn regulates the activity of response regulators, particularly DNA-binding transcriptional factors. Gram-negative bacteria utilize LuxIR QS systems, and similarly Gram-positive bacteria use specific signals of cognate receptors that are exquisitely sensitive to the signals' structures. Thus, intra-species communication confers peptide QS circuits as in LuxIR systems. Since peptide signals do not diffuse across membranes, committed oligopeptide exporters mediate cell signalizing. It has been reported that signals of peptide QS are derived from larger precursor peptides that are later modified to contain lactone and thiolactone rings, lanthionines, and isoprenyl groups. However, the biochemical processes leading to these events are not clearly understood (Ansaldi et al. 2002; Booth et al. 1996; Mayville et al. 1999; Nakayama et al. 2001). Furthermore, Gram-positive bacteria in combination with other types of QS signals also communicate with multiple peptides. With *Staphylococcus aureus* an interesting example of peptide QS exists. It is usually a benign human commensal but open penetration into host tissues causes it to become a deadly pathogen (Tenover and Gaynes 2000). A biphasic strategy is utilized by *S. aureus* to impart the disease. A low cell count leads the bacterium to express proteins that facilitate attachment by colonization; however, as the cellular density increases, the bacteria repress these traits and initiate production of toxins and proteases, which ultimately lead to dissemination (Lyon and Novick 2004). The agr QS system regulates this switching in gene expression programs. This system consists of the two-component sensor kinase-response regulator pair, agr C and agr A, respectively, and autoinducing peptides, which are encoded by agr D (Ji et al. 1995; Novick et al. 1995). It has been reported that the agr B protein is involved in the export of autoinducing peptides (AIPs) and also that it adds the thiolactone ring to these AIPs (Saenz et al. 2000). As AIPs bind to agr C it leads to agr A phosphorylation. Phosphorylated agr A leads to the expression of RNA III, which is a regulatory RNA as it blocks the synthesis of cell-adhesion factors and induces the expression of secreted factors (Novick et al. 1993). The activated agr A also leads to the expression of the agr BDCA. All of these processes increase the levels of AIPs, which

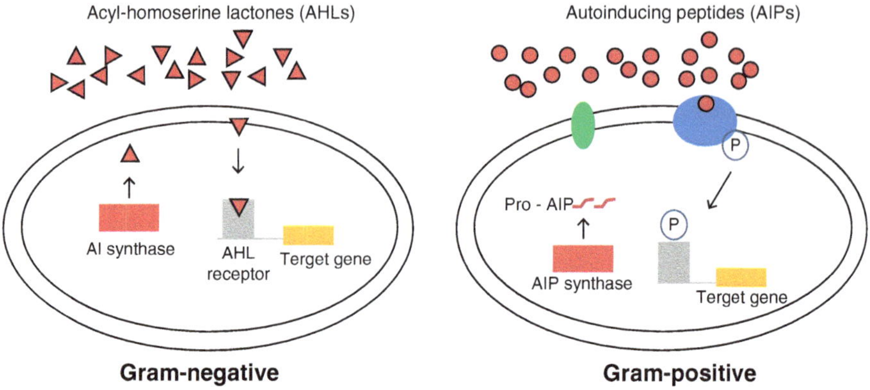

Fig. 9.1 Quorum sensing (QS) systems in bacteria. Gram-negative bacteria (left) secrete AHLs (red triangles), which in threshold concentrations penetrate into the cells and activate the cognate AHL receptor and induce the QS-regulated gene expression. Gram-positive bacteria (right) produce mature AIPs (red circles) that further interact with a transmembrane histidine kinase receptor activating the target gene expression via autophosphorylation of the trascriptional regulator (Source: Ivanova et al. 2013)

ultimately switches the entire population from a low-cell-density to the high-cell-density state (Novick et al. 1995). The *S. aureus* strains are classified on the basis of the sequence of thiolactone-containing AIP. To date four different AIPs for four different types of *S. aureus* have been identified (Dufour et al. 2002). Surprisingly, each AIP exclusively leads to the activation of agr C receptor and blocks expression of all other non-cognate receptors by competitive binding (Lyon et al. 2002). Therefore, each AIP in the other three groups of *S. aureus* inhibits activation of the virulence cascade. Intra-species competition exists in case of co-infection with two different *S. aureus* groups. The group that first establishes itself by means of its QS cascade outcompetes the other group (Mayville et al. 1999). Hence QS in *S. aureus* inhibits the dissemination of non-kin progeny while allowing dissemination of closely related progeny. Biochemical analysis has revealed that each *S. aureus* group is the primary causative agent of a specific type of *S. aureus* disease. This indicates the establishment of a specific niche for each strain in which the cell-cell communication has been instrumental (Novick 2003; Kielian et al. 2001; Kravchenko et al. 2008; Lyczak et al. 2000; Smith et al. 2002). One of the molecular mechanisms underlying the evolution of new bacterial species may be the co-divergence of the signal-receptor pairs occurring in these bacteria. The diverse family of Gram-positive soil-dwelling bacteria Streptomycetes have great clinical relevance as they are a biological systems reservoir of secondary metabolites, many of which are used as antibiotics (reviewed in Chater and Horinouchi 2003). Streptomycetes by means of QS control morphological differentiation and secondary metabolite production. They utilize γ-butyrolactones as autoinducers. These signals are fascinating as they are structurally related to AHL autoinducers. Nevertheless, there are no reports yet for cross-communication between streptomycetes and Gram-negative bacteria that communicate with AHLs (Fig. 9.1).

9.2.2 Mechanisms of Action in Gram-Negative Bacteria

The paradigm for QS in most Gram-negative bacteria is the bioluminescent marine bacterium *Vibrio fischeri* (Nealson and Hastings 1979). *V. fischeri* inhibits the Hawaiian squid *Euprymna scolopes'* light organ where it grows and increases its cell density and thereby leads to the gene expression required for bioluminescence. The *E. scolopes* utilizes this light provided by bacteria to cover its shadow and avoid predation (Visick et al. 2000). The bacteria in turn also benefit as they get nutrients from the light organ. For the expression of luciferase operon (LuxICDABE), which is involved in light production, the two proteins LuxI and LuxR are required (Fig. 9.1). The LuxI is the autoinducer synthase, which produces the acyl-homoserine lactone (AHL) (Eberhard et al. 1981; Engebrecht and Silverman 1984), and LuxR is the cytoplasmic autoinducer receptor/DNA binding transcriptional activator (Engebrecht et al. 1983). As AHLs are produced they diffuse freely within as well as outside cells and thereby increase cell density (Kaplan and Greenberg 1985). Once the signal threshold concentration is reached, it is bound to LuxR. This complex in turn activates transcription of the operon encoding luciferase (Stevens et al. 1994). Notably, LuxR-AHL complex by means of luciferase operon also induces expression of LuxI. This regulatory configuration creates a positive feedback that causes the entire population to switch into "QS mode" and produce light. A number of other Gram-negative bacteria have also been reported to possess LuxIR-type proteins and communicate by means of AHL signals (Manefield and Turner 2002). There is extreme specificity between the LuxR proteins and their cognate AHL signals, which are utilized predominantly for intra-species communication. LuxI-type proteins play an important role in lactonizing the methionine moiety from S-adenosylmethionine (SAM), besides its linkage to particular fatty acyl chains carried on acyl-acyl carrier proteins (More et al. 1996; Parsek et al. 1999). The signal specificity is maintained by means of varying lengths of side chains of fatty acyl groups (Fuqua and Eberhard 1999). Structural analysis of LuxI-type proteins indicated that they possess pockets of acyl bindings that specifically fit in a side-chain moiety (Gould et al. 2004; Watson et al. 2002). Hence LuxI proteins produce signaling molecules of high fidelity. Some of the LuxI-type proteins have also been reported to produce multiple AHLs; however, it is not clear whether they are biologically relevant (Marketon et al. 2002). Structural analysis of LuxR proteins also reveals that they possess specific acyl binding pockets that enable them to be activated by their cognate signal (Vannini et al. 2002; Zhang et al. 2002; Wright et al. 2005).

Therefore, there is a mixed-species environment wherein multiple AHL signals are present and then each species responds only to the buildup of its own signal. Significantly, the bacterium does not frequently rely on one LuxIR QS system; however, it uses many LuxIR systems, often in conjunction with other types of QS circuits. Premature activation of LuxIR-type QS circuits is prevented by mechanisms that ensure hat tboth the signal and the detector are synthesized and interact in the cytoplasm. The LuxR homologue TraR is evidence of one such type of mechanism in the plant pathogen *Agrobacterium tumefaciens*, where stability of LuxR-type proteins increases upon AI binding. It has been reported that in presence of an autoinducer, the half-life of TraR increases to over 30 min, while in its absence it has a half-life of a few minutes (Zhu and Winans 1999). The crystal structure analysis of TraR indicates that for folding of the nascent polypeptide, AHL binding is required. Furthermore, radiolabeled TraR revealed that it becomes stabilized only when its cognate AHL is added prior to labeling of the protein (Zhang et al. 2002; Zhu and Winans 2001). Thus, TraR initiates the QS cascade only when AHL accumulates in a significant concentration (both outside and inside the cell). Active export of AHL signals is another mechanism that prevents "short-circuiting" of LuxIR systems (Pearson et al. 1999). Once the signal accumulates in a significant concentration, which is the indicator of high cell density, diffusion into the cell overwhelms export and thus engages the circuit. As AHLs have long acyl side-chains they require active export to transverse the bacterial membrane (Pearson et al. 1999) (Fig. 9.1).

9.3 Role of Quorum Sensing in Bacterial Virulence

9.3.1 Gram-Positive Bacteria

Peptide QS systems are utilized by many Gram-positive bacteria to control gene expression, and in this way *S. aureus* has served as a model to study bacterial peptide signaling (Novick and Geisinger 2008). *S. aureus* is found in the human microbiota and is reported in 30 % of the adult population (George and Muir 2007; Roux et al. 2009). Despite its extensive pervasiveness in healthy subjects, it has been increasingly been found to be associated with antibiotic resistance, and thus is a very dangerous opportunistic pathogen (George and Muir 2007). *S. aureus* displays very rapid transmission as it has multiple virulence factors, hence revealing its importance as a human pathogen (Massey et al. 2006). Furthermore, *S. aureus* forms on indwelling devices such as urethral stents and biofilms on many surfaces (Kehinde et al. 2004). These biofilms and indwelling devices lead to Staphylococcus infection. *S. aureus* virulence is mediated by peptide-based QS system factors encoded by the accessory gene regulator (agr) locus (Novick and Geisinger 2008). In the agr system the autoinducer is an oligopeptide called an autoinducing peptide (AIP), which is encoded by agrD. AgrB is a membrane-bound protein trimmer and causes secretion of AIP (Ji et al. 1995; Saenz et al. 2000; Zhang et al. 2004). The active AIP is a five-membered thiolactone ring with 7–9 amino acids (Roux et al. 2009). Extracellular AIP binds to agrC, which is a membrane-bound sensor kinase. The agrC undergoes autophosphorylation and leads to activation of agrA (Ji et al. 1995; Koenig et al. 2004; Lina et al. 1998). It has been reported that agr system regulates virulence of genes predominantly from P2 and P3 promoters, which produce RNAII and RNAIII, respectively (Morfeldt et al. 1995; Novick and Geisinger 2008). From the RNAII transcript P2 promotes the transcription of the agr operon, which includes agrA, agrB, agrC, and agrD (Novick et al. 1995; Roux et al. 2009). The activated agrA is a phosphorylated homodimer that stimulates the transcription of P2 and P3 promoters, with a higher affinity for P2 (George and Muir 2007; Koenig et al. 2004). The transcription of P3 leads to the production of RNAIII, which is the effector molecule of the agr system (Roux et al. 2009). The RNAIII is a regulatory RNA (514 nt) and it also functions as messenger RNA (mRNA) of d-toxin (Balaban and Novick 1995; Benito et al. 2000; Kong et al. 2006; Novick et al. 1995). The 39 end is required for the repression of protein A synthesis while the 59 end is thought to upregulate α-haemolysin (Kong et al. 2006; Morfeldt et al. 1995). The RNAIII increases the production of capsules, toxins, and proteases, while it reduces the expression of surface adhesins (Novick and Geisinger 2008; Roux et al. 2009). Over 70 genes regulate the agr system, 23 of which are known virulence factors (George and Muir 2007). Among the virulence factors regulated by the agr system, there are two classes. The first class includes virulence factors that are involved in host and immune evasion, while the second class contain factors that lead to the production of exoproteins that are associated with invasion and toxin production (Bowden et al. 2005; Yao et al. 2006). It has been reported that the agr system essentially switches the bacteria into invasive and aggressive pathogens from an adhesive, colonizing commensal (Roux et al. 2009). On the basis of agr polymorphism, four major groups have been recognized and are categorized as I–IV (George and Muir 2007; Ji et al. 1997). Each of these groups has a distinct AIP, by means of which they bind to receptors from all groups. However, each AIP specifically activates its receptor from each group (Ji et al. 1997). All the groups are cross-inhibiting with the exception of groups I and IV, which are able to cross-activate (George and Muir 2007; Mayville et al. 1999; Otto et al. 1998). There appears to be a correlation between the relative fitness of the *S. aureus* strain and the agr group (Fleming et al. 2006). The agr impact virulence by the formation of biofilms. Biofilm formation within bacteria is a multi-step developmental process, which initiates with adhesion to a surface. The attached bacteria divide and give rise to macro-colonies, which later develop into mature biofilms, which can assume multiple topographies. The last step in biofilm development is detachment, which may be important for dissemination during an infectious process (Parsek and Tolker-Nielsen 2008). *S. aureus* is thought to possess two independent mechanisms of biofilm formation; the first involves an extracellular polysaccharide and polysaccharide intercellular adhesin (PIA). The second is thought to be PIA-independent, possibly involving adhesive proteins and the sarA and agr global regulators (Lauderdale et al. 2009; Novick and Geisinger 2008). The role of agr in biofilm formation has been explored because biofilms are thought to play a critical role in *S. aureus* infection. When agr is non-functional, *S. aureus* has enhanced adhesion abilities (Vuong et al. 2000). An agr mutant strain has a detachment defect, and the detachment of bacterial cells from biofilms was found to coincide with agr expression (Kong et al. 2006; Yarwood et al. 2004). The role of agr is thought to be due to a reduction in adhesin production and an increase in the production of both d-haemolysin and proteases (Novick and Geisinger 2008).

9.3.2 Gram-Negative Bacteria

P. aeruginosa is a Gram-negative bacterium capable of surviving in a wide range of environments. It is commonly associated with nosocomial infections and infections in severely burned individuals, and is a leading cause of death in severe respiratory infections, such as chronic lung infections in cystic fibrosis (CF) patients (Bendiak and Ratjen 2009; Bodey et al. 1983). Infections with *P. aeruginosa* are difficult to eradicate due to their high levels of antibiotic resistance and growth in biofilms (Driscoll et al. 2007). At least three intertwined QS sensing systems and one orphan autoinducer receptor affect the ability of *P. aeruginosa* to cause disease (Fig. 9.2). Two of these systems, las and rhl, rely on the production of AHLs as the signaling molecules (AIs) (De Kievit and Iglewski 2000). In the las system, N-3-oxododecanoyl-homoserine lactone (3OC12- HSL) is produced by the enzyme encoded by the lasI gene. When *P. aeruginosa* reaches a certain threshold density, 3OC12-HSL binds to the transcriptional activator LasR. LasR, in turn, dimerizes and binds to target promoters to control gene expression (De Kievit and Iglewski 2000). Similarly, in the rhl system, the rhlI gene encodes the enzyme involved in the production of N-butyryl-homoserine lactone (C4-HSL). As with 3OC12-HSL, C4-HSL binds to its cognate transcriptional regulator, RhlR, to control the activity of target promoters (De Kievit and Iglewski 2000). The rhl system is controlled by the las system at both transcriptional and post-transcriptional levels (Latifi et al. 1996). Besides LasR and RhlR, *P. aeruginosa* encodes an orphan receptor protein, QscR, which can sense 3OC12- HSL to control its own regulon (Chugani et al. 2001; Fuqua 2006; Schuster and Greenberg 2006). The las and rhl systems regulate the timing and production of multiple virulence factors, including elastase, alkaline protease, exotoxin A, rhamnolipids, pyocyanin, lectins, and superoxidase dismutase (Schuster et al. 2003; Smith and Iglewski 2003). The expression of these

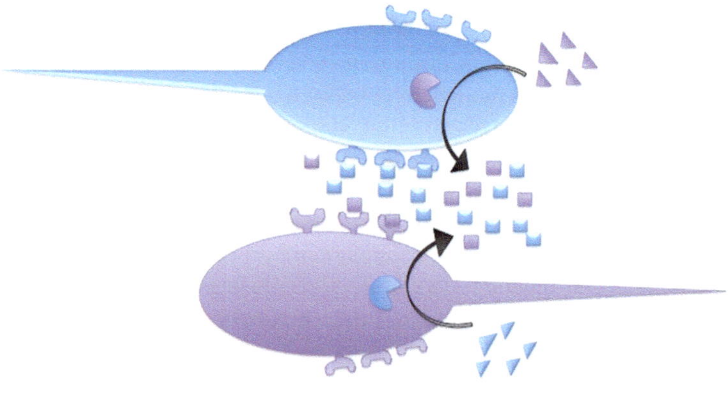

Fig. 9.2 Engineered cell consortium of two different bacterial strains: Blue cells are sensitive to molecules produced by purple cells and vice versa, depicting interdependent chemical response via quorum sensing

two QS systems has also been linked to the regulation of biofilm formation. QS signaling may start in the early stages of biofilm development, which is characterized by microcolony formation, where lasI mutants are unable to form structurally normal biofilms (Davies et al. 1998). Expression of the lasI gene is maximal at day 4 of biofilm development, decreasing between days 6 and 8. The expression of rhlI fluctuates during biofilm development and phenotypes of biofilm development with a rhlI mutant varying according to the media and model used, supposedly due to the different iron levels present (Davies et al. 1998; Yoon et al. 2002). This may indicate that QS is active during *P. aeruginosa* colonization of CF patients; however, it is important to note that QS-deficient *P. aeruginosa* strains are often isolated from CF patients (Erickson et al. 2002; Karatuna and Yagci 2010). This has spurred a major discussion in the scientific community about whether QS is really important during CF infections. It has been hypothesized that the maintenance of a functional QS system is a metabolic burden for *P. aeruginosa* and that co-colonization with QS-proficient and -deficient strains is in the best interest of this community of pathogens (Heurlier et al. 2006; Kohler et al. 2009). Moreover, social exploitation in *P. aeruginosa* communities may provide an explanation for the emergence of QS-deficient strains in human infections (Sandoz et al. 2007). Nevertheless, it has been shown that *P. aeruginosa* rhlI and lasI mutants cause less tissue destruction and decrease mortality when compared with wild-type strains in multiple animal models (Smith and Iglewski 2003), indicating an important role for QS in *P. aeruginosa* pathogenesis. Apart from regulating the expression of virulence factors, some of the AIs have been shown to directly interact with host cells. QS regulates the production of several extracellular virulence factors, promotes biofilm maturation, and regulates the expression of antibiotic efflux pumps, meaning that it has a key role in the pathogenesis of *P. aeruginosa* (Whitehead et al. 2001; Fuqua 2006).

9.4 Genetic Regulation of Quorum Sensing

Community behavior prevalent among diverse bacterial species is an example of QS. QS is the ability of a microorganism to perceive and respond to microbial population density by relying on the production of, and subsequent response to, diffusible signal molecules. The majority of Gram-negative bacteria produce acyl-HSLs that function as signaling molecules in QS. The physiological processes regulated by QS are extremely diverse, depending upon the bacterial species, ranging from bioluminescence to swarming motility. For intercellular signaling mechanisms, acyl-HSL QS has become a paradigm. Over the past decade a flurry of research has been carried out that has led to a significant understanding of many aspects of QS. These include the synthesis of acyl-HSLs, the receptors that recognize the acyl-HSL signal and transduce this information to the level of gene expression, and the interaction of these receptors with the transcriptional machinery.

The gene expression of QS within bacteria is population-density dependent. QS systems include two obligate components: a regulatory receptor protein that interacts with the regulator, and a low-molecular-weight regulator (autoinducer), readily diffusible through the cytoplasmic membrane. The abrupt activation (induction) of certain genes and operons occurs as the bacterial population reaches a critical level of density and autoinducers accumulate to a necessary threshold value. Bacteria accomplish communication between cells belonging to the same or different species, genera, and even families by means of low-molecular-weight regulators; QS systems have been shown to play a key role in the regulation of various metabolic processes in bacteria and to function as global regulators of the expression of bacterial genes.

9.5 Engineering Quorum Sensing

QS systems enable bacteria to coordinate their behavior as a function of local population density and are often used in synthetic systems that require cell–cell communication (Fig. 9.2). Many reports have documented the engineered EsaR promoter, PesaR, which is repressed by the QS regulator EsaR. EsaR-dependent gene expression from PesaR is induced by 3-oxo-hexanoyl-homoserine lactone (3OC6HSL), which is actually the set of modified PesaR promoters that contain a second EsaR binding site (Mishler et al. 2010). The changes in gene expression levels, regulatory range, 3OC6HSL sensitivity, and the regulatory role of EsaR that are dependent on the position of the second binding site were also observed. Combining the new promoters with endogenous 3OC6HSL production led to QS-dependent systems that exhibit a range of expression levels and timing. These promoters represent a new set of tools for modulating QS-dependent gene expression and may be used to tune the regulation of multiple genes in response to a single QS signal.

From the viewpoint of biotechnology, metabolic engineering mainly aims to change the natural status of a pathway in a microorganism towards the overproduction of certain bioproducts. The biochemical nature of a pathway implies that a changed pathway will often lead to the collective results of altered behavior of the metabolic enzymes encoded by corresponding genes. By finely modulating the expression of these genes or the properties of the enzyme, we can gain efficient control of the pathway. In this article, we reviewed the typical methods that have been applied to regulate the expression of genes in metabolic engineering. These methods are grouped according to the operation targets in a typical gene. The transcription of a gene is controlled by an indispensable promoter. By utilizing promoters with different strengths, expected levels of expression can be easily achieved, and screening a promoter library may find suitable mutant promoters that can provide tunable expression of a gene. Auto-responsive promoters (QS-based or oxygen-inducible) simplify the induction process by driving the expression of a gene in an automated manner. Light-responsive promoters enable reversible and noninvasive control on gene activity, providing a promising method for controlling gene expression with temporal and spatial resolution through metabolic engineering involving complicated genetic circuits. Through directed evolution and/or rational design, the encoding sequences of a gene can be altered, leading to possibly the most profound changes in the properties of metabolic enzymes. Introducing an engineered riboswitch in mRNA can make it a regulatory molecule at the same time; the ribosomal binding site is commonly engineered to be more attractive for a ribosome through design. The terminator of a gene will affect the stability of mRNA, and the intergenic region will influence the expression of many related genes. Improving the performance of these elements is generally the main goal in metabolic engineering.

9.6 Conclusions

QS has been effectively utilized in identifying inherent stress-resistance and metabolomics as well as in the bioremediation of assorted industrial and environmental pollutants. Advancing genetic engineering in QS can facilitate sensing of environmental pollutants and could be utilized in different bioremediation technologies. Integrating QS features of bacteria with biofilms can create a new platform for Ecometabolomics.

References

Ansaldi M, Marolt D, Stebe T, Mandic-Mulec I, Dubnau D (2002) Specific activation of the *Bacillus* quorum- sensing systems by isoprenylated pheromone variants. Mol Microbiol 44:1561–1573

Balaban N, Novick RP (1995) Translation of RNAIII, the Staphylococcus aureus agr regulatory RNA molecule, can be activated by a 39-end deletion. FEMS Microbiol Lett 133:155–161

Bendiak GN, Ratjen F (2009) The approach to Pseudomonas aeruginosa in cystic fibrosis. Semin Respir Crit Care Med 30:587–595

Benito Y, Kolb FA, Romby P, Lina G, Etienne J, Vandenesch F (2000) Probing the structure of RNAIII, the Staphylococcus aureus agr regulatory RNA, and identification of the RNA domain involved in repression of protein A expression. RNA 6:668–679

Bodey GP, Bolivar R, Fainstein V, Jadeja L (1983) Infections caused by Pseudomonas aeruginosa. Rev Infect Dis 5:279–313

Booth MC, Bogie CP, Sahl HG, Siezen RJ, Hatter KL, Gilmore MS (1996) Structural analysis and proteolytic activation of *Enterococcus faecalis* cytolysin, a novel lantibiotic. Mol Microbiol 21:1175–1184

Booth MC, Cheung AL, Hatter KL, Jett BD, Callegan MC, Gilmore MS (1997) Staphylococcal accessory regulator (sar) in conjunction with agr contributes to Staphylococcus aureus virulence in endophthalmitis. Infect Immun 65:1550–1556

Bowden MG, Chen W, Singvall J, Xu Y, Peacock SJ, Valtulina V, Speziale P, Hook M (2005) Identification and preliminary characterization of cell-wall-anchored proteins of Staphylococcus epidermidis. Microbiology 151:1453–1464

Chater KF, Horinouchi S (2003) Signaling early developmental events in two highly diverged Streptomyces species. Mol Microbiol 48:9–15

Cheung AL, Eberhardt KJ, Chung E, Yeaman MR, Sullam PM, Ramos M, Bayer AS (1994) Diminished virulence of a sar2/agr2 mutant of Staphylococcus aureus in the rabbit model of endocarditis. J Clin Invest 94:1815–1822
Chugani SA, Whiteley M, Lee KM, D'Argenio D, Manoil C, Greenberg EP (2001) QscR, a modulator of quorum-sensing signal synthesis and virulence in Pseudomonas aeruginosa. Proc Natl Acad Sci U S A 98:2752–2757
Cirioni O, Ghiselli R, Minardi D, Orlando F, Mocchegiani F, Silvestri C, Muzzonigro G, Saba V, Scalise G (2007) RNAIII-inhibiting peptide affects biofilm formation in a rat model of staphylococcal ureteral stent infection. Antimicrob Agents Chemother 51:4518–4520
Davies DG, Parsek MR, Pearson JP, Iglewski BH, Costerton JW, Greenberg EP (1998) The involvement of cell-to-cell signals in the development of a bacterial biofilm. Science 280:295–298
De Kievit TR, Iglewski BH (2000) Bacterial quorum sensing in pathogenic relationships. Infect Immun 68:4839–4849
DiMango E, Zar HJ, Bryan R, Prince A (1995) Diverse Pseudomonas aeruginosa gene products stimulate respiratory epithelial cells to produce interleukin-8. J Clin Invest 96:2204–2210
Driscoll JA, Brody SL, Kollef MH (2007) The epidemiology, pathogenesis and treatment of Pseudomonas aeruginosa infections. Drugs 67:351–368
Dufour P, Jarraud S, Vandenesch F, Greenland T, Novick RP et al (2002) High genetic variability of the agr locus in staphylococcus species. J Bacteriol 184:1180–1186
Eberhard A, Burlingame AL, Eberhard C, Kenyon GL, Nealson KH, Oppenheimer NJ (1981) Structural identification of autoinducer of *Photobacterium fischeri* luciferase. Biochemistry 20:2444–2449
Engebrecht J, Silverman M (1984) Identification of genes and gene products necessary for bacterial bioluminescence. Proc Natl Acad Sci U S A 81:4154–4158
Engebrecht J, Nealson K, Silverman M (1983) Bacterial bioluminescence: isolation and genetic analysis of functions from *Vibrio fischeri*. Cell 32:773–781
Erickson DL, Endersby R, Kirkham A, Stuber K, Vollman DD, Rabin HR, Mitchell I, Storey DG (2002) Pseudomonas aeruginosa quorum-sensing systems may control virulence factor expression in the lungs of patients with cystic fibrosis. Infect Immun 70:1783–1790
Fleming V, Feil E, Sewell AK, Day N, Buckling A, Massey RC (2006) Agric interference between clinical Staphylococcus aureus strains in an insect model of virulence. J Bacteriol 188:7686–7688
Fuqua C (2006) The QscR quorum-sensing regulon of Pseudomonas aeruginosa: an orphan claims its identity. J Bacteriol 188:3169–3171
Fuqua C, Eberhard A (1999) Signal generation in autoinduction systems: synthesis of acylated homoserine lactones by LuxI-type proteins. In: Dunny GM, Winans SC (eds) Cell-cell signaling in bacteria. ASM Press, Washington, DC, pp 211–230
George EA, Muir TW (2007) Molecular mechanisms of *agr* quorum sensing in virulent staphylococci. ChemBio 8:847–855
Gillaspy AF, Hickmon SG, Skinner RA, Thomas JR, Nelson CL, Smeltzer MS (1995) Role of the accessory gene regulator (agr) in pathogenesis of staphylococcal osteomyelitis. Infect Immun 63:3373–3380
Goerke C, Kraning K, Stern M, Doring G, Botzenhart K, Wolz C (2000) Molecular epidemiology of community-acquired Staphylococcus Aureus in families with and without cystic fibrosis patients. J Infect Dis 181:984–989
Gould TA, Schweizer HP, Churchill ME (2004) Structure of the *Pseudomonas aeruginosa* acylhomoserinelactone synthase LasI. Mol Microbiol 53:1135–1146
Grumbein S, Opitz M, Lieleg O (2014) Selected metal ions protect Bacillus subtilis biofilms from erosion. Metallomics 6:1441–1450
Heurlier K, Denervaud V, Haas D (2006) Impact of quorum sensing on fitness of Pseudomonas aeruginosa. Int J Med Microbiol 296:93–102

Heyer G, Saba S, Adamo R, Rush W, Soong G, Cheung A, Prince A (2002) Staphylococcus aureus agr and sarA functions are required for invasive infection but not inflammatory responses in the lung. Infect Immun 70:127–133

Ivanova K, Fernandes MM, Tzanov T (2013) Current advances on bacterial pathogenesis inhibition and treatment strategies. In: Méndez-Vilas A (eds) Microbial pathogens and strategies for combating them: science, technology and education. Formatex Research Center, pp 322–336. https://doi.org/10.13140/RG.2.1.3988.7840

Ji G, Beavis RC, Novick RP (1995) Cell density control of staphylococcal virulence mediated by an octapeptide pheromone. Proc Natl Acad Sci U S A 92:12055–12059

Ji G, Beavis R, Novick RP (1997) Bacterial interference caused by auto inducing peptide variants. Science 276:2027–2030

Kaplan HB, Greenberg EP (1985) Diffusion of autoinducer is involved in regulation of the *Vibrio fischeri* luminescence system. J Bacteriol 163:1210–1214

Karatuna O, Yagci A (2010) Analysis of the quorum sensing dependent virulence factor production and its relationship with antimicrobial susceptibility in Pseudomonas aeruginosa respiratory isolates. Clin Microbiol Infect. https://doi.org/10.1111/j.1469-0691.2010.03177.x

Kehinde EO, Rotimi VO, Al-Hunayan A, Abdul-Halim H, Boland F, Al-Awadi KA (2004) Bacteriology of urinary tract infection associated with indwelling J ureteral stents. J Endourol 18:891–896

Kielian T, Cheung A, Hickey WF (2001) Diminished virulence of an a-toxin mutant of Staphylococcus aureus in experimental brain abscesses. Infect Immun 69:6902–6911

Koenig RL, Ray JL, Maleki SJ, Smeltzer MS, Hurlburt BK (2004) Staphylococcus aureus AgrA binding to the RNAIII-agr regulatory region. J Bacteriol 186:7549–7555

Kohler T, Buckling A, Van DC (2009) Cooperation and virulence of clinical Pseudomonas aeruginosa populations. Proc Natl Acad Sci U S A 106:6339–6344

Kong KF, Vuong C, Otto M (2006) Staphylococcus quorum sensing in biofilm formation and infection. Int J Med Microbiol 296:133–139

Kravchenko VV, Kaufmann GF, Mathison JC, Scott DA, Katz AZ, Grauer DC, Lehmann M, Meijler MM, Janda KD, Ulevitch RJ (2008) Modulation of gene expression via disruption of NF-kB signaling by a bacterial small molecule. Science 321:259–263

Latifi A, Foglino M, Tanaka K, Williams P, Lazdunski A (1996) A hierarchical quorum-sensing cascade in *pseudomonas aeruginosa* links the transcriptional activators LasR and RhIR (VsmR) to expression of the stationary-phase sigma factor RpoS. Mol Microbiol 21:1137–1146

Lauderdale KJ, Boles BR, Cheung AL, Horswill AR (2009) Interconnections between sigma B, agr, and proteolytic activity in Staphylococcus aureus biofilm maturation. Infect Immun 77:1623–1635

Lina G, Jarraud S, Ji G, Greenland T, Pedraza A, Etienne J, Novick RP, Vandenesch F (1998) Transmembrane topology and histidine protein kinase activity of AgrC, the agr signal receptor in Staphylococcus aureus. Mol Microbiol 28:655–662

Lyczak JB, Cannon CL, Pier GB (2000) Establishment of Pseudomonas aeruginosa infection: lessons from a versatile opportunist. Microbes Infect 2:1051–1060

Lyon GJ, Novick RP (2004) Peptide signaling in Staphylococcus aureus and other Gram-positive bacteria. Peptides 25:1389–1403

Lyon GJ, Wright JS, Muir TW, Novick RP (2002) Key determinants of receptor activation in the *agr* autoinducing peptides of Staphylococcus aureus. Biochemistry 41:10095–10104

Manefield M, Turner SL (2002) Quorum sensing in context: out of molecular biology and into microbial ecology. Microbiology 148:3762–3764

Marketon MM, Gronquist MR, Eberhard A, Gonzalez JE (2002) Characterization of the Sinorhizobium meliloti sinR/sinI locus and the production of novel N-acyl homoserine lactones. J Bacteriol 184:5686–5695

Massey RC, Horsburgh MJ, Lina G, Hook M, Recker M (2006) The evolution and maintenance of virulence in Staphylococcus aureus: a role for host-to-host transmission. Nat Rev Microbiol 4:953–958

Mayville P, Ji G, Beavis R, Yang H, Goger M, Novick RP, Muir TW (1999) Structure-activity analysis of synthetic autoinducing thiolactone peptides from Staphylococcus aureus responsible for virulence. Proc Natl Acad Sci U S A 96:1218–1223
Mishler DM, Topp S, Reynoso CMK, Gallivan JP (2010) Engineering bacteria to recognize and follow small molecules. Curr Opin Biotechnol 21(5):653–656. https://doi.org/10.1016/j.copbio.2010.05.007
More MI, Finger LD, Stryker JL, Fuqua C, Eberhard A, Winans SC (1996) Enzymatic synthesis of a quorum-sensing autoinducer through use of defined substrates. Science 272:1655–1658
Morfeldt E, Taylor D, Von GA, Arvidson S (1995) Activation of a-toxin translation in Staphylococcus aureus by the trans encoded antisense RNA, RNAIII. EMBO J 14:4569–4577
Mulcahy H, Charron-Mazenod L, Lewenza S (2010) Pseudomonas aeruginosa produces an extracellular deoxyribonuclease that is required for utilization of DNA as a nutrient source. Environ Microbiol 12:1621–1629
Nakayama J, Cao Y, Horii T, Sakuda S, Akkermans AD, de Vos WM, Naqasawa H (2001) Gelatinase biosynthesisactivating pheromone: a peptide lactone that mediates a quorum sensing in enterococcus faecalis. Mol Microbiol 41:145–154
Nealson KH, Hastings JW (1979) Bacterial bioluminescence: its control and ecological significance. Microbiol Rev 43:496–518
Novick RP (2003) Autoinduction and signal transduction in the regulation of staphylococcal virulence. Mol Microbiol 48:1429–1449
Novick RP, Geisinger E (2008) Quorum sensing in staphylococci. Annu Rev Genet 42:541–564
Novick RP, Ross HF, Projan SJ, Kornblum J, Kreiswirth B, Moghazeh S (1993) Synthesis of staphylococcal virulence factors is controlled by a regulatory RNA molecule. EMBO J 12:3967–3975
Novick RP, Projan SJ, Kornblum J, Ross HF, Ji G, Kreiswirth B, Vandenesch F, Moghazeh S (1995) The agr P2 operon: an autocatalytic sensory transduction system in Staphylococcus aureus. Mol Gen Genet 248:446–458
Otto M, Sussmuth R, Jung G, Gotz F (1998) Structure of the pheromone peptide of the Staphylococcus epidermidis agr system. FEBS Lett 424:89–94
Parsek MR, Tolker-Nielsen T (2008) Pattern formation in Pseudomonas aeruginosa biofilms. Curr Opin Microbiol 11:560–566
Parsek MR, Val DL, Hanzelka BL, Cronan JE, Greenberg EP (1999) Acyl homoserinelactone quorum-sensing signal generation. Proc Natl Acad Sci U S A 96:4360–4365
Pearson JP, Van Delden C, Iglewski BH (1999) Active efflux and diffusion are involved in transport of Pseudomonas aeruginosa cell-to-cell signals. J Bacteriol 181:1203–1210
Roux A, Payne SM, Gilmore MS (2009) Microbial telesensing: probing the environment for friends, foes, and food. Cell Host Microbe 6:115–124
Saenz HL, Augsburger V, Vuong C, Jack RW, Gotz F, Otto M (2000) Inducible expression and cellular location of AgrB, a protein involved in the maturation of the staphylococcal quorum-sensing pheromone. Arch Microbiol 174:452–455
Sandoz KM, Mitzimberg SM, Schuster M (2007) Social cheating in *Pseudomonas aeruginosa* quorum sensing. Proc Natl Acad Sci U S A 104:15876–15881
Schu DJ, Scruggs JM, Geissinger JS, Michel KG, Stevens AM (2014) Acyl-homoserine lactone recognition and response hindering the quorum-sensing regulator EsaR. PLoS One 9:e107687
Schuster M, Greenberg EP (2006) A network of networks: quorum-sensing gene regulation in Pseudomonas aeruginosa. Int J Med Microbiol 296:73–81
Schuster M, Lostroh CP, Ogi T, Greenberg EP (2003) Identification, timing, and signal specificity of Pseudomonas aeruginosa quorum-controlled genes: a transcriptome analysis. J Bacteriol 185:2066–2079
Smith RS, Iglewski BH (2003) P. aeruginosa quorum-sensing systems and virulence. Curr Opin Microbiol 6:56–60
Smith RS, Harris SG, Phipps R, Iglewski B (2002) The Pseudomonas aeruginosa quorum-sensing molecule N-(3-oxododecanoyl) homoserine lactone contributes to virulence and induces inflammation in vivo. J Bacteriol 184:1132–1139

Steinmoen H, Teigen A, Havarstein LS (2003) Competence-induced cells of Streptococcus pneumoniae lyse competence-deficient cells of the same strain during cocultivation. J Bacteriol 185:7176–7183

Stevens AM, Dolan KM, Greenberg EP (1994) Synergistic binding of the *Vibrio fischeri* LuxR transcriptional activator domain and RNA polymerase to the lux promoter region. Proc Natl Acad Sci U S A 91:12619–12623

Tenover FC, Gaynes RP (2000) The epidemiology of *Staphylococcus aureus* infections. In: Fischetti VA, Novick RP, Ferretti JJ, Portnoy DA, Rood JI (eds) Gram- positive pathogens. ASM Press, Washington, DC, pp 414–421

Vannini A, Volpari C, Gargioli C, Muraglia E, Cortese R, De Francesco R, Neddermann P, Marco SD (2002) The crystal structure of the quorum sensing protein TraR bound to its autoinducer and target DNA. EMBO J 21:4393–4401

Visick KL, Foster J, Doino J, McFall-Ngai M, Ruby EG (2000) *Vibrio fischeri* lux genes play an important role in colonization and development of the host light organ. J Bacteriol 182:4578–4586

Vuong C, Saenz HL, Gotz F, Otto M (2000) Impact of the agr quorum-sensing system on adherence to polystyrene in Staphylococcus aureus. J Infect Dis 182:1688–1693

Watson WT, Minogue TD, Val DL, Von Bodman SB, Churchill ME (2002) Structural basis and specificity of acyl-homoserine lactone signal production in bacterial quorum sensing. Mol Cell 9:685–694

Whitehead NA, Barnard AM, Slater H, Simpson NJ, Salmond GP (2001) Quorum-sensing in Gram-negative bacteria. FEMS Microbiol Rev 25:365–404

Wright JS, Traber KE, Corrigan R, Benson SA, Musser JM, Novick RP (2005) The agr radiation: an early event in the evolution of staphylococci. J Bacteriol 187:5585–5594

Yao Y, Vuong C, Kocianova S, Villaruz AE, Lai Y, Sturdevant DE, Otto M (2006) Characterization of the Staphylococcus epidermidis accessory-gene regulator response: quorum-sensing regulation of resistance to human innate host defense. J Infect Dis 193:841–848

Yarwood JM, Bartels DJ, Volper EM, Greenberg EP (2004) Quorum sensing in Staphylococcus aureus biofilms. J Bacteriol 186:1838–1850

Yoon SS, Hennigan RF, Hilliard GM, Ochsner UA, Parvatiyar K, Kamani MC, Allen HL, DeKievit TR, Gardner PR, Schwab U, Rowe JJ, Iglewski BH, McDermott TR, Mason RP, Wozniak DJ, Hancock RE, Parsek MR, Noah TL, Boucher RC, Hassett DJ (2002) Pseudomonas aeruginosa anaerobic respiration in biofilms: relationships to cystic fibrosis pathogenesis. Dev Cell 3:593–603

Zhang RG, Pappas T, Brace JL, Miller PC, Oulmassov T, Molyneaux JM, Anderson JC, Bashkin JK, Winans SC, Joachimiak A (2002) Structure of a bacterial quorum-sensing transcription factor complexed with pheromone and DNA. Nature 417:971–974

Zhang L, Lin J, Ji G (2004) Membrane anchoring of the AgrD Nterminal amphipathic region is required for its processing to produce a quorum-sensing pheromone in Staphylococcus aureus. J Biol Chem 279:19448–19456

Zhu J, Winans SC (1999) Autoinducer binding by the quorum-sensing regulator TraR increases affinity for target promoters in vitro and decreases TraR turnover rates in whole cells. Proc Natl Acad Sci U S A 96:4832–4837

Zhu J, Winans SC (2001) The quorum-sensing transcriptional regulator TraR requires its cognate signaling ligand for protein folding, protease resistance, and dimerization. Proc Natl Acad Sci U S A 98:1507–1512

Chapter 10
From Interaction to Gene Induction: An Eco-friendly Mechanism of PGPR-Mediated Stress Management in the Plant

Yachana Jha and R.B. Subramanian

Abstract Soil bacteria living on or around the root surface that facilitates the plant's growth have been isolated from the paddy rhizosphere. Among 35 isolates, two selected isolates, *Bacillus pumilus* and *Pseudomanas pseudoalcaligenes,* have been evaluated for their use in the induction of various genes in different situations to help the paddy plant to survive in adverse conditions. The induction of defense-related pathogenesis-related protein occurs as a result of inoculation of the plant with plant growth-promoting rhizobacteria (PGPR) prior to biotic stress. Accumulation of low-molecular-weight osmoprotectants and soluble sugar has been observed in inoculated plants under abiotic stress, which helps in osmoregulation. The abiotic stress, especially salinity, results in a change in protein configuration. Molecular chaperones help these proteins to maintain their configuration under stress. Inoculation with PGPR also helps in the formation of these chaperones. These isolates also show differential induction of the stress-related gene RAB18 and catalase in paddy plants during RNA profiling. The results indicate that the ecofriendly root-associated bacteria can serve as a simple and cheap tool for regulating plant sugar concentrations and combating stress in crops, as well as for increased productivity due to their growth-promoting ability.

Keywords PGPR • PR proteins • Molecular chaperones • Osmoprotectant • Gene induction • RAB18 • β-1,3 glucanase

Y. Jha (✉)
N. V. Patel College of Pure and Applied Sciences, Sardar Patel University, Anand, Gujarat, India
e-mail: yachanajha@ymail.com

R.B. Subramanian
B R D School of Biosciences, Sardar Patel University, Anand, Gujarat, India

D. Egamberdieva, P. Ahmad (eds.), *Plant Microbiome: Stress Response*, Microorganisms for Sustainability 5, https://doi.org/10.1007/978-981-10-5514-0_10

10.1 Introduction

Plants being sessile are frequently exposed to various stress factors at the same time as they have advanced to live at fixed locations. To minimize damage caused by stress, plants have established definite mechanisms that allow them to take note of accurate ecological changes and respond to multifaceted stress conditions to protect valued resources for growth and development. When subjected to a combination of numerous stresses, plants will stimulate a precise and unique stress response. Two of the main hindrances to increasing crop growth and productivity in various parts of the world are climate change and pathogens (Smith 2011). The capability of plants to survive during stress depends on numerous mechanisms that allow them to cope. Such mechanisms are broadly divided into three categories. First is phenotypic flexibility and stress avoidance, in which plants develop the ability to sustain normal growth and development under stress conditions. Second is escaping, in which plants complete their lifecycle before the onset of the stress and undergo dormancy (Farooq et al. 2009). Third is stress tolerance, in which regular metabolic activities and plant growth are maintained even under stress.

In response to abiotic stress, plants experience numerous physiological and biochemical changes, which include osmotic adjustment, optimization of water loss, induction of antioxidant systems, morphological changes, as well as the induction of different stress-responsive genes and proteins to minimize the detrimental effects of reactive oxygen species (ROS) (Huang et al. 2014). Under dehydration stress, plants exhibit various traits, including maintenance of root viability, ionic adjustment, membrane stability, and maintenance of cell water content, as well as accumulation of proteins and other metabolites. Plants also accumulate low molecular proteins, carbohydrates, and other metabolites to maintain their osmotic balance, which work directly or indirectly in structural stabilization. Plants have the ability to respond via signal transduction pathways in maintaining their metabolic activity. A cascade of signals ranging from primary (stomatal closer, ion levels, etc.) to secondary (secondary metabolites, phytohormone production, etc.) responses act as a self-protective approach by plants against such stressful environments. Furthermore, plants must protect themselves from a huge variety of pests and pathogen attacks, including bacteria, fungi, viruses, herbivorous insects, and nematodes, (Hammond-Kosack and Jones 2000). Thus, one of the main problems associated with agrochemical control involves exploring the various naturally occurring host-plant resistance mechanisms to biotic stress agents. It is not impossible for various plant stress responsive pathways to be activated simultaneously, e.g., abiotic stress conditions, such as drought/salinity, may cause physical damage to plant tissue, which subsequently enables access of pathogens to that plant. Thus, abiotic stress can trigger defenses against biotic stress (pathogens), whereas the reverse has not often been observed. Therefore, various stress-related proteins have to be analyzed for their biochemical activities, as some of them may carry out roles that are important for both types of stresses. Furthermore, abiotic and biotic stress may cause related physiological/biochemical effects, and henceforth co-regulation of such defense-related genes may be selected during evolution for the survival of the plant. To recognize the nature of numerous responses to stress and to generate possibilities for developing multiple stress-resistant plants, there is a need to focus on plant stress research to maintain high yields. Plant systems react in order to prevent damage and ensure survival under stress; each stressor stimulates a multifaceted molecular and cellular response, but often with the loss of growth and yield. The rhizosphere is a nutrient-rich habitat and harbors a huge variety of roots associated with fungi and bacteria, which have favorable effects on the plant (Mendas et al. 2013), and improve plant growth through different mechanisms. Stresses negatively affect plant growth and productivity, but the root-associated bacteria help the plant by maintaining growth and yields under different types of stress. Plant growth-promoting rhizobacteria are among the most effective soil microorganisms that are able to promote plant performance.

10.2 Isolation, Identification, and Inoculation of Bacterial Isolates

The most active natural niche where intra- and inter-species interaction of microorganisms like bacteria, fungi, and protozoa takes place is the rhizosphere. Such microorganisms reside there due to the occurrence of varied and rich microbial nutrients (Bais et al. 2006). The plant growth-promoting rhizobacteria (PGPR) influences the induction of genes in plants under abiotic and biotic stresses and can also modulate the inhabitants of the rhizosphere around the root of the plant. The rhizosphere bacterial inhabitants have a considerable role in maintaining healthy roots by enhancing nutrient uptake and developing tolerance to environmental stress. The use of PGPR as a bio-fertilizer increases plant nutrient status by enhancing phosphorus, potassium solubilization, siderophores production, and nitrogen fixation (Jha and Subramanian 2014a). Bacterial genera such as *Pseudomonas*, *Brevibacillus*, and *Bacillus* are well known to enhance the growth, development, and yield in various non-leguminous plants, as reported by Karlidag et al. (2007). Among 35 isolates, the two selected isolates *Bacillus pumilus* and *Pseudomanas pseudoalcaligenes* have been evaluated for their ability in the induction of various genes in different situations for the survival of the paddy (unmilled rice) plant in adverse conditions (Jha et al. 2011a). The soil sample has been tested in the Sophisticated Instrumentation Centre for Applied Research and Testing (SICART) laboratory by an extracted water sample method. The soil possesses the following physio-chemical properties; pH 6.58, electrical conductivity 1480 μS/cm, salinity 8.6%, nitrate 112.5 mg kg^{-1}, chloride 128 mg kg^{-1}, sulfate 155 mg kg^{-1}, ammonia nitrogen 23.3 mg kg-1, CEC:3 cmol, organic carbon: 5500 mg kg^{-1}. Initially the bacterial stain was isolated in the semi-solid NFb medium, then a white veil-like pellicle that formed below the surface of the semi-solid NFb medium was purified and transferred to NFb agar plates. The pure culture was maintained on the nutrient agar plates. The colony formed on the NFb agar plates indicates that isolates have the ability for nitrogen fixation. NFb agar plates containing bromothymol blue, which is a pH indicator dye, are used for isolation of bacteria. The change in plate color from green to blue indicates that the pH of the medium shifts towards alkalinity due to the growth of bacteria.

Molecular identification of bacterial isolates is been done by isolation of total genomic DNA from the isolates and amplified with 16S rDNA specific primers16S F: 5′AGAGTTTGATCCTGGCTCAG3′ and 16S R: 5′AGGTTACCTTGTTACG ACTT3′ followed by sequencing as per our previously published method (Jha and Subramanian 2013a). Discrete bands of 16S rDNA amplicon of about 1500 bp are obtained in agarose gel. The phylogenetic trees were constructed using BLAST software by comparing the 16S rDNA sequence of isolates and related genera from a database using the neighbor-joining (NJ) algorithm and maximum likelihood (ML) method (Fig. 10.1a, b). The sequences with accession nos. EU921258 and EU921259 were submitted to the NCBI data bank. The isolates were identified as *P. pseudoalcaligenes* and *B. pumilus*, respectively.

Seeds of paddy variety GR11were inoculated with isolates as per our previously published methods with some modification (Jha and Subramanian 2013b). The seeds are kept in sterile distilled water after washing thoroughly with autoclaved distilled water and incubated on a rotary shaker for 5–6 h. To check possible contamination, the sterile seeds were transferred to petri dishes having tryptone glucose yeast extract agar medium, and incubated in the dark at 30 °C. The germinated seedlings without any contamination were used for inoculation experiments. The effect of the isolated inoculated bacteria on the selected paddy seedlings on various biochemical parameters has been studied by transferring the seedlings in culture tubes. The culture tube contains 400 ml of Hoagland's nutrient medium, 400 ml micronutrients, 1% agar in 40 ml distilled water, and a bacterial inoculum of the isolated bacteria, which has been added to it in a concentration of 6 × 108 cfu ml^{-1}. To obtain a mixture of both isolates for inoculation, the cultures were mixed at a concentration of 6×10^8 cfu ml^{-1} into the medium. The culture tubes were incubated for a 12-h light–dark cycle in a growth chamber at 27 °C. Confirmation of the association of PGPR with the root was done with 2, 3, 5-triphenyl tetrazolium chloride (TTC) staining, with 1.5 g of TTC and maleic acid in a sterile potassium phosphate buffer (pH 7). The surface-sterilized plant roots were incubated overnight in the TTC stain and cross-sections of the root were examined under an image analyzer microscope (Carl Zeiss) (Jha and Subramanian 2011). The presence of bacteria within the root cortex region can be noticeably visualized as red-colored cells when a thin section of the root is observed under the microscope after TTC staining (Fig. 10.2).

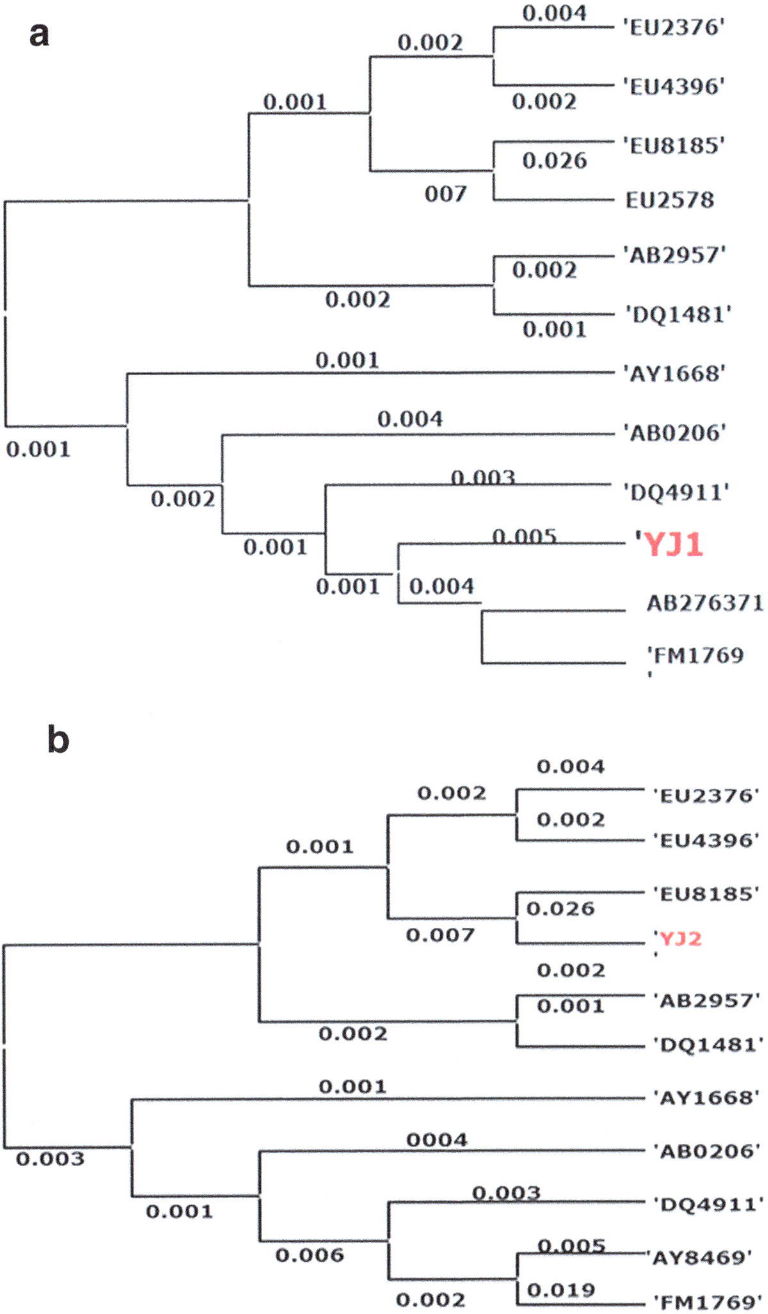

Fig. 10.1 (**a**) Phylogenetic tree for *Pseudomonas psedualcaligene* made in MEGA 3.1 software using neighbor joining method. (**b**) Phylogenetic tree for *Bacillus pumilus* made in MEGA 3.1 software using neighbor joining method

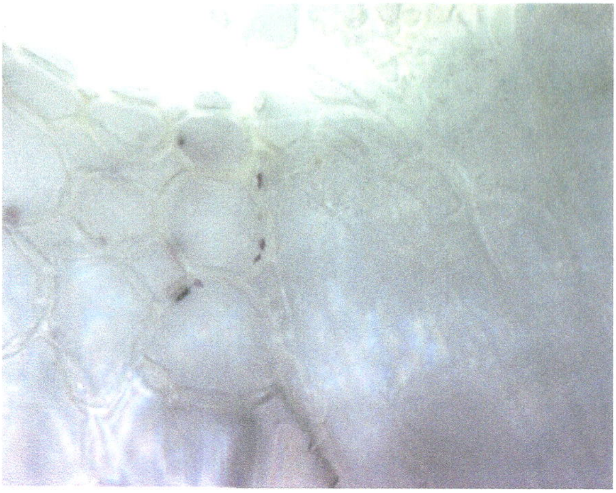

Fig. 10.2 A section of paddy plant root showing the association of bacteria in root cortex as brown sport due to TTC staining

10.3 Effect of Plant Growth-Promoting Rhizobacteria (PGPR) in the Induction of Pathogenesis-Related (PR) Proteins and β-1,3 Glucanase Genes Under Biotic Stress

Plants possess a group of chemical and physical barriers to avoid almost all unfavorable interactions encountered between plants and biotic stressors. The chemical barriers act as a constitutive defense by inhibiting the enzymatic functions of pathogens, or by rapid accumulation of secondary metabolites or proteins, which inhibit the growth of pathogens, or by toxic metabolites, which kill the pathogen. The physical barriers consist of obstacles like cuticle and cell walls to avoid harmful biotic interactions (Hanley et al. 2007). Depending on the species and the environment in which plants grow, there are different levels and forms of barriers. Plants have developed efficient mechanisms for a response to biotic stress that allows for quick and targeted responses. In comparison with the widespread use of chemical pesticides, the biological control options like non-pathogenic soil bacteria (PGPR) living in association with plant roots is a promising alternative (Aravind et al. 2009). Plants inoculated with PGPR are correlated with decreases in disease in both greenhouse and field experiments. These bacteria are native to the soil and plant rhizosphere and have an important role for plants as bio-control agents against phyto-pathogens. Bacteria that reduce the frequency or severity of plant diseases are pften referred to as biocontrol agents.

In agricultural systems, the biotics (pathogen infections) are responsible for most of the reduction that differentiates yield potential from harvestable yield. Plants when infected by pathogens illustrate a wide range of defense responses, i.e., synthesizing novel proteins that can have a direct or indirect action on the course of pathogenesis (Beneduzi et al. 2012). These proteins may comprise enzymes involved in flavonoids, peroxidases, β-I,3 glucanases, phenyl ammonia lyase (PAL), phenylpropanoid, chitinases, hydroxyproline-rich glycoproteins, and a varied group of extracellular acidic proteins collectively known as pathogenesis-related (PR) proteins. Among the PR proteins, the β-1,3-glucanases are highly important because they are developmentally and hormonally regulated in healthy plants and constantly protect plants from phyto-pathogen infection. β-1,3-glucanases and chitinase are members of the PR protein family that are either induced in response to pathogen attack or by releasing oligosaccharides from the fungal cell wall during invasion, and have the capability to degrade fungal cell walls or effectively restrict the growth of different fungi (Gupta et al. 2013). In our study of differential expression of PR proteins like polyphenol peroxidases, β-I,3 glucanases, phenyl ammonia lyase (PAL), and chitinases have been observed after 10–12 h of infection in control as well as in infected plants in the presence of PGPR (Jha et al. 2011b). The expression of different PR proteins at different levels and times indicates the differences in the type of gene expression. The pattern of PR protein expression changes at different times of infection in the presence of different isolates. Such systemic resistance induced by *Pseudomonas* is coupled with the accumulation of PAL, β-1, 3-glucanase, chitinase, and polyphenol peroxidase enzymes in the plant. The β-1, 3-glucanase and chitinase enzymes hydrolyze the β-1, 3-glucan and chitin, the main components of fungal cell walls, respectively. The higher induction of such PR proteins may be because of the enhanced association of these bacteria with the plant in the presence of *P. pseudoalcaligenes* as well as having less competition than the rhizospheric *B. pumilus* bacteria (Fig. 10.3a, b). Reports have shown that inoculation by the root-associated PGPR in the paddy plant changes the plant gene expression for β-1, 3-glucanase. The results showed that *P. pseudoalcaligenes* induces resistance to biotic stress (M*agnaporth grisea* infection) and a mixture of both *P. pseudoalcaligenes* and *B. pumilus* is more effective. With RNA dot blot assay and reverse transcription-polymerase chain reaction (RT-PCR), genes are identified whose induction intensity has been induced in a plant treated with the PGPR compared to non-inoculated controls. RNA dot blot is used to observe changes in messenger RNA (mRNA) levels in plants inoculated with PGPR under biotic stress. The observations reported that paddy plant inoculation by *P. pseudoalcaligene* and *B. pumilus* confers significant resistance to plant pathogens by differential gene expression followed by PR protein production. Such reports have highlighted their potential for both plant-growth promotion and defense against pathogen attacks. These consequences are mainly expected to be related to changes in plant gene expression and changes in plant gene profile regulated by PGPR (Poupin et al. 2013). The molecular machinery following these types of stress prevention is inducible, allowing the redirection of physiological events, which depend on a huge number of factors. However, the signal induction pathway and the molecular basis of original rhizobacteria-induced systemic resistance (ISR) differ in various aspects from pathogen-induced systemic acquired resistance (SAR).

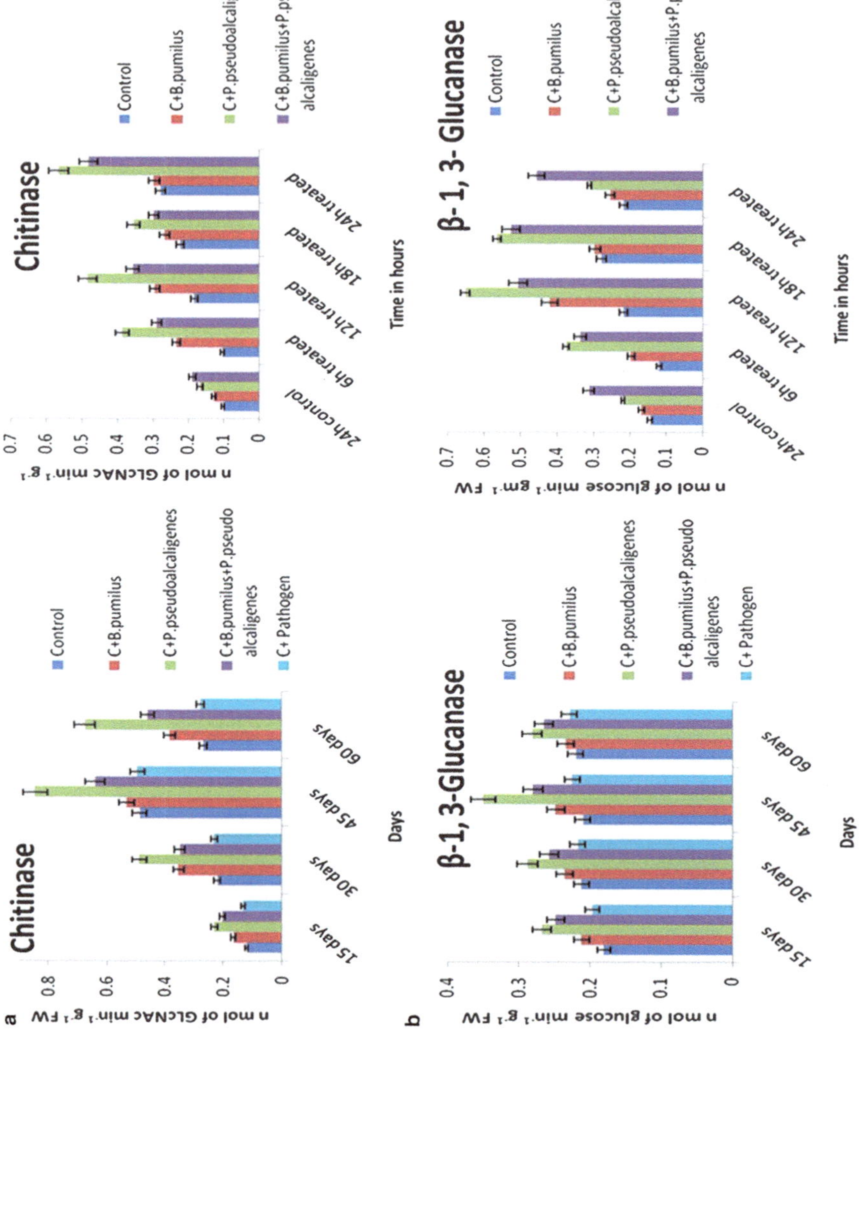

Fig. 10.3 **(a)** The graph showing the effect of PGPR β-1,3 glucanase activity at various days and different times of cultivation after pathogen treatment. **(b)** The graph showing the effect of PGPR Chitinase activity at various days and different times of cultivation after pathogen treatment

10.4 Effect of PGPR on the Induction and Accumulation of Osmoprotectant

Soil salinity is a major problem throughout the world, particularly in the semiarid and arid regions where farming is carried out under irrigation. Plant nutrient acquisition in these regions has been badly affected by salinity, resulting in a significant decline in plant metabolism and physiology. The multiple effects of salt stress results in enhanced production of ROS on the physiological scale. The ROS rise during a period of water deficiency and consequently enhance ion uptake, resulting in decreased photosynthesis (Dat et al. 2000). Plants adapt and respond to salinity-induced stress by using numerous regulatory mechanisms and composite dynamic communications, which currently are described as the plant interactome (Shavrukov 2013). Accumulation of the osmoprotectant in the plant cell takes place, which reduces the cell water potential, alleviating the detrimental effects of salt ions.

An interesting observation from our study is that plants inoculated with PGPR under non-saline as well as at different salinity levels have a greater relative water content and membrane stability index, which is in accordance with the findings of Sandhya et al. (2010). The role of osmoprotectants in preventing cell damage from salt stress-induced dehydration results in an increased level of leaf relative to water content in paddy plants growing in saline conditions, as suggested by Janska et al. (2010). Non-inoculated plants have decreased relative water content, and a lower membrane stability index with increased salinity.

Soil salinity restricts the movement of water through the plant root and causes osmotic stress to the plant cell. For protection against the immediate effects of water shortages, plant cells accumulate low-molecular-weight osmoprotectants for osmoregulation. The instant response against osmotic stress by cells is that they start to accumulate compatible solutes and the resulting efflux of cellular water facilitates the uptake of K^+ (Burg and Ferraris 2008). Such osmoprotectants are organic osmolytes, which include amino acids and their derivatives, sugars and derivatives, polyols and derivatives, betaines, and ectoines. *Pseudomonas fluorescens* helps in salt tolerance by producing glycine, alanine, glutamic acid, threonine, aspartic acid, and serine by de novo synthesis of the osmolytes in the cytosol, as reported by Paul and Nair (2008). In our study, the accumulation of osmoprotectants due to PGPR under salt stress has been observed in paddy plants inoculated with *P. pseudoalcaligenes* and *B. pumilus*. Many plant species naturally accumulate organic osmoprotectants when exposed to diverse abiotic stresses, but paddy does not have this ability. In our study, the overall expression of organic osmolytes is lower in paddy plants, but PGPR help in the accumulation of organic osmolytes in paddy leaves to some extent under saline conditions. We also focused on accumulation of sugars in plants inoculated with PGPR under stress, with the results showing different types of sugars accumulate in inoculated plants under salinity stress. To counteract the effect of osmosis, an osmoprotectant allows additional water to be taken up from the environment. The osmotic stress induced by salinity restricts the absorption of water from the soil, resulting in enhanced concentrations of potentially toxic salt ions within plant cells. The PGPR may act to confer tolerance to salt stress via regulating the transcription factors, which stimulate adaptive responses by inducing expression of ion channel and transporter genes in the cytosol. These genes are involved in the synthesis of antioxidants and osmoprotectants to remove the accumulation of toxic ions from the plant cell (Bharti et al. 2016).

10.5 Effect of PGPR in the Induction and Accumulation Chaperones Under Abiotic Stress

With global environmental changes, availability of water for irrigation is a crucial factor affecting world crop yields, and which is beyond the reach of purely socio-economic implications.

The future shortage of water required for agriculture will certainly augment the production cost of crops and constrain the use crops that use water more proficiently. Abiotic stresses, like salinity, drought, extreme temperatures, oxidative stress, and chemical toxicity, are serious problems in agriculture and are the major cause of reduced crop production. This stress is the prime reason for crop loss globally, reducing regular yields of important food crops by more than 50%. Hence the mechanism developed by the plant in achieving stress tolerance is of immense practical importance. Such important stress tolerance mechanisms include accumulation of osmoprotectants, production of late embryogenesis-abundant proteins, specificity of ion transporters, free-radical scavengers, and factors involved in transcriptional control and signaling cascades, etc. (Wang et al. 2003). The transcriptional control mechanism is the rational continuance of proteomics, which are widely used now. Proteomic analysis involves the study of multi-protein systems in an organism (Karpievitch et al. 2010). In this analysis the complete protein profile coded by the genome has been analyzed to learn about diverse proteins and their functions in such huge networked systems of an organism. This is an essential module of current systems biology approaches, the objective of which is to characterize the coordinated performance rather than the behavior of a particular component. Measurement of mRNA that codes for character alone does not directly relate to the types and level of complementary proteins in the cell and their role in organism behavior, as proteins formed on mRNA undergo various post-translational modifications and modifications by environmental agents. Proteins are the principal molecule responsible for energy production, structural organization, cell communications, cell signaling, and cell division. Therefore, a complete understanding of systems biology is extremely important. Plants are able to enhance their stress tolerance ability through numerous methods of action such as through the help of molecular chaperones. Molecular chaperones are one of the most abundant and ubiquitous proteins present in both viruses and all living organisms. Molecular

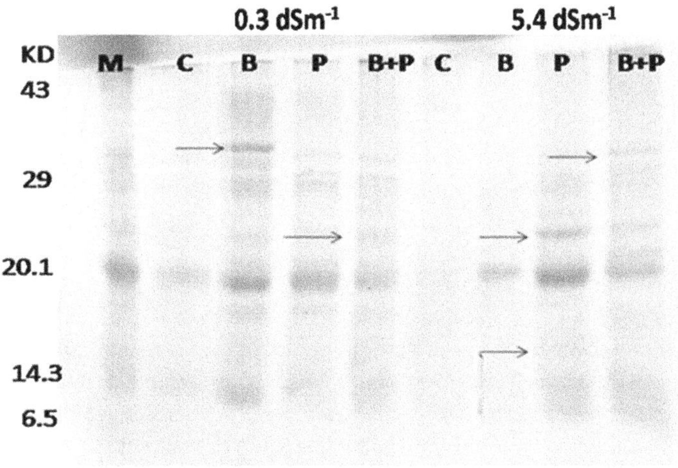

Fig. 10.4 Polyacrylamide gel of total soluble protein showing the differential expression of low molecular weight proteins due to inoculation of PGPR and salinity. Loaded samples were adjusted to a constant amount of protein (15 μg), *M* marker (KD), *C* Control, *B* Plant inoculated with *B.pumulis, P P.pseudoalcaligenes*, *B+P* inoculated with both the isolates

chaperones are molecular-binding proteins that bind with those kinetically trapped in misfolded forms of molecules to resolve the trapped structures and ensure accessibility of that molecule for their biological activity. Horn et al. (2007) reported that there is a set of small, low-molecular-weight proteins that quickly accumulate during unfavorable conditions; these proteins may play a role as molecular chaperones. The molecular chaperones bind with misfolded proteins, allowing for their proper folding (Lee and Vierling 2000). In our study, there were some new bands of low-molecular-weight proteins that occurred during the separation of leaf proteins under salinity stress in the sodium dodecyl sulfate (SDS) gel (Fig. 10.4). Abiotic stresses like salinity generally cause protein dysfunction, and preventing the accumulation of non-native proteins is significant for cell survival under stress in order to maintain proteins in their functional conformations (Jha and Subramanian 2014b). Chaperones are accountable for protein assembly, folding, translocation to a specific site, and degradation after normal cellular function. Chaperones also stabilize proteins and membranes during translocation and can help in the refolding of proteins under stress conditions. Chaperones are mostly expressed in plants when they experience high temperature stress, but are also expressed in a broad range of other environmental stresses like salinity, draught, and osmotic, oxidative, and cold stresses (Wang et al. 2004). Thus, chaperones can play an important role in defending plants under various conditions of stress by re-establishing normal protein folding and cellular homeostasis.

10.6 Effect of PGPR in the Induction of RAB 18 and Catalase Gene Under Abiotic Stress

As a result of the continuous increase in climatic global warming, additional approaches need to be developed and implemented to enhance per capita agricultural production. Among the various factors that affect agricultural production are abiotic stresses such as low temperature, drought, and salinity, which chiefly affect potential crop production compared to harvested yield (Atkinson and Urwin 2012). Salinity and salinity-induced stress is an emerging drought-related stress factor that reduces the potential crop yield in arable lands.

Plant-microbe interactions can play an important role in maintaining plant health under stress and also help in developing low-cost, sustainable agriculture methods for food and non-food crops. By understanding the actual mechanism by which microbes will interact with the host plant, researchers will successfully be able to exploit the biotechnological potential of efficient biological partnerships for a wide range of crops and applications. A most promising and ecofriendly research area for future studies is developing potential microbes to enhance the sustainable cultivation of crops under abiotic and biotic stresses. The potential of microbes to confer tolerance to plants against stress may act as a novel approach for mitigating the influence of worldwide climate change on plant populations. Once the exceptional abilities of tolerance to extremes, genetic variability, and techniques for their effective deployment of such potential microbes are known, such microbes can play an important role in controlling the effects of stress and in agriculture production. These microorganisms also deliver exceptional simulations for understanding the stress tolerance approach, and can be successively engineered into plants. Previously, microbes belonging to diverse genera, including *Pseudomonas, Paenibacillus, Rhizobium, Pantoea, Bacillus, Burkholderia, Methylobacterium, Achromobacter, Variovorax, Enterobacter Microbacterium,* and *Azospirillum,* etc., have been reported to enhance tolerance in inoculated plants under diverse abiotic stress conditions (Choudhary et al. 2011). It has also been confirmed that PGPR play a significant role in providing stress tolerance ability to the plant, and PGPR-inoculated plants show changes in transcript level and profile in response to water stress. Recent reports have shown that PGPR-mediated responses have a high intensity and rapid effect in plants under stress after inoculation and can be used through a systematic priming approach (Herman et al. 2008). To enhance previous knowledge and to analyze the differential gene expression in plants under stress as a result of inoculation with root-associated bacteria, total mRNA from the rice plant inoculated with root-associated bacteria under saline stress has been isolated after seven consecutive days (Fig. 10.5). The mRNA from the control as well as the treated samples for stress have been used for the synthesis of cDNA and are PCR-amplified using RAB18 and catalase-specific primers, respectively. The induction of catalase and RAB18 are observed in inoculated plants as intense bands (data previously communicated). The amplified cDNA sequence has been obtained and submitted to the NCBI database under the accession numbers JF495696 and JX875104,

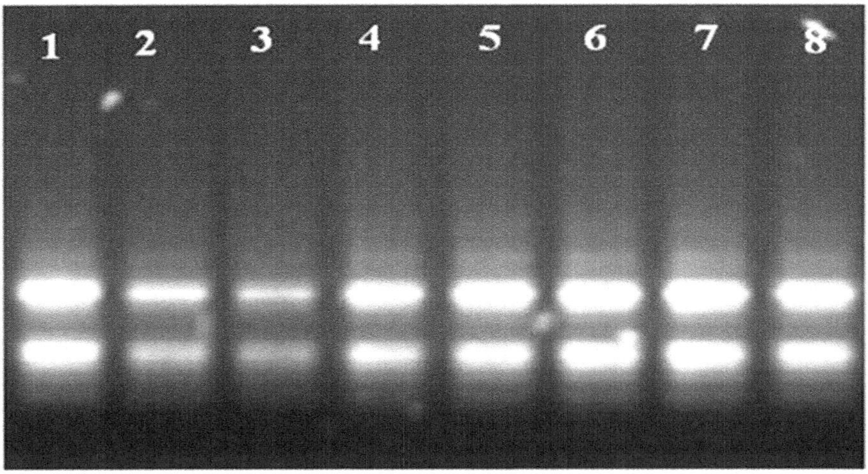

Fig. 10.5 Lane *1* = control, *2* = control + *B. pumilus,* Lane *3* = control + *P. pseudoalcaligenes*, Lane *4* = control + *B. pumilus* + *P. pseudoalcaligenes* at non-saline state. Lane *5* = stressed, Lane *6* = stressed + *B. pumilus,* Lane *7* = stressed + *P. pseudoalcaligenes* Lane *8* = stressed + *B. pumilus* + *P. pseudoalcaligenes* at 1.5% salinity

respectively. In our study, the induction of catalase and RAB18 gene in the presence of PGPR under salinity stress is the molecular basis of stress adaptation in the paddy plant, which is at present one of the chief sources of food for the world's population (Jha et al. 2014). It showed a potential benefit of plant-associated bacteria for food crops under salinity stress. A recent report showed that the association of root-associated bacteria plays a significant role in stress adaptation in a large variety of crops (Rojas-Tapias et al. 2012). The performance of habitat-adjusted symbiotic microbial association to prevent salinity stress has been previously clearly demonstrated (Rodriguez et al. 2008). In *Arabidopsis thaliana*, the model plant, the association of the PGPR *Paenibacillus polymyxa* showed an induced tolerance to abiotic stress, indicating that PGPR can be projected as an efficient stress-tolerant mechanism (Glick 2014). Such stress-signaling molecules involved in numerous aspects of cellular metabolism and physiology are supposed to play a significant role in stress interactomes. Therefore, the induction of stress-related genes by association with root-associated bacteria in plants takes place prior to abiotic and biotic stress.

10.7 Conclusion

Stress badly affects agriculture, the environment, and biodiversity. There are universal growing demands for healthy, environmentally friendly, and ecologically compatible techniques for combating all such stresses in order to increase crop productivity. Use of efficient, beneficial microorganisms such as PGPR is a viable alternative option to the traditional agricultural methods. The PGPR that reside in the rhizosphere of the plant roots help plants to overcome various types of stress and to increase crop productivity. Through its own method of stress tolerance, PGPR provides a significant advantage to plants in inducing systemic resistance to control phyto-pathogenic microorganisms. To increase crop production and yield, microbial inoculants are probably one of the most suitable approaches to agriculture, as they help in the differential expression of stress-related genes in plants under stress. Such beneficial microorganisms will allow new methods to be introduced that provide more profits from microbial inoculants while improving plant growth as well as increasing resistance to stress. A commercial application PGPR consortium with recognized actions that can act synergistically is needed as they could tender diverse modes of action.

References

Aravind R, Kumar A, Eapen SJ, Ramana KV (2009) Endophytic bacterial flora in root and stem tissues of black pepper (Piper nigrum L.) genotype: isolation, identification and evaluation against *Phytophthora capsici*. Lett Appl Microbiol 48:58–64

Atkinson NJ, Urwin PE (2012) The interaction of plant biotic and abiotic stresses: from genes to the field. J Exp Bot 63(10):3523–3543

Bais HP, Weir TL, Perry LG, Gilroy S, Vivanco JM (2006) The role of root exudates in rhizosphere interactions with plants and other organisms. Annu Rev Plant Biol 57:233–266

Beneduzi A, Ambrosini A, Passaglia LMP (2012) Plant growth-promoting rhizobacteria (PGPR): their potential as antagonists and biocontrol agents. Genet Mol Biol 35(4):1044–1051

Bharti N, Pandey SS, Barnawal D, Patel VK, Kalra A (2016) Plant growth promoting rhizobacteria *Dietzia natronolimnaea* modulates the expression of stress responsive genes providing protection of wheat from salinity stress. Sci Rep 6:34768

Burg MB, Ferraris JD (2008) Intracellular organic osmolytes: function and regulation. J Biol Chem 283:7309–7313

Choudhary DK, Sharma KP, Gaur RK (2011) Biotechnological perspectives of microbes in agroecosystems. Biotechnol Lett 33(10):1905–1910

Dat J, Vandenabeele S, Vranova E, Van Montagu M, Inze D, Van Breusegem F (2000) Dual action of the active oxygen species during plant stress responses. Cell Mol Life Sci 57:779–795

Farooq M, Wahid A, Kobayashi N, Fujita D, Basra SMA (2009) Plant drought stress: effects, mechanisms and management. Agron Sustain Dev 29:185–212

Glick BR (2014) Bacteria with ACC deaminase can promote plant growth and help to feed the world. Microbiol Res 169:30–39

Gupta P, Ravi I, Sharma V (2013) Induction of β-1,3-glucanase and chitinase activity in the defense response of *Eruca sativa* plants against the fungal pathogen *Alternaria brassicicola*. J Plant Interact 8:155–161

Hammond-Kosack KE, Jones JDG (2000) In: Buchannan B, Gruissem W, Jones R (eds) Response to plant pathogens, in biochemistry and molecular biology of plants. American Society of Plant Physiologists, Rockville, pp 1102–1157

Hanley ME, Lamont BB, Fairbanks MM, Rafferty CM (2007) Plant structural traits and their role in antiherbivore defense. Perspect Plant Ecol Evol Syst 8:157–178

Herman MAB, Nault BA, Smart CD (2008) Effects of plant growth promoting rhizobacteria on bell pepper production and green peach aphid infestations in New York. Crop Prot 2:996–1002

Horn G, Hofweber R, Kremer W, Kalbitzer HR (2007) Structure and function of bacterial cold shock proteins. Cell Mol Life Sci 64:1457–1470

Huang X, Peng X, Zhang L, Chen S, Cheng L, Liu G (2014) Bovine serum albumin in saliva mediates grazing response in *Leymus chinensis* revealed by RNA sequencing. BMC Genomics 15:1126

Janska A, Marsik P, Zelenkova S, Ovesna J (2010) Cold stress and acclimation: what is important for metabolic adjustment? Plant Biol 12:395–405

Jha Y, Subramanian RB (2011) Endophytic *Pseudomonas pseudoalcaligenes* shows better response against the *Magnaporthe grisea* than a rhizospheric *Bacillus pumilus* in *Oryza sativa* (Rice). Arch Phytopathol Plant Protect 44:592–604

Jha Y, Subramanian RB (2013a) Root associated bacteria from the rice antagonizes the growth of *Magnaporthe grisea*. J Plant Pathol Microbiol 4:164

Jha Y, Subramanian RB (2013b) Paddy inoculated with PGPR show better growth physiology and nutrient content under salinity. Child J Agric Res 73(1):213–219

Jha Y, Subramanian RB (2014a) Characterization of root associated bacteria from paddy and its growth promotion efficacy. 3 Biotech 4(3):325–330

Jha Y, Subramanian RB (2014b) Under saline stress plant growth promoting bacteria affect growth, photosynthesis and antioxidant activities in paddy. Int J Agric Environ Biotechnol 7(2):205–212

Jha Y, Subramanian RB, Patel S (2011a) Combination of endophytic and rhizospheric plant growth promoting rhizobacteria in *Oryza sativa* shows higher accumulation of osmoprotectant against saline stress. Acta Physiol Plant 33:797–802

Jha Y, Subramanian RB, Patel S (2011b) Endophytic bacteria induced enzymes against *M. grisea* in *O. sativa* under biotic stress. Afr J Basic Appl Sci 3:136–146

Jha Y, Sablok G, Naidu Subbarao N, Sudhakar R, TurabeFazil MHU, Subramanian RB, Squartini, Kumar S (2014) Bacterial-induced expression of RAB18 protein in *Orzya sativa* salinity stress and insights into molecular interaction with GTP ligand. J Mol Recognit 27:521–527

Karlidag HA, Esitken M, Turan SF (2007) Effects of root inoculation of plant growth promoting rhizobacteria (PGPR) on yield, growth and nutrient element contents of apple. Sci Hortic 114(16):20

Karpievitch YV, Polpitiya AD, Anderson GA, Smith RD, Dabney AR (2010) Liquid chromatography mass spectrometry-based proteomics: biological and technological aspects. Ann Appl Stat 4(4):1797–1823

Lee GJ, Vierling E (2000) A small heat shock protein cooperates with heat shock protein 70 systems to reactivate a heat-denatured protein. Plant Physiol 122:189–198

Mendas R, Garbeva P, Raaijmakers JM (2013) The rhizosphere microbiome: significance of plant beneficial, plant pathogenic, and human pathogenic microorganisms. FEMS Microbiol Rev 37:634–663

Paul D, Nair S (2008) Stress adaptations in a plant growth promoting rhizobacterium (PGPR) with increasing salinity in the coastal agricultural soils. J Basic Microbiol 48(5):378–384

Poupin MJ, Timmermann T, Vega A, Zuñiga A, González B (2013) Effects of the plant growth-promoting bacterium *Burkholderia phytofirmans* PsJN throughout the life cycle of *Arabidopsis thaliana*. PLOS ONE 8(7):e69435

Rodriguez RJ, Henson J, Van Volkenburgh E, Hoy M, Wright L, Beckwith F, Kim YO, Redman RS (2008) Stress tolerance in plants via habitatadapted symbiosis. ISME J 2:404–416

Rojas-Tapias D, Moreno-Galván A, Pardo-Díaz S, Obando M, Rivera D, Bonilla R (2012) Effect of inoculation with plant growth-promoting bacteria (PGPB) on amelioration of saline stress in maize (Zea mays). Appl Soil Ecol 61:264–272

Sandhya V, Ali SKZ, Minakshi G, Gopal R, Venkateswarlu B (2010) Effect of plant growth promoting *Pseudomonas spp.* on compatible solutes, antioxidant status and plant growth of maize under drought stress. Plant Growth Regul 62:21–30

Shavrukov Y (2013) Salt stress or salt shock: genes are we studying? J Exp Bot 64:119–127

Smith MD (2011) An ecological perspective on extreme climatic events: a synthetic definition and framework to guide future research. J Ecol 99:656–663

Wang WX, Vinocur B, Altman A (2003) Plant responses to drought, salinity and extreme temperatures: towards genetic engineering for stress tolerance. Planta 218:1–14

Wang WX, Vinocur B, Shoseyov O, Altman A (2004) Role of plant heat- hock proteins and molecular chaperones in the abiotic stress response. Trends Plant Sci 9(5):244–252

Chapter 11
Predisposition of Crop Plants to Stress Is Directly Related to Their DNA Health

Murat Dikilitas, Sema Karakas, and Parvaiz Ahmad

Abstract Plants have to defend themselves under abiotic and biotic stress conditions to sustain their growth and development and to transfer their genetic material into the next generation. Sometimes, plants have to face up both kinds (abiotic and biotic) of stressors at the same time. In this case, the defense of plants becomes much harder. Crop production under these circumstances has been mostly evaluated from the view of biochemical mechanism. To remediate the conditions of such plants, scientists have focused on the increase of defense mechanism via biochemical support material or via genetic transfer of nuclear material. However, health and strength of genetic material in parental plants are at prime importance before remediative studies take place. With this chapter, we evaluated the importance of DNA health and their response upon stress emergence via new and cost-effective methods.

Keywords DNA damage • DNA repair • Predisposition • Abiotic stress • Pathogen

11.1 Introduction

Abiotic and biotic stress factors are important constraints on worldwide crop production. Losses caused by diseases alone reach up to 20–40% of the global harvest each year (Savary et al. 2012). Another 20–30% of loss is caused by the abiotic stress factors. As a consequence, crop losses constitute a significant threat to global food sector. Therefore, it is a high priority to double the food production by 2050 to feed the increasing number of people globally. To increase the crop production, a fundamental

M. Dikilitas (✉)
Department of Plant Protection, Harran University, S. Urfa, Turkey
e-mail: m.dikilitas@gmail.com

S. Karakas
Department of Soil Science and Plant Nutrition, Harran University, S. Urfa, Turkey

P. Ahmad
Department of Botany and Microbiology, King Saud University, Riyadh, Riyadh, Saudi Arabia

D. Egamberdieva, P. Ahmad (eds.), *Plant Microbiome: Stress Response*, Microorganisms for Sustainability 5, https://doi.org/10.1007/978-981-10-5514-0_11

understanding of plant molecular biology, biochemistry, and genetic knowledge is needed. Under the scope of this concept, efforts for making resistant crop plants against abiotic or biotic stress factors have rapidly increased due to the reduction of fertile agricultural areas in the last two decades. Many resistant crops have been generated either by modifying the genetic structure or inducing biochemical and physiological changes. Although efforts have been made to combat against the loss of crop production resulted from abiotic or biotic stresses, crop plants are often simultaneously exposed to multiple biotic and abiotic stresses resulting in substantial yield loss (Dikilitas et al. 2016). Therefore, new focus should be diverted toward understanding DNA health and biochemical responses under combined stress conditions. Due to industrialization and global heating, abiotic and biotic stressors are now very much interconnected. Therefore, any crop plants generated for both abiotic and biotic stress factors would struggle more to stand alive. Mechanism of combined stresses of abiotic and biotic stress factors could be different from those of abiotic stresses or biotic stresses alone and sometimes become contrasting to those seen under individual stresses. For example, Rizhsky et al. (2004) reported that *Arabidopsis thaliana* plants accumulated sucrose instead of proline under drought and heat stress conditions. The enhanced transpiration rate during heat stress to cool down the leaf surface aggravated water loss and resulted in more uptake of toxic substances from the soil that put the plants under heavy pressure. This case is also realistic for abiotic and biotic interactions. Therefore, predisposition resulting from abiotic stresses during or prior to microbial infection affects susceptibility of crop plants to disease. Short- or long-term abiotic stressors either with low or high intensity predispose crop plants against pathogen attack that would be otherwise tolerated in optimal conditions. Abiotic and biotic stress responses are coordinated by complex signaling networks including various defense enzymes, phytohormones, reactive oxygen species (ROS), and oxidant/antioxidant metabolites (Rejeb et al. 2014; Pandey et al. 2017). Although abiotic stresses are able to affect the outcome of plant-pathogen interactions in terms of crop quality and yield, they also alter the resistance of crop plants by making them susceptible to further pathogen attack. Although abiotic stresses may have potential to reduce disease severity, here, we consider the negative impacts of abiotic stresses on crop plants, because the negative impact of abiotic stresses on pathogens may also negatively affect the development of crop plants via increase of pathogenic toxins, enzymes, etc. (Funari et al. 2012; Dikilitas et al. 2016). Therefore, negative effects of both stress groups have been evaluated in this chapter. It is now clearly understood that any stress-tolerant crop plants would eventually face to more aggressive pathogenic microorganisms or to harsh environmental conditions or to both than ever they had faced in the past (Dikilitas and Karakas 2012). What is probably not considered at first sight here is that relatively low or mild abiotic stress could be able to suppress the disease resistance and may not give a chance to crop plants to recover. Therefore, it is important to note that any approaches for making stress-tolerant or stress-resistant crop plants should have strong genetic backgrounds, because any further increase in the virulence of pathogens or unfavorable environmental conditions will put the so-called "resistant" plants into a great danger. For example, if a crop plant faces salinity or drought stress, the symptoms, in general, would be reduced cell enlargement, blockage of xylem, loss of turgor, reduced water potential, decline in photosynthesis, accumulation of ROS and stress-related metabolites as well as stress-responsive gene expression, etc. (Méndez-Alonzo et al. 2016; Ahanger et al. 2016). Crop plants, on the other hand, will respond to these stressors by accumulating more antioxidant metabolites and enzymes and expressing stress-related genes to the best of their genetic capacity. The components of antioxidant defense system are enzymatic and nonenzymatic antioxidants. Enzymatic antioxidants include superoxide dismutase (SOD), catalase (CAT), ascorbate peroxidase (APX), glutathione reductase (GR), etc., and nonenzymatic antioxidants are glutathione (GSH), ascorbic acid (AA), carotenoids, tocopherols, etc. (Smolinska and Szczodrowska 2017). Higher plants are also able to detoxify ROS by accumulating in different organelles or parts. They also reduce the toxic effects of chemicals by accumulating stress-related amino acids and proteins (Koobaz et al. 2016). The cell system coordinately works to protect the plant from the deleterious effects of ROS. The plant system has also genetic responses to stress. Bhatnagar-Mathur et al. (2007) stated that during stress, a number of genes and gene products have been expressed including proteins responsible for tolerance to these stresses. For example, heavy metals, irradiation, and pesticide toxicity at high doses affect gene response at different scales in different plants. They can influence DNA either directly by causing breaks in various parts of DNA and modify the chromatin structure, or they can indirectly affect DNA via oxidative stress (Hattab et al. 2009). The activation of stress-responsive genes also occurs by a complex array of signaling pathways.

During stress, plants are able to recover from the deleterious effects of stress. Recovery of plants, of course, depends on the duration, severity, and source of stress. For example, drought-stressed plants could recover from the effect of drought after rewatering or from the effect of salinity after desalinization of the habitats and so on (Dikilitas et al. 2016). If duration and severity of stress are extended considerably, the damage to DNA is inevitable (Procházková et al. 2013; Dikilitas et al. 2015). Damages to DNA could be various; deletion or insertion of a single base pair in DNA helix is accounted as damage in the native structure of DNA as well as breaks. Breaks in DNA may result from damaged DNA replication or from the oxidative destruction of deoxyribose residues. The strand breaks could be single or double. Double-strand breaks as compared to single-strand breaks are lethal as they affect both strands of DNA and lead to loss of genetic information (Kumari et al. 2008). On the other hand, there are a number of repair mechanisms such as photoreactivation, nucleotide excision repair (NER), base excision repair (BER), mismatch repair (MMR), recombinational repair, apoptosis, etc. to repair the damaged parts of DNA (Roy 2014). In humans, failure of these mechanisms could cause serious disorders in various organs and metabolic functions. These failures could be transferred to the next generations resulting in susceptibility to various diseases and stresses. In plants, these disorders could lead to a reduction in defense responses. Although damage to DNA could be recovered during or after the stress conditions, however, no recovery has been reported so far in susceptible and moderately resistant crop plants in terms of DNA damage followed by the effect of combined stresses such as salinity and pathogen or drought and pathogen, etc. The same case also applies to crop plants under severe stress conditions such as heavy metal stress or continuous drought and saline conditions (Roy 2014; Dikilitas et al. 2015). The important issue here is to determine which part of DNA has been damaged or what sorts of breaks have occurred during the stress period. Common kinds of stress metabolism and DNA damages are presented in Fig. 11.1a, b.

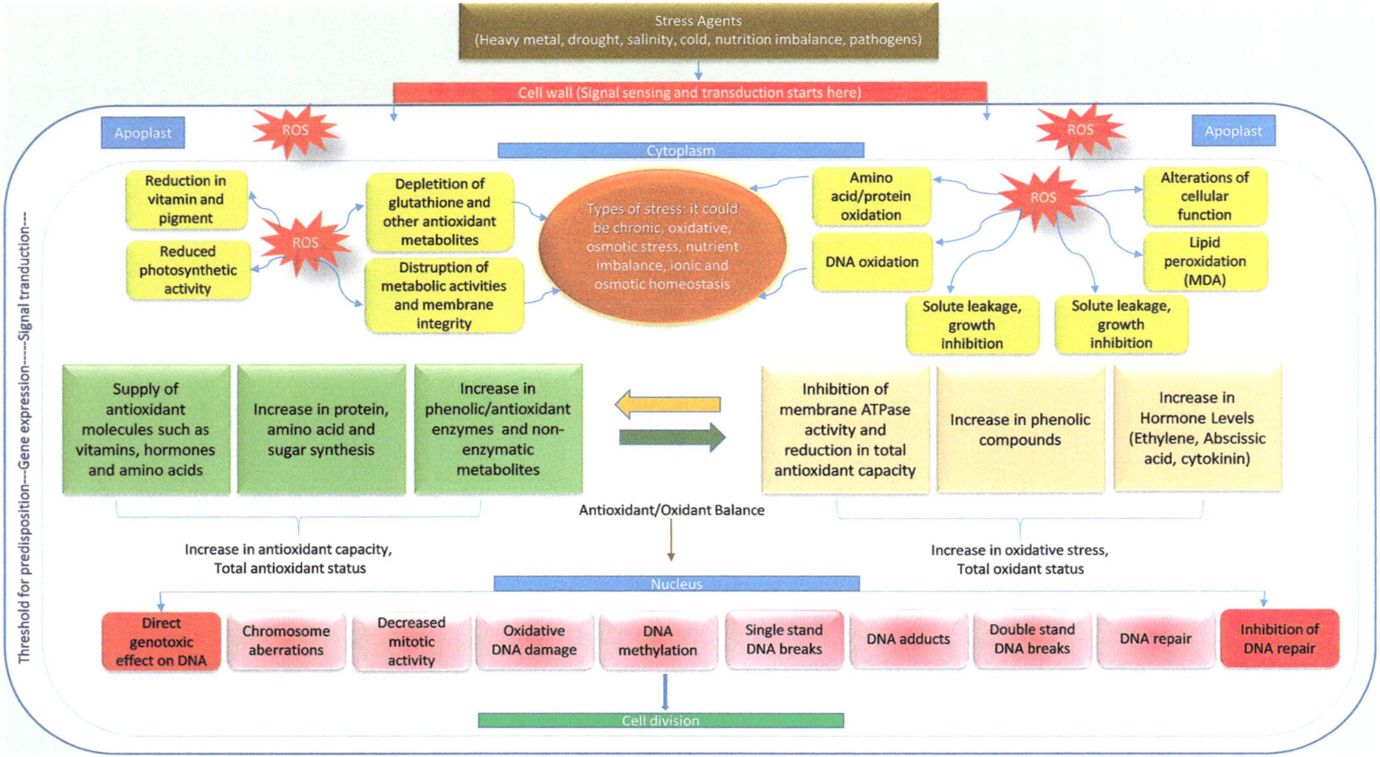

Fig. 11.1 (**a**) A general biochemical and molecular defense model for the plants under stress conditions; (**b**) status of stressed plants; they either lose their function or recover from the effects of stress via ROS detoxification
[ROS generation takes place in various cellular parts such as chloroplasts, cytoplasm, peroxisomes, mitochondria, plasma membrane, endoplasmic reticulum, and cell wall in normal and stress conditions. Stepwise downregulation of electron transport helps stress adaptation and thus reduces ROS accumulation. The explosion sign indicates the generation of ROS sites]

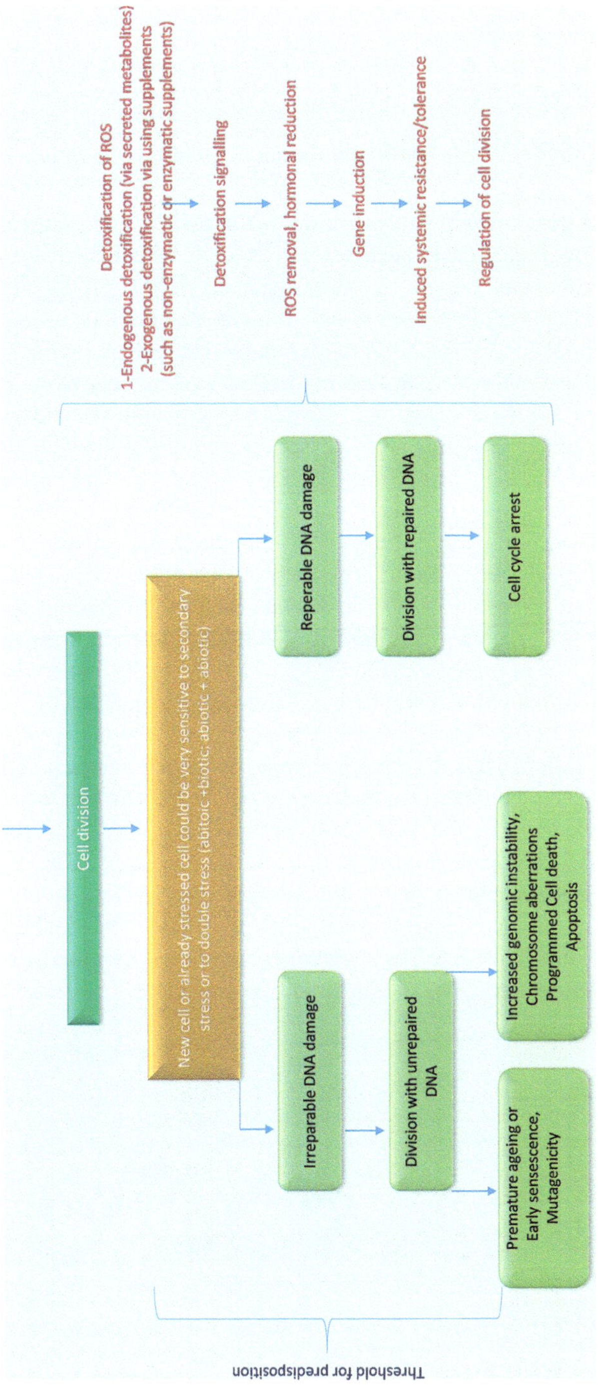

Fig. 11.1 (continued)

In this chapter, effects of stress (abiotic or biotic) in crop plants were evaluated regarding DNA damage, and a link between DNA health condition and predisposition of crop plants to stress was discussed.

11.2 Effects of Stress on DNA Damage

Living cells are known to generate ROS through physiological and biochemical processes in the plant system. Free radicals such as OH^-, O_2^-, and non-radical H_2O_2 are products of normal metabolic processes. However, the place of ROS production and the molecules to be oxidized are important issues. It is true that cell system can handle free radicals in optimal conditions, but if these radicals become excessive, it could damage the cell wall and DNA leading to abnormal cell function and eventually cell death. Unrepaired DNA damages can also lead to genomic instability, which in turn may enhance premature aging and disease susceptibility (Filipič 2012; Wang et al. 2016).

Plants experience oxidative stress following exposure to all abiotic or biotic stresses such as temperature, drought, salinity, UV radiation, pathogens, etc. Oxidative stress resulted from abiotic or biotic stress or from that of both could lead to cellular damage with overexpression of ROS as well as reactive nitrogen species (RNS), which are highly toxic species resulting in significant deteriorations in plants (Considine et al. 2015). The increase in radical species is generally associated with the induction of protein oxidation, carbohydrate degradation, lipid peroxidation, pigment breakdown and DNA strand breaks. Although plant genome is very stable and well protected, however, it is the target molecule in cellular damage. The genomic integrity of organisms is, therefore, under constant threat. DNA damage could be endogenous or exogenous, and as a result, production of ROS or RNS interferes with DNA that leads to several modifications including damaged bases, inter- and intra-strand cross-links, and single- or double-strand breaks (Fig. 11.2).

DNA damages could be divided into two categories as single- and double-stranded DNA breaks. Oxidized, alkylated base damage, base loss, DNA adducts, and intra-strand cross-links involve DNA single-strand breaks (ssB). The second

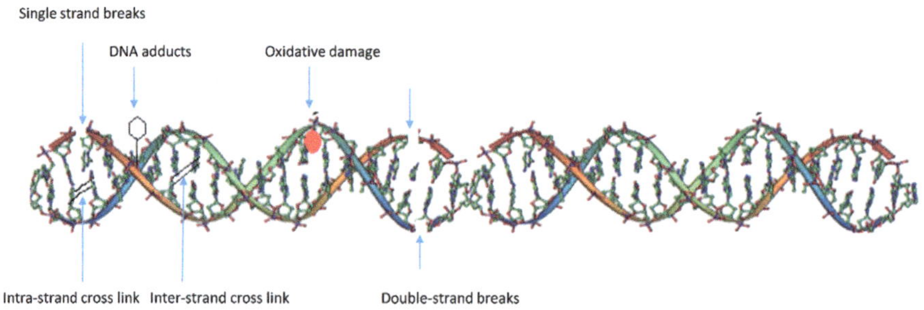

Fig. 11.2 Various DNA damages at different sites of DNA

category includes inter-strand cross-links and DNA double-strand breaks (dsB), which are the most severe type of DNA damage in the eukaryotic genome (Manova and Gruszka 2015).

DNA damages could occur via endogenous and exogenous factors. Endogenous DNA damages result from the increases in the concentration of free radicals in the cytoplasm surrounding the nucleus. Since mitochondria and chloroplasts are the main sources of ROS production sites (Sharma et al. 2012), therefore, oxidized bases could lead to DNA damages. Also, gene transmission, expression, and maintenance of genetic information have the potential to cause ssB or dsB (Montecucco and Biamonti 2013).

Exogenous DNA damages result from stressors which have alkylating potentials to methylate the DNA bases, mainly at their O- and N-positions generating small base damage as O^6-methylguanine, N^7-methylguanine, or N^3-methyladenine (Shrivastav et al. 2010). Ionizing radiation, for example, produces a large number of lesions on DNA molecule and generates dsB which leads to fragmentation (Moretton et al. 2017).

One of the most hazardous ROS is OH^- ions reacting with all organic compounds including the components of DNA molecule resulting in damages to both purine and pyrimidine bases and negatively affect the structure of the molecule (Tuteja et al. 2009). Hydroxyl ions damage to all cellular components like sugars, amino acids, phospholipids, and organic acids and produce organic radicals. The radicals attack the double bonds of heterocyclic DNA bases and remove a hydrogen atom from the methyl group of thymine and from each of the C-H bonds of 2-deoxyribose. Although ROS result in repairable DNA damages, however, in severe cases or in susceptible cultivars, repair of DNA damages could not be achieved (Dikilitas et al. 2015).

Stress not only damages the structure of DNA but also affects the contents of it. For example, Younis et al. (2009) stated that the pattern of changes in contents of RNA and DNA was evident in broad bean seedlings throughout the germination period; however, increases in NaCl or mannitol concentrations in the culture media induced a significant decrease in both RNA and DNA contents. Electrophoretic analysis of proteins of NaCl- or mannitol-treated broad beans also revealed the disappearance of some bands and the appearance of new characteristics of stress-related bands. They also stated that exogenous addition of ascorbic acid increased the contents of nucleic acids and resulted in expression of new protein bands. A close relation between the changes in nucleic acid and protein contents was also reported by Bor et al. (2003). These studies have clearly shown that the additional stress is able to result in further reduction in DNA or RNA contents and easily affect the defense mechanism of crop plants. The mechanism was explained by the fact that high concentrations of toxic substances in the culture media may affect protein synthesis through the inhibition of a number of the enzymes such as nitrate reductase, ATPase, etc. This results in the production of many metabolites including DNA and RNA contents, which are necessary for protein and enzyme synthesis (Younis et al. 2009). Although mutagenic agents have prime role in damaging DNA, but in severe cases of stress, especially under combined stress, recovery of DNA is not expected. Therefore, ROS is one of the primary casuses of DNA damages unless repaired. The generation of oxidized bases in DNA may have serious consequences for the cell affected. Unrepaired damaged DNA bases cause blocking of the DNA polymerase enzyme.

11.2.1 Effects of Abiotic Stress on DNA Damage

In normal conditions, free radicals are continuously produced as by-products and act as signaling molecules to regulate the expression of genes which play important roles for the defense system and metabolic pathways. The level of ROS in non-stress conditions remains low due to detoxification mechanisms in plants. However, exposure to environmental or pathological stressors results in an imbalance between prooxidants and antioxidants and leads to oxidative stress ((Meriga et al. 2004). If the production of oxygen species is increased or when levels of antioxidants are diminished, prooxidant and antioxidant balance in the cell is shifted toward the prooxidants. This state is called oxidative stress. It is either caused by the reduction of antioxidant concentration due to inhibition or reduction of antioxidant enzymes or an increase of ROS and RNS levels. It virtually damages all cell components. The imbalance toward oxidant levels leads to oxidative deterioration of crucial biomolecules such as nucleic acids, lipids, proteins, and carbohydrates (Aseervatham et al. 2013). For example, heavy metals like chromium, nickel, manganese, lead, mercury, and cadmium all have the capacity to damage DNA structures. Manganese, for example, ions bind with protein and form protein-mangan complex; chromium produces lesions in DNA structure (Singh et al. 2013). Not only high concentrations of toxicants such as heavy metals result in acute DNA damages, but also their low concentrations could lead to DNA damages. For example, Wang and Jia (2009) reported that DNA damages were evident in the testes of frog *Rana nigromaculata* following exposure to low concentrations of $PbNO_3$. They stated that low levels of lead resulted in DNA damages due to an increase of oxidative stress. Here, it is important to note that the chronic effects of toxicants at low doses may result in similar damages. Only the damaging effect on DNA is delayed if it is not repaired properly.

11.2.2 Effects of Biotic Stress on DNA Damage

To the best of our knowledge, there is no living organism that directly damages host DNA. However, the production of ROS and RNS produced by the host cell as a result of microbial attack would damage DNA. Here, the important issue is the interactions between pathogens and abiotic stress factors in which their combined stress would create more devastating effects on host cells than each stress factor alone would if the stress created by the pathogens is not nullified by the abiotic stress factor.

Although biotic stress factors do not cause severe DNA damages as heavy metals, pesticides, irradiators, etc., however, the combination of pathogenic agents with abiotic stressors might play more significant roles than those of either stress factor alone. At this point, recovery of the plants may not take place even if the concentration of toxic substances becomes low. Also, a pathogenic stress agent, as an additional stress agent, may secrete more toxins and pathogenic enzymes after the breakdown of the host resistance by the abiotic stress factors and may act as a more virulent pathogen (Dikilitas and Karakas 2012). Free movement of the attacking pathogen either in the root zone or in aerial parts creates more pathogenicity and may not give a chance to crop plants to recover from the effect of stress due to high accumulation of ROS (Fig. 11.3). Some abiotic and biotic stressors on DNA damage in plants are presented in Table 11.1.

11.3 Mode of Action of DNA Repair in Plants

DNA is very stable and its primary structure is hardly modified. However, DNA, unlike proteins and lipids and other higher molecules, once modified or damaged, cannot be replaced. It should either be repaired or killed through the cell system. Therefore, the stability and intactness of DNA are prerequisite for normal cellular functions. DNA normally suffers damage during biochemical process either in normal or stress conditions. Therefore, the genomic integrity of living organisms is constantly under threat by endogenous and exogenous DNA-damaging agents. Damage to DNA, in general, can occur to its components (base, sugar, and phosphate groups). The bases are the most reactive parts, and, therefore, many toxic chemicals form adducts (lesions) when attached to DNA bases (Georgieva et al. 2017). Since proteins and lipids are readily degraded and resynthesized, the most significant consequence of oxidative stress is considered to be DNA modifications. Modifications in DNA can become permanent via the formation of mutations and other types of genomic instability. The attack on DNA by ROS generates a low steady-state level of DNA adducts (Sharma et al. 2012).

Many stress agents including mutagens in high concentrations are able to cause DNA damages in a short term or in a long term with low concentrations. The stress agents, if not tolerated, could cause reduced stability in the genome and have a deleterious effect on plant development and eventually lead to crop reduction as well as quality. Therefore, the capacity of cells to prevent genotoxic damage is important to maintain genetic stability, because maintenance of DNA integrity and stability is essential for cellular transcription and biological functions (Manova and Gruszka 2015). If damaged DNA is not properly repaired, cell organization, replication, and transcription may face some difficulties to produce enzymatic and nonenzymatic metabolites for cell functioning (Polyn et al. 2015). Plants have a complex network of mechanisms to determine the damaged parts in the genome and establish a firm repair for a good function of DNA. Genomic stability is, therefore, ensured through the removal of the DNA lesions and restructuring of the original genetic information (Yoshiyama et al. 2013). However, DNA repair is not always error-free. Therefore, alterations in gene sequence could be transmitted to the next generation

Fig. 11.3 A photographic illustration of the effects of abiotic (NaCl) and biotic (*Verticillium albo-atrum*, Vaa) or interactions of both stress factors in tomato plants is visualized for the defense systems of crop plants under salt stress (Dikilitas 2003)
[Both stress factors are able affect the homeostasis of chemical signals such as Ca2+, ROS, and pH levels. Abiotic stress results in a reduction in defense mechanism of crop plants via changes in the structure of cell wall and in the properties of preformed or induced physical and chemical barriers that retard or prevent the entrance of pathogenic agents. The decrease in defense and antioxidant enzymes and increase in hormonal status have potential to cause oxidation in DNA structure]

Table 11.1 DNA-damaging effects of abiotic and biotic stress factors on plant metabolism

Stress agents	Effects on DNA	Effects on physiological and biochemical mechanisms of plants	References
Heavy metal (lead)	ssB, dsB in tobacco	Stunted growth, distorted leaves, brown root tips, decreased biomass	Gichner et al. (2008)
Heavy metal (lead)	Shortened mitotic index stage, prolonged interphase stage	Reductions of protein, chlorophyll, carbohydrate, DNA, and RNA content	Hamid et al. (2010)
Potato X virus	DNA damage through ROS accumulation	–	Cerovska et al. (2014)
Environmental pollution	DNA strand breaks	–	Akcha et al. (2008)
Air pollution	Increased DNA damage	Increased lipid peroxidation	Tai et al. (2010)
Coal contamination	DNA damage was evident	Reduction in biomass, chlorophyll, antioxidant activities	Menezes et al. (2015)
Pesticides	Micronuclei formation	Decreased in mitotic activity	Fernandes et al. (2007)
Salinity	DNA damage ssB and dsB	Increased H_2O_2 content, lipid peroxidation, and loss of chlorophyll content	Dikilitas et al. (2015)
Irradiation	DNA damage ssB and dsB	Accumulation of oxidant molecules	Ojima et al. (2009)
Drought	Genomic instability	Decrease in defense activities, chlorophyll contents	Roy (2014)
		Loss of turgor/osmotic adjustment is failed	
Drought and irradiation	Severe DNA strand breaks	Enhanced generation of ROS, damage to lipids, carbohydrates, proteins	Bandurska et al. (2013)
Pathogen and abiotic stress (salinity) combination remain elusive?	Not known in DNA structure?	Decrease in total antioxidant capacity; however, upon pathogen attack decrease further in biochemical parameters	Dikilitas (2003) and Sanogo (2006)
Drought and pathogen combination remain elusive?	Not known in DNA structure?	Drought increases in DNA damage. However, after rewatering drought-tolerant plants might recover. But under the pathogen attack, recovery in terms of DNA damage remains elusive	(Sanogo 2006)

leading to mutations. During transcription and repair, unrepaired DNA damage can lead to mutations, cellular senescence, apoptosis, progression of cancer, premature aging, and death (Filipič 2012; Dubrovina and Kiselev 2015). Once the stability of DNA is interrupted, the repair mechanism takes place immediately; however, if a cell cannot be repaired properly, transcription of the genes is negatively affected and cell functions including protein metabolism, hormones, defense enzymes, etc. are seriously affected (Holá et al. 2015).

A damaged DNA could pass various stages; it could either go into the recovery stage, or it undergoes apoptosis/programmed cell death stage, or it could undergo unregulated cell division, which can lead to cancer and cell death (Dubrovina and Kiselev 2015). DNA repair processes have been characterized in bacterial, fungal, plant, and animal cells. Our understanding of DNA repair mechanisms could play a significant role in the effective utilization of mutation technologies in future crop improvement (Manova and Gruszka 2015). Since the plant cannot escape from unfavorable conditions, they have evolved a remarkably significant mechanism to repair damaged part of DNA according to the changing environmental conditions. It has been reported that some stress factors such as light, heat, or drought lead to activation of certain DNA repair pathways. DNA repair capacity was shown to decrease with plant age mainly due to a reduction in the efficiency of DNA repair pathways (Golubov et al. 2010) although some results are controversial (Cvjetko et al. 2013).

Plants have been shown to possess various repair mechanisms. Exogenous antioxidants are the ones that are supplied from outside the cell. The free radical scavenging mechanism in the plant system under non-stress conditions is mostly sufficient; however, under stress conditions, the radical scavenging capacity may not be enough to compensate the negative effect of stress. Therefore, supplementary intake of antioxidants is needed. For example, phenolic and flavonoids are the major antioxidant compounds to protect the plants against abiotic and biotic stressors. They combine with proteins and other compounds, for example, flavonoids act as hydrogen donors and reduce the free radicals in the cell. Several mechanisms are available for repairing DNA damage both in the nucleus and in the mitochondria. Direct reversal of the damage and replacement of the base and the whole nucleotide could repair the damaged parts of DNA (Tuteja et al. 2009). The efficiency of repair mechanisms may sometimes be enhanced following exposure to ROS due to the expression of many DNA repair enzymes upregulated during stress. For example, DNA single-strand breaks could be repairable where damages occurred; however, DNA double-strand breaks are very critical for cells and usually are lethal (Georgieva and Stoilov 2015). If damages to DNA are not repaired properly, lesions or adducts can severely impair DNA synthesis and function of the cell.

Recent advances in the study of DNA repair in higher plants show that they use mechanisms similar to those present in other eukaryotes to remove and/or tolerate oxidized bases and other oxidative DNA lesions. Therefore, plants represent a valuable model system for the study of DNA oxidative repair processes in the eukaryotic cell. To deal with different kinds of DNA damages, there are a number of repair pathways for genomic stability. The base excision repair (BER), nucleotide excision repair (NER), mismatch repair (MMR), and double-strand break repair are activated upon damage at nucleotides to protect plant cell from the danger.

Base excision repair (BER) and nucleotide excision repair (NER) are commonly studied repair pathways (Balestrazzi et al. 2010). During metabolism or environmental stress, DNA is subjected to single-strand breaks and repaired by

BER. Initially, DNA glycosylases recognize and remove damaged bases to form apurinic/apyrimidinic (AP) sites. The resulting AP sites are processed by the AP endonuclease. Then, ssB can be repaired by short- or long-patch BER. To repair damaged DNA, antioxidants have significant roles. They have two dimensions as endogenous and exogenous. Endogenous antioxidants are agents that are capable of neutralizing the free radicals and have been reported to prevent oxidative damage caused by them. For example, SOD is an important endogenous antioxidant enzyme which acts as the first-line defense system against ROS by scavenging superoxide radicals to H_2O_2. On the other hand, glutathione peroxidase is present in the cytoplasm of cells that remove H_2O_2 by coupling its reduction to H_2O. Glutathione transferase (GST) is another important enzyme located in mitochondria. It plays a vital role for the detoxification of ROS and deactivation of many harmful substances (Aseervatham et al. 2013). Mismatch repair (MMR) also removes mispaired nucleotides during replication and repairs DNA adducts.

11.4 Determination of DNA Damage Through Advanced Techniques

Advances in molecular biology have led to the development of a number of selective and sensitive assays for DNA analysis in genotoxicology. Some of these methods are PCR-based; others include electrochemical, electrophoretic, etc.

DNA is constantly subjected to chemical modifications even under non-stress conditions. Genotoxic effects of toxicants not only are to be determined in the parents, but also their offspring should be tested if the stress agents have mutagenic potentials. Any stress factors, oxidative stress agents, or genotoxicants have potential to result in DNA damages, depending on the severity, duration, concentration, and defense mechanisms of the host. Therefore, we need to apply a proper method, which is reliable, cost-effective, less time-consuming, and less complicated. Commercial toxicity methods and protocols used in animal or tissue culture studies are associated with high cost and need enormous labor and sophisticated equipment. An easy, simple, cost-effective, and more importantly a reliable method to assess the status of DNA would, therefore, be more appropriate and useful for rapid determination of DNA damage, which will enable us to assess the condition of a cell. By this way, we could test the health status and transcription capacity. Also, we could determine the efficacy of chemicals applied on cells via DNA damage measuring methods. For example, the toxicity of pesticides applied on cells would tell us about the toxicity of chemicals and inform us whether the cell could recover or not from the effect of pesticides by measuring the DNA health conditions. Several types of DNA damages illustrated in Fig. 11.2 could be determined with a valid and fast method. There are various methods for measuring DNA damage and repair. The most important issue here is to measure the effect of stressors directly on DNA. The advances in molecular biology have offered many sophisticated methods for DNA analysis in the field of genotoxicology. These techniques have been constantly developed to become more sensitive, less complicated, and commonly available for many laboratories. Here, some important methods for measuring DNA damages were briefly introduced. We considered speed, reliability, time, complexity, an application on cellular or acellular DNA, and cost for the selection of methods and protocols.

Some of these methods are polymerase chain reaction (PCR), comet assay, halo assay, gas chromatography-mass spectrometry (GS-MS), fluorescence in situ hybridization (FISH), flow cytometry (FCM), immunological assay, radioimmunoassay (RIA), enzyme-linked immunosorbent assay (ELISA), electrochemical methods, etc.

11.4.1 RAPD-PCR

Random amplified polymorphic DNA (RAPD) is one of the most reliable techniques for detecting DNA damage as the amplification stops at the site of damage. It was developed by Williams et al. (1990) and Welsh and McClelland (1990). Random amplified DNA fragments of genomic DNA with single short primers of arbitrary nucleotide sequence under low annealing conditions form the basis of the method. The assay was successfully applied to detect genomic DNA alterations induced by several DNA-damaging agents, such as Cd (Liu et al. 2005) and salinity (Khan et al. 2013), in plants although more advanced techniques have been available now. Detection of genotoxic effect via RAPD involves the comparison of genetic profiles of control and treated cells. The frequency of appearance and disappearance of bands must be interpreted as polymorphism. The modification in RAPD profiles due to genotoxic exposure can be regarded as alterations in genomic template stability (GTS, a qualitative measure of genotoxic effect). The method is able to detect a wide range of DNA damage (e.g., DNA adducts, DNA breakage) as well as mutations (point mutations and large rearrangements). Many authors stated that the disappearance, appearance, changes, and intensity of the bands of RAPD products might be related to the DNA damage (e.g., single- and double-strand breaks, modified bases, abasic sites, oxidized bases, bulky adducts, DNA-protein cross-links), point mutations, and chromosomal rearrangements (Yunus et al. 2013).

11.4.2 Comet Assay

The term comet assay (single-cell gel electrophoresis) was first given by Olive et al. (1990), and numerous modifications have been reported to date to determine various types of DNA damages. This technique is used to detect mainly single-strand break, double-strand break, oxidative DNA damage, and single-strand break associated with incomplete excision repair sites caused by UV radiation, ultrasound, electromagnetic frequency radiation, etc. (Collins 2014). To improve the assay, an alkaline

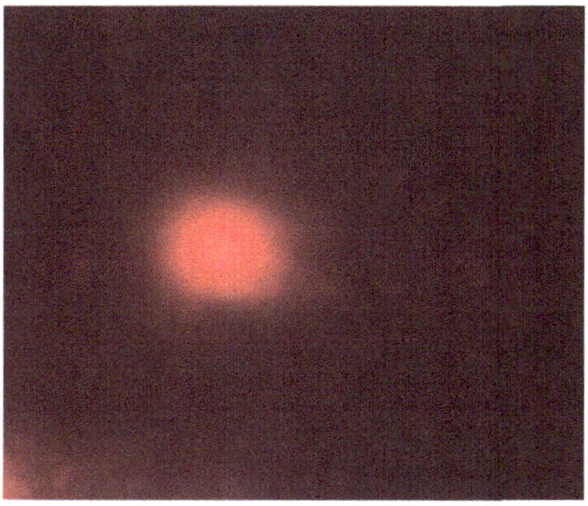

Fig. 11.4 Fragmented DNA following DNA-damaging chemicals (ethyl methane sulfonate)

lysis step was added in various studies. Recently, an enzyme formamidopyrimidine DNA glycosylase (fpg) was added to the reaction buffer to visualize the oxidative DNA damages in cells.

DNA damage is quantified by the proportion of DNA which moves out of nuclei toward the anode when individual cells or isolated nuclei, embedded in a thin layer agarose layer, are subjected to electrophoresis that results in the transport of DNA fragments out of the nucleus. After ethidium bromide or silver staining, the damage could then be quantified. Compared to other DNA assay methods, the assay is rapid, sensitive, visual, and inexpensive, and it converts oxidative damage into strand breaks using specific repair enzymes. The assay can be applied to any eukaryotic organisms and cell types. The assay can measure oxidative DNA damage in an efficient and relatively artifact-free manner. With a few cells, the assay can be performed in human, animal, fungal, and plant cells (Dikilitas et al. 2009). The assay is able to measure the damages quantitatively and qualitatively at the DNA damage level (Fig. 11.4). Recent studies have shown that the effect of low concentrations of toxicants on DNA can also be detected via this method (Pfeifer et al. 2015; Dikilitas et al. 2015). Not only the DNA health status of plant cells can be determined, but also their repair capacity and kinetics could be measured via this method. This method could determine the side effects of toxic chemicals such as pesticides on nontarget crop plants or organisms (Dikilitas et al. 2012).

11.4.3 Halo Assay

In this assay, propidium iodide (PI) or ethidium bromide (EtBr) intercalates into the DNA helix. Then, DNA can be seen as a fluorescent halo. This assay can measure single cells and does not require radioactive labeling of DNA. With this technique, cells are lysed and individual nucleoids are visualized as "halos," and, thereafter, halo area can be measured by an image analysis system which determines the chromatin fragility (Woudstra et al. 1998). For the assessment of single-strand breaks at the single-cell level, this assay was improved as an alkaline halo assay. In this modified assay, the cells are first embedded in melted agarose and spread on the microscope slides, thereafter, incubated in a high-salt alkaline lysis solution followed by another incubation in a hypotonic alkaline solution, and finally stained with ethidium bromide (Sestili et al. 2006). Under these conditions, single-stranded DNA fragments diffuse radically from the nuclear cage. Fast halo assay (FHA) is a technique similar to alkaline halo assay (AHA), but there is some modification such as simplification of the lysis, denaturation, and staining procedures (Sestili et al. 2006). With this assay, a fluorescent dye (propidium iodide) intercalates into the DNA helix and causes the change in the supercoiling status of the DNA. DNA can then be observed as a fluorescent halo after the lysis of cells. It is not sensitive as radiolabeling DNA assay. However, improved versions of alkaline and fast halo assays increased the sensitivity of the assay via improved versions of staining, denaturation, and lysis procedures. Although the AHA has similarities with comet assay, its principle is different. The most important difference is that AHA does not use electrophoresis to separate damaged DNA from undamaged DNA. This method is the most rapid, straightforward, and less expensive compared to other test systems including comet assay.

11.4.4 Gas Chromatography-Mass Spectrometry (GC-MS)

This method is commonly used to quantify DNA damage. It is capable of measuring multiple modified bases in a single DNA sample. After the hydrolysis of DNA, GC-MS converts nucleosides/bases and internal standards into measurable substances to determemine DNA damage. It is often carried out at high temperature 90 °C ± 140 °C (Dizdaroglu et al. 2015).

11.4.5 Fluorescence In Situ Hybridization (FISH)

It determines the location of damaged DNA in nuclei or chromosomes. Estimation of DNA damage is carried out on a cell basis. Chromosomes with aberrations are able to be detected efficiently (Kwasniewska and Kwasniewski 2013). This assay is also useful in detecting sister chromatid exchange, chemical adducts to DNA, and DNA strand breakage.

11.4.6 FCM (Flow Cytometry)

This assay is useful in detecting chromosomal aberrations, sister chromatid exchange, chemical adducts to DNA, and DNA strand breakage. Recently, nucleotide excision repair has been also detected by alkaline unwinding FCM assay (Monteiro et al. 2010).

11.4.7 Electrochemical Methods

The electrochemical methods offer a sensitive, selective, low cost, and miniaturized device for the detection of DNA damage. Adenine, cytosine, and guanine bases undergo redox processes at the mercury electrodes allowing for the measurement of oxidation in guanine and adenine bases. With this method, small amounts of ssDNA and dsDNA could be measured (Rahman et al. 2005).

11.4.8 DNA Damage Sensitivity Assay

This is a macroscopic method which detects the sensitivity of DNA through responses of "true leaf assay." A number of stress factors such as irradiation, heavy metals, pesticides, UV lights, etc. that are able to damage DNA allow us to determine the resistance of plants by comparing their mutant or wild-type lines by checking if true leaves have been formed (Rosa and Scheid 2014). With this method, potential damaging effects of DNA-damaging agents and repair capacity of plants could be determined in in vitro conditions.

11.4.9 Terminal Deoxyribonucleotidyltransferase-Mediated Deoxyuridine Triphosphate Nick End Labeling (TUNEL) Assay

As the name indicates, TUNEL assay detects DNA fragmentation by fluoresceinating the free ends of the DNA; therefore, with the help of fluorescence microscope, one can detect apoptosis (Bruggeman et al. 1997). It can also detect single- and double-strand breaks.

11.4.10 Enzyme-Linked Immunosorbent Assay (ELISA)

In this technique, antigens (modified DNA) bound to the plate which is blocked by the incubation of wells with a dilute protein solution. Thereafter, unknown samples are similarly mixed with antibody before addition to the plate. The bound primary antibody is quantified with enzyme-conjugated secondary antisera by the addition of appropriate substrate after incubation and washing off non-bound material (Santella 1999).

11.5 Conclusion and Future Prospects

Preservation of genomic stability and integrity is the primary task for every organism; plant genomes are constantly affected by internal metabolic processes and external stressors. DNA plays an essential role in the development and function of all living cells. Therefore, managing of a healthy DNA is crucially important. Direct damage to DNA molecule is an important criterion for genotoxicological research; it reveals the toxic level of toxicants as well as health status of an organism, including the status of the next generation of the same organism. Genotoxicants may impact DNA molecule directly through interactions with nucleotides or indirectly with DNA replication and cellular function. Induction of DNA damage is the most important step of a predisposition of living cells to stress factors. Cells with DNA damage may not be able to recover from the effects of stress unless damaged DNA is repaired. DNA molecules contain highly reactive groups; therefore, they are targeted by many genotoxic compounds. As a result, DNA molecules lose their functions. Under the abiotic stress, the defense response of the host becomes more complicated and makes the host more prone to pathogenic attacks due to changing behavior of the pathogens during the abiotic stress. The combined stress has now more damaging effects on plants. When pathogen is interacted with abiotic stress agents, the response of plants to stresses involves mechanisms of tolerance and resistance. The tolerance/resistance mechanisms involve a gene-based mechanism which hinders the invasion and the establishment of specific pathogens as well as delaying the expression of stress symptoms. There are several kinds of repair strategies against these damages in humans and other important flora and fauna. Although genotoxic studies have shown that toxic agents have highly damaging effects on DNA structures, recent studies showed that supplementation of antioxidants could improve oxidative stress and repair the damaged parts of DNA. For example, selenium was found to reverse the Cd-induced decrease in fresh mass and changes in lipid peroxidation as well as changes in the DNA methylation pattern of *Brassica napus* seedlings (Filek et al. 2010). It has also been reported that 1 mM ascorbic acid with regard to irradiation reduced the DNA damage down to 30% in cells (Ma et al. 2015). It was concluded that the addition of ascorbic acid as an antioxidant compound increased the tissue resistance.

Acknowledgments We apologize to colleagues whose work was not cited due to space limitations. Our sincere thanks to Dr. Dilfuza Egamberdieva for her patience and support.

Conflict of Interest The authors declare that there are no conflicts of interest.

References

Ahanger M, Akram N, Ashraf M, Alyemeni M, Wijaya L, Ahmad P (2016) Signal transduction and biotechnology in response to environmental stresses. Biol Plant 61(3):401–416

Akcha F, Arzul G, Rousseau S, Bardouil M (2008) Comet assay in phytoplankton as biomarker of genotoxic effects of environmental pollution. Mar Environ Res 66(1):59–61

Aseervatham G, Sivasudha T, Jeyadevi R, Arul Ananth D (2013) Environmental factors and unhealthy lifestyle influence oxidative stress in humans—an overview. Environ Sci Pollut Res 20(7):4356–4369

Balestrazzi A, Confalonieri M, Macovei A, Donà M, Carbonera D (2010) Genotoxic stress and DNA repair in plants: emerging functions and tools for improving crop productivity. Plant Cell Rep 30(3):287–295

Bandurska H, Niedziela J, Chadzinikolau T (2013) Separate and combined responses to water deficit and UV-B radiation. Plant Sci 213:98–105

Bhatnagar-Mathur P, Vadez V, Sharma K (2007) Transgenic approaches for abiotic stress tolerance in plants: retrospect and prospects. Plant Cell Rep 27(3):411–424

Bor M, Ozdemir F, Turkan I (2003) The effect of salt stress on lipid peroxidation and antioxidants in leaves of sugar beet *Beta vulgaris* L. and wild beet *Beta maritima* L. Plant Sci 164:77–84

Bruggeman LA, Dikman S, Meng C, Quaggin SE, Coffman TM, Klotman PE (1997) Neuropathy in human immunodeficiency virus-1 transgenic mice is due to renal transgene expression. J Clin Invest 100:84–92

Cerovska N, Plchova H, Vaculik P, Moravec T, Gichner T (2014) Potato virus X induces DNA damage in leaf nuclei of the host plant *Nicotiana tabacum* L. var. *xanthi*. Biol Plant 58(4):783–787

Collins A (2014) Measuring oxidative damage to DNA and its repair with the comet assay. Biochim Biophys Acta Gen Subj 1840(2):794–800

Considine M, María Sandalio L, Helen Foyer C (2015) Unravelling how plants benefit from ROS and NO reactions, while resisting oxidative stress. Ann Bot 116(4):469–473

Cvjetko P, Balen B, Peharec Štefanić P, Debogović L, Pavlica M, Klobučar G (2013) Dynamics of heat-shock induced DNA damage and repair in senescent tobacco plants. Biol Plant 58(1):71–79

Dikilitas M (2003) Effect of salinity & its interactions with *Verticillium albo-atrum* on the disease development in tomato (*Lycopersicon esculentum* Mill.) and lucerne (*Medicago sativa* & *M. media*) plants. Ph.D. Thesis, University of Wales, Swansea

Dikilitas M, Karakas S (2012) Behaviour of plant pathogens for crops under stress during the determination of physiological, biochemical, and molecular approaches for salt stress tolerance chapter 16. In: Ashraf M (ed) Crop production for agricultural improvement. Springer Publ, New York, pp 417–441

Dikilitas M, Kocyigit A, Yigit F (2009) A molecular-based fast method to determine the extent of DNA damages in higher plants and fungi. Afr J Biotechnol 8(14):3118–3127

Dikilitas M, Kocyigit A, Bilinc H, Taskin A (2012) Genotoxic effects of low doses of pesticides on mononuclear leukocyte ısolated from higher cells. Fresenius Environ Bull 21(9A):2833–2838

Dikilitas M, Collins AR, Kocyigit A, El Yamani N, Karakas S (2015) DNA damage in potato plants exposed to high level of NaCl stress. Front. Genet. Conference Abstract: ICAW 2015 - 11th International Comet Assay Workshop

Dikilitas M, Karakas S, Hashem A, Abd Allah EF, Ahmad P (2016) Oxidative stress and plant responses to pathogens under drought conditions. In: Ahmad P (ed) Water stress and crop plants: a sustainable approach, 2 volume Set. John Wiley & Sons Ltd, Chichester

Dizdaroglu M, Coskun E, Jaruga P (2015) Measurement of oxidatively induced DNA damage and its repair, by mass spectrometric techniques. Free Radic Res 49(5):525–548

Dubrovina A, Kiselev K (2015) Age-associated alterations in the somatic mutation and DNA methylation levels in plants. Plant Biol 18(2):85–196

Fernandes T, Mazzeo D, Marin-Morales M (2007) Mechanism of micronuclei formation in polyploidizated cells of *Allium cepa* exposed to trifluralin herbicide. Pestic Biochem Physiol 88(3):252–259

Filek M, Kościelniak J, Łabanowska M, Bednarska E, Bidzińska E (2010) Selenium-induced protection of photosynthesis activity in rape (*Brassica napus*) seedlings subjected to cadmium stress. Fluorescence and EPR measurements. Photosynth Res 105(1):27–37

Filipič M (2012) Mechanisms of cadmium induced genomic instability. Mutat Res/Fundam Mol Mech Mutagen 733(1–2):69–77

Funari E, Manganelli M, Sinisi L (2012) Impact of climate change on waterborne diseases. Ann Ist Super Sanità 48(4):473–487

Georgieva M, Stoilov L (2015) Neutral comet profiles: reliable system for analyses of DNA strand breaks distribution. Genet Plant Physiol 5(1):10–14

Georgieva M, Rashydov N, Hajduch M (2017) DNA damage, repair monitoring and epigenetic DNA methylation changes in seedlings of Chernobyl soybeans. DNA Repair 50:14–21

Gichner T, Žnidar I, Száková J (2008) Evaluation of DNA damage and mutagenicity induced by lead in tobacco plants. Mutat Res/Genet Toxicol Environ Mutagen 652(2):186–190

Golubov A, Yao Y, Maheshwari P, Bilichak A, Boyko A, Belzile F, Kovalchuk I (2010) Microsatellite instability in Arabidopsis increases with plant development. Plant Physiol 154(3):1415–1427

Hamid N, Bukhari N, Jawaid F (2010) Physiological responses of Phaseolus vulgaris to different lead concentrations. Pak J Bot 42(1):239–246

Hattab S, Chouba L, Ben Kheder M, Mahouachi T, Boussetta H (2009) Cadmium- and copper-induced DNA damage in *Pisum sativum* roots and leaves as determined by the Comet assay. Plant Biosyst 143(1):S6–S11

Holá M, Vágnerová R, Angelis K (2015) Mutagenesis during plant responses to UVB radiation. Plant Physiol Biochem 93:29–33

Khan F, Hakeem K, Siddiqi T, Ahmad A (2013) RAPD markers associated with salt tolerance in soybean genotypes under salt stress. Appl Biochem Biotechnol 170(2):257–272

Koobaz P, Ghanati F, Hosseini Salekdeh G, Moradi F, Hadavand H (2016) Drought tolerance in four-day-old seedlings of a drought-sensitive cultivar of wheat. J Plant Nutr 40(4):574–583

Kumari S, Rastogi RP, Singh KL, Singh SP, Sinha RP (2008) DNA damage: detection strategies. EXCLI J 7:44–62

Kwasniewska J, Kwasniewski M (2013) Comet-FISH for the evaluation of plant DNA damage after mutagenic treatments. J Appl Genet 54(4):407–415

Liu W, Li P, Qi X, Zhou Q, Zheng L, Sun T, Yang Y (2005) DNA changes in barley (*Hordeum vulgare*) seedlings induced by cadmium pollution using RAPD analysis. Chemosphere 61(2):158–167

Ma Y, Ogawa N, Yoshikawa Y, Mori T, Imanaka T, Watanabe Y, Yoshikawa K (2015) Protective effect of ascorbic acid against double-strand breaks in giant DNA: marked differences among the damage induced by photo-irradiation, gamma-rays and ultrasound. Chem Phys Lett 638:205–209

Manova V, Gruszka D (2015) DNA damage and repair in plants – from models to crops. Front Plant Sci 6:1–26

Méndez-Alonzo R, López-Portillo J, Moctezuma C, Bartlett M, Sack L (2016) Osmotic and hydraulic adjustment of mangrove saplings to extreme salinity. Tree Physiol 36(12):1562–1572

Menezes A, Da Silva J, Rossato R, Santos M, Decker N, Da Silva F, Cruz C, Dihl R, Lehmann M, Ferraz A (2015) Genotoxic and biochemical changes in *Baccharis trimera* induced by coal contamination. Ecotoxicol Environ Saf 114:9–16

Meriga B, Krishna Reddy B, Rajender Rao K, Ananda Reddy L, Kavi Kishor P (2004) Aluminium-induced production of oxygen radicals, lipid peroxidation and DNA damage in seedlings of rice (*Oryza sativa*). J Plant Physiol 161(1):63–68

Montecucco A, Biamonti G (2013) Pre-mRNA processing factors meet the DNA damage response. Front Genet 4:1–14

Monteiro M, Rodriguez E, Loureiro J, Mann R, Soares A, Santos C (2010) Flow cytometric assessment of Cd genotoxicity in three plants with different metal accumulation and detoxification capacities. Ecotoxicol Environ Saf 73(6):1231–1237

Moretton A, Morel F, Macao B, Lachaume P, Ishak L, Lefebvre M, Garreau-Balandier I, Vernet P, Falkenberg M, Farge G (2017) Selective mitochondrial DNA degradation following double-strand breaks. PLoS One 12(4):e0176795

Ojima Y, Nishioka M, Matsumoto M, Taya M (2009) Quantification of DNA damage by the comet assay in radish sprouts exposed to excess light irradiation. Biochem Eng J 46(1):69–72

Olive PL, Banath JP, Durand RE (1990) Heterogeneity in radiation-induced DNA damage and repair in tumor and normal cells measured using the 'comet' assay. Radiat Res 122:86–94

Pandey P, Irulappan V, Bagavathiannan MV, Senthil-Kumar M (2017) Impact of combined abiotic and biotic stresses on plant growth and avenues for crop improvement by exploiting physio-morphological traits. Front Plant Sci 8:1–15

Pfeifer D, Chung Y, Hu M (2015) Effects of low-dose bisphenol a on DNA damage and proliferation of breast cells: the role of c-Myc. Environ Health Perspect 123(12):1271–1279

Polyn S, Willems A, DeVeylder L (2015) Cell cycle entry, maintenance, and exit during plant development. Curr Opin Plant Biol 23:1–7

Procházková D, Wilhelmová N, Pavlíková D, Száková J, Gichner T (2013) Zinc induces DNA damage in tobacco roots. Biol Plant 57(4):783–787

Rahman MA, Kwon NH, Won MS, Choe ES, Shim YB (2005) Functionalized conducting polymer as an enzyme-immobilizing substrate: an amperometric glutamate microbiosensor for *in vivo* measurements. Anal Chem 77:4854–4860

Rejeb IB, Pastor V, Mauch-Mani B (2014) Plant responses to simultaneous biotic and abiotic stress: molecular mechanisms. Plants 3:458–475

Rizhsky L, Liang H, Shuman J, Shulaev, Davletova S, Mittler R (2004) When defense pathways collide. The response of *Arabidopsis* to a combination of drought and heat stress. Plant Physiol 134:1683–1696

Rosa M, Scheid OM (2014) DNA damage sensitivity assays with arabidopsis seedlings. Bio-protocol 4(7):e1093

Roy S (2014) Maintenance of genome stability in plants: repairing DNA double strand breaks and chromatin structure stability. Front Plant Sci 5:1–5

Sanogo S (2006) Predispositional effect of soil water saturation on infection of Chile pepper by *Phytophthora capsici*. Hortscience 41(1):172–175

Santella RM (1999) Immunological methods for detection of carcinogen-DNA damage in humans. Cancer Epidemiol Biomark Prev 8:733–739

Savary S, Ficke A, Aubertot JN, Hollier C (2012) Crop losses due to diseases and their implications for global food production losses and food security. Food Sec 4:519–537

Sestili P, Martinelli C, Stocchi V (2006) The fast halo assay: an improved method to quantify genomic DNA strand breakage at the single-cell level. Mutat Res 607:205–214

Sharma P, Jha A, Dubey R, Pessarakli M (2012) Reactive oxygen species, oxidative damage, and antioxidative defense mechanism in plants under stressful conditions. J Bot 2012:1–26

Shrivastav N, Li D, Essigmann J (2010) Chemical biology of mutagenesis and DNA repair: cellular responses to DNA alkylation. Carcinogenesis 31(1):59–70

Singh H, Mahajan P, Kaur S, Batish D, Kohli R (2013) Chromium toxicity and tolerance in plants. Environ Chem Lett 11(3):229–254

Smolinska B, Szczodrowska A (2017) Antioxidative response of *Lepidium sativum* L. during assisted phytoremediation of Hg contaminated soil. New Biotechnol 38:74–83

Tai H, Percy K, Karnosky D (2010) DNA damage in *Populus tremuloides* clones exposed to elevated O3. Environ Pollut 158(4):969–976

Tuteja N, Ahmad P, Panda B, Tuteja R (2009) Genotoxic stress in plants: shedding light on DNA damage, repair and DNA repair helicases. Mutat Res/Rev Mutat Res 681(2–3):134–149

Wang MZ, Jia XY (2009) Low levels of lead exposure induce oxidative damage and DNA damage in the testes of the frog Rana nigromaculata. Ecotoxicology 18:94–99

Wang H, He L, Song J, Cui W, Zhang Y, Jia C, Francis D, Rogers H, Sun L, Tai P, Hui X, Yang Y, Liu W (2016) Cadmium-induced genomic instability in Arabidopsis: molecular toxicological biomarkers for early diagnosis of cadmium stress. Chemosphere 150:258–265

Welsh J, McClelland M (1990) Fingerprinting genomes using PCR with arbitrary primers. Nucl Acids Res 18:7213–7218

Williams JGK, Kubelik AR, Livak KJ, Rafalski JA, Tingey SV (1990) DNA polymorphisms amplified by arbitrary primers are useful as genetic markers. Nucl Acids Res 18:6531–6535

Woudstra EC, Driessen C, Konings AWT, Kampinga HH (1998) DNA damage induction and tumour cell radiosensitivity: PFGE and halo measurements. Int J Radiat Biol 73:495–502

Yoshiyama K, Sakaguchi K, Kimura S (2013) DNA damage response in plants: conserved and variable response compared to animals. Biology 2:1338–1356

Younis M, Hasaneen M, Kazamel A (2009) Plant growth, metabolism and adaptation in relation to stress conditions. XXVII. Can ascorbic acid modify the adverse effects of NaCl and mannitol on amino acids, nucleic acids and protein patterns in *Vicia faba* seedlings? Protoplasma 235(1–4):37–47

Yunus M, Aziz M, Kadir M, Daud S, Rashid A (2013) *In vitro* mutagenesis of *Etlingera elatior* (Jack) and early detection of mutation using RAPD markers. Turkish J Biol 37:716–725

Chapter 12
Legume, Microbiome, and Regulatory Functions of miRNAs in Systematic Regulation of Symbiosis

Syed Sarfraz Hussain, Meeshaw Hussain, Muhammad Irfan, and Kadambot H.M. Siddique

Abstract Legumes represent the most-valued food after cereals for humans and animals. They are grown extensively in the dry/semiarid tropics worldwide, mostly under rainfed conditions. Legumes have the potential to establish symbiotic relationships with both rhizobial bacteria and arbuscular mycorrhizal fungi (AMF). This cooperation leads to atmospheric nitrogen fixation in nodules and phosphorus in arbuscules. Recent advances in high-throughput sequencing and other molecular technologies have provided opportunities to study the molecular basis of symbiosis in legumes. Several important components of the gene networks involved in legume symbiosis have been identified, including microRNAs (miRNAs), which have emerged as key players in gene expression, developmental processes, and stress in legumes. To date, a plethora of conserved and legume-specific miRNAs have been reported that are associated with symbiotic interactions by experimental and bioinformatic approaches. In this chapter, we combine data from published literature—especially genomic and deep sequencing data on miRNAs involved in symbiosis, biological nitrogen fixation, and phosphorus availability through nodules and

S.S. Hussain (✉)
School of Agriculture, Food & Wine, Waite Campus, University of Adelaide, Adelaide, SA, Australia

Department of Biological Sciences, Forman Christian College (A Chartered University), Lahore, Pakistan
e-mail: syedsarfraz_2000@yahoo.com

M. Hussain
Institute of Molecular Biology & Biotechnology, Bahauddin Zakariya University, Multan, Pakistan

M. Irfan
Department of Biological Sciences, Forman Christian College (A Chartered University), Lahore, Pakistan

K.H.M. Siddique
The UWA Institute of Agriculture, The University of Western Australia, Crawley, WA, Australia

D. Egamberdieva, P. Ahmad (eds.), *Plant Microbiome: Stress Response*, Microorganisms for Sustainability 5, https://doi.org/10.1007/978-981-10-5514-0_12

arbuscules—to address the specificity functions of miRNA in establishing symbiosis in legumes. Furthermore, we highlight the interaction of the legume microbiome and miRNA in particular, establishing symbiosis for environmentally sustainable agriculture and increased global crop productivity. However, due to the complex nature of xxx, a concerted effort is required to fully understand the roles of miRNAs in the development of symbiosis in legumes.

Keywords Legumes • MicroRNAs • Plant microbiome • Nutritional deficiency • Symbiosis

12.1 Legumes, microRNAs, and Microbiome: A Triangle for Future Agriculture

Legumes are a diverse and biologically significant group of plants, which belong to the Fabaceae family, and this family has 20,000 species. Legumes constitute the second largest group of important food and feed crops grown worldwide (Iantecheva et al. 2013). Members of this family fill critical niches in most terrestrial biomes and have a significant impact on global agriculture, environment, human/animal nutrition, and health. Legumes stand third behind cereals and oilseeds (Popelka et al. 2004; Ashraf et al. 2010), accounting for 27% of primary food production around the globe (Graham and Vance 2003). Annually, around 250 million tons of grain legumes are produced, consisting of major crops used for food, feed, and vegetable oil, and these alone constitute 33% of the dietary protein and nitrogen needs of humans (Zhu et al. 2005; Shrama et al. 2010; Kudapa et al. 2012). A key contribution of legumes to a sustainable environment is their ability to fix atmospheric molecular nitrogen in most agricultural ecosystems in symbiotic association with rhizobia through a process called symbiotic nitrogen fixation (SNF) or root nodule symbiosis (RNS) (Urvardi and Scheible 2005; El-Enany et al. 2013; Mantri et al. 2013). Legumes, working as green manure, fix around 200 million tons of nitrogen (Graham and Vance 2003) in soil, thus reducing the need for chemical nitrogen fertilizer (Arrese-Igor et al. 2011; Valentine et al. 2011) and improving the productivity of cereals and other crops in agricultural rotations. Therefore, legumes play a key role in sustainable intensified agriculture, when used in common farming practices as an intercrop in crop rotation systems by dramatically improving organic contents in the field.

Legumes have some unique features, including nitrogen fixation, protein-rich physiology, and secondary metabolites; it is critical to develop genomic tools in legumes to understand these features (Cook 1999). Systematic research in legumes has been introduced in the selection of two model systems, *Lotus japonicus* (Handberg and Stougaard 1992) and *Medicago truncatula* (Barker et al. 1990). Intensive genome sequencing efforts of model and legume plants are under way which have made available genome sequences of *L. japonicus*, *M. truncatula*, pigeon pea, and chickpea (Sato et al. 2008; Young et al. 2011; Varshney et al. 2012,

2013) and are available for future research exploitation. Similarly, genome sequencing of other important legume plants such as peanut, pea, and lentils are near completion. In addition, significant progress has been made on expression profiling of legume genes (Lamblin et al. 2003; Oldroyd 2005; Thompson et al. 2005; Opdenakker et al. 2012). As part of this comprehensive approach, de novo transcriptome assemblies have been developed in different legumes (Cheung et al. 2006; Deschamps and Campbell 2010; Cramer et al. 2011; Garg et al. 2011; Hiremath et al. 2011; Kudapa et al. 2012). The availability of this sequencing data will serve as a useful resource for research on legumes at molecular and genomic levels (Garg et al. 2011; Jogaiah et al. 2012).

Various studies have yielded 87 novel and 42 conserved miRNAs in soybean (Joshi et al. 2010; Subramanian et al. 2008; Wang et al. 2009). Similarly, more than 100 novel miRNAs have been identified in *M. truncatula* (Szittya et al. 2008; Jagadeeswaran et al. 2009; Lelandais-Briere et al. 2009). De Luis et al. (2012a) reported 35 miRNA families explicitly present in *L. japonicus*. This vast diversity compared to rice, maize, and wheat is probably due to more interest in legumes which resulted in a great deal of small RNA deep sequencing and analysis. Similarly, high-throughput sequencing has been used to systematically identify stress-related miRNAs in legumes (Abdel-Ghany and Pilon 2008; Li et al. 2010, 2011; Barrera-Figueroa et al. 2011; Chen et al. 2012a, b; Wang et al. 2011a, b; Zhou et al. 2012a, b). An increased interest in miRNA research in legumes has resulted in the collective discovery of 229 and 179 miRNA families in *M. truncatula* and soybean, respectively, in the last few years.

In legumes, miRNA research is in its infancy, but the past few years have witnessed significant progress, resulting in the accumulation of both conserved and species-specific miRNAs in different plants including soybean, common bean, peanut, chickpea, cowpea, and *M. truncatula* (Subramanian et al. 2008; Szittya et al. 2008; Jagadeeswaran et al. 2009; Lelandais-briere et al. 2009; Wang et al. 2009; Joshi et al. 2010; Turner et al. 2012). In addition, the availability of high-throughput techniques, such as computational prediction and sequencing approaches, has resulted in a plethora of miRNA families, including miRNA precursors and mature miRNAs in different legume species under abiotic stress (Valdes-Lopez et al. 2008; Barrera-Figueroa et al. 2011; Chen et al. 2012a, b; Dong et al. 2013; Xu et al. 2013).

Molecular and omics tools combined with advanced microscopic techniques have contributed several significant, diverse, and unexpected discoveries in the domain of the plant-associated microbiome in the past few decades (Mendes et al. 2011; Bulgarelli et al. 2012; Lundberg et al. 2012; Berg et al. 2016; Timmusk et al. 2017). A wide range of agriculturally important microbiomes has been extensively exploited for plant growth, development, and stress and disease management. Several studies have shown beneficial effects of plant-associated microbiomes, and this partnership can significantly contribute to establishing novel solutions under nutrient deficiency and alleviating other biotic and abiotic stresses using a variety of mechanisms (Bayuelo-Jimenez et al. 2002; Hayat et al. 2010; Mapelli et al. 2013; Vejan et al. 2016). The sustainability of crop plants challenged by soil nutrient deficiency becomes more important and needs non-conventional solutions such as the use of microbiomes (Schaeppi and Bulgarelli 2015). Plant-rhizobium and plant arbuscular mycorrhizal fungi (AMF) symbioses represent two of the most researched plant-associated microbiomes (Bazin et al. 2012; Wu et al. 2016). Recent evidence suggests that miRNAs play pivotal roles in the development of both types of symbiosis (Subramanian et al. 2008; Lelandais-Briere et al. 2009; Simon et al. 2009; Bazin et al. 2012; Bustos-Sanmamed et al. 2013; Yan et al. 2013). Taming microbiomes beneficial to both plants and the environment offers a promising strategy for the development of sustainable future agriculture. In this chapter, we signify the advantages of the plant-associated microbiome approach, along with miRNAs and their diverse functions toward coping with different abiotic stresses including nutrient deficiency in legumes, which poses a serious threat to global crop productivity.

12.2 MicroRNAs (miRNAs): Master Regulators with Diverse Functions

Plants, being sessile in nature, are constantly challenged by a complex array of abiotic stresses and must respond to these stresses at different levels to adapt and survive (Suzaki et al. 2013). Therefore, both morphological and physiological adaptations to overcome these abiotic stresses require many complex rearrangements of gene expression networks at the transcriptional and posttranscriptional level. Several microRNAs have been responsive to various abiotic stresses, and recent data have shown a crucial regulatory role of microRNAs in the plant response to environmental cues.

Currently, miRNAs have been reported in 64 plant species including different crop plants of economic importance, such as rice, barley, wheat, sugarcane, legumes, tomato, potato, and in many other plant species (Kruszka et al. 2012, 2014; Naya et al. 2014; Pandey et al. 2014). The miRNA database contains 24,521 mature miRNA sequences (Release 20.0, June 2013) including 6843 miRNAs from plant species. However, the rate of miRNA identification in crop plants has increased rapidly due to high-throughput sequencing methods, and the availability of complete genome sequences has improved computational and experimental protocols (Yao et al. 2007; Subramanian et al. 2008; Jagadeeswaran et al. 2009; Lelandais-Briere et al. 2009; Joshi et al. 2010; Zhao et al. 2010; Schreiber et al. 2011; Kim et al. 2012; Li et al. 2012; Wang et al. 2012; Zhang et al. 2012; Liang et al. 2013; Lin and Lai 2013).

By 2009, the Arabidopsis miRNA atlas contained more than 184 miRNAs, which were predicted to regulate more than 600 genes including 225 known targets (Griffiths-Jones et al. 2008; Alves et al. 2009). Some of these miRNAs have been analyzed at the molecular level for their roles in the regulation of target genes (Llave et al. 2002; Reinhart et al. 2002; Chen et al. 2004; Laufs et al. 2004; Duan et al. 2006). However, the number of newly discovered Arabidopsis miRNAs is continually growing. For example, miRBase release 18 contained 291 known miRNAs for *A.*

thaliana while release 21 contained 427 miRNAs (Griffiths-Jones 2004; Kozomara and Griffiths-Jones 2014).

The regulatory roles of miRNAs in plants have been established primarily through overexpression or by generating plants that express miRNA-resistant versions (Chen et al. 2004; Park et al. 2005; Schwab et al. 2005; Gandikota et al. 2007; Li et al. 2010; Khan et al. 2011; Bustos-Sanmamed et al. 2013; Turner et al. 2013).

Numerous studies have revealed that plant miRNAs are involved in almost all biological and metabolic processes (Comai and Zhang 2012; Khraiwesh et al. 2012; Sun 2012; Turner et al. 2013) including seed germination, morphogenesis (Reyes and Chau 2007), flower initiation, development, and sex determination (Chen 2004; Chuck et al. 2009), plant growth and development including vegetative and reproductive phase change (Chen et al. 2004; Chen 2005; Willmann and Poethig 2005; Jones-Rhoades et al. 2006; Mallory and Vaucheret 2006; Nogueira et al. 2006; Yang and Xue 2007; Lelandais-Briere et al. 2010; Rubio-Somoza and Weigel 2011; Wu et al. 2016; Yan et al. 2016), root development (Zhang et al. 2005; Gutierrez et al. 2009), and phytohormone signaling (Achard et al. 2004; Guo et al. 2005; Reyes and Chau 2007; Meng et al. 2009). In addition, miRNAs have been identified as potent regulators of biological processes such as metabolism, biotic and abiotic stresses, signal transduction, protein degradation, and maintenance of genome integrity in plants (Sunkar and Zhu 2004a, b; Bari et al. 2006; Mallory and Vaucheret 2006; Sunkar et al. 2006; Liu et al. 2008; Pant et al. 2008; Zhou et al. 2008; Ding et al. 2009; Trindale et al. 2010; Chitwood and Timmermans 2010; Liang et al. 2010; Meng et al. 2010; Wang et al. 2011a, b; Khraiwesh et al. 2012; Sunkar et al. 2012; Hussain and Shi 2014, 2016; Ferdous et al. 2015). Given that miRNAs are master regulators and serve as the core of gene regulatory networks (Jones-Rhoades et al. 2006), miRNA research provides countless opportunities to unravel the mechanisms underlying challenging plant traits (Sun 2012; Liang et al. 2013). This chapter mainly focuses on the plethora of miRNAs and their functions in legumes.

12.3 MiRNA Identification and Functional Diversity in Legumes

Regulation of gene expression for proper functioning is crucial for all organisms. Over 10 years, noncoding RNAs have emerged as master regulators of gene expression in living things including plants (Hussain and Shi 2014; Ferdous et al. 2015). Plant noncoding RNAs (miRNAs) function as a mediator to guide AGO, the effector nuclease of the RNA-induced silencing complex (RISC), to cleave target transcripts and/or repress translation (Bustos-Sanmamed et al. 2013). In the last decade, miRNA functions have been extensively researched in plant development and responses to biotic and abiotic stresses in legumes (Chen et al. 2009; Simon et al. 2009; Khraiwesh et al. 2012; Hussain and Shi 2014, 2016; Ferdous et al. 2015; Leladais-Briere et al. 2016). In plants, miRNA can be identified by either high-throughput sequencing like miRNA library sequencing (Lu et al. 2007; Szittya et al. 2008; Ramesh et al. 2013) or computer-based prediction software (Sunkar and Jagadeeswaran 2008; Arenas-Huertero et al. 2009; Xie et al. 2010). Hence, hundreds and thousands of miRNAs have been identified in many plant species including legumes such as *M. truncatula*, *L. japonicus*, soybean, common bean, peanut, chickpea, cowpea (Subramanian et al. 2008; Szittya et al. 2008; Arenas-Huertero et al. 2009; Lelandais-Briere et al. 2009; Libault et al. 2010a, b; Lu and Yang 2010; Paul et al. 2011; Bazin et al. 2012, 2013; De Luis et al. 2012a; Turner et al. 2012; Bustos-Sanmamed et al. 2013; Formey et al. 2014, 2015, 2016; Yan et al. 2015, 2016). In the miRBase (http://www.mirbase.org), an archive of miRNA sequences and annotations, there are 1256 sequences belonging to 285 legume families. These miRNAs have the potential to regulate species-specific biological processes in legumes (Subramanian et al. 2008; Szittya et al. 2008).

Conserved miRNAs are generally encoded by multigenic families (Allen et al. 2004). A comparative study showed that soybean has a higher gene number per miRNA family than *M. truncatula*. This is probably due to the large genome of soybean (1115 Mbp) and genome duplication (Schmutz et al. 2010). However, for some miRNA families, an opposite profile has been revealed. For example, the miRBase reveals that miR395 and miR399 have more members in *M. truncatula* than in soybean. Besides conserved miRNAs, non-conserved miRNAs—which were revealed as crop-specific (soybean: Subramanian et al. 2008; Wang et al. 2009; Joshi et al. 2010/*M. truncatula*: Szittya et al. 2008; Jagadeeswaran et al. 2009, Lelandais-Briere et al. 2009)—have now been declared as legume-specific.

12.4 Exploring miRNA Functions Under Various Abiotic Stresses in Legumes

Abiotic stresses including drought, salinity, extreme temperatures, and nutrition deficiency are major factors contributing to the reduction or total loss of crop production (Hussain et al. 2011a, b). Historically, plants have evolved morphological, physiological, and molecular adaptations to cope with abiotic stresses (Hussain et al. 2011a). It has been widely reported that miRNAs are involved in the regulation of a wide variety of physiological processes in plants (Sunkar et al. 2007). Concomitantly, growing evidence has suggested that miRNAs act as important regulators of gene expression under abiotic stress in plants (Sunkar and Zhu 2004a, b; Fuji et al. 2005; Aung et al. 2006; Chiou et al. 2006; Okamura et al. 2008; Khraiwesh et al. 2008, 2010; Ding et al. 2011; Dong et al. 2012). Recent studies have revealed more than 40 miRNA families associated with abiotic stresses in plants, 13 of which were responsive to drought and salinity stresses (Barrera-Figuero et al. 2012; Nageshbabu et al. 2013). As a result of extensive plant genome sequencing coupled with miRNA annotation, it has been revealed that almost all stress-related miRNAs are conserved, which suggests that miRNA-mediated regulatory roles may be evolutionarily conserved for specific stresses in plants (Allen et al. 2004; Szittya et al. 2008). However, one miRNA, which responds to a specific abiotic stress in one species, may not necessarily have the same function in another species. For example, Barrera-Figuero et al. (2012) demonstrated that at least 10 miRNAs involved in stress have shown the opposite expression in rice and *Arabidopsis* under drought stress. Scientists have employed numerous computational, experimental, and traditional Sanger sequencing methods in their studies that have provided low coverage in the pioneering efforts. However, with technical advancements in high-throughput technologies, particularly next-generation sequencing (NGS) and more advanced computational techniques, it has become much easier and more cost-effective to perform genome-wide profiling for the identification of low abundant stress-responsive miRNAs (Rajagopalan et al. 2006; Fahlgren et al. 2007; Yao et al. 2007; Moxon et al. 2008). As a result, the discovery of stress-related miRNAs has expanded from a few model plants such as *Arabidopsis*, tomato, and rice (Zhao et al. 2007; Li et al. 2008, 2010; Liu et al. 2008; Zhou et al. 2010) to other non-model plants (Ding et al. 2009; Jia et al. 2009; Song et al. 2011; Wang et al. 2011a, b; Barakat et al. 2012; Eldem et al. 2012; Li et al. 2013; Ozhuner et al. 2013; Shuai et al. 2013; Yanik et al. 2013) including orphan plants such as legumes (Bari et al. 2006; Liu et al. 2008; Pant et al. 2008, 2009; Subramanian et al. 2008; Szittya et al. 2008; Zhou et al. 2008; Arenas-Huertero et al. 2009; Jagadeeswaran et al. 2009, Lelandais-Briere et al. 2009; Wang et al. 2009; Jin et al. 2010; Joshi et al. 2010; Liang et al. 2010; Lu and Yang 2010; Trindale et al. 2010; Wang and long 2010; Xin et al. 2010; Barrera-Figueroa et al. 2011; Kulcheski et al. 2011; Lima et al. 2011; Paul et al. 2011; Yu et al. 2011; Wong et al. 2011; Barrera-Figuero et al. 2012; Chen et al. 2012a, b; Turner et al. 2012; Zhou et al. 2012a, b; Dong et al. 2013; Mantri et al. 2013; Nageshbabu and Jyothi 2013; Nageshbabu et al. 2013; Xu et al. 2013; Zhang et al. 2013; Formey et al. 2014, 2015, 2016; Yan et al. 2015, 2016).

12.5 MiRNAs Regulate Nutritional Balance in Plants

Higher plants require several mineral elements for the completion of plant growth, development, and successful reproduction (Nath and Tuteja 2016). These include macronutrients like nitrogen, phosphorus, and sulfur that are required in relatively large quantities (Maathuis 2009; Ohkama-Ohtsu and Wasaki 2010) along with some micronutrients (Haensch and Mendel 2009). In field conditions, plants often face the depletion of one or more essential nutrients which can cause growth retardation and other severe physiological disorders leading to reduced crop productivity and yield. Being sessile organisms, plants must be able to sense and respond to variation in the availability of different mineral nutrients to adapt to a wide range of environmental conditions. To do this, plants employ a broad spectrum of metabolic, physiological, and developmental adaptations (Kehr 2013). Regulatory miRNAs have emerged as major players in the response to nutrient stresses; specific miRNAs have been found to react to nutrient deficiencies. Growing bodies of research support the role of miRNAs under nutrient stresses where they have been identified as key regulators (Kehr 2012a, b, 2013; Sunkar et al. 2012; Baek et al. 2013) and are involved in the orchestration of adaptive responses to other stress conditions (Hussain et al. 2015, Hussain and Shi 2014). Furthermore, results from other plant models revealed that plants use miRNA-mediated posttranscriptional cleavage of target genes to coordinate the regulation of complex plant processes including the maintenance of N and P nutrient homeostasis (Pant et al. 2008, 2009; Buhtz et al. 2008, 2010; Kawashima et al. 2009; Varkonyi-Gasic et al. 2010; Hu et al. 2011 Kuo and Chiou 2011; Scheible et al. 2011; Zhao et al. 2011; Liang et al. 2012; Wang et al. 2012; Hackenberg et al. 2013a, b; Vidal et al. 2013; Xu et al. 2013; He et al. 2014; Khan et al. 2014; Zeng et al. 2014; Nath and Tuteja 2016).

12.6 MiRNAs and Nutritional Deficiency in Legumes

Expression profiles of several miRNAs have been reported in response to a nutrient deficiency in legumes. However, only a few attempts have been made to comprehensively examine the molecular mechanisms where miRNAs are important for the adaptive response.

12.7 MiRNAs and Nitrogen Starvation

Nitrogen represents a key element among macronutrients and plays a significant role in plant growth, development, and productivity in the extensive crop production system (Vance et al. 2003; Simon et al. 2009; Sinha et al. 2015; Nath and Tuteja 2016). Approximately 85–90 million tons of N are required annually worldwide (Lopez-Arredondo et al. 2014). However, plants can consume up to 40% of applied N, while the rest is lost to the environment through various processes such as surface runoff, leaching, denitrification, and microbial consumption (Good et al. 2004; Good and Beatty 2011; Kant et al. 2011; McAllister et al. 2012; Nguyen et al. 2015). Kant et al. (2011) estimated that approximately $1.1 billion can be saved annually by increasing the nitrogen use efficiency (NUE) in just 1% of crop plants. Therefore, engineering to increase fertilizer use efficiency in crop plants may give birth to a revolution in agriculture.

MiRNAs have been implicated in nutrient sensing, signaling, uptake, transport, and assimilation (Simon et al. 2009; Kehr 2013; Nacry et al. 2013; Frank et al. 2014). Under N-limiting conditions, miRNAs can be up- or downregulated. Expression profiles of different miRNA families have been observed in various crop species such as maize (Xu et al. 2011; De Luis et al. 2012b; Trevisan et al. 2012; Zhao et al. 2013), rice (Cai et al. 2012; Yan et al. 2014), soybean (Wang et al. 2013), and *Arabidopsis* (Pant et al. 2009; Liang et al. 2012). Considerable data have demonstrated that changes in the expression patterns of miRNAs play crucial roles in modulating adaptive responses under low nitrogen. MiR167 was associated with lateral root growth under N deficiency, miR169 and miR398a were repressed upon N-limitation (Pant et al. 2009; Zhao et al. 2010; Wang et al. 2013), and miR393 was induced by high nitrate and controls root architecture (Gifford et al. 2008; Vidal et al. 2010). Vidal et al. (2013) identified miR5090 and miR826 in *Arabidopsis* under N starvation. These miRNAs were negatively correlated with alkenyl hydroxyalkyl producing 2 (AOP2) gene and were involved in glucosinolate synthesis in response to changes in N availability (He et al. 2014). In agreement with the above, several studies have demonstrated the regulatory roles of miRNAs in root architecture, flowering time, and growth under limited N supply (Xing et al. 2010; Khan et al. 2011; Fu et al. 2012; Liang et al. 2012; Shikata et al. 2012; Yamaguchi and Abe 2012; Zhao et al. 2012; Wang et al. 2013; Spanudakis and Jackson 2014; Vidal et al. 2014).

12.8 MiRNAs and Phosphorus Deficiency

Phosphorus (P) is another essential macronutrient which plays a significant role in plant growth and development (Hackenberg et al. 2013a). P deficiency severely affects plant growth because the plant cannot access 30–80% of the organic and insoluble forms of P (Abel et al. 2002; Sha et al. 2012). To address the nutritive deprivation condition, plants have evolved a broad spectrum of morphological, physiological, and metabolic adaptations including reducing growth rates, modifying root system architecture (RSA) for increased surface area, establishing mycorrhizal symbiosis for environmental P fixation and utilization, enhancing expression of phosphorus transporter genes, and avoiding P requiring steps (Shulaev et al. 2008; Yuan and Liu 2008; Kehr 2013). In the past decade, collective studies have revealed the roles of several metabolic genes, transcription factors, riboregulators, plant hormones, and ubiquitin-related proteins under a low P situation (Chen et al. 2007a, b; Devaiah et al. 2007; Morcuende et al. 2007; Mueller et al. 2007; Nilsson et al. 2007; Yano et al. 2008; Yuan and Liu 2008; Rubio et al. 2009; Capoen et al. 2011; Gobbato et al. 2012; Soyano et al. 2013; Takeda et al. 2013; Laporte et al. 2014; Lopez-Arredondo et al. 2014; Gobbato 2015; Rich et al. 2015; Xue et al. 2015).

Diverse plant species have shown differential expression of miRNAs under P limitation including soybean (Zeng et al. 2010), barley (Hachenberg et al. 2013b), *Arabidopsis* (Hsieh et al. 2009; Lundmark et al. 2010), common bean (Valdes-Lopez et al. 2010), white lupin (Zhu et al. 2010), and rice (Zhou et al. 2008; Lin et al. 2010; Yan et al. 2014). Several miRNAs have been involved in low P signaling (Fuji et al. 2005; Aung et al. 2006; Bari et al. 2006; Chiou et al. 2006). Therefore, miRNAs were considered a vital element in gene regulatory networks under nutrient deprivation conditions, particularly P-starvation stress (Schachtman and Shin 2007; Chen et al. 2008; Guo et al. 2008; Valdes-Lopez and Hernandez 2008; Yuan and Liu 2008; Hackenberg et al. 2013a). In *Arabidopsis*, P homeostasis is posttranscriptionally regulated by miR399 and forms an essential component of the PHR1 signaling pathway (Franco-Zorrilla et al. 2004; Fuji et al. 2005; Aung et al. 2006; Bari et al. 2006; Chiou et al. 2006). MiRNA399 was positively regulated by *AtPHR1* (Aung et al. 2006; Bari et al. 2006) but negatively regulated by *IPS1/At4* using target mimicry mechanism (Franco-Zorrilla et al. 2007a, b). A similar case was observed in barley (Hackenberg et al. 2013b) where miR399 targeted UBC24, which encodes a ubiquitin-conjugating E2 enzyme, also known as PHOSPHATE 2 (PHO2) (Fuji et al. 2005; Aung et al. 2006; Bari et al. 2006). The *PHO2* gene negatively regulates phosphate uptake and root-to-shoot allocation (Bari et al. 2006). MiR399 is also involved in P-deficiency signaling in rapeseed and pumpkin phloem sap and common bean roots (Pant et al. 2008; Valdes-Lopez et al. 2008). Overexpression of miR399 in mutant plants rescued the phenotype and resulted in high phosphate (Pi) levels in shoots and significantly reduced the target transcript (Chiou et al. 2006). Similarly, Pant et al. (2008) demonstrated that miR399 accumulated at a high level in phloem sap of rapeseed and pumpkin under P deficiency, indicating a role in long-distance communication. This hypothesis was verified using reciprocal grafting experiments between transgenic *Arabidopsis* expressing miR399 and wild-type plants. This experiment revealed that miR399 can move from shoots to roots but cannot move in the opposite direction. Furthermore, a reduced level of PHO2 mRNA in the rootstock was observed. This confirms that translocated miR399 is also functional (Lin et al. 2008; Pant et al. 2008; Buhtz et al. 2010).

Recently, Xu et al. (2013) identified an additional miRNA (miR2111) which was induced under P deficiency in soybean. This miRNA was almost undetectable under normal conditions (full nutrition) but became highly abundant under limited P. In addition, miR2111 accumulated in phloem sap, like miR399, during P deficiency, indicating that it might also be involved in long-distance communication (Pant et al. 2009). MiR2111 targets the Kelch repeat-containing F-box protein in *Arabidopsis* and soybean, suggesting that it has a possible role in the control of protein abundance under low P (Hsieh et al. 2009). Moreover, several miRNAs such as miR156, miR778, and miR827 were upregulated in *Arabidopsis* under P deficiency, while miR169, miR395, and miR398 showed downregulation (Hsieh et al. 2009; Pant et al. 2009; Kant et al. 2011; Lin et al. 2013; Liu et al. 2014; Park et al. 2014). In contrast, many miRNAs (e.g., miR157, miR160, miR165, miR166, miR169 miR393, pvu-miR2118, gma-miR1524, gma-miR1526, and gma-miR1532) have shown differential expression with P deficiency in common beans (Valdes-Lopez et al. 2010). Interestingly, 167 miRNAs belonging to 35 miRNA families showed differential expression in white lupin in response to P deficiency (Zhu et al. 2010). Furthermore, Zeng et al. (2010) demonstrated that 27 miRNA families representing at least 57 miRNA members showed significant changes in expression level under P starvation in soybean.

12.9 MiRNAs Play Proactive Roles in the Establishment of Symbiosis in Legumes

Plants have the ability to establish mutual relationships with soil organisms to assist in the absorption of important nutrients such as phosphates, nitrogen, and metal elements from the soil (Oldroyd 2013). Plant-rhizobium symbiosis and plant AMF symbiosis represent two of the most researched plant-microbe symbioses (Bazin et al. 2012; Wu et al. 2016), leading to the development of nitrogen-fixing nodules and mycorrhizal arbuscules, respectively. Rhizobial-based symbiosis is restricted to legume plants while AMF symbiosis occurs in more than 80% of land plants (Wang and Qiu 2006; Oldroyd 2013; Leladais-Briere et al. 2016). The available evidence suggests the involvement of miRNAs in the establishment of legume-rhizobia nitrogen-fixing symbiosis (Bazin et al. 2012; Bustos-Sanmamed et al. 2013). Rhizobial infection of legumes occurs primarily through root hairs (Findley et al. 2016). Atmospheric nitrogen fixation occurs during legume-rhizobia symbiosis in specialized structures called nodules. The development of nodules represents one of the complex processes that results from tight communication between the two partners through molecular signals (Ferguson et al. 2010). In a nutshell, symbiosis is initiated by the exchange of molecular/chemical signals between the host plant and compatible bacteria species, which in turn release lipochitooligosaccharides called the "Nod" signal. Subsequently, the bacteria colonize root hairs and gain access to nodule cells through a specialized structure called the infection thread (Oldroyd and Downie 2008; Oldroyd 2013). This triggers cell division in the cortex and pericycle cell layer, which gives rise to nodule primordia leading to nodule formation (Oldroyd and Downie 2008). Ultimately, an endocytosis-like process releases bacteria from the infection thread, which results in a specialized structure in the nodule called the symbiosome, also known as the functional unit of biological nitrogen fixation (Kereszt et al. 2011; Yan et al. 2015). Various studies have identified the significance of different plant hormones in the initiation, development, and ultimate establishment of symbiosis in legumes (Suzaki et al. 2013). Recent studies have revealed that some legume-specific miRNAs play significant roles in nodulation and the establishment of symbiosis (Lelandais-Briere et al. 2009; Simon et al. 2009; Subramanian et al. 2008; Yan et al. 2013). Several miRNAs (miR156, miR160, miR166, miR167, miR169, miR319, miR393) involved in auxin signaling are regulated in the early events of nodulation and symbiosis (Subramanian et al. 2008; Li et al. 2010).

Conserved and legume-specific miRNA families that are differentially expressed, particularly in the nodulation process, have been reported in *M. truncatula*, *L. japonicum*, *Glycine max*, and *Phaseolus vulgaris* using experimental, sequencing, and bioinformatic approaches (Qiu et al. 2007; Xie et al. 2007; Subramanian et al. 2008; Zhang et al. 2008; Jagadeeswaran et al. 2009; Lelandais-Briere et al. 2009; Wang et al. 2009; Joshi et al. 2010; Li et al. 2010; Lu and Yang 2010; De Luis et al. 2012a; Turner et al. 2012; Barros-Carvalho et al. 2014; Formey et al. 2014, 2015). However, most of these studies only characterized conserved miRNAs involved in the different stages of legume-rhizobia symbiosis (Combier et al. 2006; Boualem et al. 2008; Subramanian et al. 2008; Nogueira et al. 2009; Devers et al. 2011; Wang et al. 2011a, b; Barros-Carvahlo et al. 2014; Yan et al. 2013). A few studies have revealed the involvement and significant roles played by three partners (host plant, rhizobia, and miRNA) in the process of symbiosis (Subramanian et al. 2008; Lelandais-Briere et al. 2009; De Luis et al. 2012a; Turner et al. 2013; Yan et al. 2015). Formey et al. (2014) studied differentially expressed miRNAs in soybean roots in response to Bradyrhizobium inoculation (3 h) and identified 120 novel miRNAs expected to play important roles in the establishment of symbiosis. Recently, another study analyzed miRNAs that respond to *B. japonicum* inoculation in soybean root hairs (Yan et al. 2016). In addition, many conserved miRNAs including miR156, miR160, miR167, miR172, miR398, and miR399 are expressed in other nodule tissues (Lelandais-Briere et al. 2009; Yan et al. 2013).

System-based approaches revealed 114 miRNAs in soybean including 22 novel miRNAs with 405 soybean miRNA targets, which have potentially relevant roles in the early stages of legume-rhizobia symbiosis (Wan et al. 2005; Brechenmacher et al. 2010, 2012; Libault et al. 2010; Yan et al. 2016). However, the dynamics of miRNA expression in the later stages of nodule development and symbiosis have not been investigated.

Proper plant growth and development depends on the coordinated regulation of Pi uptake and allocation between different plant organs, particularly under P deficiency. Therefore, the development of symbiosis with AMF represents a common P-starvation response, which enables the plant to receive increased P assisted by symbiotic fungi (Jovat et al. 2007; Parniske 2008; Devers et al. 2011, 2013). Collective studies have demonstrated significant roles of miRNAs in phosphate homeostasis in plants. It is of interest to explore the interaction between AMFs, miRNAs, and hosts in Pi uptake, relocation, and remobilization under P deprivation. Devers et al. (2011) reported mRNA cleavage mediated by miRNA in root cell reprogramming during AMF symbiosis in *M. truncatula*. This study based on high-throughput sequencing of sRNA and dendrogram tags identified 243 novel miRNAs in *M. truncatula*, which suggests potential miRNA roles in AMF symbiosis. Similarly, bioinformatic analysis revealed that several identified miRNA targets are involved in root symbiosis, which led to the conclusion that miRNAs play potentially significant roles in the establishment of AMF symbiosis. Similarly, the role of miR399 in the regulation of P homeostasis via Pi acquisition and translocation is well characterized in different plants (Aung et al. 2006; Bari et al. 2006; Chiou et al. 2006; Chiou 2007; Liu et al. 2010, 2012, 2014). Several studies have highlighted the potential role of miR827 in maintaining Pi homeostasis in plants (Kant et al. 2011; Lin et al. 2013; Liu et al. 2014; Park et al. 2014). Both miR399 and miR827 negatively regulate their targets PHO2 (UBC24) and NLA (nitrogen limitation adaptation) which ultimately results in Pi uptake in plants under P deprivation (Branschied et al. 2010; Kant et al. 2011). Another study reported that miR171h is involved in the AM colonization pattern in *M. truncatula* by targeting the NSP2 gene (Lauressergues et al. 2012). Furthermore, the same study reported induction of miR169d/I in mycorrhizal roots (Lauressergues et al. 2012).

Along with conserved miRNA families (miR169, miR171, miR398, miR399, miR408, miR778), many novel miRNAs such as miR5229a/b, miR5282, and miR5204 have potential roles in P homeostasis regulation in arbuscule-containing cells/mycorrhizal roots during AMF symbiosis (Pumplin and Harrison 2009; Mica et al. 2009; Devers et al. 2011; Gobbato et al. 2012).

12.10 Conclusions and Future Prospects

The significant roles of miRNAs in various physiological processes have drawn considerable attention since being discovered in 2002 (Reinhart et al. 2002), and rigorous research has identified a number of miRNAs. The past 5 years have witnessed an explosion in miRNA knowledge due to newly cloned plant miRNAs (15,041 in total). Subsequently, this resulted in an increase in target transcripts to 178,138 in 46 species (Yi et al. 2015). Several miRNA-based transgenic plants have shown improved tolerance to different biotic and abiotic stresses (Kamthan et al. 2015). Current research is directed toward exploring the essential roles of miRNAs in gene regulation, representing the largest families of gene regulatory molecules in legumes. These efforts provide a foundation for the evaluation of individual roles of miRNAs in posttranscriptional regulation of developmental processes and stress responses in economically important legumes.

Available data shows that the characterization and experimental validation of a few miRNAs and their target genes have been completed, and there are plenty of miRNAs with unclear functions in the queue. This chapter reviewed the complex mechanisms between various miRNAs, the microbiome and plants in the development of symbiosis for the regulation of N and P uptake, assimilation and utilization, and plant adaptation to both N and P limited conditions. It is vital to explore further the regulatory roles of miRNAs and the microbiome in crosstalk between N and P in plants under nutrient deprivation. Future efforts should be directed toward supplementary experimental approaches for answers to specific questions and in turn to understand the complex gene regulatory networks of miRNAs.

References

Abdel-Ghany SE, Pilon M (2008) MicroRNA-mediated systemic down-regulation of copper protein expression in response to low copper availability in *Arabidopsis*. J Biol Chem 283:15932–15945

Abel S, Tocconi CA, Delatorre CA (2002) Phosphate sensing in higher plants. Physiol Plant 115:1–8

Achard P, Herr A, Baulcombe DC, Harberd NP (2004) Modulation of floral development by a gibberellins-regulated microRNA. Development 13:3357–3365

Allen RS, Millgate AG, Chitty JA, Thisleton J, Miller JA, Fist AJ, Gerlach WL, Larkin PJ (2004) RNAi-mediated replacement of morphine with the nonnarcotic alkaloid reticuline in opium poppy. Nat Biotechnol 22:1559–1566

Alves L Jr, Niemeier S, Hauenschild A, Rehsmeier M, Merkle T (2009) Comprehensive prediction of novel microRNA targets in *Arabidopsis thaliana*. Nucleic Acids Res 37:4010–4021

Arenas-Huertero C, Pérez B, Rabanal F, Blanco-Melo D, De la Rosa C, Estrada-Navarrete G, Sanchez F, Covarrubias AA, Reyes JL (2009) Conserved and novel miRNAs in the legume *Phaseolus vulgaris* in response to stress. Plant Mol Biol 70:385–401

Arrese-Igor C, Gonzalez EM, Marino D, Ladrera R, Larrainzer E, Gil-Quintana E (2011) Physiological response of legume nodules to drought. Plant Stress 5(1):24–31

Ashraf MY, Ashraf M, Arshad M (2010) Major nutrient supply in legume crops under stress environments. In: Yadav SS, McNeil DL, Redden R, Patil SA (eds) Climate change and management of cool season grain legume crops. Springer, Dordrecht, pp 155–170

Aung K, Lin SI, Wu CC, Huang YT, Su CL, Chiou TJ (2006) PHO2, a phosphate over accumulator, is caused by a nonsense mutation in a microRNA399 target gene. Plant Physiol 141:1000–1011

Baek D, Kim MC, Chun HJ, Kang S, Park HC, Shin G, Park J, Shen M, Hong H, Kim WY, Kim DH, Lee SY, Bressan RA, Bohnert HJ, Yun DJ (2013) Regulation of miR399f transcription by *AtMYB2* affects phosphate starvation responses in *Arabidopsis*. Plant Physiol 161:362–373

Barakat A, Siram A, Park J, Zhebentyayeva T, Main D, Abbott A (2012) Genome-wide identification of chilling responsive microRNAs in *Prunus persica*. BMC Genomics 13:481

Bari R, Pant B, Stitt M, Scheible WR (2006) microRNA399 and PHR1 define a phosphate-signalling pathway in plants. Plant Physiol 141:988–999

Barker D, Bianchi S, Blondon F, Dattee Y, Duc G, Essad S, Flament T, Gallusci T, Genier G, Guy P, Muel X, Tourner J, Denarie J, Huguet T (1990) Medicago truncatula: a model plant for studying the molecular genetics of the rhizobium-legume symbiosis. Plant Mol Biol Report 8:40–49

Barrera-Figuero BE, Gao L, Wu Z, Zhou X, Zhu J, Jin H, Liu R, Zhu JK (2012) High throughput sequencing reveals novel and abiotic stress-regulated microRNAs in the inflorescence of rice. MBC Plant Biol 12:132

Barrera-Figueroa BE, Gao L, Diop NN, Wu Z, Ehlers JD, Roberts PA, Close TJ, Zhu JK, Liu R (2011) Identification and comparative analysis of drought-associated microRNAs in two cowpea genotypes. BMC Plant Biol 11:127

Barros-Carvalho GA, Paschoal AR, Marcelino-Guimaraes FC, Hungria M (2014) Prediction of potential novel microRNAs in soybean when in symbiosis. Genet Mol Res 13:8519–8529

Bayuelo-Jimenez JS, Craig R, Lynch JP (2002) Salinity tolerance of Phaseolus species during germination and early seedling growth. Crop Sci 42:1584–1594

Bazin J, Bustos-Sanmamed P, Hartmann C, Lelandais-Brière C, Crespi M (2012) Complexity of miRNA-dependent regulation in root symbiosis. Philos Trans R Soc B 367:1570–1579

Bazin J, Khan GA, Combier JP, Bustos-Sanmamed P, Debernardi JM, Rodriguez R, Sorin C, Palatnik J, Hartmann C, Crespi M, Lelandais-Brière C (2013) miR396 affects mycorrhization and root meristem activity in the legume *Medicago truncatula*. Plant J 74:920–934

Berg G, Rybakova D, Grube M, Koberl M (2016) The plant microbiome explored: implications for experimental botany. J Exp Bot 67:995–1002

Boualem A, Laporte P, Jovanovic M, Laffont C, Plet J, Combier JP, Niebel A, Crespi M, Fruqier F (2008) MicroRNA166 controls root and nodule development in *Medicago truncatula*. Plant J 54:876–887

Branschied A, Sieh D, Pant BD, May P, Devers EA, Elkrog A (2010) Expression pattern suggests role of miR399 in the regulation of the cellular response to local Pi increase during arbuscular mycorrhizal symbiosis. Mol Plant-Microbe Interact 23:915–926

Brechenmacher L, Lei ZT, Libault M, Findley S, Sugawara M, Sadowsky MJ, Sumner LW, Stacey G (2010) Soybean metabolites regulated in root hairs in response to the symbiotic bacterium *Bradyrhizobium japonicum*. Plant Physiol 153:1808–1822

Brechenmacher L, Nguyen THN, Hixson K, Libault M, Aldrich J, Pasa-Tolic L, Stacey G (2012) Identification of soybean proteins from a single cell type: the root hair. Proteomics 12:3365–3373

Buhtz A, Springer F, Chappell L, Baulcombe DC, Kehr J (2008) Identification and characterization of small RNAs from the phloem of *Brassica napus*. Plant J 53:739–749

Buhtz A, Pieritz J, Springer F, Kehr J (2010) Phloem small RNAs, nutrient stress responses and systemic mobility. BMC Plant Biol 10:64

Bulgarelli D, Rott M, Schlaeppi K, Ver L, van Themaat Ahmadinejad N, Assenza F, Rauf P, Huettel B, Reinhardt R, Schmelzer E, Peplies J, Gloeckner FO, Amann R, Eickhorst T, Schulze-Lefert P (2012) Revealing structure and assembly cues for *Arabidopsis* root-inhabiting bacterial microbiota. Nature 488:91–95

Bustos-Sanmamed P, Bazin J, Hartmann C, Crespi M, Lelandais-Brière C (2013) Small Bustos-Sanmamed P, Bazin J, Hartmann C, Crespi M, Lelasdais-Brierre C (2013) small RNA pathways and diversity in model legumes: lessons from genomics. Front Plant Sci 4:236

Cai H, Lu Y, Xie W, Zhu T, Lian X (2012) Transcriptome response to nitrogen starvation in rice. J Biosci 37:731–747

Capoen W, Sun J, Wysham D, Otegui MS, Venkateshwaran M, Hirsch S, Miwa H, Downie JA, Morris RJ, Ané JM, Oldroyd GE (2011) Nuclear membranes control symbiotic calcium signaling of legumes. Proc Natl Acad Sci U S A 108:14348–14353

Chen X (2004) A microRNA as a translational repressor of APETALA2 in *Arabidopsis* flower development. Science 303:2022–2025

Chen X (2005) microRNA biogenesis and function in plants. FEBS Lett 579:5923–5931

Chen Y, Lohuis D, Goldbach R, Prins M (2004) High frequency induction of RNA-mediated resistance against cucumber mosaic virus using inverted repeat constructs. Mol Breed 14:215–226

Chen C, Tao C, Peng H, Ding Y (2007a) Genetic analysis of salt stress responses in Asparagus bean {(*Vigna unguiculata* L.) ssp. Sesquipedalis Verds.}. J Hered 98:655–665

Chen ZH, Nimmo GA, Jenkins G, Nimmo HG (2007b) BHLH32 modulates several biochemical and morphological process that respond to Pi starvation in *Arabidopsis*. Biochem J 405:191–198

Chen YF, Wang Y, Wu WH (2008) Membrane transporters for nitrogen, phosphate and potassium uptake in plants. J Integr Plant Biol 50:835–848

Chen R, Hu Z, Zhang H (2009) Identification of microRNAs in wild soybean (*Glycine soja*). J Integr Plant Biol 51:1071–1079

Chen L, Wang T, Zhao M, Tian Q, Zhang WH (2012a) Identification of aluminium-responsive microRNAs in *Medicago truncatula* by genome-wide high throughput sequencing. Planta 235:375–386

Chen L, Wang T, Zhao M, Zhang W (2012b) Ethylene-responsive microRNAs in roots of *Medicago truncatula* identified by high throughput sequencing at whole genome level. Plant Sci 184:14–19

Cheung F, Hass BJ, Goldberg SMD, May GD, Xiao Y, Town CD (2006) Sequencing Medicago truncatula expressed sequenced tags using 454 life sciences technology. BMC Genomics 7:272

Chiou TJ (2007) The role of microRNA in sensing nutrient stress. Plant Cell Environ 30:323–332

Chiou TJ, Aung K, Lin SI, Wu CC, Chiang SF, Su CL (2006) Regulation of phosphate homeostasis by microRNAs in *Arabidopsis*. Plant Cell 18:412–421

Chitwood DH, Timmermans MCP (2010) Small RNAs are on the move. Nature 467:415–419

Chuck G, Candela H, Hake S (2009) Big impact by small RNAs in plant development. Curr Opin Plant Biol 12:81–86

Comai L, Zhang B (2012) MicroRNAs: key gene regulators with versatile functions. Plant Mol Biol 80:1

Combier JP, Frugier F, De Billy F, Boualem A, El-Yahyaoui F, Moreau S, Verni ET, Ott T, Gamas P, Crespi M, Niebel A (2006) *MtHAP2-1* is a key transcriptional regulator of symbiotic nodule development regulated by microRNA169 in *Medicago truncatula*. Genes Dev 20:3084–3088

Cook DR (1999) *Medicago truncatula*-a model in the making. Curr Opin Plant Biol 2:301–304

Cramer GR, Urano K, Delrot S, Pezzotti M, Shinozaki K (2011) Effects of abiotic stress on plants: a system biology perspective. BMC Plant Biol 11:163

De Luis A, Markmann K, Cognat V, Holt BD, Charpentier M, Parniske M, Stouggard J, Voinnet O (2012a) Two microRNAs linked to nodule infection and nitrogen-fixing ability in the legume *Lotus japonicus*. Plant Physiol 160:2137–2154

De Luis A, Markmann K, Cognat V, Holt DB, Charpentier M, Parniske M, Stougaard J, Voinnet O (2012b) Two microRNAs linked to nodule infection and nitrogen-fixing ability in the legume *Lotus japonicus*. Plant Physiol 160:2137–2154

Deschamps S, Campbell MA (2010) Utilization of next-generation sequencing platforms in plant genomics and genetic variant discovery. Mol Breed 25:553–570

Devaiah B, Karthikeyan AS, Ragothama KG (2007) WRKY75 transcription factor is a modulator of phosphate acquisition and root development in Arabidopsis. Plant Physiol 143:1789–1801

Devers EA, Branschield A, May P, Krajinski F (2011) Stars and symbiosis: microRNA and microRNA*-mediated transcript cleavage involved in *Arbuscular mycorrhizal* symbiosis. Plant Physiol 156:1990–2010

Devers EA, Teply J, Reinert A, Gaude N, Krajinski F (2013) An endogenous artificial microRNA system for unravelling the function of root endosymbiosis related genes in *Medicago truncatela*. BMC Plant Biol 13:82

Ding D, Zhang L, Wang H, Liu Z, Zhang Z, Zheng Y (2009) Differential expression of miRNAs in response to salt stress in maize roots. Annals Bot 103:29–38

Ding Y, Chen Z, Zhu C (2011) Microarray based analysis of cadmium-responsive microRNAs in rice (*Oryza sativa*). J Exp Bot 62:3563–3573

Dong QH, Han J, Yu HP, Wang C, Zhao MZ, Liu H, Ge AJ, Fang JG (2012) Computational identification of microRNAs in strawberry expressed sequence tags and validation of their precise sequences by miR-RACE. J Hered 103:268–277

Dong Z, Shi L, Wang Y, Chen L, Cai Z, Wang Y, Jin J, Li X (2013) Identification and dynamic regulation of microRNAs involved in salt stress responses in functional soybean nodules by high throughput sequencing. Int J Mol Sci 14:2717–2738

Duan CG, Wang CH, Guo HS (2006) Regulation of microRNA on plant development and viral infection. Chin Sci Bull 51:269–278

Eldem V, Celikkol-Akcay U, Ozhuner E, Bakr Y, Uranbey S, Unver T (2012) Genome-wide identification of miRNAs responsive to drought in peach (*Prunus persica*) by high throughput deep sequencing. PLoS One 7:e50298

El-Enany AE, Al-Anazi AD, Dief N, Taisan WA (2013) Role of antioxidant enzymes in amelioration of water deficit and water logging stresses on *Vigna sinensis* plants. J Biol Earth Sci 3:B144–B153

Fahlgren L, Howell MD, Kasschau KD, Chapman EJ, Sullivan CM, Cumbie JS, Givan SA, Law TF, Grant SR, Dangl JL, Carrington JC (2007) High-throughput sequencing of *Arabidopsis* microRNAs: evidence for frequent birth and death of MIRNA genes. PLoS ONE 2:e219

Ferdous J, Hussain SS, Shi BJ (2015) Role of microRNAs in plant drought tolerance. Plant Biotech J 13:293–305

Ferguson BJ, Indrasumunar A, Hayashi S, Lin MH, Lin YH, Reid DE, Gresshoff PM (2010) Molecular analysis of legume nodule development and autoregulation. J Intgr Plant Biol 52:61–76

Findley S, Joshi T, Qiu L, Sherrier DJ, Ji T, Meyers BC, Xu D, Stacey G (2016) Identification and functional characterization of soybean root hair microRNAs expressed in response to *Bradyrhizobium japonicum* infection. Plant Biotechnol J 14:1–10

Formey D, Sallet E, Lelandais-Brière C, Ben C, Bustos-Sanmamed P, Niebel A, Frugier F, Combier JP, Debellé F, Hartmann C, Poulain J, Gavory F, Wincker P, Roux C, Gentzbittel L, Gouzy J, Crespi M (2014) The small RNA diversity from *Medicago truncatula* roots under biotic interactions evidences the environmental plasticity of the miRNAome. Genome Biol 15:457

Formey D, Iñiguez LP, Peláez P, Li YF, Sunkar R, Sánchez F, Reyes JL, Hernández G (2015) Genome-wide identification of the *Phaseolus vulgaris* sRNAome using small RNA and degradome sequencing. BMC Genomics 16:423

Formey D, Martin-Rodriguez JA, Leija A, Santana O, Quinto C, Cardenas L, Hernandez G (2016) Regulation of small RNAs and corresponding targets in Nod factor-induced *Phaseolus vulgaris* root hair cells. Int J Mol Sci 17:887

Franco-Zorrilla JM, Gonzalez E, Bustos R, Linhares F, Leyva A, Paz-Ares J (2004) The transcriptional control of plant responses to phosphate limitation. J Exp Bot 55:285–293

Franco-Zorrilla JM, Valli A, Todesco M, Mateos I, Puga MI, Rubio-Somoza I, Leyva A, Weigel D, Garcia JA, Paz-Ares J (2007a) Target mimicry provides a new mechanism for regulation of microRNA activity. Nat Genet 39:1033–1037

Franco-Zorrilla JM, Valli A, Todesco M, Mateos I, Puga MI, Rubio-Somoza I, Leyva A, Weigel D, Garcia JA, Paz-Ares J (2007b) Target mimicry provides a new mechanism for regulation of microRNA activity. Nat Genet 39:1033–1037

Frank S, Tiemeyer B, Gelbrecht J, Freibauer A (2014) High soil solution carbon and nitrogen concentrations in a drained Atlantic bog are reduced to natural levels by 10 years of rewetting. Biogeosciences 11:2309–2324

Fu C, Sunkar R, Zhou C, Shen H, Zhang JY, Matts J, Wolf J, Mann DG, Stewart CN Jr, Tang Y, Wang ZY (2012) Overexpression of miR156 in switchgrass (*Panicum virgatum* L.) results in various morphological alterations and leads to improved biomass production. Plant Biotechnol J 10:443–452

Fuji H, Chiou TJ, Lin SI, Aung K, Zhu JK (2005) A miRNA involved in phosphate starvation response in *Arabidopsis*. Curr Biol 15:2038–2043

Gandikota M, Birkenbihl RP, Hohmann S, Cardon GH, Saedler H, Huijser P (2007) The miRNA156/157 recognition element in the 30 UTR of the *Arabidopsis* SBP box gene SPL3 prevents early flowering by translational inhibition in seedlings. Plant J 49:683–693

Garg R, Patel RK, Tyagi AK, Jain M (2011) DE novo assembly of chickpea transcriptome using short reads for gene discovery and marker identification. DNA Res 18:53–63

Gifford ML, Dean A, Gutierrez RA, Coruzzi GM, Birnbaum KD (2008) Cell specific nitrogen responses mediate developmental plasticity. Proc Natl Acad Sci U S A 105:803–808

Gobbato E (2015) Recent developments in arbuscular mycorrhizal signaling. Curr Opin Plant Biol 26:1–7

Gobbato E, Marsh JF, Vernié T, Wang E, Maillet F, Kim J, Miller JB, Sun J, Bano SA, Ratet P, Mysore KS, Dénarié J, Schultze M, Oldroyd GE (2012) A GRAS-type transcription factor with a specific function in mycorrhizal signaling. Curr Biol 22:2236–2241

Good AG, Beatty PH (2011) Fertilizing nature: a tradegy of excess in thecommons. PLoS Biol 9:e1001124

Good AG, Shrawat AK, Muench DG (2004) Can less yield more? Is reducing nutrient input into the environment compatible with maintaining crop production? Trends Plant Sci 9:597–605

Graham PH, Vance CP (2003) Legumes: importance and constraints for greater use. Plant Physiol 131:872–877

Griffiths-Jones S (2004) The microRNA registry. Nucleic Acids Res 32:109–111

Griffiths-Jones S, Saini HK, van Dongen S, Enright AJ (2008) miRBase: tools for microRNA genomics. Nucleic Acids Res 36:D154–D158

Guo HS, Xie Q, Fei JF, Chua NH (2005) MicroRNA directs mRNA cleavage of the transcription factor *NAC1* to downregulate auxin signals for *Arabidopsis* lateral root development. Plant Cell 17:1376–1386

Guo WB, Zhang LN, Zhao J, Liao H, Zhuang CX, Yan XL (2008) Identification of temporally and spatially phosphate-starvation responsive genes in *Glycine max*. Plant Sci 175:574–584

Gutierrez L, Bussell JD, Pacurar DI, Schwambach J, Pacurar M, Bellini C (2009) Phenotypic plasticity of adventitious rooting in *Arabidopsis* is controlled by complex regulation of AUXIN RESPONSE FACTOR transcripts and microRNA abundance. Plant Cell 21:3119–3132

Hackenberg M, Huang PJ, Huang CY, Shi BJ, Gustafson P, Langridge P (2013a) A comprehensive expression profile of microRNAs and other classes of non-coding small RNAs in barley under phosphorous-deficient and -sufficient conditions. DNA Res 20:109–125

Hackenberg M, Shi BJ, Gustafson P, Langridge P (2013b) Characterization of phosphorus-regulated miR399 and miR827 and their isomirs in barley under phosphorus-sufficient and phosphorus-deficient conditions. BMC Plant Biol 13:214

Haensch R, Mendel RR (2009) Physiological functions of mineral micronutrients (Cu, Zn, Mn, Fe, Ni, Mo, B, Cl). Curr Opin Plant Biol 12:259266

Handberg K, Stougaard J (1992) Lotus japonica, an autogamous, diploid legume species for classical and molecular genetics. Plant J 2:487–496

Hayat R, Ali S, Amara U, Khalid R, Ahmed I (2010) Soil beneficial bacteria and their role in plant growth promotion: a review. Ann Microbiol 60:579–598

He H, Liang G, Li Y, Wang F, Yu D (2014) Two young microRNAs originating from target duplication mediate nitrogen starvation adaptation via regulation of glucosinolate synthesis in *Arabidopsis thaliana*. Plant Physiol 164:853–865

Hiremath PJ, Farmer A, Cannon SB, Woodward J, Kupada H, Tuteja R, Kumar A, Bhanuprakash A, Mulaosmanovic B, Gujaria N, Krishnamurthy L, Gaur PM, Kavikishor PB, Shah T, Srinivasan R, Lohse M, Xiao Y, Town CD, Cook RD, May GD, Varshney RK (2011) Large scale transcriptome analysis in chickpea (*Cicer arietinum* L.), an orphan legume crop of the semi-arid tropics of Asia and Africa. Plant Biotechnol J 9:922–931

Hsieh LC, Lin SI, Shih ACC, Chen JW, Lin WY, Tseng CY, Lin WH, Chiou TJ (2009) Uncovering small RNA-mediated responses to phosphate deficiency in Arabidopsis by deep sequencing. Plant Physiol 151:2120–2132

Hu B, Zhu C, Li F, Tang J, Wang Y, Lin A, Liu L, Che R, Chu C (2011) LEAF TIP NECROSIS1 plays a pivotal role in the regulation of multiple phosphate starvation responses in rice. Plant Physiol 156:1101–1115

Hussain SS, Shi B (2014) Role of miRNAs in abiotic and biotic stresses in plants. In: Ahmad P, Rasool S (eds) Emerging technologies and management of crop stress tolerance, Biological techniques, vol 1. Elsevier, USA, pp 181–207

Hussain SS, Ali M, Ahmad M, Siddique KHM (2011a) Polyamines: natural and engineered abiotic stress tolerance in plants. Biotechnol Adv 29:300–3111

Hussain SS, Kayani MA, Amjad M (2011b) Transcription factors as tools to engineer enhanced drought stress tolerance in plants. Biotechnol Prog 27:297–306

Hussain SS, Ali M, Shi B (2015) MiRNA mediated regulatory functions under abiotic stresses in legumes. In: Ahmad P, Azoozor M (eds) Legumes under environmental stress: yield, improvement and adaptation. Wiley, UK, pp 231–245

Hussain SS, Asif MA, Sornaraj P, Ali M, Shi BJ (2016) Towards integration of system based approach for understanding drought stress in plants. In: Ahmad P, Rasool S (eds) Water stress and crop plants: a sustainable approach. Elsevier, USA, pp 227–247

Iantecheva A, Mysore KS, Ratet P (2013) Transformation of leguminous plants to study symbiotic interactions. Int J Dev Biol 57:577–586

Jagadeeswaran G, Zhang Y, Li YF, Shukla LI, Matts J, Hoyt P, Macmil SL, Wiley GB, Roe BA, Zhang W, Sunkar R (2009) Cloning and characterization of small RNAs from *Medicago truncatula* reveals four novel legume specific miRNA families. New Phytol 184:85–98

Jia X, Wang WX, REN L, Chen QJ, Mendu V (2009) Differential and dynamic regulation of miR398 in response to ABA and salt stress in *Populus tremula* and *Arabidopsis thaliana*. Plant Mol Biol 71:51–59

Jin T, Chang Q, Li W, Yin D, Li Z, Wang D, Liu B, Liu L (2010) Stress-inducible expression of *GmDREB1* conferred salt tolerance in transgenic alfalfa. Plant Cell Tissue Organ Cult 100:219–227

Jogaiah S, Govind SR, Tran LSP (2012) Systems biology-based approaches towards understanding drought tolerance in food crops. Crit Rev Biotech 33:23–39

Jones-Rhoades MW, Bartel DP, Bartel DP (2006) MicroRNAs and their regulatory roles in plants. Annu Rev Plant Biol 57:19–53

Joshi T, Yan Z, Libault M, Jeong DH, Park S, Green PJ, Sherier DJ, Farmer A, May G, Meyers BC, Xu D, Stacey G (2010) Prediction of novel miRNAs and associated target genes in *Glycine max*. BMC Bioinformatics 1:S14

Jovat H, Pumplin N, Harrison MJ (2007) Phosphate in the arbuscular mycorrhizal symbiosis: transport properties and regulatory roles. Plant Cell Environ 30:310–322

Kamthan A, Chaudhuri A, Kamthan M, Datta A (2015) Small RNAs in plants:recent development and application for crop improvement. Front Plant Sci 6:208

Kant S, Peng M, Rothstein SJ (2011) Genetic regulation by NLA and microRNA827 for maintaining nitrate-dependent phosphate homeostasis in *Arabidopsis*. PLoS Genet 7:e1002021

Kawashima CG, Yoshimoto N, Maruyama-Nakashita A, Tsuchiya YN, Saito K, Takahashi H, Dalmay T (2009) Sulphur starvation induces the expression of microRNA-395 and one of its target genes but in different cell types. Plant J 57:313–321

Kehr J (2012a) Roles of miRNAs in nutrient signaling and homeostasis. In: Sunkar R (ed) MicroRNAs in plant development and stress responses, signaling and communication in plants. Springer, Berlin, pp 197–217

Kehr J (2012b) Long-distance signaling by small RNAs through the phloem. In: Kragler F, Huelskamp M (eds) Short and long distance signaling, Advances in plant biology series. Springer, New York, p 131149

Kehr J (2013) Systemic regulation of mineral homeostasis by microRNAs. Front Plant Sci 4:145

Kereszt A, Mergaert P, Kondorosi E (2011) Bacterioid development in legume nodules: evolution of mutual benefits or of sacrificial victims? Mol Plant-Microbe Interact 24:1300–1309

Khan GA, Declerck M, Sorin C, Hartmann C, Crespi M, Leladais-Briere C (2011) MicroRNAs as regulators root development and architecture. Plant Mol Biol 77:47–58

Khan GA, Bouraine S, Wege S, Li Y, de Carbonnel M, Berthomieu P, Poirier Y, Rouached H (2014) Coordination between zinc and phosphate homeostasis involves the transcription factor PHR1, the phosphate exporter PHO1, and its homologue PHO1;H3 in *Arabidopsis*. J Exp Bot 65:871–884

Khraiwesh B, Ossowski S, Weigel D, Reski R, Frank W (2008) Specific gene silencing by artificial microRNAs in *Physcomitrella patens*: an alternative to targeted gene knockouts. Plant Physiol 148:684–693

Khraiwesh B, Arif MA, Seumel GI, Ossowski S, Weigel D, Reski R, Frank W (2010) Transcriptional control of gene expression by microRNAs. Cell 140:111–122

Khraiwesh B, Zhu JK, Zhu J (2012) Role of miRNAs and siRNAs in biotic and abiotic stress responses of plants. Biochem Biophys Acta 1819:137–148

Kim B, Yu HJ, Park SG, Shin JY, Oh M, Kim N, Mun JH (2012) Identification and profiling of novel microRNAs in the *Brassica rapa* genome based of small RNA deep sequencing. BMC Plant Biol 12:218

Kozomara A, Griffiths-Jones S (2014) miRBase: annotating high confidence microRNAs using deep sequencing data. Nucleic Acids Res 42:D68–D73

Kruszka K, Pieczynski M, Windels D, Bielewicz D, Jarmolowski A, Szweykowska-Kulinska Z, Vazquez F (2012) Role of microRNAs and other sRNAs of plants in their changing environments. J Plant Physiol 169:1664–1672

Kruszka K, Pacak A, Swida-Barteczka A, Nuc P, Alaba S, Wroblewska Z, Karlowski W, Jarmolowski A, Szweykowska-Kulinska Z (2014) Transcriptionally and post-transcriptionally regulated microRNAs in heat stress response in barley. J Exp Bot 65:6123–6135

Kudapa H, Bharti AK, Cannon SB, Farmer AD, Mulaosmanovic B, Kramer R, Bohra A, Weeks NT, Crow JA, Tuteja R, Shah T, Dutta S, Gupta DK, Singh A, Gaikwad K, Sharma TR, May GD, Singh NK, Varshney RK (2012) A comprehensive transcriptome assembly of pigeonpea (*Cajanus cajan* L.) using Sanger and second generation sequencing platforms. Mol Plant 5:1020–1028

Kulcheski FR, de Oliveira LF, Molina LG, Almerao MP, Rodrigues FA, Marcolino J, Barbosa JF, Stolf-Moreira R, Nepomuceno AL, Marcelino-Guimaraes FC, Abdelnoor RV, Nascimento LC, Carazolle MF, Pereira GAG, Margis R (2011) Identification of novel soybean microRNAs involved in abiotic and biotic stresses. BMC Genomics 12:307

Kuo HF, Chiou TJ (2011) The role of microRNAs in phosphorus deficiency signaling. Plant Physiol 156:1016–1024

Lamblin AF, Crow JA, Johnson JE, Silverstein KA, Kunau TM, Kilian A, Benz D, Stromvik M, Endré G, VandenBosch KA, Cook DR, Young ND, Retzel EF (2003) MtDB: a database for personalized data mining of the model legume *Medicago truncatula* transcriptome. Nucleic Acids Res 31:196–201

Laporte P, Lepage A, Fournier J, Catrice O, Moreau S, Jardinaud MF, Mun JH, Larrainzar E, Cook DR, Gamas P, Niebel A (2014) The CCAAT box-binding transcription factor NF-YA1 controls rhizobial infection. J Exp Bot 65:481–494

Laufs P, Peaucelle A, Morin H, Traas J (2004) MicroRNA regulation of the CUC genes is required for boundary size control in *Arabidopsis* meristems. Development 131:4311–4322

Lauressergues D, Delaux PM, Formey D, Leladaid-Briere C, Fort S, Cottaz S, Becard G, Niebel A, Roux C, Combier JP (2012) The microRNA miR171h modulates arbuscular mycorrhizal colonization of *Medicago truncatula* by targeting NSP2. Plant J 72:512–522

Leladais-Briere C, Moreau J, Hartmann C, Crespi M (2016) Noncoding RNAs, emerging regulators in root endosymbioses. Mole Plant-Microbe Interact 29:170–180

Lelandais-Briere C, Naya L, Sallet E, Calenge F, Frugier F, Hartmann C, Gouzy J, Crespi M (2009) Genome-wide *Medicago truncatula* small RNA analysis revealed novel microRNAs and isoforms differentially regulated in roots and nodules. Plant Cell 21:2780–2796

Lelandais-Briere C, Sorin C, Declerck M, Benslimane A, Crespi M, Hartmann C (2010) Small RBA diversity in plants and its impact in development. Curr Genomics 11:14–23

Li WX, Oono Y, Zhu J, He XJ, Wu JM, Iida K, Lu XY, Cui X, Jin H, Zhu JK (2008) The *Arabidopsis* NFYA5 transcription factor is regulated transcriptionally and post-transcriptionally to promote drought resistance. Plant Cell 20:2238–2251

Li H, Deng Y, Wu T, Subramaniun S, Yu O (2010) Misexpression of miR482, miR1512, and miR1515 increases soybean nodulation. Plant Physiol 153:1759–1770

Li H, Dong Y, Wang Y, Yang J, Liu X (2011) Characterization of the stress associated microRNAs in *Glycine max* by deep sequencing. BMC Plant Biol 11:170–181

Li T, Chen J, Qiu S, Zhang Y, Wang P, Yang L, Lu Y, Shi J (2012) Deep sequencing and microarray hybridization identify conserved and species specific microRNAs during somatic embryogenesis in hybrid yellow poplar. PLoS One 7:e43451

Li B, Duan H, Li H, Deng XW, Yin W, Xia X (2013) Global identification of miRNAs and targets in *Populus euphratica* under salt stress. Plant Mol Biol 81:525–539

Liang G, Yang F, Yu D (2010) MicroRNA395 mediates regulation of sulfate accumulation and allocation in *Arabidopsis thaliana*. Plant J 62:1046–1057

Liang G, He H, Yu D (2012) Identification of nitrogen starvation-responsive microRNAs in *Arabidopsis thaliana*. PLoS One 7:e48951

Liang L, Li Y, He H, Wang F, Yu D (2013) Identification of miRNAs and miRNA-mediated regulatory pathways in *Carica papaya*. Planta 238:739–752

Libault M, Brechenmacher L, Cheng J, Xu D, Stacey G (2010a) Root hair systems biology. Trends Plant Sci 15:641–650

Libault M, Farmer A, Joshi T, Takahashi K, Langley RJ, Franklin LD, He J, Xu D, May G, Stacey G (2010b) An integrated transcriptome atlas of the crop model *Glycine max*, and its use in comparative analysis in plants. Plant J 63:86–99

Lima JC, Arenhart RA, Margis-Pinheiro M (2011) Aluminum triggers broad changes in microRNA expression in rice roots. Genet Mol Res 10:2817–2832

Lin Y, Lai ZX (2013) Comparative analysis reveals dynamic changes in miRNA and their targets and expression during somatic embryogenesis in longan (*Dimocarpus longan* Lour.) PLoS One 8:e60337

Lin SI, Chiang SF, Lin WY, Chen JW, Tseng CY, Wu PC, Chiou TJ (2008) Regulatory network of microRNA399 and PHO2 by systemic signaling. Plant Physiol 147:732–746
Lin SI, Santi C, Jobet E, Lacut E, Kholti NE, Karlowski WM, Verdeil JL, Breitler JC, Perin C, Ko SS, Guiderdoni E, Chiou TJ, Echeverria M (2010) Complex regulation of two target genes encoding SPX-MFS proteins by rice miR827 in response to phosphate starvation. Plant Cell Physiol 51:2119–2131
Lin WY, Huang TK, Chiou TJ (2013) NITROGEN LIMITATION ADAPTATION, a target of microRNA827, mediates degradation of plasma membrane-localized phosphate transporters to maintain phosphate homeostasis in *Arabidopsis*. Plant Cell 25:4061–4074
Liu HH, Tian X, Li YJ, Wu CA, Zheng CC (2008) Microarray-based analysis of stress-regulated microRNAs in *Arabidopsis thaliana*. RNA 14:836–843
Liu JQ, Allan DL, Vance CP (2010) Systemic signaling and local sensing of phosphate in common bean: cross-talk between photosynthate and microRNA399. Mol Plant 3:428–437
Liu TY, Huang TK, Tseng CY, Lai YS, Lin SI, Lin WY, Chen JW, Chiou TJ (2012) PHO2-dependent degradation of PHO1 modulates phosphate homeostasis in Arabidopsis. Plant Cell 24:2168–2183
Liu T, Lin W, Huang T, Chiou T (2014) MicroRNA-mediated surveillance of phosphate transporters on the move. Trends Plant Sci 19:647–655
Llave C, Kasschau KD, Rector MA, Carrington JC (2002) Endogenous and silencing-associated small RNAs in plants. Plant Cell 14:1605–1619
Lopez-Arredondo DL, Leyva-Gonzalez MA, Gonzalez-Morales SI, Lopez-Bucio J, Herrera-Estrella L (2014) Phosphate nutrition: improving low-phosphate tolerance in crops. Annu Rev Plant Biol 65:95–123
Lu Y, Yang X (2010) Computational identification of novel microRNAs and their targets in *Vigna unguiculate*. Comp Func Genomics Article # 128297
Lu C, Meyers BC, Green PJ (2007) Construction of small RNA cDNA libraries for deep sequencing. Methods 43:110–117
Lundberg DS, Lebeis SL, Paredes SH, Yourstone S, Gehring J, Malfatti S, Tremblay J, Engelbrektson A, Kunin V, del Rio TG, Edgar RC, Eickhorst T, Ley RE, Hugenholtz P, Tringe SG, Dangl JL (2012) Defining the core *Arabidopsis thaliana* root microbiome. Nature 488:86–90
Lundmark M, Korner CJ, Nielsen TH (2010) Global analysis of microRNA in *Arabidopsis* in response to phosphate starvation as studied by locked nucleic acid-based microarrays. Physiol Plant 140:57–68
Maathuis FJ (2009) Physiological functions of mineral macronutrients. Curr Opin Plant Biol 12:250–258
Mallory AC, Vaucheret H (2006) Functions of microRNAs and related small RNAs in plants. Nat Genet 38:31–36
Mantri N, Basker N, Ford R, Pang E, Pardeshi V (2013) The role of micro-ribonucleic acids in legumes with a focus on abiotic stress response. Plant Genome 6:1–14
Mapelli F, Marasco R, Rolli E, Barbato M, Cherif H, Guesmi A, Ouzari I, Daffonchio D, Borin S (2013) Potential for plant growth promotion of rhizobacteria associated with Salicornia growing in Tunisian hypersaline soils. Biomed Res 2013:248078
McAllister CH, Beatty PH, Good AG (2012) Engineering nitrogen use efficient crop plants: the current status. Plant Biotechnol J 10:1011–1025
Mendes R, Kruijt M, de Bruijn I, Dekkers E, van der Voort M, Schneider JH, Piceno YM, DeSantis TZ, Andersen GL, Bakker PA, Raaijmakers JM (2011) Deciphering the rhizosphere microbiome for disease-suppressive bacteria. Science 332:1097–1100
Meng Y, Huang F, Shi Q, Cao J, Chen D, Zhang J, Ni J, Wu P, Chen M (2009) Genome-wide survey of rice microRNAs and microRNA-target pairs in the root of a novel auxin-resistant mutant. Planta 230:883–898
Meng YJ, Ma XX, Chen DJ, Wu P, Chen M (2010) Micro RNA mediated signaling involved in plant root development. Biochem Biophys Res Commun 393:345–349

Mica E, Piccolo V, Delledonnne M, Ferrarini A, Pezzotti M, Casati C, Del Fabbro C, Valle G, Policriti A, Morgante M, Pesole G, Pe ME, Horner DS (2009) High throughput approaches reveal splicing of primary microRNA transcripts and tissue specific expression of mature microRNAs in *Vitis vinifera*. BMC Genomics 10:558

Morcuende R, Bari R, Gibon Y, Zheng W, Pant BD, Blasing O, Udasel B, Czechowski T, Urvardi MK, Stitt M, Scheible WR (2007) Genome-wide, reprogramming of metabolism and regulatory networks of *Arabidopsis* in response to phosphorus. Plant Cell Environ 30:85–112

Moxon S, Jing R, Szittya G, Schwach F, Rusholme Pilcher RL, Moulton V, Dalmay T (2008) Deep sequencing of tomato short RNAs identifies microRNAs targeting genes involved in fruit ripening. Genome Res 18:1602–1609

Mueller R, Morant M, Jarmer H, Nilsson L, Nielsen TH (2007) Genome-wide analysis of the *Arabidopsis* leaf transcriptome reveals interaction of phosphate and sugar metabolism. Plant Physiol 143:156–171

Nacry P, Bouguyon E, Gojon A (2013) Nitrogen acquisition by roots: physiological and developmental mechanisms ensuring plant adaptation to a fluctuating resource. Plant Soil 370:1–29

Nageshbabu R, Jyothi MN (2013) Profile of small interfering RNAs from French bean Phyaeolus vulgaris under abiotic stress conditions. Int J Pharm Bio Sci 4B:176–185

Nageshbabu R, Jyothi MN, Sharadamma N, Sarika S, Rai DV, Devaraj VR (2013) Expression of miRNAs regulates growth and development of French bean (*Physeolus vulgaris*) under salt and drought stress conditions. Int Res J Boil Sci 2:52–56

Nath M, Tuteja N (2016) NPKS, sensing, and signalling and miRNAs in plant nutrient stress. Protoplasma 253:767–786

Naya L, Paul S, Valdez-Lopez O, Mendoza-Soto AB, Nova-Franco B, Sosa-Valencia G, Reyes JL, Hernandez G (2014) Regulation of copper homeostasis and biotic interactions by microRNA 398b in common bean. PLoS ONE 9:e84416

Nguyen GN, Rothstein SJ, Spangenberg G, Surya K (2015) Role of microRNAs involved in plant response to nitrogen and phosphorous limiting conditions. Front Plant Sci 6:629

Nilsson L, Mueller R, Nielson TH (2007) Increased expression of the MYB-related transcription factor PHR1, leads to enhanced phosphate uptake in *Arabidopsis thaliana*. Plant Cell Environ 30:1499–1512

Nogueira FT, Sarkar AK, Chitwood DH, Timmermanns MC (2006) Organ polarity in plants is specified through the opposing activity of two distinct small regulatory RNAs. Cold Spring Harbor Sym Quan Biol 71:157–164

Nogueira FT, Chitwood DH, Madi S, Ohtsu K, Schnable PS, Scanlon MJ, Timmermans MC (2009) Regulation of small RNA accumulation in the maize shoot apex. PLoS Genet 5:e1000320

Ohkama-Ohtsu N, Wasaki J (2010) Recent progress in plant nutrition research: cross-talk between nutrients, plant physiology, and soil microorganisms. Plant Cell Physiol 51:1255–1264

Okamura K, Phillips MD, Tyler DM, Duan H, Chou YT, Lai EC (2008) The regulatory activity of microRNA* species has substantial influence on microRNA and 3′ UTR evolution. Nat Struct Mol Biol 15:354–363

Oldroyd GE (2005) Sequencing the model legume *Medicago truncatula*. Grain Legumes 41:23

Oldroyd GE (2013) Speak, friend and enter: signaling systems that promote beneficial symbiotic associations in plants. Nat Rev Microbiol 11:252–263

Oldroyd GE, Downie JA (2008) Coordinating nodule morphogenesis with rhizobial infection in legumes. Annu Rev Plant Biol 59:519–546

Opdenakker K, Remans T, Vangronsveld J, Cuypers A (2012) Mitogen-activated protein (MAP) kinases in plant metal stress: regulation and responses in comparison to other biotic and abiotic stresses. Int J Mol Sci 13:7828–7852

Ozhuner E, Eldem V, Ipek A, Okay S, Sakcali S, Zhang B, Boke H, Unver T (2013) Boron stress responsive microRNAs and their targets in barley. PLoS One 8:e59543

Pandey R, Joshi G, Bhardwaj AR, Agarwal M, Katiyar-Agarwal S (2014) A comprehensive genome-wide study on tissue-specific and abiotic stress-specific miRNAs in *Triticum aestivum*. PLoS ONE 9:e95800

Pant BD, Buhtz A, Kehr J, Scheible WR (2008) MicroRNA399 is a long-distance signal for the regulation of plant phosphate homeostasis. Plant J 53:31–38

Pant BD, Musialak-Lange M, Nuc P, May P, Buhtz A, Kehr J, Walther D, Scheible WR (2009) Identification of nutrient-responsive *Arabidopsis* and rapeseed microRNAs by comprehensive real-time polymerase chain reaction profiling and small RNA sequencing. Plant Physiol 150:1541–1555

Park MY, Wu G, Gonzalez-Sulser A, Vaucheret H, Poethig RS (2005) Nuclear processing and export of microRNAs in *Arabidopsis*. Proc Natl Acad Sci 102:3691–3696

Park BS, Seo JS, Chua NH (2014) NITROGEN LIMITATION ADAPTATION recruits PHOSPHATE2 to target the phosphate transporter PT2 for degradation during the regulation of *Arabidopsis* phosphate homeostasis. Plant Cell 26:454–464

Parniske M (2008) Arbuscular mycorrhiza: the mother of plant root endosymbiosis. Nat Rev Microbiol 6:763–775

Paul S, Kundu A, Pal A (2011) Identification and validation of conserved microRNAs along with their differential expression in roots of *Vigna unguiculata* grown under salt stress. Plant Cell Tiss Organ Cult 105:233–242

Popelka JC, Terryn N, Higgins TJV (2004) Gene technology for grain legumes: can it contribute to the food challenge in developing countries? Plant Sci 167:195–206

Pumplin N, Harrison MJ (2009) Live-cell imaging reveals periarbuscular membrane domains and organelle location in *Medicago truncatula* roots during arbuscular mycorrhizal symbiosis. Plant Physiol 151:809–819

Qiu CX, Xie FL, Zhu YY, Guo K, Huang SQ, Nie L, Yang ZM (2007) Computational identification of microRNAs and their targets in *Gossypium hirsutum* expressed sequence tags. Gene 395:49–61

Rajagopalan R, Vaucheret H, Trejo J, Bartel DP (2006) A diverse and evolutionarily fluid set of microRNAs in *Arabidopsis thaliana*. Genes Dev 20:3407–3425

Ramesh V, Admane N, Hussain SM (2013) Small RNAs landscape (sRNAome) of soybean (*Glycine max* L.): biogenesis, vital functions and potential applications. Plant Knowledge J 2:24–37

Reinhart BJ, Weinstein EG, Rhoades MW, Bartel B, Bartel DP (2002) MicroRNAs in plants. Genes Dev 16:1616–1626

Reyes JL, Chau NH (2007) ABA induction of miR 159 controls transcript levels of two MYB factors during *Arabidopsis* seed germination. Plant J 49:592–606

Rich MK, Schorderet M, Bapaume L, Falquet L, Morel P, Vandenbussche M, Reinhardt D (2015) The petunia GRAS transcription factor ATA/RAM1 regulates symbiotic gene expression and fungal morphogenesis in Arbuscular mycorrhiza. Plant Physiol 168:788–797

Rubio V, Bustos R, Irigoyen ML, Cardona-Lopez X, Rojas-Triana M, Paz-Ares J (2009) Plant hormones and nutrient signaling. Plant Mol Biol 69:361–373

Rubio-Somoza I, Weigel D (2011) MicroRNA networks and developmental plasticity in plants. Trends Plant Sci 16:258–264

Sato S, Nakamura Y, Kaneko T, Asamizu E, Kato T, Nakao M, Sasamoto S, Watanabe A, Ono A, Kawashima K, Fujishiro T, Katoh M, Kohara M, Kishida Y, Minami C, Nakayama S, Nakazaki N, Shimizu Y, Shinpo S, Takahashi C, Wada T, Yamada M, Ohmido N, Hayashi M, Fukui K, Baba T, Nakamichi T, Mori H, Tabata S (2008) Genome structure of the legume, *Lotus japonicus*. DNA Res 15:227–239

Schachtman DP, Shin R (2007) Nutrient sensing and signaling: NPKS. Annu Rev Plant Biol 58:47–69

Schaeppi K, Bulgarelli D (2015) The plant microbiome at work. Mol Plant-Microbe Interact 28:212–217

Scheible WR, Pant B, Musialak-Lange M, Nuc P (2011) Nutrient-responsive plant microRNAs. In: Erdmann VA, Barciszewski J (eds) Non coding RNAs in plants. Springer, Berlin, pp 313–337

Schmutz J, Cannon SB, Schlueter J, Ma J, Mitros T, Nelson W, Hyten DL, Song Q, Thelen JJ, Cheng J, Xu D, Hellsten U, May GD, Yu Y, Sakurai T, Umezawa T, Bhattacharyya MK, Sandhu D, Valliyodan B, Lindquist E, Peto M, Grant D, Shu S, Goodstein D, Barry K, Futrell-Griggs

M, Abernathy B, Du J, Tian Z, Zhu L, Gill N, Joshi T, Libault M, Sethuraman A, Zhang XC, Shinozaki K, Nguyen HT, Wing RA, Cregan P, Specht J, Grimwood J, Rokhsar D, Stacey G, Shoemaker RC, Jackson SA (2010) Genome sequence of the palaeopolyploid soybean. Nature 463:178–183

Schreiber AW, Shi BJ, Huang CY, Langridge P, Baumann U (2011) Discovery of barley miRNAs through deep sequencing of short reads. BMC Genomics 12:129

Schwab R, Palatnik JF, Riester M, Schommer C, Schmid M, Weigel D (2005) Specific effects of microRNAs on the plant transcriptome. Dev Cell 8:517–527

Sha A, Chen Y, Ba H, Shan Z, Zhang X, Wu X, Qiu D, Chen S, Zhou X (2012) Identification of *Glycine max* microRNAs in response to phosphorus deficiency. J Plant Biol 55:268–280

Shikata M, Yamaguchi H, Sasaki K, Ohtsubo N (2012) Overexpression of *Arabidopsis* miR157b induces bushy architecture and delayed phase transition in *Torenia fournieri*. Planta 236:1027–1035

Shrama HC, Srivastava CP, Durairaj C, Gowda CLL (2010) Pest management in grain legumes and climate change. In: Yadav SS, McNeil DL, Redden R, Patil SA (eds) Climate change and management of cool season grain legume crops. Springer, Dordrecht, pp 115–140

Shuai P, Liang D, Zhang Z, Yin W, Xia X (2013) Identification of drought responsive and novel *Populus trichocarpa* microRNAs by high throughput sequencing and their targets using degradome analysis. BMC Genomics 14:233

Shulaev V, Cortes D, Miller G, Mittler R (2008) Metabolomics for plant stress responses. Physiol Plant 132:199–208

Simon SA, Meyers BC, Sherrier DJ (2009) MicroRNAs in rhizobia legume symbiosis. Plant Physiol 151:1002–1008

Sinha S, Srinivasan R, Mandal PK (2015) MicroRNA-based approach to improve nitrogen use efficiency in crop plants. In: Rakshit A, Singh HB, Sen A (eds) Nutrient use efficiency: from basics to advances. Springer, New Delhi, pp 221–235

Song QX, Liu YF, Hu XY, Zhang WK, Ma B, Chen SY, Zhang JS (2011) Identification of microRNAs and their target genes in developing soybean seeds by deep sequencing. BMC Plant Biol 11:5

Soyano T, Kouchi H, Hirota A, Hayashi M (2013) Nodule inception directly targets NF-Y subunit genes to regulate essential processes of root nodule development in *Lotus japonicus*. PLoS Genet 9:e1003352

Spanudakis E, Jackson S (2014) The role of microRNAs in the control of flowering time. J Exp Bot 65:365–380

Subramanian S, Fu Y, Sunkar R, Barbazuk WB, Zhu JK, Yu O (2008) Novel and nodulation regulated microRNAs in soybean roots. MBC Genomics 9:160

Sun G (2012) MicroRNAs and their diverse functions in plants. Plant Mol Biol 80:17–36

Sunkar R, Jagadeeswaran G (2008) *In silico* identification of conserved microRNAs in large number of diverse plant species. BMC Plant Biol 8:37

Sunkar R, Zhu JK (2004a) Novel and stress-regulated microRNAs and other small RNAs from *Arabidopsis*. Plant Cell 16:2001–2019

Sunkar R, Zhu JK (2004b) Novel and stress-regulated microRNAs and other small RNAs from *Arabidopsis*. Plant Cell 16:2001–2019

Sunkar R, Kapoor A, Zhu JK (2006) Posttranscriptional induction of two Cu/Zn superoxide dismutase genes in Arabidopsis is mediated by down-regulation of miR398 and important for oxidative stress tolerance. Plant Cell 18:2051–2065

Sunkar R, Chinnusamy V, Zhu J, Zhu JK (2007) Small RNAs as big players in plant abiotic stress responses and nutrient deprivation. Trends Plant Sci 12:301–309

Sunkar R, Li YF, Jagadeeswaran G (2012) Functions of microRNAs in plant stress responses. Trends Plant Sci 17:196–203

Suzaki T, Ito M, Kawaguchi M (2013) Genetic basis of cytokinin and auxin functions during root nodule development. Front Plant Sci 4:42

Szittya G, Moxon S, Santos DS, Jing R, Fevereiro MPS, Moulton V, Dalmay T (2008) High throughput sequencing of *Medicago truncatula* short RNAs identifies eight new miRNA families. BMC Genomics 9:593

Takeda N, Tsuzuki S, Suzaki T, Parniske M, Kawaguchi M (2013) CERBERUS and NSP1 of *Lotus japonicus* are common symbiosis genes that modulate *Arbuscular mycorrhiza* development. Plant Cell Physiol 54:1711–1723

Thompson R, Ratet P, Kuster H (2005) Identification of gene functions by applying TILLING and insertional mutagenesis strategies on microarray based expression data. Grain Legumes 41:20–22

Timmusk S, Behers L, Muthoni J, Aronsson AC (2017) Perspectives and challenges of microbe application for crop improvement. Front Plant Sci 8:49

Trevisan S, Nonis A, Begheldo M, Manoli A, Palme K, Caporale G, Ruperti B, Quaggiotti S (2012) Expression and tissue-specific localization of nitrate-responsive miRNAs in roots of maize seedlings. Plant Cell Environ 35:1137–1155

Trindale I, Capitao C, Dalmay T, Fevereiro MP, Santos DM (2010) miR398 and miR408 are up-regulated in response to water deficit in *Medicago truncatula*. Plant 231:705–716

Turner M, Yu O, Subramanian S (2012) Genome organization and characteristics of soybean microRNAs. BMC Genom 13:169

Turner M, Nizampatnam NR, Baron M, Coppin S, Damodaran S, Adhikari S, Arunachalam S, Yu O, Subramaniun S (2013) Ectopic expression of miR160 results in auxin hypersensitivity, cytokinin hyposensitivity, and inhibition of symbiotic nodule development in soybean. Plant Physiol 162:2042–2055

Urvardi MK, Scheible WR (2005) GRAS genes and the symbiotic green revolution. Science 308:1749–1750

Valdes-Lopez O, Hernandez G (2008) Transcriptional regulation and signaling in phosphorus starvation: what about legumes? J Integr Plant Biol 10:1213–1222

Valdes-Lopez O, Arena-Huertero C, Ramirez M, Girard L, Sanchez F, Vance CP, Reyes JL, Hernandez G (2008) Essential role of MYB transcription factor: PvPHR1 and microRNA: PvmiR399 in phosphorus-deficiency signaling in common bean roots. Plant Cell Environ 31:1834–1843

Valdes-Lopez O, Yang S, Aparicio-Fabre R, Graham PH, Reyes JL, Vance CP, Hernandez G (2010) MicroRNA expression profiles in common bean (*Phaseolus vulgaris*) under nutrient deficiency stresses and manganese toxicity. New Phytol 187:805–818

Valentine AJ, Benedito VA, Kang Y (2011) Legume nitrogen fixation and soil abiotic stress: from physiology to genomic and beyond. Annual Plant Rev 42:207–248

Vance CP, Udhe-Stone C, Allen DI (2003) Phosphorus acquisition and use: critical adaptation by plants for securing a non-renewable resource. New Phytol 157:423–447

Varkonyi-Gasic E, Gould N, Sandanayaka M, Sutherland P, MacDiarmid RM (2010) Characterization of microRNAs from apple (*Malus domestica* 'Royal Gala') vascular tissue and phloem sap. BMC Plant Biol 10:159

Varshney RK, Chen W, Li Y, Bharti AK, Saxena RK, Schlueter JA, Donoghue MTA, Azam S, Fan G, Whaley AM, Farmer AD, Sheridan J, Iwata Tuteja AR, Penmetsa AV, Wu W, Upadhyaya HD, Yang SP, Shah T, Saxena KB, Michael T, McCombie WR, Yang B, Zhang G, Yang H, Wang J, Spillane C, Cook DR, May GD, Xu X, Jackson SA (2012) Draft genome sequence of pigeonpea (*Cajanus cajan*), an orphan legume crop of resource-poor farmers. Nat Biotechnol 30:83–89

Varshney RK, Song C, Saxena RK, Azam S, Yu S, Sharpe AG, Cannon S, Baek J, Rosen BD, Tar'an B, Millan T, Zhang X, Ramsay LD, Iwata A, Wang Y, Nelson W, Farmer AD, Gaur PM, Soderlund C, Penmetsa RV, Xu C, Bharti AK, He W, Winter P, Zhao S, Hane JK, Garcia NC, Condie JA, Upadhyaya HD, Luo MC, Thudi M, Gowda CLL, Singh NP, Lichtenzveig J, Gali KK, Rubio J, Nadarajan N, Dolezel J, Bansal KC, Xu X, Edwards D, Zhang G, Kahl G, Gil J, Singh KB, Datta SK, Jackson SA, Wang J, Cook DR (2013) Draft genome sequence of chickpea (*Cicer arietinum*) provides a resource for trait improvement. Nat Biotechnol 31:240–246

Vejan P, Abdullah R, Khadiran T, Ismail S, Nusrulhaq BA (2016) Role of plant growth-promoting rhizobacteria in agricultural sustainability: a review. Molecules:21–573

Vidal EA, Araus V, Lu C, Parry G, Green PJ, Coruzzi GM, Gutierrez RA (2010) Nitrate-responsive miR393/ABF3 regulatory module controls root system architecture in *Arabidopsis thaliana*. Proc Natl Acad Sci U S A 107:4477–4482

Vidal EA, Moyano TC, Riveras E, Contreras-López O, Gutiérrez RA (2013) Systems approaches map regulatory networks downstream of the auxin receptor AFB3 in the nitrate response of *Arabidopsis thaliana* roots. Proc Natl Acad Sci U S A 110:12840–12845

Vidal EA, Moyano TC, Canales J, Gutiérrez RA (2014) Nitrogen control of developmental phase transitions in *Arabidopsis thaliana*. J Exp Bot 65:5611–5618

Wan J, Torres M, Ganapathy A, Thelen J, Dague BB, Mooney B, Xu D, Stacey G (2005) Proteomic analysis of soybean root hairs after infection by *Bradyrhizobium japonicum*. Mol Plant-Microbe Interact 18:458–467

Wang Y, Long L (2010) Identification and isolation of the cold-resistance related miRNAs in *Pisum sativum* Linn. Liaoning 2:231–236

Wang B, Qiu YL (2006) Phylogenetic distribution and evolution of mycorrhizas in land plants. Mycorrhiza 16:299–363

Wang Y, Li P, Cao X, Wang X, Zhang A, Li X (2009) Identification and expression analysis of miRNAs from nitrogen-fixing soybean nodules. Biochem Biophys Res Commun 378:799–803

Wang T, Chen L, Zhao M, Tian Q, Zhang WH (2011a) Identification of drought-responsive microRNAs in *Medicago truncatula* by genome-wide high throughput sequencing. BMC Genomics 12:367

Wang XB, Jovel J, Udomporn P, Wang Y, Wu Q, Li WX, Gasciolli Y, Vaucheret H, Ding SW (2011b) The 21-nucleotide, but not 22-nucleotide, viral secondary small interfering RNAs direct potent antiviral defense by two cooperative argonautes in *Arabidopsis thaliana*. Plant Cell 23:1625–1638

Wang C, Han J, Kibet KN, Kayesh E, Shangguan L, Li X, Fang J (2012) Identification of microRNAs from Amur grapes (*Vitis amurensis* Rupr.) by deep sequencing and analysis of microRNA variations with bioinformatics. BMC Genomics 13:122

Wang Y, Zhang C, Hao Q, Sha A, Zhou R, Zhou X, Yuan L (2013) Elucidation of miRNAs-mediated responses to low nitrogen stress by deep sequencing of two soybean genotypes. PLoS One 8:e67423

Willmann MR, Poethig RS (2005) Time to grow up: the temporal role of small RNAs in plants. Curr Opin Plant Biol 8:548–552

Wong CE, Zhao YT, Wang XJ, Croft L, Wang ZH, Haerizadeh F, Mattick JS, Singh MB, Carroll BJ, Bhalla PL (2011) MicroRNAs in the shoot apical meristem of soybean. J Exp Bot 62:2495–2506

Wu P, Wu Y, Liu CC, Liu LW, Ma FF, Wu XY, Wu M, Hang YY, Chen JQ, Shao ZQ, Wang B (2016) Identification of arbuscular mycorrhiza (AM)-responsive microRNAs in tomato. Front Plant Sci 7:429

Xie FL, Huang SQ, Guo K, Xiang AL, Xhu YY, Nie L, Yang ZM (2007) Computational identification of novel microRNA and targets in *Brassica napus*. FEBS Lett 58:1464–1474

Xie F, Frazier TP, Zhang B (2010) Identification and characterization of microRNAs and their targets in the bioenergy plant switchgrass (*Panicum virgatum*). Planta 232:417–434

Xin M, Wang Y, Yao Y, Xie C, Peng H, Ni Z, Sun Q (2010) Diverse set of microRNAs are responsive to powdery mildew infection and heat stress in wheat (*Triticum aestivum* L.) BMC Plant Biol 10:123

Xing S, Salinas M, Höhmann S, Berndtgen R, Huijser P (2010) MiR156-targeted and non targeted SBP-Box transcription factors act in concert to secure male fertility in *Arabidopsis*. Plant Cell 22:3935–3950

Xu Z, Zhong S, Li X, Li W, Rothstein SJ, Zhang S, Bi Y, Xie C (2011) Genome-wide identification of microRNAs in response to low nitrate availability in maize leaves and roots. PLoS One 6:e28009

Xu F, Liu Q, Chen L, Kuang J, Walk T, Wang J, Hong L (2013) Genome-wide identification of soybean microRNAs and their targets reveals their organ-specificity and responses to phosphate starvation. BMC Genom 14:66

Xue L, Cui H, Buer B, Vijayakumar V, Delaux PM, Junkermann S, Bucher M (2015) Network of GRAS transcription factors involved in the control of arbuscule development in *Lotus japonicus*. Plant Physiol 167:854–871

Yamaguchi A, Abe M (2012) Regulation of reproductive development by non-coding RNA in *Arabidopsis*: to flower or not to flower. J Plant Res 125:693–704

Yan Z, Hossain MS, Wang J, Valdés-López O, Liang Y, Libault M, Qiu L, Stacey G (2013) miR172 regulates soybean nodulation. Mol Plant-Microbe Interact 26:1371–1377

Yan Y, Wang H, Hamera S, Chen X, Fang R (2014) MiR444a has multiple functions in the rice nitrate-signaling pathway. Plant J 78:44–55

Yan Z, Hossain MS, Arikit S, Valdés-López O, Zhai J, Wang J, Libault M, Ji T, Qiu L, Meyers BC, Stacey G (2015) Identification of microRNAs and their mRNA targets during soybean nodule development: functional analysis of the role of miR393j-3p in soybean nodulation. New Phytol 207:748–759

Yan Z, Hossain MS, Valdés-López O, Hoang NT, Zhai J, Wang J, Libault M, Brechenmacher L, Findley S, Joshi T, Qiu L, Sherrier DJ, Ji T, Meyers BC, Xu D, Stacey G (2016) Identification and functional characterization of soybean root hair microRNAs expressed in response to *Bradyrhizobium japonicum* infection. Plant Biotechnol J 14:332–341

Yang T, Xue L (2007) Functional diversity of miRNAs in plants. Plant Sci 172:423–432

Yanik H, Turktas M, Dundar E, Hernendez P, Dorado G, Unver T (2013) Genome wide identification of alternative bearing-associated microRNAs (miRNAs) in olive (Olea europaea L.) BMC Plant Biol 13:10

Yano K, Yoshida S, Müller J, Singh S, Banba M, Vickers K, Markmann K, White C, Schuller B, Sato S, Asamizu E, Tabata S, Murooka Y, Perry J, Wang TL, Kawaguchi M, Imaizumi-Anraku H, Hayashi M, Parniske M (2008) CYCLOPS, a mediator of symbiotic intracellular accommodation. Proc Natl Acad Sci U S A 105:20540–20545

Yao Y, Guo G, Ni Z, Sunkar R, Du J, Zhu JK, Sun Q (2007) Cloning and characterization of microRNAs from wheat (*Triticum aestivum L.*) Genome Biol 8:R96

Yi X, Zhang Z, Ling Y, Xu W, Su Z (2015) PNRD: a plant non-coding RNA database. Nucleic Acids Res 43:D982–D989

Young ND, Debelle F, GED O, Geurts R, Cannon SB, Udvardi M, Benedito VA, KFX M, Gouzy J, Schoof H, Van de Peer Y, Proost S, Cook DR, Meyers BC, Spannag M, Cheung F, Mita SD, Krishnakumar V, Gundlach H, Zhou S, Mudge J, Bharti AK, Murray JD, Naoumkina MA, Rosen B, Silverstein KAT, Tang H, Rombauts S, Zhao PX, Zhou P, Barbe V, Bardou P, Bechner M, Bellec A, Berger A, Berges H, Bidwell S, Bisseling T, Choisne N, Couloux A, Denny R, Deshpande S, Dai X, Doyle JJ, Dudez AM, Farmer AD, Fouteau S, Franken C, Gibelin C, Gish J, Goldstein S, Gonzalez AJ, Green PJ, Hallab A, Hartog M, Hua A, Humphray SJ, Jeong DH, Jing Y, Jocker A, Kenton SM, Kim DJ, Klee K, Lai H, Lang C, Lin S, Macmil SL, Magdelenat G, Matthews L, McCorrison J, Monaghan EL, Mun JH, Najar FZ, Nicholson C, Noirot C, O'Bleness M, Paule CR, Poulain J, Florent Prion F, Qin B, Qu C, Retzell EF, Riddle C, Sallet E, Samain S, Samson N, Sanders I, Scarpelli SC, Schiex T, Segurens B, Severin AJ, Sherrier DJ, Shi R, Sims S, Singer SR, Sinharoy S, Sterck L, Viollet A, Wang BB, Wang K, Wang M, Wang X, Warfsmann J, Weissenbach J, White DJ, White JD, Wiley GB, Wincker P, Xing Y, Yang L, Yao Z, Ying F, Zhai J, Zhou L, Zuber A, Denarie J, Dixon RA, May GD, Schwartz DC, Rogers J, Quetier F, Town CD, Roe BA (2011) The Medicago genome provides insight into the evolution of rhizobial symbioses. Nature 480:520–524

Yu H, Song C, Jia Q, Wang C, Li F, Nicholas KK, Zhang X, Fang J (2011) Computational identification of microRNAs in apple expressed sequence tags and validation of their precise sequences by miR-RACE. Plant Physiol 141:56–70

Yuan H, Liu D (2008) Signaling components involved in plant responses to phosphate starvation. J Integr Plant Biol 50:849–859

Zeng HQ, Zhu YY, Huang SQ, Yang ZM (2010) Analysis of phosphorus deficient responsive microRNAs and cis-elements from soybean (*Glycine max* L.) J Plant Physiol 167:1289–1297
Zeng H, Wang G, Hu X, Wang H, Du L, Zhu Y (2014) Role of microRNAs in plant responses to nutrient stress. Plant Soil 374:1005–1021
Zhang BH, Pan XP, Wang QL, Cobb GP, Anderson TA (2005) Identification and characterization of new plant microRNAs using EST analysis. Cell Res 15:336–360
Zhang B, Pan X, Stellwag EJ (2008) Identification of soybean microRNAs and associated target genes. Planta 229:161–182
Zhang J, Zhang S, Han S, Wu T, Li X, Li W, Qi L (2012) Genome-wide identification of microRNAs in larch and stage-specific modulation of 11 conserved microRNAs and their targets during somatic embryogenesis. Planta 236:647–657
Zhang LW, Song JB, Shu XX, Zhang Y, Yang ZM (2013) miR395 is involved in detoxification of cadmium in *Brassica napus*. J Hazard Mater 250-251:204–211
Zhao L, KimYJ DT, Chen X (2007) MiR172 regulates stem cell fate and defines the inner boundary of APETALA3 and PISTILLATA expression domain in *Arabidopsis* floral meristem. Plant J 51:840–849
Zhao CZ, Xia H, Frazier TP, Yao YY, Bi YP, Li AQ, Li MJ, Li CS, Zhang BH, Wang XJ (2010) Seep sequencing identifies novel and conserved microRNAs in peanuts (*Arachis hypogaea* L.) BMC Plant Biol 10:3
Zhao M, Ding H, Zhu JK, Zhang F, Li XW (2011) Involvement of miR169 in the nitrogen-starvation responses in *Arabidopsis*. New Phytol 190:906–915
Zhao M, Tai H, Sun S, Zhang F, Xu Y, Li WX (2012) Cloning and characterization of maize miRNAs involved in responses to nitrogen deficiency. PLoS One 7:e29669
Zhao Y, Xu Z, Mo Q, Zou C, Li W, Xu Y, Xie C (2013) Combined small RNA and degradome sequencing reveals novel miRNAs and their targets in response to low nitrate availability in maize. Ann Bot 112:633–642
Zhou X, Wang G, Sutoh K, Zhu J, Zhang W (2008) Identification of cold-inducible microRNAs in plants by transcriptome analysis. BBA-Gene Regul Mech 1779:780–788
Zhou LG, Liu YH, Liu ZC, Kong DY, Duan M, Luo LJ (2010) Genome-wide identification and analysis of drought-responsive microRNAs in *Oryza sativa*. J Exp Bot 61:4157–4168
Zhou ZS, Song JB, Yang ZM (2012a) Genome-wide identification of *Brassica napus* microRNAs and their targets in response to cadmium. J Exp Bot 63:4597–4613
Zhou ZS, Zeng HQ, Liu ZP, Yang ZM (2012b) Genome-wide identification of *Medicago truncatula* microRNAs and their targets reveals their different regulation by heavy metal. Plant Cell Environ 35:86–99
Zhu H, Choi H, Cook DR, Shoemaker RC (2005) Bridging model and crop legumes through comparative genomics. Plant Physiol 137:1189–1196
Zhu YY, Zeng HQ, Dong CX, Yin XM, Shen QR, Yang ZM (2010) microRNA expression profiles associated with phosphorus deficiency in white lupin (Lupinus albus L.) Plant Sci 178:23–29

Chapter 13
Plant-Microbe-Metal Interactions: Basics, Recent Advances, and Future Trends

Rahul Mahadev Shelake, Rajesh Ramdas Waghunde, Eugene Hayato Morita, and Hidenori Hayashi

Abstract All organisms require metal ions to complete their life cycle. Excess or shortage of essential metal ions is toxic to plants. Also, some heavy metals are toxic at all concentrations and hinder the functioning of plants. Therefore, plants including microbes have evolved metal homeostatic machineries to tackle toxic levels of metals inside the cell. Since a long time, scientists have investigated metal homeostasis mechanisms in plants. In last few decades, anthropogenic activities together with natural catastrophic events have increased the bioavailable concentration of heavy metals in the biosphere. Heavy metals are persistent in nature and cannot be biodegraded. Thus, heavy-metal pollution is becoming a threat to environment, agriculture, and human health. The microbes are the most sensitive creature to metal stress than the rest of soil fauna. Some plant-microbe interactions are beneficial under stress induced by heavy metal thereby enhancing uptake, translocation, distribution, and detoxification by either or both the partners, i.e., plant or microbe. The rapid progress in the research about the molecular and physiological mechanisms of plant-associated microbes is helping us to understand the factors influencing plant-microbe-metal interactions under heavy-metal stress. In this chapter, we have summarized various aspects and recent updates of three major interactions, i.e., plant-metal, plant-microbe, and plant-microbe-metal interactions. Further, we have assessed recent updates in beneficial plant-microbe interactions and their application in the management of metal-induced abiotic stress in plants.

Rahul Mahadev Shelake and Rajesh Ramdas Waghunde contributed equally to this work.

R.M. Shelake (✉) • H. Hayashi
Proteo-Science Center, Ehime University, Matsuyama, Japan
e-mail: rahultnau@gmail.com

R.R. Waghunde
Department of Plant Pathology, College of Agriculture, Navsari Agricultural University, Bharuch, India

E.H. Morita
Department of Chemistry, Faculty of Science, Josai University, Saitama, Japan

D. Egamberdieva, P. Ahmad (eds.), *Plant Microbiome: Stress Response*, Microorganisms for Sustainability 5, https://doi.org/10.1007/978-981-10-5514-0_13

Keywords Plant-microbe interactions • Metal bioavailability • Heavy metal • Metal uptake • Remediation • Abiotic stress

13.1 Introduction

Metal ions are essential for life because one-quarter to one-third of all proteins require metals in order to function (Mahadev et al. 2013). Metalloenzymes are reported in all six classes established by the International Union of Biochemistry (Waldron et al. 2009). All the organisms actively maintain a beneficial intracellular concentration of essential metal ions by delicately balancing the expression of proteins involved in specific metal uptake and export/storage (Outten and O'Halloran 2001). Conversely, essential metal ions can also be poisonous to the plant viability if the intracellular concentration is more than physiological requirements (Touati 2000). Some of the metal elements are toxic to plants at all concentrations and commonly known as heavy metals such as cadmium (Cd), arsenic (As), lead (Pb), and mercury (Hg) (Chetia et al. 2011).

The term "heavy metal" is rather inexact (Duffus 2002). However, naturally abundant transition metal elements, chromium (Cr), manganese (Mn), iron (Fe), cobalt (Co), nickel (Ni), copper (Cu), zinc (Zn), molybdenum (Mo), cadmium (Cd), and mercury (Hg), together with the metallic element lead (Pb), are often designated as "heavy metals" (Table 13.1). The term "heavy metal" is defined in the refereed literature as metal elements having a density higher than water (Fergusson 1990), specifically more than 5 g/cm^3 (Jarup 2003).

Heavy metal ions have serious effects on terrestrial and aquatic ecosystems. Metal toxicity increases health risks for humans and plants altering physiological processes. Heavy metal ions can cause serious health issues in human once ingested into the body through a food chain. Major health risks due to heavy-metal toxicity consist of cardiovascular dysfunctions, chronic anemia, cancer, skin allergies, tooth

Table 13.1 List of heavy metals

Heavy metal	Density (g/cm^3)	Form acquired
Chromium (Cr)	7.19	Cr^{3+}, (Cr^{2+}), CrO^4, CrO^{2-}
Manganese (Mn)	7.43	Mn^{2+}
Iron (Fe)	7.87	Fe^{2+}
Nickel (Ni)	8.908	Ni^{2+}
Cobalt (Co)	8.9	Co^{2+}
Copper (Cu)	7.87	Cu^+, Cu^{2+}
Zinc (Zn)	7.14	Zn^{2+}
Molybdenum (Mo)	10.28	MoO^{2-}, MoO_4
Cadmium (Cd)	8.7	Cd^{2+}
Mercury (Hg)	13.56	Hg
Lead (Pb)	11.34	Pb^{2+}

decay, bone damage, kidney damage, cognitive impairment, abnormal functioning of the nervous system and brain, etc. (summarized in Ullah et al. 2015). Therefore, to reduce these health risks, the reduction, immobilization, and possibly total exclusion of heavy metals from the food and environment are necessary.

To maintain metal homeostasis inside the plant tissues, plants have evolved several mechanisms that regulate the metal uptake, translocation, accumulation, and detoxification of essential and nonessential metal ions (Zitka et al. 2013). These mechanisms can be influenced by several factors such as soil properties, plant growth stage, climate, root exudates, and chemicals released by associated microbes inside the plants or in the rhizosphere (DalCorso et al. 2014). Phytochelatins (PCs) and metallothioneins (MTs) are the most studied heavy-metal-binding ligands associated with metal homeostasis in plants (Cobbett and Goldsbrough 2002). Plant and microbes coexist in nature, and plant-microbe interaction is either beneficial (mutual or beneficial to only one partner) or harmful (pathogenic or parasitic) to plants (Rosenblueth and Martínez-Romero 2006). Therefore, the study of plants along with microbes living in close association, also known as plant microbiome or plant microbiota has evolved. The plant microbiome beneficial to plant health can classify into two subcategories:

1. The microbes residing belowground that include free-living rhizobacteria (rhizosphere), epiphytes (rhizoplane), and endophytes (internal endosymbionts).
2. The microbes residing aboveground plant parts that may include epiphytes and endophytes.

Recent reports have suggested vast potential in an exploitation of plant-microbe interactions for plant growth-promoting (PGP) activity along with abiotic stress management in plants, specifically microbe-assisted phytoremediation of toxic metals (Ma et al. 2016). In this chapter, we have summarized the factors influencing the bioavailability, uptake, and translocation mechanisms of metal ions in plants. Further, the effects of metal toxicity on plant physiology and defense strategies adopted by the plant species discussed. Also, the role of plant-associated microbes in immobilization of toxic metal, mobilization of essential metal ions to plants, and recent updates of plant-microbe-metal interactions in the process of tolerance or remediation of toxic metals are summarized.

13.2 Factors Affecting Bioavailability of Metal Ions

Heavy metals are found everywhere in the soil, but concentrations are variable at different locations (Luo et al. 2014). The soil properties are the most studied natural factors influencing the bioavailability of metals such as pH, cation exchange capacity, clay and organic matter composition, etc. Over centuries, industrial, mining/smelting, and military activities, as well as farming and waste practices implemented by human beings, have contaminated large areas of developed countries with high concentrations of heavy metals and organic pollutants (Ali et al. 2014). Additionally, natural disasters like volcano eruptions alter the metal bioavailability directly or indirectly in the affected regions.

13.2.1 Anthropogenic Activities

Anthropogenic activities such as mining or smelting of metal ores are one of the sources of heavy metals in soils. It is also responsible for increased prevalence and occurrence of heavy-metal contamination at the Earth's surface. Mining activities have a negative impact on soil and water bodies in the environment and produce many sulfide-rich tailings (Bhattacharya et al. 2006). Higher deposition of metal ions is recorded in soils around the mining area of Jequetepeque river basin of Peru (Yacoub et al. 2012), tungsten mines of Panasqueira in Portugal (Candeias et al. 2014), Katanga Copperbelt in Democratic Republic of Congo (Pourret et al. 2016), and Bayan Obo rare earth element mine in Mongolia of China (Pan and Li 2016).

Disposal of industrial wastewater also contaminates the soils, and it is documented in different parts of the world. Some examples include Etang de Berre in France (Georgeaud et al. 1997), Damascus in Syria (Moller et al. 2005; Song et al. 2000), Hangzhou city in China (Lu and Bai 2006), Kurang River in Pakistan (Zahra et al. 2014), and Lucknow city in India (Gupta et al. 2015). If these metals are in bioavailable form, they can be accumulated in some plants and may pose a potential threat to humans and grazing animals. Some agricultural soils contain high concentrations of metals from natural geochemical sources (Adriano 1986), while others contaminated by metals from anthropogenic activity such as phosphate fertilizer application, fuel spills, industrial effluents derived from the mining process, and wastewater sludge.

13.2.2 Natural Processes

In addition to human activities, natural processes such as volcanic eruptions and continental dust also lead to emission and accumulation of heavy metals in the ecosystem. For example, the analysis of soils in Kanagawa area of Japan, located near Mount Fuji showed considerably higher levels of Ni, Cu, Zn, Cd, and Pb present in the dry soil (Okamoto et al. 1997). Similar results observed for samples collected from the vicinity of Masaya volcano in Nicaragua (Central America) thereby confirming volcano smoke causes metal deposition into soil (Hinrichs et al. 2011). Heavy metal ions released into the atmosphere by anthropogenic activities and natural processes can be easily transported to a distant location through soil, air, or water during natural catastrophic events such as the volcano, flooding, tsunami, hurricane, and forest fire (Jovanovic et al. 2015).

13.3 Metal Ions in Plant Life

Essential nutrients are defined as the substances that cannot be replaced with any other nutrients, and its absence prevents the completion of the life cycle of the plant. Besides carbon dioxide, water, and oxygen, all plants require total 14 essential elements that divided into two groups: macronutrients (requirement more than 1000 mg/kg of dry weight; N, P, K, Ca, Mg, and S) and micronutrients (requirement less than 100 mg/kg of dry weight; Cl, Fe, B, Ni, Cu, Mn, Zn, and Mo) (Mengel and Kirkby 2012). As mentioned above, some of the heavy metals are essential nutrients because they are integral parts of many enzymes and other proteins (Fig. 13.1).

As shown in Fig. 13.1, metals are important in different physiological processes. Metal ions are integral components of chemical reactions involved in plant metabolism and take part either directly (e.g., Fe in electron transport) or indirectly (as cofactors of regulatory proteins, e.g., Ni in urease). Functions of these six

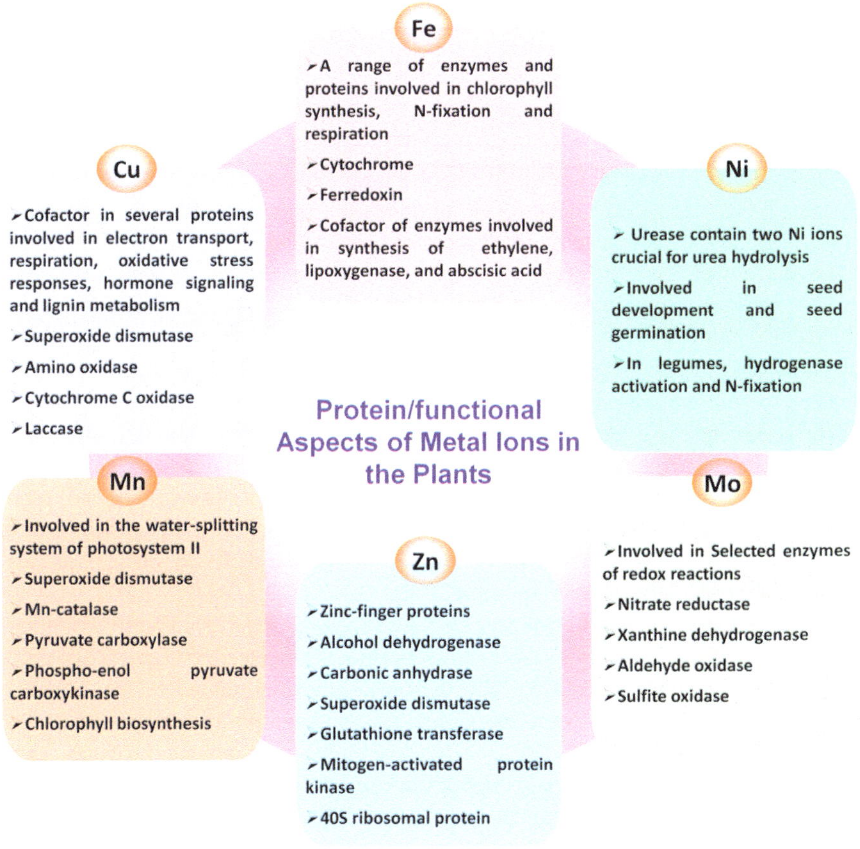

Fig. 13.1 Role of micronutrient-metal ions (*Fe*, *Ni*, *Cu*, *Mn*, *Zn*, and *Mo*) in various plant protein/processes briefly explained

micronutrient-metals in plants are well studied and reviewed in earlier reports (Fe, Rout and Sahoo 2015; Ni, Seregin and Kozhevnikova 2006; Cu, Yruela 2005; Mn, Millaleo et al. 2010; Zn, Broadley et al. 2007; and Mo, Kaiser et al. 2005). Some metal elements are an integral part of redox reactions that are fundamental to energy conversion processes and cellular activities. For example, Fe is vital component of cytochromes, catalase, and ferredoxin. Cu is a component of proteins involved in electron transfer chain in photosynthesis and respiration. Mn plays a key role in photosynthesis specifically in oxygen evolution. Zn is a vital component in structure and function of several proteins as well as enzymes.

13.4 Metal Uptake and Translocation in Plants

Metals are accessible to plants in air, water, or soil. Plants can take up metal ions through any of the media depending on the environment, metal type, and the nature of plant species. In addition to the metal uptake, plants also release metals back into the environment during several physiological processes. Consequently, metal accumulation inside the plant body depends on both uptake and release of metal across the plant body (Greger 1999).

Although some nutrients such as NH_3, SO_2, and NO_2 can be absorbed in the gaseous form via stomata and used directly by leaves (Yang et al. 2006; Vallano and Sparks 2008), plants take most of the minerals through the roots. Due to charged state of metal ions, they cannot make an easy entry into the plant roots. To facilitate the metal movement from soil to root surface and then penetration of cell wall plus cell membrane into root tissues, plants require several membrane trafficking proteins that can transport metal ion. These proteins are commonly known as transporters that consist of an extracellular domain to carry metal ion during transport across the cell membrane (Tak et al. 2013). There are numbers of metal transporter families of proteins in plant genomes suggesting highly developed metal homeostasis mechanisms to adjust with dynamic changes in metal availability (Migeon et al. 2010). Metal transporter families are divided into two categories by functional properties: families involved in metal sequestration into the plant cell, i.e., influx system and families involved in metal transportation across the different plant parts, i.e., efflux system (Colangelo and Guerinot 2006).

The members of metal transporter families involved in influx and efflux in plants have reviewed in the past (Colangelo and Guerinot 2006; DalCorso et al. 2014). ZIP (ZRT zinc-regulated transporter, IRT-like protein, iron-regulated transporter) and the NRAMP (natural resistance-associated macrophage protein) family members are major players in influx and efflux system of plants metal uptake (DalCorso et al. 2014). Besides, yellow stripe and copper transporter families are important in metal distribution and compartmentalization in different tissues. The metal efflux system consists of the cation diffusion facilitator family, the multidrug and toxic compound extrusion family, the cation exchanger family, heavy-metal-transporting ATPases, the plant cadmium resistance family, and ferroportin families.

Metals are first taken into the root apoplast (passive entry) and then into the cell wall (through adsorption). Several strategies have been known to contribute metal absorption in plant rhizospheres such as the secretion of chelating molecules, acidification, and production of high-affinity metal transporters (Rauser 1999). Plants take almost all the metal ions in the cationic forms except Mo (Greger 1999). Once entered in the apoplast, metal needs to be taken actively into the symplast because of the Casparian strip in the root stele (Chen et al. 2011). Functional roles of several metal-transporter family members regarding substrate affinity are far more complex, and ongoing research is exploring novel facets of complex functioning at the molecular level (Rajkumar et al. 2013; Akhter et al. 2012; Visioli et al. 2015; Ma et al. 2016).

Absorbed metal ions can be sequestered by root cells in the apoplast or translocated into the root stele. Then the metal ions loaded into the xylem to get transported to the shoot portion are promoted by transpiration process (DalCorso et al. 2014). The phloem transport of metals is not an easy process because the phloem tissue consists of several metal-interacting components. To avoid metal intrusion in xylem, plant systems chelate the metal ions with phytochelatins (metallothioneins, amino acids, phytosiderophores, nicotianamine, organic acids) during translocation and distribution. Chelation helps to evade an oxidative stress. The transport through xylem tissue, long-distance mobilization, and distribution into different tissues in aerial parts such as vacuole, chloroplast, and mitochondria is not yet well known. Also, unlike some microbes such as *Helicobacter*, studies focused on storage of essential metal ions in plants are limited, and further research is required (Shelake et al. 2017).

13.5 Metal Toxicity and Defense Responses in Plants

13.5.1 Metal Toxicity in Plants

In most plants, an average amount of each heavy metal is variable and consists of 50, 10, 200, 0.05, 0.1, 1, and 1.5 (μg per g of weight) for Zn, Cu, Mn, Cd, As, Pb, and Ni, respectively (Van der Ent et al. 2013). Metal-induced toxicity can result from any of the following three modes: replacing crucial components in biomolecules such as proteins and enzymes, blocking functions of biomolecules, modifying structure or function cellular components activating reactive oxygen species (ROS) production and stress signaling in plants. Production of ROS in plant cell induces a series of reactions that produces toxic chemicals, for instance, hydrogen peroxide, superoxide, hydroxyl radicals, and singlet oxygen causing protein denaturation, lipid peroxidation, and DNA mutations, and also impairs the chlorophyll function by damaging pigments (Sudo et al. 2008). Plants have developed effective antioxidant systems to scavenge the ROS toxicity induced by abiotic or biotic stress (Skorzynska-Polit et al. 2010).

13.5.2 Plant-Defense Responses

Plants have evolved many mechanisms to deal with metal toxicity comprising strategies either to avoid or tolerate the metal stress. Avoidance strategies include immobilization of metal or by altering soil pH to reduce metal solubility by excreting several chemicals such as plant exudates, organic acids, etc. (Costa et al. 1997; Yang et al. 2001). The tolerance mechanisms include the production of phytochelatins (PCs), metallothioneins (MTs), and organic acids and compartmentalizing these complexes within metabolically inactive sites, such as the vacuole (Rajkumar et al. 2013; Akhter et al. 2012). Another such strategy followed by plants is to regulate the influx of metal ion via lowering the transporter activity or extruding ions from inside the cell to outside environment. Two major compounds synthesized by plants to deal the metal ion toxicity are PCs and MTs. Even though both are cysteine-rich polypeptides, the main difference is in the synthesis process (Joshi et al. 2016). PCs are synthesized by enzymatic process, whereas MTs are proteins encoded by genes and produced by translation.

13.5.2.1 Phytochelatins

The plants synthesize PCs from reduced glutathione without translation, a group of heavy-metal chelating molecules in plants, and transpeptidation reaction is catalyzed by the PC synthase enzyme (Anjum et al. 2014). PCs are structurally related to glutathione (GSH) and highly similar in chemical configuration with GSH. GSH or related compounds are the main substrate for biosynthesis of PC (Zenk 1996; Joshi et al. 2016). The γ-Glu-Cys dipeptidyl transpeptidase (EC 2.3.2.15), also known as PC synthase (tetramer with molecular mass of 95,000), was originally identified by Grill et al. (1989) in *Silene cucubalus*, and then the enzyme activity was reported in several other plants like pea, *Arabidopsis*, and tomato (Klapheck et al. 1995; Howden et al. 1995; Chen et al. 1997).

PCs play a central role in the inactivation of metal ions and form PC-metal complexes to store in vacuoles. Additionally, they provide some of the essential metal ions such as Cu and Zn to apoenzymes or nucleic acid structures, like zinc fingers (Pinter and Stillman 2014).

13.5.2.2 Metallothioneins

The MTs are Cys-rich low-molecular-weight (4–8 kDa), highly heterogeneous metal-binding proteins encoded by a specific gene family. MTs are ubiquitously found in bacteria, plants, and animals (Koszucka and Dąbrowska 2006). First plant metallothionein was reported in 1983 from wheat embryos (Hanley-Bowdoin and Lane 1983) and further characterized in 1987. Since then several types of plant MTs have been reported to have several functional roles in plants such as metal ion homeostasis, detoxification, metal storage, and also in the case of protection against oxidative stress (Blindauer and Leszczyszyn 2010; Hassinen et al. 2011).

Plant MTs consist of four distinct subgroups (Type 1, Type 2, Type 3, and Type 4) depending on the Cys arrangement in amino acid sequence (Cobbett and Goldsbrough 2002). Specific details about distribution, diversity, and structural and functional features of plant MTs are well documented in recent reports (Freisinger 2011; Leszczyszyn et al. 2013; Joshi et al. 2016). MTs are mainly involved in metal homeostasis, but their production in plants is not limited to heavy-metal toxicity. MTs expression is found to be altered in various biotic and abiotic stresses suggesting a complex role for MTs in plant defense and not only in metal stress (Joshi et al. 2016).

13.6 Plant-Microbe-Metal Interactions

Multiple factors affect heavy-metal tolerance in plants when exposed to alleviated bioavailable metal concentrations. Thus, it is difficult to elucidate the precise mechanisms responsible for tolerance of each metal ion. To improve plant tolerance under metal stress and bioremediation, two promising strategies are emerging for revegetation of heavy-metal-contaminated sites; first is the plant growth enhancement, and second is the reduced metal translocation, both of which can achieve by soil amendments and/or inoculations of plant-associated microbes.

In general, plants have the ability to choose their root microflora from surroundings soil, and each plant species has a distinctive group of associated microbes (Hartmann et al. 2009). We know that the bacteria communicate with each other via diffusible chemical signals, commonly known as quorum sensing (Greenberg 1997). Also, plants can interact with soil microorganisms and form symbiotic relationship such as N-fixation. This information about microbe-microbe and plant-microbe interaction suggests that the plant and microbes have evolved signaling mechanisms to sense and act in response to each other. Plants also excrete several chemicals through the root, known as root exudates that comprise sugars, amino acids, and organic acids (Bayliss et al. 1997; Penrose and Glick 2001). These root exudates are a rich source of nutrients for soil microbes that could be one of the major factors for higher population number of microbes in the rhizosphere. Root exudates can alter the metal availability through metal mobility, solubility, and bioavailability in soil (Chiang et al. 2011; Luo et al. 2014). In turn, microbes, primarily PGP rhizobacteria, can improve plant growth by producing phytohormones (auxins, gibberellins, ABA, cytokinins), vitamins, enzymes, siderophores, and antibiotics (Noordman et al. 2006; Ahmad et al. 2008; Ma et al. 2016).

Plant-microbe interactions have been investigated with reference to plant-pathogen interactions (Tak et al. 2013). Plant and microbes have been studied independently under metal stress. The microbes are the oldest life on the Earth with different habitats. They have characteristics of utilization of metal from soil; hence,

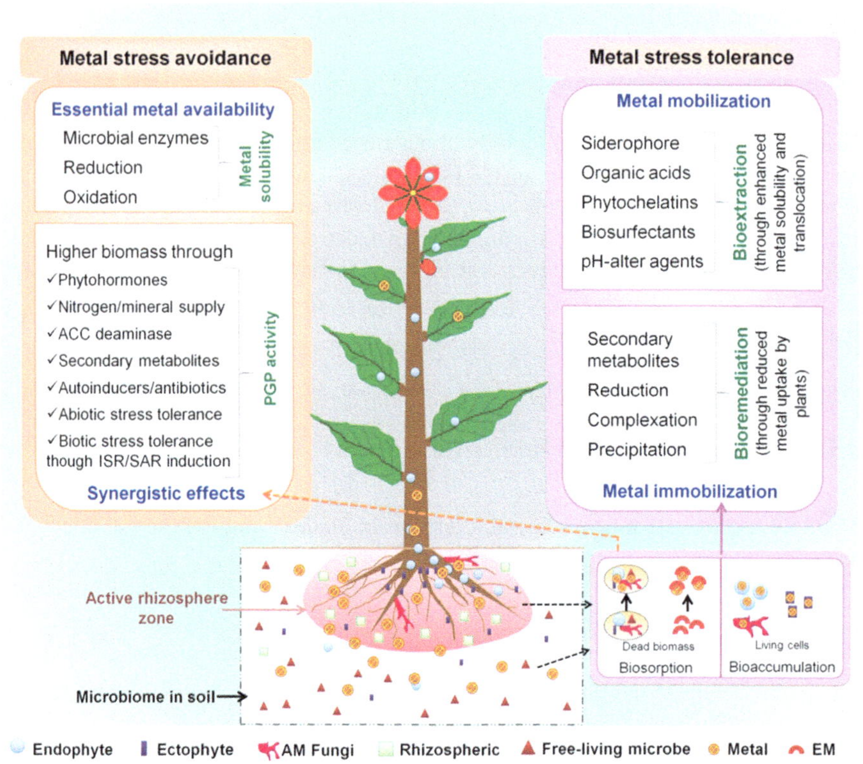

Fig. 13.2 Plant-microbe-metal interactions and possible mechanisms for direct or indirect detoxification of metals are summarized. *EM* extracellular substances released by microbes, *ACC* 1-ami nocyclopropane-1-carboxylate, *AM fungi* arbuscular mycorrhizal fungi, *ISR/SAR* systemic acquired resistance/induced systemic resistance

they can be best research materials to study the resistance mechanism against metal. Until the last two decades, the understanding of plant-microbe interactions under metal stress was limited. Recent reports on the microbial application in reclamation of metal-polluted soils provide more ideas about plant-microbe-metal interactions (Nies 1999). PGP microorganisms such as rhizobacteria including free-living microbes, endophytes, and ectophytes have evolved several mechanisms to tackle metal toxicity and succeed to have normal physiological growth in polluted sites (Fig. 13.2). These mechanisms may include one of the following ways: exclusion of metal from metabolic pathways, extrusion through cells, immobilization with metal-binding proteins or chelating agents, oxidation or reduction of metal to convert into less toxic form, demethylation, and methylation (Tak et al. 2013).

13.7 Application of Plant Microbiome in Metal Stress Management

Plant-associated microbes have a huge potential to improve direct extraction or stabilization of toxic metal ions in contaminated soils. In Fig. 13.2, the mechanisms of metal tolerance, PGP activity, and the probable metal transforming abilities of the plant-microbe-metal ion interactions presented. Also, recent data about beneficial plant-microbe interactions under metal stress conditions summarized in Table 13.2. Plant-microbial association can be exploited to improve tolerance and remediation efficiency by phytostabilization of toxic metals not only in polluted soil; it can also be useful for cleanup of water bodies (Rajkumar et al. 2013). The metal detoxification mechanisms can be divided into two broad categories (Fig. 13.2). First one includes the strategies that allow plants to evade the metal stress though higher biomass production or effective uptake of essential nutrients. The second category involves the strategies that help the plants to tolerate metal stress by either immobilizing the toxic metal or enhancing the solubility, mobility, and translocation of metal during the process of detoxification.

Use of plant-microbe interactions for detoxification of metal ions including other pollutants is defined as bioremediation (USDA 2001), but sometimes these processes are not clear enough to assign them to a specific category. Definitions of some important concepts/mechanisms and their relevance to plant-microbe-metal interactions are provided here (Table 13.3).

13.7.1 Plant Growth-Promoting (PGP) Microbes and Metal Stress in Plants

In some cases, plant species benefit from the symbiotic relationship with microbes, producing a higher amount of biomass owing to metal-stress-induced physiological changes. For example, a symbiotic association of nitrogen fixer *Frankia* sp. with alder (*Alnus glutinosa* L. Gaertn.) plants has shown a promising plant-microbe-metal interaction that allows the plants to grow in the presence of heavy metal by producing higher biomass. This strategy of *Frankia*-alder interaction can allow recultivating the metal-contaminated sites with little risk of metal dispersal such as Cu, Ni, Zn, Pb, and Cd (Belanger et al. 2015). In some studies about plant-associated microbes, PGP activity is a direct or indirect result of metal-induced response mechanisms to retain relatively normal growth by plants, possibly through siderophore and phytohormone production. For instance, *Cicer arietinum* L. associated *Pseudomonas aeruginosa* strain OSG41 under Cr stress (Oves et al. 2013), and *Hibiscus cannabinus* associated *Enterobacter* sp. strain EG16 under Cd stress (Chen et al. 2016b).

Table 13.2 Beneficial plant-microbe-metal interactions reported in recent years are summarized

Plant	Microbe	Metal	References
Plant uptake, phytoextraction, phytostabilization			
Medicago sativa	*Rhizophagus irregularis*	Cd, Ni	Mnasri et al. (2017)
Maize	AM fungi	Hg	Kodre et al. (2017)
Solanum nigrum	Endophytic strains RSF-4L and RSF-6L	Cd	Khan et al. (2017)
Noccaea tymphaea and *Alyssum murale*	68 rhizobacterial strains	Ni	Durand et al. (2016)
Brassica juncea, *Lupinus albus*	13 indigenous bacterial strains	As, Hg	Franchi et al. (2016)
Arthrocnemum macrostachyum	48 different cultivable isolates from the aerial part, roots, and rhizosphere	As, Cu, Pb, Zn	Navarro-Torre et al. (2016)
Phytolacca americana	*Bacillus* sp., *Sphingomonas* sp., *Pantoea* sp.	Mn	Zhang et al. (2015a)
Intercropping *Sedum alfredii* with *Brassica napus*	*Acinetobacter calcoaceticus*	Cd	Chen et al. (2015)
Lens culinaris	*Agrobacterium tumefaciens*, *Rahnella aquatilis*, *Pseudomonas* sp.	Pb	Jebara et al. (2015)
Eruca sativa	*Pseudomonas putida* (ATCC 39213)	Cd	Kamran et al. (2015)
Prosopis juliflora	5 rhizospheric, 21 endophytic bacteria (mainly *Bacillus*, *Staphylococcus*, *Aerococcus*)	Cr, Cd, Cu, Pb, Zn	Khan et al. (2015)
Helianthus annuus	Root endophytic strains isolated from *Agrostis capillaris*	Cu	Kolbas et al. (2015)
Medicago lupulina	*Sinorhizobium meliloti* CCNWSX0020	Cu	Kong et al. (2015)
Solanum nigrum	*Glomus versiforme*	Cd	Liu et al. (2015a)
Sedum plumbizincicola	9 Cd-tolerant PGP bacteria isolated from roots	Cd	Liu et al. (2015b)
Brassica juncea, *Ricinus communis*	*Psychrobacter* sp. SRS8 (rhizosphere), *Pseudomonas* sp. A3R3 (endophytic)	Ni, Zn, Fe	Ma et al. (2015)
Spartina maritima	25 endophytic bacterial strains	As, Cd, Cu, Ni, Pb, Zn	Mesa et al. (2015)
Sedum plumbizincicola	*Trichoderma reesei* FS10-C	Cd	Teng et al. (2015)
Noccaea caerulescens	*Arthrobacter*, *Microbacterium* sp.	Ni	Visioli et al. (2015)

(continued)

Table 13.2 (continued)

Plant	Microbe	Metal	References
Boehmeria nivea (L.) Gaud	*Pseudomonas aeruginosa* ATCC 9027	Cd	Xie et al. (2015)
Elsholtzia splendens	*Pseudomonas putida* CZ1	Cu	Xu et al. (2015)
Pennisetum	*Bacillus megaterium 1Y31*	Mn	Zhang et al. (2015b)
Salix variegata Franch.	Several root endophytic fungi	Cd	An et al. (2015)
Miscanthus sinensis	*Pseudomonas koreensis* AGB-1	As, Cd, Cu, Pb, Zn	Babu et al. (2015)
Pinus sylvestris	*Trichoderma* sp. PDR1-7	Pb	Babu et al. (2014)
Sedum alfredii	*Sphingomonas* SaMR12	Cd	Chen et al. (2014)
Pteris cretica	*Pseudomonas aeruginosa*	As	Jeong et al. (2014)
Several dicot and cotyledon plants	*Pseudomonas sp. PDMZnCd2003*	Zn, Cd	Taboonma et al. (2014)
Oryza sativa L.	Endophytic microbes isolated from *Sedum alfredii* H.	Zn	Wang et al. (2014)
Alnus firma	*Bacillus thuringiensis* GDB-1	Pb	Babu et al. (2013)
Sedum plumbizincicola	*Phyllobacterium myrsinacearum* RC6b	Cd, Zn, Pb	Ma et al. (2013)
Sinapis alba L.	3 *Pseudomonas putida* and 2 *P. fluorescens* strains	Zn, Cd, Cu	Płociniczak et al. (2013)
Brassica juncea, *Luffa cylindrica*, *Sorghum halepense*	*Bacillus megaterium* SR28C	Ni	Rajkumar et al. (2013)
Metal-stress tolerance through PGP activity			
Capsicum annuum	*Neorhizobium huautlense* T1–17	Cd, Pb	Chen et al. (2016a)
Hibiscus cannabinus	*Enterobacter* sp. strain EG16	Cd	Chen et al. (2016b)
Lonicera japonica	AM fungi	Cd	Jiang et al. (2016)
Wheat cultivars UP-2565, Kalyan Sona S-227	*Dietzia Maris*, *Lysinibacillus* strains	Cd	Gusain et al. (2017)
Hieracium pilosella, *Medicago sativa*	Combination of AMF and N_2-fixing bacteria	Zn, Pb	Ogar et al. (2015)
Alnus glutinosa L. Gaertn.	*Frankia alni* ACN14a	Cu, Ni, Zn, Pb, Cd	Belanger et al. (2015)
Agrostis capillaris, *Festuca rubra*	21 bacterial strains isolated from rhizosphere	Cu, Zn	Nicoară et al. (2014)

(continued)

Table 13.2 (continued)

Plant	Microbe	Metal	References
Calopogonium mucunoides	AM fungus *Glomus etunicatum*	Pb	Souza et al. (2014)
Sorghum bicolor subsp. Drummondii	*Pseudomonas fluorescens* JH 70-4	Pb	Shim et al. (2014)
Cicer arietinum L.	*Pseudomonas aeruginosa* strain OSG41	Cr	Oves et al. (2013)
Sorghum bicolor L., *Phytolacca acinosa*, *Solanum nigrum* L.	*Bacillus* sp. SLS18	Mn, Cd	Luo et al. (2012)
Arabidopsis thaliana	*Geobacillus*, *Ralstonia*, *Bacillus*, *Sphingomonas*, *Burkholderia* sp.	Cd	Remans et al. (2012)

Table 13.3 Important processes involved in metal stress tolerance and remediation of toxic metal ions. Even though some of the terms mentioned are used to describe broad ideas, but here those processes are explained in relation to metal

Process/mechanism	Details
Precipitation	Enzymatic processes induced by plant or microbial compounds that convert toxic metal form to insoluble precipitate
Chelation	Scavenging of metal ions by chemical compounds produced by plant or microbes and making it unavailable for other chemical interactions
Bioleaching	Microbial dissolution of metals from their mineral source through various metabolic processes
Biosorption	Immobilization of metals in dead or alive biomass
Bioaccumulation	Metal accumulation inside living cells
Bioexclusion	Export of toxic metal ions from cell cytoplasm through active efflux systems
Complexation	Metal biosorption by substances secreted by plant-associated microbes in extracellular environment
Phytovolatilization	The use of plants to absorb and volatilize metal ions into atmosphere
Phytofiltration	Adsorption or absorption of heavy metals in plant roots (rhizofiltration) or seedlings (blastofiltration) from water
Phytotransformation	Chemical modification of organic or complex forms into less toxic or nontoxic forms by chemicals produced by plants
Hyperaccumulator	Plants capable to grow in heavy-metal contaminated soils and accumulate comparatively higher amount of toxic metals in their tissues

Chemicals produced by root-colonizing endophytes can alter the bioavailable metal levels in rhizosphere through diverse biogeochemical mechanisms like immobilizing toxic metal ions, transformation, translocation, chelation by chemical compounds, solubilization of unavailable forms into bioavailable one, precipitation, and volatilization (Rajkumar et al. 2013). Such processes facilitate the metal uptake in plants directly conferring metal stress tolerance. Also, owing to induced signaling mechanisms under metal stress in plants results in higher biomass production indirectly. Even though siderophores have a high affinity for Fe, they also possess high affinity for several other metal ions such as Zn Cd, Ga, Al, Cu, and Pb (Schalk et al. 2011). An example of plant metal uptake induced by endophytes includes Mn-resistant endophytic bacteria isolated from a Mn-hyperaccumulator species *Phytolacca americana* (Zhang et al. 2015a). Higher Mn uptake and biomass production were observed in *P. americana* inoculated with Mn-resistant bacterial strains.

13.7.2 Plant-Microbe-Metal Interactions and Phytoremediation

Application of living plants for the remediation of contaminated sites is called as phytoremediation. Sometimes, phytoremediation and bioremediation terms are used alternatively due to the closeness of microbes with plants and their influence on plant growth and development (Reichenauer and Germida 2008). Also, phytoremediation is not only the remediation of heavy metals but it also includes the other toxic substances such as organic pollutants and petroleum products. This process is subdivided into phytoextraction and phytostabilization. The phytoextraction and phytostabilization processes can be improved greatly by plant-associated microbes through several biochemical processes either solubilizing or immobilize metal ions (Table 13.3). Sometimes these terms are alternatively used to explain the same concept governing metal detoxification.

13.7.2.1 Phytoextraction

The phytoextraction is a branch of phytoremediation and comprises the use of plants to extract metals from the soil and transport them in different plant parts including roots, shoots, and leaves (Garbisu and Alkorta 2001). Plant-associated microbes have been reported to enhance the metal translocation and promote the phytoextraction process (Rajkumar et al. 2013). For example, a metal-resistant rhizobacterial strain *Pseudomonas putida* CZ1 isolated from *Elsholtzia splendens* induces PGP activity in response to Cu stress and promotes the Cu accumulation and root-to-shoot translocation (Xu et al. 2015).

Two bacterial strains of *Variovorax paradoxus* isolated from the rhizosphere of *Noccaea tymphaea* and *Alyssum murale* showed improved Ni phytoextraction (Durand et al. 2016). Plants inoculated with mesocosms of *V. paradoxus* showed higher Ni accumulation in roots and shoots. AM fungi boost the phytoextraction potential of the host (*Medicago sativa*) as well as nonhost (*Sesuvium portulacastrum*) plants under Ca and Ni stress conditions through stimulated absorption in roots and translocation to shoots (Mnasri et al. 2017).

13.7.2.2 Phytostabilization

The plant applications for reduced bioavailability of pollutants define as phytostabilization (Garbisu and Alkorta 2001). Unlike phytoextraction, this process focuses on immobilization of metal ions in soil and not in plant body that reduces its bioavailability to other living forms (Mendez and Maier 2008). A broader definition of phytostabilization also includes the vegetation cover of contaminated sites as a way to prevent the spread of pollutants to other places (Bolan et al. 2011). Plant-mediated immobilization of metals in soil generally achieve by releasing root exudates. Microbial role in metal immobilization in soil and improved phytostabilization is well established.

One of the examples of phytostabilization is *Miscanthus sinensis* associated endophyte *Pseudomonas koreensis* AGB-1 in heavy-metal-contaminated soil (Babu et al. 2015). The *M. sinensis* is a mine-tailing growing plant species contaminated with several toxic metals such as As, Cd, Cu, Pb, and Zn. Fungal endophytes are also good candidates for improving metal stress such as RSF-4 L and RSF-6 L strains isolated from leaves of Cd-hyperaccumulator *Solanum nigrum* (Khan et al. 2017). An assessment of plant physiological responses and Cd accumulation in inoculated plants showed higher biomass due to PGP activity of fungal endophytes even without accumulating Cd by microbe. Fungal endophytes protected the host plants by altering oxidative stress responsive enzyme activities (lower peroxidase and polyphenol peroxidase activities, high catalase activity) under Cd stress. This study suggests the right combination of PGP microbes and heavy-metal hyperaccumulator plants can enhance phytostabilization or immobilization ability of host plants.

13.8 Conclusion and Future Trends

In this chapter, we have given a brief introduction about the role of metal ions in plant life and how altered bioavailable concentration is dangerous for the environment, human, and plants. We address the natural and anthropogenic factors influencing the bioavailability of heavy metals to the living cells in the environment. Plants and microbes have evolved several defense mechanisms to tackle the changing metal concentrations and to escape or tolerate the higher metal content. Abiotic stress in plants induced by higher metal levels can be countered by the use of plant microbiome altering phyto-availability or enhancing plant-defense responses. Metal-contaminated sites can be used for revegetation through PGP and metal-tolerant microbes by providing key nutrients required for plant growth.

Finally, further laboratory studies on the appropriate combination of plant and microbe than can be used for reclamation of metal-contaminated sites is important for field application and phytoremediation. The focus of future research on the following points may be valuable for effective utilization of plant-microbe-metal interactions for phytoremediation and management of abiotic stress induced by metals in the plant:

1. Research on the elucidation of molecular mechanisms of the plant-microbe interactions under metal stress is required.
2. Finding the right combination of plant-microbes for enhanced PGP activity for specific sites is recommended.
3. Conversion of lab technology or pilot studies into practical application at field level will be encouraging to further applied research.

Maintenance of up-to-date genetic and genomic data of plant microbiome and its accessibility to all microbiologist and plant scientists is vital.

References

Adriano DC (1986) https://doi.org/10.1002/food.19870310321

Ahmad F, Ahmad I, Khan MS (2008) Screening of free-living rhizospheric bacteria for their multiple plant growth promoting activities. Microbiol Res 163:173–181

Akhter MF, McGarvey B, Macfie SM (2012) Reduced translocation of cadmium from roots is associated with increased production of phytochelatins and their precursors. J Plant Physiol 169:1821–1829

Ali M, Chernova TA, Newnam GP, Yin L, Shanks J, Karpova TS, Lee A, Laur O, Subramanian S, Kim D, McNally JG, Seyfried NT, Chernoff YO, Wilkinson KD (2014) Stress-dependent proteolytic processing of the actin assembly protein Lsb1 modulates a yeast prion. Biol Chem 289(40):27625–27639

An H, Liu Y, Zhao X, Huang Q, Yuan S, Yang X, Dong J (2015) Characterization of cadmium-resistant endophytic fungi from *Salix variegata* Franch. in three Gorges Reservoir Region, China. Microbiol Res 176:29–37

Anjum NA, Gill SS, Gill R, Hasanuzzaman M, Duarte AC, Pereira E, Ahmad I, Tuteja R, Tuteja N (2014) Metal/metalloid stress tolerance in plants: role of ascorbate, its redox couple, and associated enzymes. Protoplasma 251:1265–1283

Babu AG, Kim JD, Oh BT (2013) Enhancement of heavy metal phytoremediation by *Alnus firma* with endophytic *Bacillus thuringiensis* GDB-1. J Hazard Mater 250-251:477–483

Babu AG, Shea PJ, Oh BT (2014) *Trichoderma* sp. PDR1-7 promotes *Pinus sylvestris* reforestation of lead-contaminated mine tailing sites. Sci Total Environ 476–477:561–567

Babu AG, Shea PJ, Sudhakar D, Jung IB, Oh BT (2015) Potential use of *Pseudomonas koreensis* AGB-1 in association with *Miscanthus sinensis* to remediate heavy metal (loid)-contaminated mining site soil. J Environ Manag 151:160–166

Bayliss C, Bent E, Culham DE, MacLellan S, Clarke AJ, Brown GL, Wood JM (1997) Bacterial genetic loci implicated in the *Pseudomonas putida* GR12-2R3-canola mutualism: identification of an exudate-inducible sugar transporter. Can J Microbiol 43:809–818

Belanger PA, Bellenger JP, Roy S (2015) Heavy metal stress in alders: tolerance and vulnerability of the actinorhizal symbiosis. Chemosphere 138:300–308

Bhattacharya A, Routh J, Gunnar J, Bhattacharya P, Mörth M (2006) Environmental assessment of abandoned mine tailings in Adak, Västerbotten district (Northern Sweden). Appl Geochem 21:1760–1780

Blindauer CA, Leszczyszyn OI (2010) Metallothioneins: unparalleled diversity in structures and functions for metal ion homeostasis and more. Nat Prod Rep 27:720–741

Bolan NS, Park JH, Robinson B, Naidu R, Huh KY (2011) Phytostabilization: a green approach to contaminant containment. In: Donald LS (ed) . Adv, Academic Press, Agron, pp 145–204
Broadley MR, White PJ, Hammond JP, Zelko I, Lux A (2007) Zinc in plants. New Phytol 173:677–702
Candeias C, Melo R, Ávila PF, da Silva EF, Salgueiro AR, Teixeira JP (2014) Heavy metal pollution in mine-soil-plant system in S. Francisco de Assis–Panasqueira mine (Portugal). Appl Geochem 44:12–26
Chen J, Zhou J, Goldsbrough PB (1997) Characterization of phytochelatin synthase from tomato. Physiol Plant 101:165–172
Chen T, Cai X, Wu X, Karahara I, Schreiber L, Lin J (2011) Casparian strip development and its potential function in salt tolerance. Plant Signal Behav 6:1499–1502
Chen B, Zhang Y, Rafiq MT, Khan KY, Pan F, Yang X, Feng Y (2014) Improvement of cadmium uptake and accumulation in *Sedum alfredii* by endophytic bacteria *Sphingomonas* SaMR12: effects on plant growth and root exudates. Chemosphere 117:367–373
Chen B, Ma X, Liu G, Xu X, Pan F, Zhang J, Tian S, Feng Y, Yang X (2015) An endophytic bacterium *Acinetobacter calcoaceticus* Sasm3-enhanced phytoremediation of nitrate-cadmium compound polluted soil by intercropping *Sedum alfredii* with oilseed rape. Environ Sci Pollut Res 22:17625–17635
Chen L, He L y, Wang Q, Sheng X f (2016a) Synergistic effects of plant growth-promoting *Neorhizobium huautlense* T1-17 and immobilizers on the growth and heavy metal accumulation of edible tissues of hot pepper. J Hazard Mater 312:123–131
Chen Y, Chao Y, Li Y, Lin Q, Bai J, Tang L, Wang S, Ying R, Qiu R (2016b) Survival strategies of the plant-associated bacterium *Enterobacter* sp. strain EG16 under cadmium stress. Appl Environ Microbiol 82:1734–1744
Chetia M, Chatterjee S, Banerjee S, Nath MJ, Singh L, Srivastava RB, Sarma HP (2011) Groundwater arsenic contamination in Brahmaputra river basin: a water quality assessment in Golaghat (Assam), India. Environ Monit Assess 173:371–385
Chiang PN, Chiu CY, Wang MK, Chen BT (2011) Low-molecular-weight organic acids exuded by Millet (*Setaria italica* (L.) Beauv.) roots and their effect on the remediation of cadmium-contaminated soil. Soil Sci 176:33–38
Cobbett C, Goldsbrough P (2002) Phytochelatins and metallothioneins: roles in heavy metal detoxification and homeostasis. Annu Rev Plant Biol 53:159–182
Colangelo EP, Guerinot ML (2006) Put the metal to the petal: metal uptake and transport throughout plants. Curr Opin Plant Biol 9:322–330
Costa G, Michaut JC, Guckert A (1997) Amino acids exuded from axenic roots of lettuce and white lupin seedlings exposed to different cadmium concentrations. J Plant Nutr 20:883–900
DalCorso G, Manara A, Piasentin S, Furini A (2014) Nutrient metal elements in plants. Metallomics 6:1770–1788
Duffus JH (2002) "Heavy metals" a meaningless term? (IUPAC technical report). Pure Appl Chem 74:793–807
Durand A, Piutti S, Rue M, Morel JL, Echevarria G, Benizri E (2016) Improving nickel phytoextraction by co-cropping hyperaccumulator plants inoculated by plant growth promoting rhizobacteria. Plant Soil 399:179–192
Fergusson JE (1990) The heavy elements: chemistry, environmental impact and health effects. Pergamon Press, Oxford
Franchi E, Rolli E, Marasco R, Agazzi G, Borin S, Cosmina P, Pedron F, Rosellini I, Barbafieri M, Petruzzelli G (2016) Phytoremediation of a multi contaminated soil: mercury and arsenic phytoextraction assisted by mobilizing agent and plant growth promoting bacteria. J Soils Sediments 1–13
Freisinger E (2011) Structural features specific to plant metallothioneins. J Biol Inorg Chem 16:1035–1045
Garbisu C, Alkorta I (2001) Phytoextraction: a cost-effective plant-based technology for the removal of metals from the environment. Bioresour Technol 77:229–236

Georgeaud VM, Rochette P, Ambrosi JP, Vandamme D, Williamson D (1997) Relationship between heavy metals and magnetic properties in a large polluted catchment: the Etang de Berre (south of France). Phys Chem Earth 22(1–2 Spec Iss):211–214

Greenberg EP (1997) Quorum sensing in Gram-negative bacteria. ASM News 63:371–377

Greger M (1999) Metal availability and bioconcentration in plants. In: Heavy metal stress in plants. Springer, Berlin/Heidelberg, pp 1–27

Grill E, Loffler S, Winnacker E-L, Zenk MH (1989) Phytochelatins, the heavy-metal-binding peptides of plants, are synthesized from glutathione by a specific γ-glutamylcysteine dipeptidyl transpeptidase (phytochelatin synthase). Proc Natl Acad Sci U S A 86:6838–6842

Gupta VK, Nayak A, Bhushan B, Agarwal S (2015) A critical analysis on the efficiency of activated carbons from low-cost precursors for heavy metals remediation. Crit Rev Environ Sci Technol 45:613–668

Gusain P, Paliwal R, Singh V (2017) Rhizoremediation of cadmium-contaminated soil associated with hydroxamate siderophores isolated from Cd-resistant plant growth-promoting *Dietzia maris* and *Lysinibacillus* strains. Int J Phytoremediation 19:290–299

Hanley-Bowdoin L, Lane BG (1983) A novel protein programmed by the mRNA conserved in dry wheat embryos. Eur J Biochem 135:9–15

Hartmann A, Schmid M, Van Tuinen D, Berg G (2009) Plant-driven selection of microbes. Plant Soil 321:235–257

Hassinen VH, Tervahauta AI, Schat H, Kärenlampi SO (2011) Plant metallothioneins-metal chelators with ROS scavenging activity? Plant Biol 13:225–232

Hinrichs M, Rymer H, Gillman M et al. (2011) Characterisation and distribution of heavy metals at Masaya volcano, Nicaragua. Paper presented at AGU Fall Meeting abstracts Dec 2011

Howden R, Goldsbrough PB, Andersen CR, Cobbett CS (1995) Cadmium-sensitive, cad1, mutants of *Arabidopsis thaliana* are phytochelatin deficient. Plant Physiol 107:1059–1066

Jarup L (2003) Hazards of heavy metal contamination. Br Med Bull 68:167–182

Jebara SH, Saadani O, Fatnassi IC, Chiboub M, Abdelkrim S, Jebara M (2015) Inoculation of *Lens culinaris* with Pb-resistant bacteria shows potential for phytostabilization. Environ Sci Pollut Res 22:2537–2545

Jeong S, Moon HS, Nam K (2014) Enhanced uptake and translocation of arsenic in Cretan brake fern (*Pteris cretica* L.) through siderophore-arsenic complex formation with an aid of rhizospheric bacterial activity. J Hazard Mater 280:536–543

Jiang Q-Y, Zhuo F, Long S-H, Zhao H-D, Yang D-J, Ye Z-H, Li S-S, Jing Y-X (2016) Can arbuscular mycorrhizal fungi reduce Cd uptake and alleviate Cd toxicity of *Lonicera japonica* grown in Cd-added soils? Sci Rep 6:21805

Joshi R, Pareek A, Singla-Pareek SL (2016) Plant metallothioneins: classification, distribution, function, and regulation. In: Ahmad P (ed) Plant metal interaction: emerging remediation techniques. Elsevier, Amsterdam, pp 239–261

Jovanovic VS, Mitić V, Mandić SN, Ilić M, Simonović S (2015) Heavy metals in the post-catastrophic soils. In: Heavy metal contamination of soils. Springer, pp 3–21

Kaiser BN, Gridley KL, Brady JN, Phillips T, Tyerman SD (2005) The role of molybdenum in agricultural plant production. Ann Bot 96:745–754

Kamran MA, Syed JH, Samas E, Munis MFH, Chaudhary HJ (2015) Effect of plant growth-promoting rhizobacteria inoculation on cadmium (Cd) uptake by *Eruca sativa*. Environ Sci Pollut Res 22:9275–9283

Khan MU, Sessitsch A, Harris M, Fatima K, Imran A, Arslan M, Shabir G, Khan QM, Afzal M (2015) Cr-resistant rhizo- and endophytic bacteria associated with Prosopis juliflora and their potential as phytoremediation enhancing agents in metal-degraded soils. Front Plant Sci 5:755

Khan AR, Ullah I, Waqas M, Park G-S, Khan AL, Hong S-J, Ullah R, Jung BK, Park CE, Ur-Rehman S, Lee I-J, Shin J-H (2017) Host plant growth promotion and cadmium detoxification in *Solanum nigrum*, mediated by endophytic fungi. Ecotoxicol Environ Saf 136:180–188

Klapheck S, Schlunz S, Bergmann L (1995) Synthesis of phytochelatins and homo-phytochelatins in *Pisum sativum* L. Plant Physiol 107:515–521

Kodre A, Arcon I, Debeljak M, Potisek M, Likar M, Vogel-Mikus K (2017) Arbuscular mycorrhizal fungi alter Hg root uptake and ligand environment as studied by X-ray absorption fine structure. Environ Exp Bot 133:12–23
Kolbas A, Kidd P, Guinberteau J, Jaunatre R, Herzig R, Mench M (2015) Endophytic bacteria take the challenge to improve Cu phytoextraction by sunflower. Environ Sci Pollut Res 22:5370–5382
Kong Z, Mohamad OA, Deng Z, Liu X, Glick BR, Wei G (2015) Rhizobial symbiosis effect on the growth, metal uptake, and antioxidant responses of *Medicago lupulina* under copper stress. Environ Sci Pollut Res 22:12479–12489
Koszucka AM, Dąbrowska G (2006) Plant metallothioneins. Adv Cell Biol 33:285–302
Leszczyszyn OI, Imam HT, Blindauer CA (2013) Diversity and distribution of plant metallothioneins: a review of structure, properties and functions. Metallomics 5:1146–1169
Liu W, Wang Q, Wang B, Hou J, Luo Y, Tang C, Franks AE (2015a) Plant growth-promoting rhizobacteria enhance the growth and Cd uptake of *Sedum plumbizincicola* in a Cd-contaminated soil. J Soils Sediments 15:1191–1199
Liu H, Yuan M, Tan S, Yang X, Lan Z, Jiang Q, Ye Z, Jing Y (2015b) Enhancement of arbuscular mycorrhizal fungus (*Glomus versiforme*) on the growth and Cd uptake by Cd-hyperaccumulator *Solanum nigrum*. Appl Soil Ecol 89:44–49
Lu SG, Bai SQ (2006) Study on the correlation of magnetic properties and heavy metals content in urban soils of Hangzhou city, China. J Appl Geophys 60:1–12
Luo S, Xu T, Chen L, Chen J, Rao C, Xiao X, Wan Y, Zeng G, Long F, Liu C, Liu Y (2012) Endophyte-assisted promotion of biomass production and metal-uptake of energy crop sweet sorghum by plant-growth-promoting endophyte *Bacillus* sp. SLS18. Appl Microbiol Biotechnol 93:1745–1753
Luo ZB, Wu C, Zhang C, Li H, Lipka U, Polle A (2014) The role of ectomycorrhizas in heavy metal stress tolerance of host plants. Environ Exp Bot 108:47–62
Ma Y, Rajkumar M, Luo Y, Freitas H (2013) Phytoextraction of heavy metal polluted soils using *Sedum plumbizincicola* inoculated with metal mobilizing *Phyllobacterium myrsinacearum* RC6b. Chemosphere 93:1386–1392
Ma Y, Rajkumar M, Rocha I, Oliveira RS, Freitas H (2015) Serpentine bacteria influence metal translocation and bioconcentration of *Brassica juncea* and *Ricinus communis* grown in multi-metal polluted soils. Front Plant Sci 5:757
Ma Y, Oliveira RS, Freitas H, Zhang C (2016) Biochemical and molecular mechanisms of plant-microbe-metal interactions: relevance for phytoremediation. Front Plant Sci 7:1–19
Mahadev SR, Hayashi H, Ikegami T, Abe S, Morita EH (2013) Improved protein overexpression and purification strategies for structural studies of cyanobacterial metal-responsive transcription factor, SmtB from marine *Synechococcus* sp. PCC 7002. Protein J 32:626–634
Mendez MO, Maier RM (2008) Phytostabilization of mine tailings in arid and semiarid environments-an emerging remediation technology. Environ Health Perspect 116:278
Mengel K, Kirkby EA (2012) Principles of plant nutrition. Springer, Berlin
Mesa J, Mateos-Naranjo E, Caviedes MA, Redondo-Gomez S, Pajuelo E, Rodriguez-Llorente ID (2015) Endophytic cultivable bacteria of the metal bioaccumulator *Spartina maritima* improve plant growth but not metal uptake in polluted marshes soils. Front Microbiol 6:1–15
Migeon A, Blaudez D, Wilkins O, Montanini B, Campbell MM, Richaud P, Thomine S, Chalot M (2010) Genome-wide analysis of plant metal transporters, with an emphasis on poplar. Cell Mol Life Sci 67:3763–3784
Millaleo R, Reyes-Díaz M, Ivanov AG, Mora ML, Alberdi M (2010) Manganese as essential and toxic element for plants: transport, accumulation and resistance mechanisms. J Soil Sci Plant Nutr 10:470–481
Mnasri M, Janoušková M, Rydlová J, Abdelly C, Ghnaya T (2017) Comparison of arbuscular mycorrhizal fungal effects on the heavy metal uptake of a host and a non-host plant species in contact with extra-radical mycelial network. Chemosphere 171:476–484
Moller A, Muller HW, Abdullah A, Abdelgawad G, Utermann J (2005) Urban soil pollution in Damascus, Syria: concentrations and patterns of heavy metals in the soils of the Damascus Ghouta. Geoderma 124:63–71

Navarro-Torre S, Mateos-Naranjo E, Caviedes MA, Pajuelo E, Rodriguez-Llorente ID (2016) Isolation of plant-growth-promoting and metal-resistant cultivable bacteria from *Arthrocnemum macrostachyum* in the Odiel marshes with potential use in phytoremediation. Mar Pollut Bull 110:133–142

Nicoară A, Neagoe A, Stancu P, de Giudici G, Langella F, Sprocati AR, Iordache V, Kothe E (2014) Coupled pot and lysimeter experiments assessing plant performance in microbially assisted phytoremediation. Environ Sci Pollut Res 21:6905–6920

Nies DH (1999) Microbial heavy metal resistance. Appl Microbiol Biotechnol 51:730–750

Noordman WH, Reissbrodt R, Bongers RS, Rademaker ILW, Bockelmann W, Smit G (2006) Growth stimulation of *Brevibacterium* sp. by siderophores. J Appl Microbiol 101:637–646

Ogar A, Sobczyk Ł, Turnau K (2015) Effect of combined microbes on plant tolerance to Zn-Pb contaminations. Environ Sci Pollut Res 22:19142–19156

Okamoto T, Kamiyama K, Wachi K (1997) Background levels of heavy metals in Kanagawa prefecture farm soils impacted by volcanic ash from Mt. Fuji. In: Ando T, Fujita K, Mae T, Matsumoto H, Mori S, Sekyia J (eds) Plant nutrition for sustainable food production and environment developments in plant and soil sciences, vol 78. Springer, Dordrecht, pp 553–554

Outten CE, O'halloran TV (2001) Femtomolar sensitivity of metalloregulatory proteins controlling zinc homeostasis. Science 292:2488–2492

Oves M, Khan MS, Zaidi A (2013) Chromium reducing and plant growth promoting novel strain Pseudomonas aeruginosa OSG41 enhance chickpea growth in chromium amended soils. Eur J Soil Biol 56:72–83

Pan Y, Li H (2016) Investigating heavy metal pollution in mining brown field and its policy implications: a case study of the Bayan Obo rare earth mine, Inner Mongolia, China. Environ Manag 57:879–893

Penrose DM, Glick BR (2001) Levels of 1-aminocyclopropane-1-carboxylic acid (ACC) in exudates and extracts of canola seeds treated with plant growth-promoting bacteria. Can J Microbiol 47:368–372

Pinter TB, Stillman MJ (2014) The zinc balance: competitive zinc metalation of carbonic anhydrase and metallothionein 1A. Biochemistry 53:6276–6285

Płociniczak T, Kukla M, Wątroba R, Piotrowska-Seget Z (2013) The effect of soil bioaugmentation with strains of *Pseudomonas* on Cd, Zn and Cu uptake by *Sinapis alba* L. Chemosphere 91:1332–1337

Pourret O, Lange B, Bonhoure J, Colinet G, Decrée S, Mahy G, Séleck M, Shutcha M, Faucon MP (2016) Assessment of soil metal distribution and environmental impact of mining in Katanga (Democratic Republic of Congo). Appl Geochem 64:43–55

Rajkumar M, Ma Y, Freitas H (2013) Improvement of Ni phytostabilization by inoculation of Ni resistant *Bacillus megaterium* SR28C. J Environ Manag 128:973–980

Rauser WE (1999) Structure and function of metal chelators produced by plants. Cell Biochem Biophys 31:19–48

Reichenauer TG, Germida JJ (2008) Phytoremediation of organic contaminants in soil and groundwater. Chem Sus Chem 1:708–717

Remans T, Thijs S, Truyens S, Weyens N, Schellingen K, Keunen E, Gielen H, Cuypers A, Vangronsveld J (2012) Understanding the development of roots exposed to contaminants and the potential of plant-associated bacteria for optimization of growth. Ann Bot 110:239–252

Rosenblueth M, Martínez-Romero E (2006) Bacterial endophytes and their interactions with hosts. Mol Plant-Microbe Interact 19:827–837

Rout GR, Sahoo S (2015) Role of iron in plant growth and metabolism. Rev Agric Sci 3:1–24

Schalk IJ, Hannauer M, Braud A (2011) New roles for bacterial siderophores in metal transport and tolerance. Environ Microbiol 13:2844–2854

Seregin IV, Kozhevnikova AD (2006) Physiological role of nickel and its toxic effects on higher plants. Russ J Plant Physiol 53:257–277

Shelake RM, Ito Y, Masumoto J, Morita EH, Hayashi H (2017) A novel mechanism of "metal gel-shift" by histidine-rich Ni^{2+}-binding Hpn protein from *Helicobacter pylori* strain SS1. PLoS One 12:e0172182

Shim J, Babu AG, Velmurugan P, Shea PJ, Oh B-T (2014) *Pseudomonas fluorescens* JH 70-4 promotes Pb stabilization and early seedling growth of sudan grass in contaminated mining site soil. Environ Technol 35:2589–2596

Skorzynska-Polit E, Drazkiewicz M, Krupa Z (2010) Lipid peroxidation and antioxidative response in *Arabidopsis thaliana* exposed to cadmium and copper. Acta Physiol Plant 32:169–175

Song Z, Williams CJ, Edyvean RGJ (2000) Sedimentation of tannery wastewater. Water Res 34:2171–2176

Souza LA, Camargos LS, Schiavinato MA, Andrade SAL (2014) Mycorrhization alters foliar soluble amino acid composition and influences tolerance to Pb in Calopogonium mucunoides. Theor Exp Plant Physiol 26:211–216

Sudo E, Itouga M, Yoshida-Hatanaka K, Ono Y, Sakakibara H (2008) Gene expression and sensitivity in response to copper stress in rice leaves. J Exp Bot 59:3465–3474

Taboonma P, Nakbanpote W, Sangdee A (2014) Resistance and plant growth-promoting properties under Zn / Cd stress of Pseudomonas sp. ZnCd 2003. In: Kumar R (ed) Proc. Intl. Conf. on future trends in bio-informatics and environmental science—FTBES. Bangkok, pp 15–20

Tak HI, Ahmad F, Babalola OO (2013) Advances in the application of plant growth-promoting rhizobacteria in phytoremediation of heavy metals. Reviews of Rev Environ Contam Toxicol 223:33–52

Teng Y, Luo Y, Ma W, Zhu L, Ren W, Luo Y, Christie P, Li Z (2015) *Trichoderma reesei* FS10-C enhances phytoremediation of Cd-contaminated soil by *Sedum plumbizincicola* and associated soil microbial activities. Front Plant Sci 9:1–10

Touati D (2000) Iron and oxidative stress in bacteria. Arch Biochem Biophys 373:1–6

Ullah A, Heng S, Munis MF, Fahad S, Yang X (2015) Phytoremediation of heavy metals assisted by plant growth promoting (PGP) bacteria: a review. Environ Exp Bot 117:28–40

United States Environmental Protection Agency (2001) Citizen's guide to bioremediation. U.S. Environmental Protection Agency, Washington, DC

Vallano DM, Sparks JP (2008) Quantifying foliar uptake of gaseous nitrogen dioxide using enriched foliar $\delta^{15}N$ values. New Phytol 177:946–955

Van der Ent A, Baker AJ, Reeves RD, Pollard AJ, Schat H (2013) Hyperaccumulators of metal and metalloid trace elements: facts and fiction. Plant Soil 362:319–334

Visioli G, Vamerali T, Mattarozzi M, Dramis L, Sanangelantoni AM (2015) Combined endophytic inoculants enhance nickel phytoextraction from serpentine soil in the hyperaccumulator *Noccaea caerulescens*. Front Plant Sci 6:1–12

Waldron KJ, Rutherford JC, Ford D, Robinson NJ (2009) Metalloproteins and metal sensing. Nature 460:823–830

Wang Y, Yang X, Zhang X, Dong L, Zhang J, Wei Y, Feng Y, Lu L (2014) Improved plant growth and Zn accumulation in grains of rice (*Oryza sativa* L.) by inoculation of endophytic microbes isolated from a Zn hyperaccumulator, *Sedum alfredii* H. J Agric Food Chem 62:1783–1791

Xie J, Liu Y, Zeng G, Liu H, Zheng B, Tang H, Xu W, Sun Z, Tan X, Nie J, Jiang Z, Gan C, Wang S (2015) The effects of *P. aeruginosa* ATCC 9027 and NTA on phytoextraction of Cd by ramie (*Boehmeria nivea* (L.) Gaud). RSC Adv 5:67509–67517

Xu C, Chen X, Duan D, Peng C, Le T, Shi J (2015) Effect of heavy-metal-resistant bacteria on enhanced metal uptake and translocation of the Cu-tolerant plant, *Elsholtzia splendens*. Environ Sci Pollut Res 22:5070–5081

Yacoub C, Pérez-Foguet A, Miralles N (2012) Trace metal content of sediments close to mine sites in the Andean region. Sci World J 2012:732519

Yang J, Mosby D, Casteel S, Blanchar R (2001) Lead immobilization using phosphoric acid in a smelter-contaminated urban soil. Environ Sci Technol 35:3553–3559

Yang L, Stulen I, De Kok LJ (2006) Impact of sulfate nutrition on the utilization of atmospheric SO_2 as sulfur source for Chinese cabbage. J Plant Nutr Soil Sci 169:529–534

Yruela I (2005) Copper in plants. Braz J Plant Physiol 17:145–156

Zahra A, Hashmi MZ, Malik RN, Ahmed Z (2014) Enrichment and geo-accumulation of heavy metals and risk assessment of sediments of the Kurang Nallah—feeding tributary of the Rawal Lake Reservoir, Pakistan. Sci Total Environ 470:925–933

Zenk MH (1996) Heavy metal detoxification in higher plants: a review. Gene 179:21–30

Zhang WH, Chen W, He LY, Wang Q, Sheng XF (2015a) Characterization of Mn-resistant endophytic bacteria from Mn-hyperaccumulator *Phytolacca americana* and their impact on Mn accumulation of hybrid penisetum. Ecotoxicol Environ Saf 120:369–376

Zhang WH, He LY, Wang Q, Sheng XF (2015b) Inoculation with endophytic *Bacillus megaterium* 1Y31 increases Mn accumulation and induces the growth and energy metabolism-related differentially-expressed proteome in Mn hyperaccumulator hybrid pennisetum. J Hazard Mater 300:513–521

Zitka O, Krystofova O, Hynek D et al (2013) Metal transporters in plants. In: Gupta DK, Corpas FJ, Palma JM (eds) Heavy metal stress in plants. Springer, Berlin, pp 19–41

Chapter 14
Potential of Endophytic Bacteria in Heavy Metal and Pesticide Detoxification

Anket Sharma, Vinod Kumar, Neha Handa, Shagun Bali, Ravdeep Kaur, Kanika Khanna, Ashwani Kumar Thukral, and Renu Bhardwaj

Abstract Heavy metal (HM) and pesticide contamination in the soil is of major concern in the present era. Both of these contaminants disturb soil microflora and adversely affect the growth and development of plants. The soil contamination can be reduced by ecofriendly techniques. The use of endophytic bacteria (EB) in the rhizosphere is one such technique where EB reduce the HM and pesticide contaminants in the soil. They can efficiently reduce the HM and pesticide concentration in the soil by enhancing the phytoremediating efficiency of plants. Moreover, EB can also degrade the pesticides in soil by producing various hormones and enzymes which ultimately result in promotion of the growth of plants. Hence, keeping in mind the efficiency of EB in reducing the HM and pesticide contamination in soil, the present review gives a detailed view of HM and pesticide detoxification by these bacteria.

Keywords Bioremediation • Endophytic bacteria • Heavy metal detoxification • Pesticide degradation • Soil contamination

14.1 Introduction

Plants are exposed to various abiotic factors like temperature, drought, heavy metals, and pesticides. Any distraction from the normal levels of these factors results in origin of stress (Kumar et al. 2008; Parvaiz et al. 2008). The indiscriminate release

A. Sharma (✉) • V. Kumar
Plant Stress Physiology Lab, Department of Botanical and Environmental Sciences, Guru Nanak Dev University, Amritsar, Punjab, India

Department of Botany, DAV University, Jalandhar, India
e-mail: anketsharma@gmail.com

N. Handa • S. Bali • R. Kaur • K. Khanna • A.K. Thukral • R. Bhardwaj (✉)
Plant Stress Physiology Lab, Department of Botanical and Environmental Sciences, Guru Nanak Dev University, Amritsar, India
e-mail: renubhardwaj82@gmail.com

D. Egamberdieva, P. Ahmad (eds.), *Plant Microbiome: Stress Response*, Microorganisms for Sustainability 5, https://doi.org/10.1007/978-981-10-5514-0_14

of heavy metals (HM) in the environment is a serious problem (Kumar et al. 2017). The stress of HM to plants results in oxidative burst which leads to lipid degradation, disturbance in electron transport mechanism, and damage to biomolecules (Gill 2014; Sharma et al. 2016a). At the same time, pesticides are regularly used to protect crops from pests, but their application also causes toxicity to crop plants by generation of reactive oxygen species (ROS), which result in impaired growth and reduced photosynthetic efficiency of plants (Xia et al. 2009; Sharma et al. 2016b, 2017). Pesticides also persist in soil and plant parts for longer times in the form of pesticide residues (Xia et al. 2009; Sharma et al. 2016c).

Endophytic bacteria (EB) colonize inside the tissues of plants without causing harmful consequences or symptomatic infections inside their host plants (Schulz and Boyle 2006). These inhabit in apoplasm or symplasm and have the ability of stimulating physiological changes that assist the plant growth and development (Conrath et al. 2006). These bacteria provide vitamins and phytohormones for nutrient accumulation and metabolism in plants (Shi et al. 2009). They also provide tolerance against heavy metal stress and may trigger the growth of host plant involving various mechanisms such as synthesis of growth regulators, increased uptake of mineral nutrients and water, nitrogen fixation, and systemic resistance against pathogens (Ryan et al. 2008). Inoculation of endophytic bacteria in plants causes several physiological changes like decreased membrane potentials, stomatal regulation, osmolyte accumulation, osmotic adjustment, and altered level of phospholipids in cell membranes (Compant et al. 2005). Isolation of EB from metal hyperaccumulators shows the capacity to tolerate high concentrations of metal (Idris et al. 2004).

EB help the plants to encounter metal stress by reducing the toxicity of heavy metals by transforming metal ions to less toxic or nontoxic forms (Zhu et al. 2014). They enhance the resistance of plants against heavy metals (Rajkumar et al. 2009; Ma et al. 2015) and degrade pesticides in soil (Clive 2003; Nawaz et al. 2011). Additionally they produce various enzymes which help in degradation of pesticides (Singh 2008). Extracellular precipitation of heavy metals is also one of the mechanisms followed by these bacteria with which they reduce the toxicity of heavy metals and enhance bioremediation (Babu et al. 2015). The bacteria also have the ability to adsorb and desorb metal ions which is helpful in biological remediation of polluted soils (Guo et al. 2010; Luo et al. 2011). These plant-microbe associations thus aid in bioremediation of heavy metals as well as other pollutants of the soils and waters.

One of the major concerns dealing with pesticides is that they have the tendency to persist in the environment and enter the food chain through plants (Liu and Xiong 2001). Their complete biodegradation by bacteria converts them to carbon dioxide and water. Microbial degradation requires the application of efficient microorganism to degrade toxic pesticides such as organochlorines, organophosphates, and carbamates into nontoxic substance through enzymatic action. The present review is planned to give detailed information about the roles of EB in the detoxification of HM and pesticides. An attempt has been made to review the latest developments made toward potential of EB in HM and pesticide detoxification strategies.

14.2 Sources of Heavy Metal and Pesticide Contamination

Heavy metal pollution is one of the major concerns of the environment which has caused significant changes in their biogeochemical balance. There is an excessive release of different heavy metals like lead, cadmium, chromium, zinc, copper, nickel, etc., into the soil and other natural resources which causes harmful effects on biotic life (Dixit et al. 2015). Moreover, pesticides are exposed to environment during their application, manufacturing, and formulation. However, the use of pesticides in agriculture is very common practice which leads to their accumulation in fruits, food, vegetables, fodder, environmental pollution, and ecological imbalance (Parween et al. 2016). Use of pesticides in agriculture purposes is the main source of pesticide pollution in soil. Many of the heavy metals such as nickel, cadmium, mercury, arsenic, chromium, lead, zinc, copper, silver, etc. possess carcinogenic, cytotoxic, and mutagenic properties (Ahmad et al. 2016). Heavy metals and pesticides within soils are originated from different sources like use of various chemical pesticides and insecticides, fertilizers, sewage irrigation, atmospheric deposition, mining activities, and improper waste management (industrial, nuclear, agricultural, hospital, and domestic waste) (Zhang et al. 2011a). The main sources of heavy metals can be broadly classified as natural and anthropogenic. The natural sources are those which occur through natural activities like erosion, volcanic eruption, forest fires, weathering of rocks, etc., and anthropogenic sources are those which depend upon the human activities like industrial and domestic discharges, sewage discharge, electronic wastes, and agricultural practices (Dixit et al. 2015). Details about HM sources have been given in Table 14.1.

Table 14.1 Sources of heavy metals

Source category	Heavy metals	Source type	References
Anthropogenic (industrial)	As	Industrial wastes	Tripathi et al. (2007)
	Cd, Ni, Zn	Sewage/sludge	Keller et al. (2002)
			McLaren et al. (2004)
	Cr, Pb, Zn	Mining/smelting	Sumner (2000)
	Zn	Mining/coal waste, combustion, and steel processing	Greany (2005)
	Ni	Corroded metal pipes, containers	Cempel and Nikel (2006)
	Ni	Metal plating industries, electroplating, combustion of fossil fuels, mining	Khodadoust et al. (2004)
	Pb	Paint and petrol additives, smelting, coal combustion, cement production industries	Li et al. (2014)

(continued)

Table 14.1 (continued)

Source category	Heavy metals	Source type	References
	Cd	Galvanized pipe breakdown	Terry and Stone (2002)
	Cr	Metallurgical refractories, chemical industries	Kotasa and Stasicka (2000)
	Cu	Industrial waste additives, pipes	Mohod and Dhote (2013)
	Fe, Mn	Effluent, acid mine drainage, landfill, leachate, sewage	Kotasa et al. (2000)
	Zn	Mining	Huffmeyer et al. (2009)
Anthropogenic (agricultural)	Cd	Fertilizers	Jarup (2014)
	As	Pesticides, swine feed additives, poultry	Mukherjee et al. (2006)
	Cd	Agricultural crops	Campbell (2006)
	Cu, Cr, As	Agriculture fields	McLaughlin et al. (2000)
	Hg, As, Cu, Zn	Pesticides	Arao et al. (2010)
	Cd	Fertilizers	Boyd (2010)
Natural	Hg	Surface runoff	Kowalski et al. (2014)
	As, Cd, Cu, Zn	Weathering (acid mine drainage)	Razo et al. (2003)
	As	Mineral deposits	Nordstrom (2002)
	Cd, Co, Mn	Earth crust	Jarup et al. (2014)
	Fe, Cr, As, Zn, Mn	Weathering of sedimentary rocks like sandstone, limestone, shale, dolomite, etc.	Camacho and Armienta (2000)
			Ball and Izbick (2004)
			Viers et al. (2007)
	Ni	Atmospheric aerosols, volcanic emissions, vegetation, forest fires, windblown dust	Cempel and Nikel (2006)
	Pb		Li et al. (2014)
	Hg		Kang et al. (2011)
			Hsu et al. (2010)

14.3 Toxic Effects of Pesticides and HM

The toxicity caused by pesticides and HM toward plants and microbes is as follows:

14.3.1 Pesticides

The pesticide use for the removal of plant weeds and pests is the most important practice applied in agriculture. It causes adverse effects on human health when applied in higher quantities and causes ecological imbalance, residues in water, soil, fodder, food, vegetables, and environmental pollution. Pesticide application also causes adverse effects on plants such as changes in the germination, physiological, and biochemical activities that ultimately affect the yield of the plants (Parween et al. 2016). The germination in *Zea mays* L. cv NAAC-6002 decreases drastically after the treatment of pendimethalin (Rajashekhar et al. 2012). Moreover, many adverse effects of pesticides have been seen on beneficial soil microorganisms (Aktar et al. 2009). They show direct lethal effects toward beneficial microbes (Staley et al. 2015). A summary of toxic effects caused by pesticides on plants is given in Table 14.2 and toxic effects of pesticides on microbes in Table 14.3.

Table 14.2 Toxic effects of pesticides on plants

S.NO	Pesticides	Plants	Effects	References
1.	Diuron, atrazine, hexazinone, tebuthiuron	*Zostera muelleri, Halodule uninervis*	Damage to PS II system	Flores et al. (2013)
			Reduction in photosynthesis and inhibition of quantum yield	
2.	Pendimethalin	*Zea mays*	Decrease in length of radical and plumule along with germination rate	Rajashekhar et al. (2012)
			Suppression in seed reserve mobilization	
			Impaired degradation of seed reserve	
3.	Chlorpyrifos	*Vigna radiata*	Decrease in plant height, total surface area, number of leaf branches	Parween et al. (2011)
			Impairment in photosynthetic pigments such as Chl a, Chl b, total Chl and carotenoid contents	
			Number of pods and seeds decreased	
4.	Chlorpyrifos, cypermethrin, fenvalerate	*Cenchrus setigerus, Pennisetum pedicellatum*	Reduction in seed germination	Dubey and Fulekar (2011)
			Stunted plant growth	
			Inhibitory effect on bacterial population within soil	
5.	Acetochlor (AC), bensulfuron-methyl (BSM)	*Oryza sativa*	Decrease in the fresh weight, nitrate reductase (NR), glutamine synthetase (GS) activities	Huang and Xiong (2009)
			Increase in free amino acid content	
			Reduced soluble protein and nitrate content	

(continued)

Table 14.2 (continued)

S.NO	Pesticides	Plants	Effects	References
6.	Dimethoate	*Momordica charantia*	Reduction in photosynthetic pigments	Mishra et al. (2009)
			ROS accumulation in leaves causing electrolyte leakage and lipid peroxidation	
			Increase in the activities of SOD, CAT, POD	
7.	Diuron, atrazine, hexazinone	*Navicula, Nephroselmis pyriformis*	Inhibition in PS II components	Magnusson et al. (2008)
			Growth rate and photosynthesis inhibited	
8.	Hexaconazole (HEX), paclobutrazol (PBZ)	*Daucus carota* L.	Increase in nonenzymatic antioxidant, reduced glutathione, and ascorbate peroxidase (APX) activities	Gopi et al. (2007)
			Increase in fresh weight, biomass, dry weight, and carotenoid content	
			Increase in starch-hydrolyzing activities, anthocyanin, proline, starch sugar content, amino acids	
9.	Fusilade (fluazifop-p-butyl)	*Lens culinaris Medik*	Reduction in shoot and lateral root growth	Aksoy and Dane (2007)
			Chlorosis, curling, asymmetry, and leaf expansion occurs	
			Inhibition in lipid synthesis	
10.	Chlorotoluron	*Triticum aestivum*	O^{2-} and H_2O_2 accumulation leads to peroxidation of PM lipids	Song et al. (2007)
			Proline accumulation	
			Increase in the activities of POD, APX in roots and leaves	
			Decreases in CAT activity	
11.	MT-101(naproanilide), NOP(2-(2-naphthyloxy) propionanilide	*Sesbania, Oryza sativa*	Decrease in germination	Hirase and Molin (2002)
			Decrease in plant height, density and dry weight	

Chl chlorophyll, *NR* nitrate reductase, *GS* glutamine synthetase, *AC* acetochlor, *BSM* bensulfuron-methyl, *ROS* reactive oxygen species, *SOD* superoxide dismutase, *POD* peroxidase, *CAT* catalase, *GSH* glutathione, *POX* peroxidase, *HEX* hexaconazole, *PBZ* paclobutrazol, *APX* ascorbate peroxidase

Table 14.3 Toxic effects of pesticides on microbes

S.NO	Pesticides	Microorganisms	Effects	References
1.	Ethoxyquin, ortho-phenylphenol	Nitrifying bacteria, ammonia-oxidizing bacteria	Inhibition in nitrification	Papadopoulou et al. (2015)
			Ethoxyquin transformation into QI (quinone imine) inhibited nitrification	
2.	Carbaryl, glyphosate, thiophanate methyl	*Batrachochytrium dendrobatidis*	Inhibition in zoospore production	Hanlon and Parris (2012)
3.	Carbamate (carbaryl)	Cyanobacteria (*Anabaena flos-aquae, Microcystis flos- aquae, M. aeruginosa*), green algae (*Chlorella vulgaris, Chlorella pyrenoidosa, Selenastrum capricosnutum, S. quadricauda, S. obliquus*)	Green algae showed more sensitivity than cyanobacteria	Ma et al. (2006)
4.	Atrazine, isoproturon, metribuzin, sulfosulfuron	*Bradyrhizobium*	Inhibition in photosynthesis due to reduction in photosynthetic pigments and blocking of electron transfer from compound Q to PQ (plastoquinone) in PS II	Khan et al. (2006)
			Seed yield decreases	
5.	Atrazine	Cyanobacteria	Decreased growth rate and cell density effected	Lockert et al. (2006)
			Reduction in Chl content	

QI quinone imine, *Chl* chlorophyll, *PQ* plastoquinone

Table 14.4 Toxic effects of heavy metals on plants

S.NO	Heavy metals	Plants	Effects	References
1.	Pb	*Jatropha curcas L.*	Reduced contents of carotenoids and chlorophyll	Shu et al. (2012)
			Induced membrane damages and reduced leaf growth, root length, and photosynthesis	
			Increased activities of CAT and POD and decrease in SOD activity	
2.	Pb	*Pistia stratiotes*	Decrease in chlorophyll content	Vesely et al. (2011)
			Accumulation of lead in roots and leaves	
3.	Cd, Cu, Pb, Zn	*Fontinalis antipyretica*	Increased levels of MDA and lipid peroxidation	Dazy et al. (2009)
			Increased activities of SOD, CAT, GR APX, GPX	
4.	Hg	*Cucumis sativus*	Elevated levels of lipid peroxidases, protein oxidation	Cargnelutti et al. (2006)
			Reduced chlorophyll content and catalase activity	
5.	Hg	*Sesbania drummondii*	Disturbances in photosynthesis	Israr et al. (2006)
6.	Co	*Phaseolus vulgaris*	Growth suppression due to chlorosis and necrosis	Chatterjee et al. (2006)
			Decrease in number of seeds, flowers	
			Reduction in chlorophyll content, biomass, hill reaction, catalase activity, sugar, starch, and protein nitrogen	
			Increased enzymatic concentrations of ribonucleases, peroxidases, phosphatases, and phenols	
7.	Cd	*Glycine max*	Oxidative stress due to thiobarbituric acid reactive substances	Balestrasse et al. (2004)
			Decreased levels of leghemoglobin, nitrogenase, and peroxidases	
			Increased ethylene production, ammonium, and protease activity leads to senescence	
			Nodular senescence due to loss of nitrogen fixing area and bacteroids	
8.	Cd	*Glycine max*	Reduction in spermidine levels, DAO activity, GO/GAT system	Balestrasse et al. (2005)
			Elevation in Put, ammonium, and proline contents in nodules and roots	

SOD superoxide dismutase, *CAT* catalase, *GRD* glutathione reductase, *APX* ascorbate, *GPX* guaiacol peroxidase, *POD* peroxidase, *MDA* malondialdehyde, *GR* glutathione reductase, *DAO* diamine oxidase, *GO/GAT* glutamine oxoglutarate aminotransferase, *Put* putrescine

14.3.2 Heavy Metals

Heavy metal contamination in soil is one of the major problems caused because of chemical fertilizers, industrial wastes, and other methods involved in agriculture. Heavy metals such as Cu, Cr, Cd, Ni, Cu, Zn, Pb, etc. negatively affect plants and microflora present within soil. Plants respond toward HM toxicity by the generation of reactive oxygen species (ROS) thereby causing oxidative stress to plants (Mithofer et al. 2004). The microbial communities living within soils have also been shown to decline under heavy metal contamination (Abaye et al. 2005). A summary on toxic effects of heavy metals on plants has been given in Table 14.4 and toxic effects of heavy metals on microbes in Table 14.5.

Table 14.5 Toxic effects of heavy metals on microbes

S.NO	Heavy metals	Microorganisms	Effects	References
1.	Cd, Zn	*Anabaena, Microcystis, Nostoc*	NH_4^+ and PO_4^{3-} uptake inhibited	Yadav et al. (2016)
			Na^+ and K^+ efflux adversely effected	
			Reduction in glutamine synthetase and alkaline phosphatase activity	
2.	Cu, Pb, Cd, Li	*Cyanothece* COY 0110	Metabolism effected due to change in photosynthesis, nitrogen and carbon metabolism, translation, amino acid metabolism, CO_2 metabolism	Mota et al. (2015)
3.	Pb	Ammonia-oxidizing bacteria	Loss in viability of microbes	Yuan et al. (2015)
			Reduction in bacterial community	
4.	Cu	*Plasmopara viticola*	Inhibition of enzymatic activities such as phosphatases, ureases, arylsulfatases, invertases, xylanases	Mackie et al. (2013)
			Carbon and nitrogen metabolism adversely effected	
5.	Fe, Zn, Pb, Cr, Mn, Cd, Ni, Cu	*Azotobacter*	Growth rate affected	Lenart and Koladka (2013)
6.	Cd, Hg	*Arthrospira platensis*	APA activity inhibited	Tekaya et al. (2013)
7.	Cu	*Acidobacteria, Actinobacteria, Gammaproteobacteria*	Microbial metabolism affected due to suppression in respiratory rate of bacterium	de Boer et al. (2012)

(continued)

Table 14.5 (continued)

S.NO	Heavy metals	Microorganisms	Effects	References
8.	Cr, Ag	*Spirulina platensis*	Inhibition in Hill reaction activity	Babu et al. (2010)
			Disruption in ETC components along with phycobilisomes and phycocyanin	
9.	Zn, Cu, Cd	*Rhizobium leguminosarum*	Reduction in microbial number in soil due to toxic effects of heavy metals	Chaudri et al. (2008)

NH_4^+ ammonium ion, *PO_4^{3-}* phosphate, *Na^+* sodium ion, *K^+* potassium ion, *APA* alkaline phosphatase, *ETC* electron transport chain

14.4 Endophytic Bacteria and Their Biodiversity

EB inhabit the internal tissues of the plants. Endophytes are often classified as either obligate or facultative depending upon their life strategies. Obligate endophytes are strictly dependent upon their host for their growth, survival, and transmission, whereas facultative endophytes live a part of their life cycle outside the host plant (Hardoim et al. 2008) and can develop different relationships, e.g., mutualistic, symbiotic, commensalistic, and trophobiotic with the host plant (Ryan et al. 2008). Plants are normally associated with diverse endophytic bacteria, and they constitute vast niches for them. Likely, majority of the plant species are associated with endophytic bacteria (Rosenblueth and Martinez-Romero 2006). Plants and their associated endophytic bacteria have been review by many authors indicating huge diversity of bacteria harboring inside plants (Hallmann et al. 1997; Lodewyckx et al. 2002; Rosenblueth and Martinez-Romero 2006). Endophytes can originate from the natural environment, i.e., rhizosphere or phyllosphere (Ryan et al. 2008). The dominant phylum of endophytic bacteria found in plants belongs to *Proteobacteria* followed by *Firmicutes* and *Actinobacteria*. Other less common phyla include *Acidobacteria*, *Bacteroidetes*, *Planctomycetes*, and *Verrucomicrobia* (Santoyo et al. 2016). EB can be beneficial to their host in many ways as they can act as plant growth regulators and biomass promoters. They can synthesize important natural products which may have potential use in agriculture, medicine, and industry (Ryan et al. 2008). Endophytes harboring contaminated habitats are contaminant resistant and are successfully used in phytoremediation (Li et al. 2012).

14.4.1 Heavy Metal-Resistant Endophytic Bacterial Diversity

Isolation of heavy metal-resistant endophytic bacteria studies reveal that wide ranges of genera are metal resistant that include *Acinetobacter*, *Agreia*, *Agrobacterium*, *Arthrobacter*, *Bacillus*, *Burkholderia*, *Caulobacter*,

Chryseobacterium, *Clostridium*, *Curtobacterium*, *Enterobacter*, *Herbaspirillum*, *Kocuria*, *Kushneria*, *Methylobacterium*, *Micrococcus*, *Microbacterium*, *Moraxella*, *Paenibacillus*, *Pantoea*, *Paracoccus*, *Plantibacter*, *Proteobacteria*, *Pseudoalteromonas*, *Pseudomonas*, *Rahnella*, *Rhizobium*, *Rhodococcus*, *Salinicola*, *Sanguibacter*, *Serratia*, *Sphingomonas*, *Stenotrophomonas*, *Streptomyces*, *Staphylococcus*, *Stenotrophomonas*, *Variovorax*, *Vibrio*, *Xanthomonadaceae*, etc. These HM-resistant EB have been listed in Table 14.6.

Table 14.6 Metal-resistant endophytic bacterial diversity

S. No.	Endophytic bacteria	Plant species	Isolated from	Resistant heavy metal	Reference
1.	*Methylobacterium mesophilicum*, *M. extorquens*, *Sphingomonas* sp., *Curtobacterium* sp., *Plantibacter flavus*, *Rhodococcus* sp.,	*Thlaspi goesingense*	Stem	Ni	Idris et al. (2004)
2.	*Pseudomonas fluorescens* G10 and *Microbacterium* sp. G16	*Brassica napus*	Roots	Pb	Sheng et al. (2008)
3.	*Clostridium aminovalericum*, *Enterobacter* sp., *Sanguibacter* sp., *Stenotrophomonas* sp., *Pseudomonas* sp., and *Xanthomonadaceae*	*Nicotiana tabacum*	Seeds	Cd	Mastretta et al. (2009)
4.	*Bacillus sp.* (EB L14)	*Solanum nigrum*	Leaves	Multi-metal resistant	Guo et al. (2010)
5.	*Exiguobacterium aurantiacum*, *Burkholderia* sp., *Bacillus cereus*, *Serratia marcescens*, *Acinetobacter calcoaceticus*, *Acinetobacter junii*, *Micrococcus luteus*, *Bacillus firmus*, *Bacillus megaterium*, *Moraxella* sp., *Paracoccus* sp., *Arthrobacter* sp., *Bacillus pumilus*, *Sphingomonas* sp., *Arthrobacter* sp., *Herbaspirillum* sp., *Microbacterium kitamiense*, *Bacillus* sp.	*Elsholtzia splendens*, *Commelina communis*	Roots, stems, and leaves	Cu	Sun et al. (2010)
6.	*Pseudomonas* sp., *Bacillus* sp., *Stenotrophomonas* sp., *Acinetobacter* sp.	*Sedum alfredii*	Leaves, stems, and roots	Zn, Cd	Xinxian et al. (2011)
7.	*Caulobacter* sp., *Rhizobium* sp., and *Bacillus* sp.	Belgisch rood (*Salix × rubens* var. basfordiana)	Roots	Cd, Zn, Pb	Weyens et al. (2013)
	Paenibacillus sp., *Bacillus* sp.,				
	Pantoea sp., and *Pseudomonas* sp.		Shoots		

(continued)

Table 14.6 (continued)

S. No.	Endophytic bacteria	Plant species	Isolated from	Resistant heavy metal	Reference
8.	*Pseudomonas* sp., *Bacillus* sp., *Streptomyces* sp., *Staphylococcus* sp., *Arthrobacter* sp., and *Caulobacter* sp.	Tora (*Salix schwerinii* × *Salix viminalis*)	Roots	Cd, Zn, Pb	Weyens et al. (2013)
	Gammaproteobacteria, *Bacillus* sp., *Staphylococcus* sp., *Rhodococcus* sp., *Acinetobacter* sp., *Micrococcus* sp., and *Chryseobacterium* sp.		Shoots		
9.	*Pseudomonas*, *Microbacterium*, *Methylobacterium*, and *Burkholderia*	*Alyssum serpyllifolium*	Roots, stems, and leaves	Ni	Ma et al. (2011)
10.	*Agrobacterium tumefaciens* Q2BJ3, *Bacillus* spp. Q2BJ1, Q2CJ3, and Q2BG1, *Acinetobacter* sp. Q2BJ2, *Bacillus* subtilis Q2CJ5, and *Bacillus megaterium* Q2BG4	*Commelina communis*	Roots, stems, and leaves	Pb	Zhang et al. (2011b)
11.	*Rahnella* sp. JN6	*Polygonum pubescens*	Roots and stems	Lead, cadmium, zinc,	He et al. (2013)
12.	*Rahnella* sp. JN27	*Zea mays*	Roots	Cd	Yuan et al. (2014)
13.	*Microbacterium* sp., *Arthrobacter* sp., *Agreia* sp., *Bacillus* sp., *Stenotrophomonas* sp., *Kocuria* sp., and *Variovorax* sp.	*Noccaea caerulescens*	Roots	Ni	Visioli et al. (2014)
14.	*Burkholderia* sp. *SaZR4*, *Burkholderia* sp. *SaMR10*, *Sphingomonas* sp. *SaMR12*, and *Variovorax* sp. *SaNR1*,	*Sedum alfredii Hance*	Roots	Zn, Cd	Zhang et al. (2013)
15.	*Micrococcus yunnanensis*, *Vibrio sagamiensis*, and *Salinicola peritrichatus*	*Spartina maritime (MA)*	Leaves, stems, and roots	As, Cu, Zn	Mesa et al. (2015)
16.	*Kushneria* sp.	*Arthrocnemum macrostachyum*	Shoots	Ni, Pb	Navarro-Torre et al. (2016)
	Micrococcus sp.			As, Pb	
	Pseudoalteromonas sp.		Roots	Zn	
	Vibrio sp.			Co	
17.	*Enterobacter* sp. K3-2	*Sorghum sudanense*	Roots	Cu	Li et al. (2016)

14.4.2 Pesticide-Resistant Endophytic Bacterial Diversity

Wide range of plant-associated bacteria mainly belongs to α, β, and γ classes of *Proteobacteria* which assist their host in degradation of toxic organic pesticide compounds (Tetard-Jones and Edwards 2016). Inoculation of *Pisum sativum* with *Pseudomonas putida* POPHV6, isolated from poplar trees, enhanced removal of 2,4-dichlorophenoxyacetic acid from the soil, and aerial parts of pea showed no 2,4-D accumulation (Germaine et al. 2006). Eight genera of these bacteria including *Bacillus* sp., *Microbacterium* sp., *Paenibacillus* sp., *Aeromonas* sp., *Flavobacterium* sp., *Klebsiella terrigena*, *Pantoea* sp., and *Pseudomonas* sp. isolated from different parts of *Phragmites communis*, *Nymphaea tetragona*, and *Najas marina* possessed high rates of pesticide degradation (Chen et al. 2012). *Streptomyces* sp. atz2 isolated from leaves of sugarcane, decreased atrazine concentration, and resulted in appearance of nontoxic metabolite (Mesquini et al. 2015).

14.5 Plant Growth Promotion by EB

EB are universally present in the plant kingdom. They play an important role in horticulture, silviculture, and agriculture as well as in phytoremediation (Santoyo et al. 2016). EB promote the growth of plants by repressing the plant pathogens indirectly. This enhanced state of resistance is protective against a wide range of parasites and pathogens, i.e., bacteria, viruses, fungi, nematodes, etc. (Vauterin and Swings 1997; Murphy et al. 2003; Ryu et al. 2004). Plant growth-promoting bacteria are used as inoculants for enhancing the yield and growth of agricultural crops by replacing the use of chemical fertilizers, pesticides, etc. (Stefan et al. 2008; Ashrafuzzaman et al. 2009; Saharan 2011). Plant growth-promoting bacteria such as *Corynebacterium*, *Bacillus*, and *Enterobacter* have been reported to enhance the growth and improve the health through direct and indirect mechanisms which are helpful to the plant (Gupta et al. 1998, El-Banna and Winkelmann 1988; Idriss et al. 2002). Ji et al. (2014) isolated and characterize the endophytic bacteria from Korean rice varieties which promote growth. When seeds of rice were treated with bacteria, there was improvement in the growth, height, and weight of the plants. These provide many advantages to the host plant such as promotion of plant growth and protection against pathogens. Under such broad environmental conditions, EB interact and communicate with the plants in a better way as compared to the rhizospheric bacteria (Ali et al. 2012; Coutinho et al. 2015). Mechanisms used by EB for facilitating the growth of plant are well known and understood (Glick 2012; Gamalero and Glick 2011). These bacteria affect the growth of the plant either directly or indirectly. Direct promotion of plant growth takes place when EB increase the addition of resources from the environment such as N, P, Fe, etc. or regulating the plant growth by providing plant hormones such as auxin, ethylene, or cytokinin. EB promotes the growth of plants directly through various mechanisms such as nitrogen fixation, phytohormone production, and solubilization of minerals. Indirect promotion of plants by EB takes place when bacterium reduces or inhibits the loss to plants that might be caused by pathogenic bacteria, i.e., fungi, nematodes, and bacteria. There are various mechanisms that EB use indirectly to promote the growth of the plants including the antibiotic production, enzymes involved in cell wall degradation, decreasing the content of ethylene, reducing the concentration of iron available to pathogens, and synthesis of pathogen-preventing volatile compounds (Glick 2015). Certain EB reduce the toxicity of metal and increase the plant growth by one or more of these mechanisms (Rajkumar et al. 2009; Pereira and Castro 2014). EB are natural biocontrol agents and produce substances which may effectively reduce the phytopathogen-caused diseases by the production of antibiotics, hydrolytic enzymes, and chitinases and antimicrobial volatile organic compounds (Sheoran et al. 2015). EB directly or indirectly promote the growth in plants by improving the nutritional status of plants, production of phytohormones, etc. Bacterial strains have been isolated from the root nodules of *Mimosa pudica* and showed better results suggesting its importance in use as biofertilizers, biostimulant, and biopesticide as a sustainable agricultural approach (Nivya 2015).

14.6 Effects of EB on Heavy Metals

Even though heavy metals are toxic to plants and microbes, metal-accumulating plants are widely reported to harbor metal-resistant plant growth-promoting endophytic (PGPE) bacteria in their interiors (Ma et al. 2011). Endophytic bacteria isolated from metal-hyperaccumulating plant species are resistant to metal toxicity and play an important role in the survival and growth of plants (Rajkumar et al. 2009). Inoculation of metal-resistant PGPE endophytic bacteria is recently attained due attention for phytoremediation of metal-contaminated soils because they not only promote plant growth and detoxify metal toxicity but also enhance metal accumulation in plants (Rajkumar et al. 2009; Ma et al. 2016). For instance, inoculation of indigenous Pb-resistant bacteria, *Pseudomonas fluorescens* G10 and *Microbacterium sp.* G16, promoted biomass and Pb uptake in *Brassica napus* raised in Pb-spiked soils (Sheng et al. 2008). Plant growth-promoting and Ni-resistant strain *Pseudomonas* A3R3 isolated from tissues of *Alyssum serpyllifolium* increased biomass and enhanced Ni content by 10% in *A. serpyllifolium* and by 15% in *B. juncea* (Ma et al. 2011). Inoculation of Pb-resistant endophytic bacteria isolated from *Commelina communis* to rape increased Pb uptake in the rape from 58% to 62% (Zhang et al. 2011b). Bio-inoculation of Cu-tolerant and ACC deaminase-producing endophytic bacteria (*Ralstonia* sp. J1-22-2, *Pantoea agglomerans* Jp3-3, and *Pseudomonas thivervalensis* Y1-3-9) isolated from plants growing in Cu mine area promoted growth and Cu accumulation in *Brassica napus* (Zhang et al. 2011c). Endophytic bacteria (*Variovorax* sp. SaNR1, *Burkholderia* sp. SaZR4, *Burkholderia* sp. SaMR10, and *Sphingomonas* sp. SaMR12) isolated from the roots of *S. alfredii* growing in metal-contaminated soils, when inoculated to Zn- and Cd-spiked perlite, SaMR12 and SaNR1 bacterial strains significantly enhanced growth and metal phytoextraction in *S. alfredii*. SaMR10 had least effect on growth and metal phytoextraction, while SaZR4 strain significantly improved Zn phytoextraction but not Cd extraction (Zhang et al. 2013). Further hydroponic experiments conducted by Chen et al. (2014) showed that inoculation of SaM12 significantly increased biomass, root exudates, and Zn uptake in *S. alfredii*. SaM12 inoculation to nonhost (*Brassica napus*) raised in Cd-spiked soils resulted in successful colonization of SaM12 in rape roots. Cd uptake and translocation to leaves were enhanced in inoculated plants (Pan et al. 2016). Inoculation of *Rahnella* sp. JN6, metal-tolerant PGPE bacteria isolated from *Polygonum pubescens*, improved the efficiency of Cd, Pb, and Zn phytoextraction by *Brassica napus*. *Rahnella* significantly stimulated growth and metal uptake by producing growth-promoting substances like IAA, ACC deaminase, siderophores, and/or increasing bioavailability of metals (He et al. 2013). Endophytic bacterial strain *Rahnella* sp. JN27 is also reported to enhance growth and Cd phytoextraction in *Amaranthus hypochondriacus* and *Amaranthus mangostanus* which are well-established Cd hyperaccumulators (Yuan et al. 2014). Mn-resistant *Bacillus megaterium* 1Y31 promoted growth and Mn accumulation in Mn hyperaccumulator hybrid pennisetum through strengthening energy metabolism and improving photosynthetic efficiency and proportion of PGPE bacteria in pennisetum (Zhang et al. 2015). Integrated community of Ni-tolerant endophytic bacteria rather than individual strains proves to be a more effective strategy in improving Ni phytoextraction in *N. caerulescens* (Visioli et al. 2015). Bioaugmentation of multi-metal-contaminated soils sited with PGPE bacterial strains (*Stenotrophomonas sp.* E1L, *Bacillus* sp. E1S2, *Bacillus pumilus* E2S2, *Bacillus* sp. E4S1, and *Achromobacter* sp. E4L5) isolated from tissues of *Sedum plumbizincicola* improved phytoextraction efficiency of *S. plumbizincicola* indicating potential of plant's own PGPE bacteria in enhancing phytoextraction of multi-metal-contaminated soils (Ma et al. 2015). Inoculation of metal-resistant and PGPE bacteria stains (*Bacillus megaterium* JL35 and *Burkholderia* sp. GL12) isolated from *Elsholtzia splendens* to its host improved Cu uptake by 223%, whereas improved Cu uptake by 31.3% in nonhost plant *Brassica napus* when grown in Cu-contaminated soil (Sun et al. 2015). Inoculation of multi-metal-resistant strain endophytic bacterial strain E6S isolated from stems of *Sedum plumbizincicola* (Zn/Cd hyperaccumulator) improved root accumulation in the host plant. Whereas, the root to shoot translocation of Cd and Zn was reduced suggesting the role of endophytic bacterial strain E6S in rhizoaccumulation of metal in plants (Ma et al. 2016). Root endophytic isolate of *Sorghum sudanense* and *Enterobacter* sp. K3-2 increased biomass of host grown in Cu mine area and also increased Cu uptake in roots by 83–86% than non-inoculated plants (Li et al. 2016). Few studies have also reported noneffectiveness of PGPE bacteria for improving metal accumulation by plants. Mesa et al. (2015) observed reduction in metal uptake by inoculation of indigenous metal-resistant endophytic bacteria to *Spartina maritime*, whereas indigenous bacteria enhanced growth of the plants under metal stress.

14.7 Bioremediation Potential of EB

EB have been reported to secrete hormones which enhance the nutrient uptake, increased root growth, and plant biomass (Gravel et al. 2007; Shi et al. 2009; Phetcharat and Duangpaeng 2012). Mitigation of heavy metal stress was caused by endophytic bacteria through an association of nutritional and biochemical assistances. In plants, IAA produced by these bacteria is entailed in different physiological processes like plant development, stimulation of plant defense system, and to signaling mechanisms (Gravel et al. 2007; Spaepen et al. 2007; Navarro et al. 2006).

Gibberellins synthesized by EB play an important role in plant tissue extension, in particular of stem region, thereby enhancing growth of plants (Salisbury 1994). In *Graminaceae* species, endophytic bacteria such as *Acetobacter diazotrophicus* and *Herbaspirillum seropedicae* were capable to synthesize plant hormone indole-3-acetic acid and gibberellins which further enhanced growth and yield of plants (Bastian et al. 1998).

In plants, ethylene plays a crucial role in regulating growth and metabolism and participates in plant-microbe association and plant nutrient cycle and also provides resistance against biotic and abiotic stresses (Ping and Boland 2004). Bacteria are proficient in decreasing stress-mediated effects on plants through enzymatic hydrolysis of ACC (1-aminocyclopropane-1-carboxylic acid). ACC is participated in biosynthetic of ethylene which acts as intermediate that converts methionine to ethylene (Adams and Yang 1979). ACC is discharged through roots or seeds and then employed by ACC-utilizing bacteria prior to its oxidation by ACC oxidase and then cleaved into α-ketobutyrate (αKB) and ammonia by ACC deaminase (Contesto et al. 2008). Bacteria uses ammonia formed by ACC deaminase as the only source of nitrogen and thus decline ACC within plants and simultaneously decrease the level of ethylene (Glick et al. 1998; Belimov et al. 2002). The reduced concentration of ethylene in ACC-utilizing bacteria host plants obtain advantage by decreasing stress and improving plant productivity (Cheng et al. 2007; Dell'Amico et al. 2008; Hardoim et al. 2008).

Heavy metals can stimulate the production of ethylene in plants, and high concentration of this phytohormone may hinder the plant metabolism (Jackson 1991; Weckx et al. 1993). Various soil bacteria promote root growth in plants on the basis of capability of bacterial ACC deaminase to hydrolyze and decline the quantity of ACC (precursor of ethylene) which further reduces the production of ethylene in plants (Glick et al. 1994, 1998; Hall et al. 1996). In canola seedlings exposed to high concentration of nickel, it was demonstrated that inoculation of *Kluyvera ascorbata* SUD 165 reduced the biosynthesis of ethylene (Burd et al. 1998).

It has been demonstrated that *Variovorax paradoxus* was the prevalent strain found in rhizosphere of *Brassica juncea* raised on Cd-contaminated site (Belimov et al. 2005). This strain was observed to synthesize ACC deaminase and had the ability to utilize ACC as a source of energy. In vitro studies revealed that there was a positive correlation between the activity of ACC deaminase and influence of bacteria on elongation of root in plants. From this study it has been elucidated that *B.*

juncea inoculated with *V. paradoxus* in Cd-contaminated soil may regulate the root growth, which could be used in the phytoremediation processes. PGPR (ACC deaminase producers) were isolated from *Graminaceae* grasses raised on heavy metal-contaminated water meadow, and these strains have an important role in plant growth (Dell'Amico et al. 2005). Inoculation of *Pseudomonas* sp. in wheat plants under chromium stress improved root growth and auxin synthesis and simultaneously declined the level of chromium in the plants (Hasnain and Sabri 1997).

Microorganisms play an important role in the detoxification and in situ removal of toxic substances from contaminated sites (Desaint et al. 2000; Wang et al. 2005; Wood 2008). PGPR comprising symbiotic N2 fixers assist growth of plant indirectly by avoiding the harmful effects of pesticides or directly by producing plant growth regulators (Figueiredo et al. 2007; Jeon et al. 2003; Lopez et al. 2005). In plants, the regulation of root system depends on the activity of auxins which may enhance or reduce the radicle cell size, depending on the level and cross talk with other plant hormones like cytokinins (Evans 1984). IAA triggers the enzyme H^+-ATPase which is elementary for synthesis of energy in the nodule of leguminous plant roots (Rosendahl and Jochimsen 1995). IAA synthesis was observed in both rhizobacteria and symbiotic bacteria like genus *Bradyrhizobium* (Boddey and Hungria 1994). Inoculation of *Mesorhizobium* sp. (MRC4) in chickpea provides protection from pesticides like fipronil and pyriproxyfen but also enhanced the performance of chickpea raised in insecticide soils. Improved chemical and biological characteristics of chickpea plants may be due to influence of plant hormones or siderophores produced by the isolate MRC4 (Ahemad and Khan 2009). Remediation of pesticides from agriculture lands is one of the essential problems as these chemicals are very costly and challenging as they form venomous chemicals by reaction of diverse organic and inorganic substances present in soils (Jain et al. 2005). Proficient microbial technology is helpful for pesticide degradation/removal from the agricultural lands. Microbial degradation (utilize of bacteria, fungi, viruses, and actinomycetes) can efficiently eradicate pesticides from the polluted soils such as organochlorines, organophosphates, and carbamates by enzymatic action (Porto et al. 2011).

14.7.1 Mechanism of HM and Pesticide Detoxification

14.7.1.1 HM Detoxification

Bacteria have developed resistance to toxic heavy metals chiefly by two mechanisms, viz., detoxification of heavy metal by transformation of their toxic forms to nontoxic or unavailable forms and efflux pumping of toxic metal from the cells which is an active mechanism (Evanko and Dzombak 1997). Detoxification of heavy metals through basic redox reactions occurs in soils, and microbes reduce heavy metal ions by acting as oxidizing agents (Dixit et al. 2015). Microbial activities involve both aerobic and anaerobic activities, wherein former uses oxygen as an electron acceptor and, in latter, alternative electron acceptors (sulfates, nitrates, ferric oxides, etc.) are reduced by microorganisms. Another mechanism of efflux pumping has been reported to be much developed in EB. *Bacillus* sp. MN3-4, which is a strain of endophytic bacteria, has been reported to develop a P-type ATPase efflux pump that has the ability to utilize energy from ATP and use it to transfer metal ions across the biomembranes against the concentration gradients (Shin et al. 2012). Apart from the above two mechanisms, endophytic bacteria are also involved in alteration of phenotype and functional characteristics of the host plants and ultimately enable them to resist heavy metal stress (Li et al. 2011). It has been established that endophytic bacteria can strengthen the antioxidative defense system of plants by reducing lipid peroxidation and enhancing the activities of enzymes such as superoxide dismutase, catalase, glutathione peroxidase, and ascorbate peroxidase and hence equip the plant to fight against oxidative stress caused by heavy metals (Zhang et al. 2010; Wan et al. 2012). Some endophytic bacteria also contain certain genes which encode enzymes responsible for metal detoxification. *MerB* and *MerA* are the genes present in some mercury-resistant gram-negative bacteria in a mercury resistance (*Mer*) operon (Brown et al. 2003). The former encodes the enzyme organomercurial lyase involved in conversion of organomercurials to mercuric ions, while *MerA* expresses to form mercuric reductase to convert mercuric ions to elemental mercury which is less toxic (Cursino et al. 2000; Brown et al. 2003).

The EB also express 1-aminocyclopropane-1-carboxylase (ACC) deaminase enzyme that cleaves ACC to α-ketobutyrate and ammonia. The bacteria thus aid in lowering the ACC levels of host plants and ultimately inhibit the production of ethylene that induces growth inhibition. Therefore, the plants inhabiting endophytic bacteria with ACC deaminase when exposed to biotic or abiotic stresses are able to resist growth inhibition due to ethylene (Glick 2014, 2015). This property of bacterial endophytes has also been reported to promote endophyte-assisted phytoremediation of polluted soils. A study conducted on *S. plumbizincicola* showed that this hyperaccumulator plant contains strains of bacteria exhibiting siderophore production, indole-3-acetic acid production, phosphorus solubilization, ACC deaminase activity, and metal resistance (Zn, Cd, and Pb). These bacterial strains when inoculated to *S. plumbizincicola* plants are made to grow in soils contaminated with Zn, Cd, and Pb, and a significant increase in metal uptake, root and shoots lengths, and fresh and dry biomass was observed (Ma et al. 2015; Ullah et al. 2015). Another study on *Medicago lupulina* infested with *Sinorhizobium meliloti* strain CCNWSX0020 was reported to promote accumulation of Cu in both roots and shoots. The antioxidative defense system of the plant was also reported to increase with bacterial symbiosis and helped in combating Cu stress (Kong et al. 2015).

14.7.1.2 Pesticide Detoxification

The process of bacterial degradation of pesticides is catalyzed by enzymes as well as controlled by environmental factors such as temperature, water potential, pH, and nutrient composition of the soil (Singh 2008). The plasmids of these bacteria encode for pesticide-degrading enzymes and referred as catabolic plasmids. The presence of catabolic plasmids has been seen in several species like *Pseudomonas*, *Actinobacter*, *Flavobacterium*, *Arthrobacter*, etc. (Sayler et al. 1990). Both gram-positive and gram-negative bacteria have been reported to be involved in biological degradation of a wide variety of pesticides. Studies conducted on *Stenotrophomonas maltophilia* showed that this bacterial strain has the ability to detoxify polyaromatic hydrocarbons as well as xenobiotic compounds (Juhasz et al. 2000; Lee et al. 2002). The M1 strain of this bacterium can also degrade methomyl which is an oxime carbamate and is declared as one of the most toxic pesticides (Clive 2003; Nawaz et al. 2011). The M1 strain contains two plasmids out of which one (PMb) is responsible of methomyl degradation. This plasmid can be transformed into other bacterial strains and thus can be used to enhance the ability of methomyl degradation (Nawaz et al. 2011). Another pesticide aldicarb, which is toxic for human health, can be degraded by esterase and amidase which are bacterial enzymes (Burgess et al. 1994; Lifshitz et al. 1997; Turan et al. 2008; Nawaz et al. 2011). The bacterium *S. maltophilia* has also been reported to be involved in degradation of aldicarb by producing esterase enzyme (Turan et al. 2008). Other bacteria responsible for degradation of this pesticide are *Alcaligenes denitrificans*, *Enterobacter gergoviae*, *Flavimonas oryzihabitans*, and *Bacillus subtilis* (Turan et al. 2008).

14.7.2 Detoxification of Other Soil Pollutants

Bioremediation technique provides a cheap and safe alternative method for removal of contaminants from the environment. Bacterial process is the main activity which is involved in organic pollutant hydrolysis (Vasileva-Tonkova and Galabova 2003). Enzymes have the great capacity to detoxify the toxic polluting substances because they have been identified to convert the pollutants at a detectable limit and have the suitable capacity to restore the polluted environment (Rao et al. 2010). Bioaugmentation with a selected combination of two endophytic bacteria enhanced the growth of host plant and decreased the content of crude oil in the soil (Fatima et al. 2016). In another experiment, bulk soil, rhizosphere, and endophytic strains of *Acer pseudoplatanus* were chosen according to their capability for plant growth and trinitrotoluene conversion potential and formed a combination. By the inoculation of this combination in the *Acer capillaries,* it prevents the grass from oxidative stress and contributes to the conversion of trinitrotoluene (Thijs et al. 2014). Various enzymes from fungi, plants, and bacteria are involved in biodegradation of toxic organic pollutants, and it is a cost-effective and nature-friendly technology (Karigar and Rao 2011). Detoxification of toxic organic compounds by various bacteria, fungi, and higher plants by oxidative coupling is done by oxidoreductase (Gianfreda et al. 1999; Bollag and Dec 1998). Oxidoreductases are involved in humification of many phenolic substances which are manufactured by the decomposition of lignin in a soil environment. In the similar manner, they can also detoxify toxic xenobiotics, i.e., aniline or phenolic compounds via the polymerization and copolymerization with other substances or binding with humic substances (Park et al. 2006).

Catechol dioxygenase degrades the aromatic molecules in the environment. EB are screened for cytokinin like compounds. They are known to produce growth-promoting compounds. Cytokinin is one of the hormones which can be used in agriculture for management of leafy vegetables and fruits during preharvest and post-harvest. All the plants shelter various groups of endophytic bacteria. EB produce a number of compounds which affect the growth of host plants positively. EB import nutrients to plants directly effecting the growth of plant by the synthesis of phytohormones (Miliute and Buzaite 2011). Strains of EB were able to produce indole-3-acetic acid (IAA). Better understanding of the plant growth-enhancing activity of these strains provides significant contribution in enhancing the environmental sustainability in agriculture (Abbamondi et al. 2016). It was observed that EB strains are bio-prospective for enhancing the growth of plant which may be a good strategy to boost the growth of crops in marginal lands (Khan et al. 2016). EB, i.e., *Azospirillum lipoferum*, promotes growth in plants by the production of abscisic acid (ABA), indole-3-acetic acid (IAA), and gibberellins (GA). Cohen et al. (2009) analyzed the effects of *Azospirillum lipoferum* in maize plants treated with inhibitors of ABA and GA and subjected to drought stress or provided adequate water. The inhibitors reduced the growth of plant under drought stress as well as adequate supply of water, but treatment of *Azospirillum lipoferum* reversed this effect. These results indicate that ABA and GA alleviate water stress of plant by *Azospirillum*.

14.8 Conclusions and Future Prospects

EB are able to detoxify the heavy metals and pesticides in soil which ultimately results in reducing the soil contamination. The ability of these bacteria to secrete various growth-promoting hormones also helps plants to counterattack the negative effects of heavy metals and pesticides. Moreover, their presence in soil also enhances the phytoremediation efficiency of plants. The heavy metal- and pesticide-resistant strains of these bacteria can be helpful in cleaning up the contaminated soils. Additionally, the identification of heavy metal and pesticide detoxification genes in these bacteria followed by incorporation of those identified genes in hyperaccumulator plants can be beneficial to reduce the adverse effects of heavy metals and pesticides on plants.

References

Abaye DA, Lawlor K, Hirsch PR, Brookes PC (2005) Changes in the microbial community of an arable soil caused by long-term metal contamination. Eur J Soil Sci 56:93–102

Abbamondi GR, Tommonaro G, Weyens N, Thijs S, Sillen W, Gkorezis P, Iodice C, Rangel WM, Nicolaus B, Vangronsveld J (2016) Plant growth-promoting effects of rhizospheric and endophytic bacteria associated with different tomato cultivars and new tomato hybrids. Chem Biol Technol Agric 3:1. https://doi.org/10.1186/s40538-015-0051-3

Adams DO, Yang SF (1979) Ethylene biosynthesis: identification of 1-aminocyclopropanecarboxylic acid as an intermediate in the conversion of methionine to ethylene. Proc Natl Acad Sci U S A 76:170–174

Ahemad M, Khan MS (2009) Effect of insecticide-tolerant and plant growth-promoting *Mesorhizobium* on the performance of chickpea grown in insecticide stressed alluvial soils. J Crop Sci Biotechnol 12:217–226

Ahmad S, Tabassum H, Alam A (2016) Role of microbial bioremediation of heavy metal from contaminated soils: an update. Int J Biol Pharm Allies Sci 5(7):1605–1622

Aksoy O, Dane F (2007) The effects of fusillade (Fluazifop-p-butyl) on root and shoot growth of lentil (*Lens culinaris Medik.*) seedlings. J Appl Environ Biol Sci 1(3):9–13

Aktar W, Sengupta D, Chowdhury A (2009) Impact of pesticides use in agriculture: their benefits and hazards. Interdiscip Toxicol 2(1):1–12

Ali S, Charles TC, Glick BR (2012) Delay of flower senescence by bacterial endophytes expressing 1-aminocyclopropane-1-carboxylate deaminase. J Appl Microbiol 113:1139–1144

Arao T, Ishikawa S, Murakam IM (2010) Heavy metal contamination of agricultural soil and counter measures in Japan. Paddy Water Environ 8(3):247–257

Ashrafuzzaman M, Hossen FA, Ismail MR, Hoque A, Islam MZ, Shahidullah SM, Meon S (2009) Efficiency of plant growth-promoting rhizobacteria (PGPR) for the enhancement of rice growth. Afr J Biotechnol. 6 8(7):1247–1252

Babu NG, Sarma PA, Attitalla IH, Murthy SDS (2010) Effect of selected heavy metal ions on the photosynthetic electron transport and energy transfer in the thylakoid membrane of the cyanobacterium, *Spirulina platensis*. Acad J Plant Sci 3:46–49

Babu AG, Shea PJ, Sudhakar D, Jung IB, Oh BT (2015) Potential use of *Pseudomonas koreensis* AGB-1 in association with *Miscanthus sinensis* to remediate heavy metal(loid)-contaminated mining site soil. J Environ Manag 151:160–166

Balestrasse KB, Gallego SM, Tomaro ML (2004) Cadmium-induced senescence in nodules of soybean (*Glycine max* L.) plants. Plant Soil 262:373–381

Balestrasse KB, Gallego SM, Benavides MP, Tomaro ML (2005) Polyamines and proline are affected by cadmium stress in nodules and roots of soybean plants. Plant Soil 27:343–353

Ball JW, Izbick JA (2004) Occurrence of hexavalent chromium in ground water in the western Mojave Desert, California. Appl Geochem 19:1123–1135

Bastián F, Cohen A, Piccoli P, Luna V, Bottini R, Baraldi R, Bottini R (1998) Production of indole-3-acetic acid and gibberellins A_1 and A_3 by *Acetobacter diazotrophicus* and *Herbaspirillum seropedicae* in chemically-defined culture media. Plant Growth Regul 24:7–11

Belimov AA, Safronova VI, Mimura T (2002) Response of spring rape to inoculation with plant growth-promoting rhizobacteria containing 1-aminocyclopropane-1-carboxylate deaminase depends on nutrient status of the plant. Can J Microbiol 48:189–199

Belimov AA, Hontzeas N, Safronova VI, Demchinskaya SV, Piluzza G, Bullitta S, Glick BR (2005) Cadmium-tolerant plant growth-promoting bacteria associated with the roots of Indian mustard (*Brassica juncea* L. Czern.) Soil Biol Biochem 37:241–250

Boddey LH, Hungria M (1994) Classificacao de estirpes de Bradyrhizobium japonicum em genotipo I e II baseada nas caracterısticas fenotıpicas e genotıpicas. In: Simposio Brasileiro Sobre Microbiologia de Solo, Londrina. Resumos. LAPAR, Londrina, p 66

Bollag JM, Dec J (1998) Use of plant material for the removal of pollutants by polymerization and binding to humic substances, Tech. Rep. R-82092. Center for Bioremediation and Detoxification Environmental Resources Research Institute The Pennsylvania State University, University Park

Boyd RS (2010) Heavy metal pollutants and chemical ecology: exploring new frontiers. J Chem Ecol 36:46–58

Brown NL, Stoyanov JV, Kidd SP, Hobman JL (2003) The MerR family of transcriptional regulators. FEMS Microbiol Rev 27:145–163

Burd GI, Dixon DG, Glick BR (1998) A plant growth-promoting bacterium that decreases nickel toxicity in seedlings. Appl Environ Microbiol 64:3663–3668

Burgess JL, Bernstein JN, Hurlbut K (1994) Aldicarb poisoning. A case report with prolonged cholinesterase inhibition and improvement after pralidoxime therapy. Arch Intern Med 154(2):221–224
Camacho JR, Armienta MA (2000) Natural chromium contamination of groundwater at Leon Valley Mexico. J Geochem Explor 68:167–181
Campbell PGC (2006) Cadmium-A priority pollutant. Environ Chem 3:387–388
Cargnelutti D, Tabaldi LA, Spanevello RM, Jucoski GO, Battisti V, Redin M, Linares CEB, Dressler VL, Flores MM, Nicoloso FT, Morsch VM, Schetinger MRC (2006) Mercury toxicity induces oxidative stress in growing cucumber seedlings. Chemosphere 65:999–1106
Cempel M, Nikel G (2006) Nickel: a review of its sources and environmental toxicology. Pol J Environ Stud 15:375–382
Chatterjee C, Gopal R, Dube BK (2006) Physiological and biochemical responses of French bean to excess cobalt. J Plant Nutr 29:127–136
Chaudri AM, McGrath SP, Gibbs P, Chambers BC, Carlton-Smith C, Bacon J, Campbell C, Aitken A (2008) Population size of indigenous *Rhizobium leguminosarum biovar trifolii* in long-term field experiments with sewage sludge cake, metal-amended liquid sludge or metal salts: effects of zinc, copper and cadmium. Soil Biol Biochem 40:1670–1680
Chen WM, Tang YQ, Mori K, Wu XL (2012) Distribution of culturable endophytic bacteria in aquatic plants and their potential for bioremediation in polluted waters. Aquat Biol 15:99–110
Chen B, Shen J, Zhang X, Pan F, Yang X, Feng Y (2014) The endophytic bacterium, *Sphingomonas* SaMR12, improves the potential for zinc phytoremediation by its host, *Sedum alfredii*. PLoS One 9(9):e106826. https://doi.org/10.1371/journal.pone.0106826
Cheng Z, Park E, Glick BR (2007) 1-aminocyclopropane-1-carboxylate deaminase from Pseudomonas putida UW4 facilitates the growth of canola in the presence of salt. Can J Microbiol 53:912–918
Clive DS (2003) The pesticide manual, 13th edn. BCPC Publications, Farnham, p 1344. ISBN 978-1-90139613-3
Cohen AC, Travaglia CN, Bottini R, Piccoli PN (2009) Participation of abscisic acid and gibberellins produced by endophytic Azospirillum in the alleviation of drought effects in maize. Botany 87:455–462
Compant S, Reiter B, Sessitsch A, Nowak J, Clement C, Barka EA (2005) Endophytic colonization of *Vitis vinifera* L. by a plant growth-promoting bacterium, Burkholderia sp. strain PsJN. Appl Environ Microbiol 71:1685–1693
Conrath U, Beckers GJM, Flors V, García-Agustín P, Jakab G, Mauch F et al (2006) Priming: getting ready for battle. Mol Plant-Microbe Interact 19:1062–1071
Contesto C, Desbrosses G, Lefoulon C, Béna G, Borel F, Galland M et al (2008) Effects of rhizobacterial ACC deaminase activity on Arabidopsis indicate that ethylene mediates local root responses to plant growth-promoting rhizobacteria. Plant Sci 175:178–189
Coutinho BG, Licastro D, Mendonça-Previato L, Cámara M, Venturi V (2015) Plant-influenced gene expression in the rice endophyte Burkholderia kururiensis M130. Mol Plant Microbe Interact 28:10–21
Cursino L, Mattos SV, Azevedo V, Galarza F, Bücker DH, Chartone-Souza E, Nascimento AM (2000) Capacity of mercury volatilization by *mer* (from *Escherichia coli*) and glutathione S-transferase (from *Schistosoma mansoni*) genes cloned in *Escherichia coli*. Sci Total Environ 261(1):109–113
Dazy M, Masfaraud JF, Ferard JF (2009) Induction of oxidative stress biomarkers associated with heavy metal stress in *Fontinalis antipyretica* Hedw. Chemosphere 75:297–302
de Boer TE, Tas N, Braster M, Temminghoff EJ, Roling WF, Roelofs D (2012) The influence of long-term copper contaminated agricultural soil at different pH levels on microbial communities and springtail transcriptional regulation. Environ Sci Technol 46(1):60–68
Dell'Amico E, Cavalca L, Andreoni V (2005) Analysis of rhizobacterial communities in perennial Graminaceae from polluted water meadow soil, and screening of metal-resistant, potentially plant growth-promoting bacteria. FEMS Microbiol Ecol 52:153–162

Dell'Amico E, Cavalca L, Andreoni V (2008) Improvement of Brassica napus growth under cadmium stress by cadmium-resistant rhizobacteria. Soil Biol Biochem 40:74–84

Desaint S, Hartmann A, Parekh NR, Fournier JC (2000) Genetic diversity of carbofuran-degrading soil bacteria. FEMS Microbiol Ecol 34:173–180

Dixit R, Malaviya D, Pandiyan K, Singh UB, Sahu A, Shukla R, Singh BP, Rai JP, Sharma PK, Lade H, Paul D (2015) Bioremediation of heavy metals from soil and aquatic environment: an overview of principles and criteria of fundamental processes. Sustainability 7(2):2189–2212

Dubey KK, Fulekar MH (2011) Effect of pesticides on the seed germination of *Cenchrus setigerus* and *Pennisetum pedicellatum* as monocropping and co-cropping system: implications for Rhizospheric bioremediation. Rom Biotechnol Lett 16(1):5909–5918

El-Banna N, Winkelmann G (1988) Pyrrolnitrin from Burkholderia cepacia: antibiotic activity against fungi and novel activities against streptomycetes. J Appl Microbiol 85:69–78

Evanko CR, Dzombak DA (1997) Remediation of metals-contaminated soil and groundwater. Environ Sci 412:1–45

Evans ML (1984) Functions of hormones at the cellular level of organization. In: Scott TK (ed) Hormonal regulation of development, II. Springer, Berlin, pp 23–62. ISBN 10:0387101969

Fatima K, Imran A, Amin I, Khan QM, Afzal M (2016) Plant species affect colonization patterns and metabolic activity of associated endophytes during phytoremediation of crude oil contaminated soil. Environ Sci Pollut Res 23:6188–6196

Figueiredo MVB, Martinez CR, Burity HA, Chanway CP (2007) Plant growth promoting rhizobacteria for improving nodulation and nitrogen fixation in the common bean (*Phaseolus vulgaris* L). World J Microbiol Biotechnol. https://doi.org/10.1007/s11274-007-9591-4

Flores F, Collier CJ, Mercurio P, Negri AP (2013) Phytotoxicity of four photosystem II herbicides to tropical seagrasses. PLoS One 8(9):e75798

Gamalero E, Glick BR (2011) Mechanisms used by plant growth-promoting bacteria. In: Maheshwari MK (ed) Bacteria in agrobiology: plant nutrient management. Springer-Verlag, Berlin, pp 17–46

Germaine KJ, Liu X, Cabellos GG, Hogan JP, Ryan D, Dowling DN (2006) Bacterial endophyte-enhanced phytoremediation of the organochlorine herbicide 2,4-dichlorophenoxyacetic acid. FEMS Microbiol Ecol 57:302–310

Gianfreda L, Xu F, Bollag JM (1999) Laccases: a useful group of oxidoreductive enzymes. Biorem J 3:1–26

Gill M (2014) Heavy metal stress in plants: a review. Int J Adv Res 2:1043–1055

Glick BR (2012) Plant growth-promoting bacteria: mechanisms and applications. Scientifica. Article ID 963401. https://doi.org/10.6064/2012/963401

Glick BR (2014) Bacteria with ACC deaminase can promote plant growth and help to feed the world. Microbiol Res 169(1):30–39

Glick BR (2015) Beneficial plant-bacterial interactions. Springer, Heidelberg

Glick BR, Jacobson CB, Schwarze MMK, Pasternak JJ (1994) 1-Aminocyclopropane-1-carboxylic acid deaminase mutants of the plant growth promoting rhyzobacterium Pseudomonas putidaGR 12-2 do not stimulate canola root elongation. Can J Microbiol 40:911–915

Glick BR, Penrose DM, Li J (1998) A model for the lowering of plant ethylene concentrations by plant growth-promoting bacteria. J Theor Biol 190:63–68

Gopi R, Jaleel CA, Sairam R, Lakshmanan GM, Gomathinayagam M, Panneerselvam R (2007) Differential effects of hexaconazole and paclobutrazol on biomass, electrolyte leakage, lipid peroxidation and antioxidant potential of *Daucus carota* L. Colloids Surf B: Biointerfaces 60:180–186

Gravel V, Antoun H, Tweddell RJ (2007) Growth stimulation and fruit yield improvement of greenhouse tomato plants by inoculation with Pseudomonas putida or Trichoderma atroviride: possible role of indole acetic acid (IAA). Soil Biol Biochem 39:1968–1977

Greany KM (2005) An assessment of heavy metal contamination in the marine sediments of Las Perlas Archipelago, Gulf of Panama. M.S. thesis, School of Life Sciences Heriot-Watt University, Edinburgh, Scotland

Guo H, Luo S, Chen L, Xiao X, Xi Q, Wei W, Zeng G, Liu C, Wan Y, Chen J, He Y (2010) Bioremediation of heavy metals by growing hyperaccumulator endophytic bacterium *Bacillus* sp. L14. Bioresour Technol 101:8599–8605

Gupta A, Saxena AK, Gopal M, Tilak KV (1998) Effect of plant growth promoting rhizobacteria on competitive ability of introduced Bradyrhizobium sp. (Vigna) for nodulation. Microbiol Res 153:113–117

Hall JA, Peirson D, Ghosh S, Glick BR (1996) Root elongation in various agronomic crops by the plant growth promoting rhizobacterium Pseudomonas putida GR12-2. Isr J Plant Sci 44:37–42

Hallmann J, Quadt-Hallmann A, Mahaffee WF, Kloepper JW (1997) Bacterial endophytes in agricultural crops. Can J Microb 43:895–914

Hanlon SM, Parris MJ (2012) The impact of pesticides on the pathogen *Batrachochytrium dendrobatidis* independent of potential hosts. Arch Environ Contam Toxicol 63:137–143

Hardoim PR, van Overbeek LS, van Elsas JD (2008) Properties of bacterial endophytes and their proposed role in plant growth. Trends Microbiol 16:463–471

Hasnain S, Sabri AN (1997) Growth stimulation of Triticum aestivum seedlings under Cr-stresses by non-rhizospheric pseudomonad strains. Environ Pollut 97:265–273

He H, Ye Z, Yang D, Yan J, Xiao L, Zhong T, Yuan M, Cai X, Fang Z, Jing Y (2013) Characterization of endophytic *Rahnella* sp. JN6 from *Polygonum pubescens* and its potential in promoting growth and Cd, Pb, Zn uptake by *Brassica napus*. Chemosphere 90:1960–1965

Hirase K, Molin WT (2002) Effects of MT-101 and NOP on germination and seedling growth of hemp sesbania and rice. Weed Sci 50:386–391

Hsu SC, Wong GTF, Gong GC, Shiah FK, Huang YT, Kao SJ, Tsai F, Lung SCC, Lin FJ, Lin II, Hung CC, Tseng CM (2010) Sources, solubility, and dry deposition of aerosol trace elements over the East China Sea. Mar Chem 120:116–127

Huang H, Xiong ZT (2009) Toxic effects of cadmium, acetochlor and bensulfuron-methyl on nitrogen metabolism and plant growth in rice seedlings. Pestic Biochem Physiol 94:64–67

Huffmeyer N, Klasmeier J, Matthies M (2009) Geo-referenced modeling of zinc concentrations in the Ruhr river basin (Germany) using the model GREAT-ER. Sci Total Environ 407:2296–2305

Idris R, Trifonova R, Puschenreiter M, Wenzel WW, Sessitsch A (2004) Bacterial communities associated with flowering plants of the Ni hyperaccumulator *Thaspi goesingense*. Appl Environ Microbiol 70:2667–2677

Idriss EE, Makarewicz O, Farouk A, Rosner K, Greiner R, Bochow H, Richter T, Borriss R (2002) Extracellular phytase activity of Bacillus amyloliquefaciens FZB45 contributes to its plant-growth-promoting effect. Microbiology 148:2097–2109

Israr M, Sahi S, Datta R, Sarkar D (2006) Bioaccumulation and physiological effects of mercury in *Sesbania drummonii*. Chemosphere 65:591–598

Jackson MR (1991) Ethylene in root growth and development. In: Mattoo AK, Suttle JC (eds) The plant hormone ethylene. CRC Press, Boca Raton, pp 151–189

Jain RK, Kapur M, Labana S, Lal B, Sarma PM, Bhattacharya D, Thakur IS (2005) Microbial diversity: application of microorganisms for the biodegradation of xenobiotics. Curr Sci 89:101–112

Jarup L (2014) Hazards of heavy metal contamination. Br Med Bull 68:167–182

Jeon JS, Lee SS, Kim HY, Ahn TS, Song HG (2003) Plant growth promotion in soil by some inoculated microorganisms. J Microbiol 41:271–276

Ji SH, Gururani MA, Chun SC (2014) Isolation and characterization of plant growth promoting endophytic diazotrophic bacteria from Korean rice cultivars. Microbiol Res 169:83–98

Juhasz AL, Stanley GA, Britz ML (2000) Microbial degradation and detoxification of high molecular weight polycyclic aromatic hydrocarbons by *Stenotrophomonas maltophilia* strain VUN 10,003. Lett Appl Microbiol 30:396–401

Kang J, Choi MS, Yi HI, Song YH, Lee D, Cho JH (2011) A five-year observation of atmospheric metals on Ulleung Island in the East/Japan Sea: temporal variability and source identification. Atmos Environ 45:4252–4262

Karigar CS, Rao SS (2011) Role of microbial enzymes in the bioremediation of pollutants: a review. Enzym Res 7:2011. https://doi.org/10.4061/2011/805187

Keller C, McGrath SP, Dunham SJ (2002) Trace metal leaching through a soil-grassland system after sewage sludge application. J Environ Qual 31(5):1550–1560

Khan MS, Chaudhary P, Wani PA, Zaidi A (2006) Biotoxic effects of the herbicides on growth, seed yield and grain protein of green gram. J Appl Sci Environ Manag 10(3):141–146

Khan AL, Halo BA, Elyassi A, Ali S, Al-Hosni K, Hussain J, Al-Harrasi A, Lee IJ (2016) Indole acetic acid and ACC deaminase from endophytic bacteria improves the growth of *Solanum lycopersicum*. Electron J Biotechnol 21:58–64

Khodadoust AP, Reddy KR, Maturi K (2004) Removal of nickel and phenanthrene from kaolin soil using different extractants. Environ Eng Sci 21(6):691–704

Kong Z, Mohamad OA, Deng Z, Liu X, Glick BR, Wei G (2015) Rhizobial symbiosis effect on the growth, metal uptake, and antioxidant responses of *Medicago lupulina* under copper stress. Environ Sci Pollut Res 22(16):12479–12489

Kotasa J, Stasicka Z (2000) Chromium occurrence in the environment and methods of its speciation. Environ Pollut 107:263–283

Kowalski A, Siepak M, Boszke L (2014) Mercury contamination of surface and ground waters of Poznań, Poland. Pol J Environ Stud 16:67–74

Kumar N, Soni JI, Kumar H, Bhatt I (2008) Assessing heavy metal hyperaccumulation and mobility in selected vegetable crops: a case study of organic farm, Gujarat, India. Nat Environ Pollut Tech 7:203–210

Kumar V, Sharma A, Dhunna G, Chawla A, Bhardwaj R, Thukral AK (2017) A tabulated review on distribution of heavy metals in various plants. Environ Sci Pollut Res 24:2210–2260

Lee EY, Jun YS, Cho KS, Ryu HW (2002) Degradation characteristics of toluene, benzene, ethylbenzene, and xylene by *Stenotrophomonas maltophilia* T3-c. J Air Waste Manage Assoc 52:400–406

Lenart A, Koładka KW (2013) The effect of heavy metal concentration and soil pH on the abundance of selected microbial groups within Arcelor Mittal Poland Steelworks in Cracow. Bull Environ Contam Toxicol 90:85–90

Li T, Liu MJ, Zhang XT, Zhang HB, Sha T, Zhao ZW (2011) Improved tolerance of maize (Zea mays L.) to heavy metals by colonization of a dark septate endophyte (DSE) *Exophiala pisciphila*. Sci Total Environ 409(6):1069–1074

Li HY, Wei DQ, Shen M, Zhou ZP (2012) Endophytes and their role in phytoremediation. Fungal Divers 54:11–18. https://doi.org/10.1007/s13225-012-0165-x

Li Y, Yang R, Zhang A, Wang S (2014) The distribution of dissolved lead in the coastal waters of the East China Sea. Mar Pollut Bull 85:700–709

Li Y, Wang Q, Wang L, He LY, Sheng XF (2016) Increased growth and root Cu accumulation of *Sorghum sudanense* by endophytic *Enterobacter* sp. K3-2: implications for *Sorghum sudanense* biomass production and phytostabilization. Ecotoxicol Environ Saf 124:163–168. https://doi.org/10.1016/j.ecoenv.2015.10.012

Lifshitz M, Shahak E, Bolotin A, Sofer S (1997) Carbamate poisoning in early childhood and in adults. Clin Toxicol 35(2):5–27

Liu YY, Xiong Y (2001) Purification and characterization of a dimethoate-degrading enzyme of *Aspergillus niger* ZHY256 isolated from sewage. Appl Environ Microbiol 67:3746–3749

Lockert CK, Hoagland KD, Siegfrie BD (2006) Comparative sensitivity of freshwater algae to atrazine. Bull Environ Contam Toxicol 76:73–77

Lodewyckx C, Vangronsveld J, Porteous F, Moore ERB, Taghavi S, Mezgeay M, Lelie DVD (2002) Endophytic bacteria and their potential applications. Crit Rev Plant Sci 21:583–606

Lopez L, Pozo C, Rodelas B, Calvo C, Juarez B, MartinezToledo MV, Gonzalez-Lopez J (2005) Identification of bacteria isolated from an oligotrophic lake with pesticide removal capacities. Ecotoxicology 14:299–312

Luo SL, Wan Y, Xiao X, Guo H, Chen L, Xi Q, Zeng G, Liu C, Chen J (2011) Isolation and characterization of endophytic bacterium LRE07 from cadmium hyperaccumulator *Solanum nigrum* L. and its potential for remediation. Appl Microbiol Biotechnol 89:1637–1644

Ma J, Lu N, Qin W, Xu R, Wang Y, Chen X (2006) Differential responses of eight cyanobacterial and green algal species to carbamate insecticides. Ecotoxicol Environ Saf 63:268–274

Ma Y, Rajkumar M, Luo Y, Freitas H (2011) Inoculation of endophytic bacteria on host and non-host plants—effects on plant growth and Ni uptake. J Hazard Mater 195:230–237

Ma Y, Oliveira RS, Nai FJ, Rajkumar M, Luo YM, Rocha I, Freitas H (2015) The hyperaccumulator *Sedum plumbizincicola* harbors metal-resistant endophytic bacteria that improve its phytoextraction capacity in multi-metal contaminated soil. J Environ Manag 156:62–69

Ma Y, Zhang C, Oliveira RS, Freitas H, Luo Y (2016) Bioaugmentation with endophytic bacterium E6S homologous to *Achromobacter piechaudii* enhances metal Rhizoaccumulation in host *Sedum plumbizincicola*. Front Plant Sci 7:75. https://doi.org/10.3389/fpls.2016.00075

Mackie KA, Müller T, Zikeli S, Kandeler E (2013) Long-term copper application in an organic vineyard modifies spatial distribution of soil micro-organisms. Soil Biol Biochem 65:245–253

Magnusson M, Heimann K, Negri AP (2008) Comparative effects of herbicides on photosynthesis and growth of tropical estuarine microalgae. Mar Pollut Bull 56:1545–1552

Mastretta C, Taghavi S, Lelie DVD, Mengoni A, Galardi F, Gonnelli C, Barac T, Boulet J, Weyens N, Vangronsveld J (2009) Endophytic bacteria from seeds of *Nicotiana tabacum* can reduce cadmium phytotoxicity. Int J Phytoremediation 11:251–267

McLaren RG, Clucas LM, Taylor MD, Hendry T (2004) Leaching of macronutrients and metals from undisturbed soils treated with metal-spiked sewage sludge. 2. Leaching of metals. Aust J Soil Res 42(4):459–471

McLaughlin MJ, Hamon RE, McLaren RG, Speir TW, Rogers SL (2000) Review: a bioavailability-based rationale for controlling metal and metalloid contamination of agricultural land in Australia and New Zealand. Aust J Soil Res 38(6):1037–1086

Mesa J, Mateos-Naranjo E, Caviedes MA, Redondo-Gómez S, Pajuelo E, Rodríguez-Llorente ID (2015) Endophytic cultivable bacteria of the metal bioaccumulator *Spartina maritime* improve plant growth but not metal uptake in polluted marshes soils. Front Microbiol 6:1450. https://doi.org/10.3389/fmicb.2015.01450

Mesquini JA, Sawaya ACHF, Lopez BGC, Oliveira VM, Miyasaka NRS (2015) Detoxification of atrazine by endophytic *Streptomyces* sp. isolated from sugarcane and detection of non-toxic metabolite. Bull Environ Contam Toxicol 95:803–809. https://doi.org/10.1007/s00128-015-1673-7

Miliūtė I, Buzaitė O (2011) IAA production and other plant growth promoting traits of endophytic bacteria from apple tree. Biologija 57:98–102

Mishra V, Srivastava G, Prasad SM (2009) Antioxidant response of bitter gourd (Momordica charantia L.) seedlings to interactive effect of dimethoate and UV-B irradiation. Hortic Sci 120:373–378

Mithofer A, Schulze B, Boland W (2004) Biotic and heavy metal stress response in plants: evidence for common signals. FEBS Lett 566:1–5

Mohod CV, Dhote J (2013) Review of heavy metals in drinking water and their effect on human health. Int J Innov Res Sci Eng Technol 2(7):2992–2996

Mota R, Pereiraa SB, Meazzinid M, Fernandesa R, Santosa A, Evanse CA, Philippisd RD, Wrighte PC, Tamagnini P (2015) Effects of heavy metals on *Cyanothece* sp. CCY 0110 growth, extracellular polymeric substances (EPS) production, ultrastructure and protein profiles. J Proteome 120:75–94

Mukherjee A, Sengupta MK, Hossain MA, Ahamed S, Das B, Nayak B, Lodh D, Rahman M, Chakraboti D (2006) Arsenic contamination in groundwater: a global perspective with emphasis on the Asian scenario. J Health Popul Nutr 24:142–163

Murphy JF, Reddy MS, Ryu CM, Kloepper JW, Li R (2003) Rhizobacteria-mediated growth promotion of tomato leads to protection against Cucumber mosaic virus. Phytopathology 93:1301–1307

Navarro L, Dunoyer P, Jay F, Arnold B, Dharmasiri N, Estelle M, Voinnet O, Jones JDG (2006) A plant miRNA contributes to antibacterial resistance by repressing auxin signaling. Science 312:436–439

Navarro-Torre S, Mateos-Naranjo E, Caviedes MA, Pajuelo E, Rodríguez-Llorente ID (2016) Isolation of plant-growth-promoting and metal-resistant cultivable bacteria from *Arthrocnemum macrostachyum* in the Odiel marshes with potential use in phytoremediation. Mar Pollut Bull 110(1):133–142. https://doi.org/10.1016/j.marpolbul.2016.06.070

Nawaz K, Hussain K, Choudary N, Majeed A, Ilyas U, hani A, Lin F, Ali K, Afghan S, Raza G, Lashar MI (2011) Eco-friendly role of biodegradation against agricultural pesticides hazards. Afr J Microbiol Res 5(3):177–183

Nivya RM (2015) A study on plant growth promoting activity of the endophytic bacteria isolated from the root nodules of *Mimosa Pudica* plant. Int J Innov Res Sci Eng Tech 4:6959–6968

Nordstrom DK (2002) Worldwide occurrences of arsenic in ground water. Science 296:2143–2145

Pan F, Meng Q, Luo S, Shen J, Chen B, Khan KY, Japenga J, Ma X, Yang X, Feng Y (2016) Enhanced cd extraction of oilseed rape (*Brassica napus*) by plant growth promoting bacteria isolated from Cd Hyperaccumulator *Sedum alfredii* Hance. Int J Phytoremediation. https://doi.org/10.1080/15226514.2016.1225280

Papadopoulou ES, Tsachidou B, Sławomir Sułowicz S, Menkissoglu-Spiroudi U, Karpouzas DG (2015) Land spreading of wastewaters from the fruit-packaging industry and potential effects on soil microbes: effects of the antioxidant Ethoxyquin and its metabolites on ammonia oxidizers. Appl Environ Microbiol 82:747–755

Park JW, Park BK, Kim JE (2006) Remediation of soil contaminated with 2, 4-dichlorophenol by treatment of minced shepherd's purse roots. Arch Environ Contam Toxicol 50:191–195

Parvaiz KV, Singh N, Behl HM, Srivastava S (2008) Influence of plant growth promoting bacteria and its mutant on heavy metal toxicity in *Brassica juncea* grown in fly ash amended soil. Chemosphere 72:678–683

Parween T, Jan S, Mahmooduzzafar, Fatma T (2011) Assessing the impact of Chlorpyrifos on growth, photosynthetic pigments and yield in Vigna radiata L. at different phenological stages. Afr J Agric Res 6:4432–4440

Parween T, Jan S, Mahmooduzzafar S, Fatma T, Siddiqui ZH (2016) Selective effect of pesticides on plant—a review. Crit Rev Food Sci Nutr 56(1):160–179

Pereira SI, Castro PM (2014) Diversity and characterization of culturable bacterial endophytes from Zea mays and their potential as plant growth-promoting agents in metal-degraded soils. Environ Sci Pollut Res 21:14110–14123

Phetcharat P, Duangpaeng A (2012) Screening of endophytic bacteria from organic rice tissue for indole acetic acid production. Procedia Eng 32:177–183

Ping L, Boland W (2004) Signals from the underground: bacterial volatiles promote growth in Arabidopsis. Trends Plant Sci 9:263–266

Porto AM, Melgar GZ, Kasemodel MC, Nitschke M (2011) Biodegradation of pesticides, pesticides in modern world-pesticides use and management. [online]. http://www.intechopen.com/books/pesticides-in-the-modern-world-pesticidesuse-and-management/biodegradation-of-pesticides

Rajashekhar N, Prakasha, Murthy TCS (2012) Seed germination and physiological behavior of Maize (cv. Nac-6002) seedlings under abiotic stress (Pendimethalin) condition. Asian J Crop Sci 4(2):80–85

Rajkumar M, Ae N, Freitas H (2009) Endophytic bacteria and their potential to enhance heavy metal phytoextraction. Chemosphere 77:153–160

Rao MA, Scelza R, Scotti R, Gianfreda L (2010) Role of enzymes in the remediation of polluted environments. J Soil Sci Plant Nutr 10:333–353

Razo I, Carrizales L, Castro J, Barriga FD, Manroy M (2003) Arsenic and heavy metal pollution of soil, water and sediments in a semi-arid climate mining area in Mexico. Water Air Soil Pollut 152:129–152

Rosenblueth M, Martínez-Romero E (2006) Bacterial endophytes and their interactions with hosts. Mol Plant-Microbe Interact 19(8):827–837. https://doi.org/10.1094/MPMI-19-0827

Rosendahl L, Jochimsen BV (1995) Uptake of indoleacetic acid in symbiosomes from soybean (*Glycine max* L.) root nodules. In: Nitrogen fixation: fundamentals and applications. Proceedings. 10. International Congress of Nitrogen Fixation, 11, St. Petersburg. Resumos. St. Petersburg, Russian Academy of Sciences, Russia, p 336

Ryan RP, Germaine K, Franks A, Ryan DJ, Dowling DN (2008) Bacterial endophytes: recent developments and applications. FEMS Microbiol Lett 278:1–9

Ryu CM, Farag MA, Hu CH, Reddy MS, Kloepper JW, Paré PW (2004) Bacterial volatiles induce systemic resistance in Arabidopsis. Plant Physiol 134:1017–1026

Saharan BS (2011) Plant growth promoting rhizobacteria: a critical review. Life Sci Med Res 21:1–30

Salisbury FB (1994) Role of plant hormones. In: Wilkinson RE (ed) Plant-environment interactions. Marcel Dekker, New York, pp 39–81

Santoyo G, Moreno-Hagelsieb G, del Carmen Orozco-Mosqueda M, Glick BR (2016) Plant growth-promoting bacterial endophytes. Microbiol Res 183:92–99

Sayler GS, Hooper SW, Layton AC, King JMH (1990) Catabolic plasmids of environmental and ecological significance. Microb Ecol 19:1–20

Schulz B, Boyle C (2006) What are endophytes? In: Schulz BJE, Boyle CIC, Sieber TN (eds) Microbial root endophytes. Springer-Verlag, Berlin, pp 1–13

Sharma P, Kumar A, Bhardwaj R (2016a) Plant steroidal hormone Epibrassinolide regulate-heavy metal stress tolerance in *Oryza sativa* L. by modulating antioxidant defense expression. Environ Exp Bot 122:1–9

Sharma A, Kumar V, Singh R, Thukral AK, Bhardwaj R (2016b) Effect of seed pre-soaking with 24-epibrassinolide on growth and photosynthetic parameters of *Brassica juncea* L. in imidacloprid soil. Ecotoxicol Environ Saf 133:195–201

Sharma A, Bhardwaj R, Kumar V, Thukral AK (2016c) GC-MS studies reveal stimulated pesticide detoxification by brassinolide application in *Brassica juncea* L. plants. Environ Sci Pollut Res 23:14518–14525

Sharma A, Thakur S, Kumar V, Kesavan AK, Thukral AK, Bhardwaj R (2017) 24-epibrassinolide stimulates imidacloprid detoxification by modulating the gene expression of Brassica juncea L. BMC Plant Biol 17(1):56

Sheng X-F, Xia J-J, Jiang C-Y, He L-Y, Qian M (2008) Characterization of heavy metal-resistant endophytic bacteria from rape (*Brassica napus*) roots and their potential in promoting the growth and lead accumulation of rape. Environ Pollut 156:1164–1170

Sheoran N, Nadakkakath AV, Munjal V, Kundu A, Subaharan K, Venugopal V, Rajamma S, Eapen SJ, Kumar A (2015) Genetic analysis of plant endophytic Pseudomonas putida BP25 and chemo-profiling of its antimicrobial volatile organic compounds. Microbiol Res 173:66–78

Shi YW, Lou K, Li C (2009) Promotion of plant growth by phytohormone producing endophytic microbes of sugar beet. Biol Fertil Soils 45:645–653

Shin MN, Shim J, You Y, Myung H, Bang KS, Cho M, Kamala-Kannan S, Oh BT (2012) Characterization of lead resistant endophytic *Bacillus* sp. MN3-4 and its potential for promoting lead accumulation in metal hyperaccumulator *Alnus firma*. J Hazard Mater 199:314–320

Shu X, Yin L, Zhang Q, Wang W (2012) Effect of Pb toxicity on leaf growth, antioxidant enzyme activities, and photosynthesis in cuttings and seedlings of *Jatropha curcas* L. Environ Sci Pollut Res 19:893–902

Singh DK (2008) Biodegradation and bioremediation of pesticide in soil: concept, method and recent developments. Indian J Microbiol 48:35–40

Song NH, Yin XL, Chen GF, Yang H (2007) Biological responses of wheat *(Triticum aestivum)* plants to the herbicide chlorotoluron in soils. Chemosphere 68(9):1779–1787

Spaepen S, Vanderleyden J, Remans R (2007) Indole-3-acetic acid in microbial and microorganism-plant signaling. FEMS Microbiol Rev 31:425–448

Staley ZR, Harwood VJ, Rohr JR (2015) A synthesis of the effects of pesticides on microbial persistence in aquatic ecosystems. Crit Rev Toxicol 45(10):813–836

Stefan M, Mihasan M, Dunca S (2008) Plant growth promoting rhizobacteria can inhibit the in vitro germination of *Glycine max* L seeds. Scientific annals of university "Alexandru Ioan Cuza" Iasi. Sect Genet Mol Biol 3:105–110

Sumner ME (2000) Beneficial use of effluents, wastes, and biosolids. Commun Soil Sci Plant Anal 31:1701–1715

Sun LN, Zhang YF, He LY, Chen ZJ, Wang QY, Qian M, Sheng XF (2010) Genetic diversity and characterization of heavy metal-resistant-endophytic bacteria from two copper-tolerant plant species on copper mine wasteland. Bioresour Technol 101(2):501–509. https://doi.org/10.1016/j.biortech.2009.08.011

Sun L, Wang X, Li Y (2015) Increased plant growth and copper uptake of host and non-host plants by metal-resistant and plant growth-promoting endophytic bacteria. Int J Phytoremediation. https://doi.org/10.1080/15226514.2015.1115962

Tekaya N, Saiapina O, Ouada HB, Lagarde F, Ouada HB, Renault NJ (2013) Ultra-sensitive conductometric detection of heavy metals based on inhibition of alkaline phosphatase activity from *Arthrospira platensis*. Bioelectrochemistry 90:24–29

Terry PA, Stone W (2002) Biosorption of cadmium and copper contaminated water by Scenedesmus abundans. Chemosphere 47:249–255

Tétard-Jones C, Edwards R (2016) Potential roles for microbial endophytes in herbicide tolerance in plants. Pest Manag Sci 72(2):203–209. https://doi.org/10.1002/ps.4147

Thijs S, Van Dillewijn P, Sillen W, Truyens S, Holtappels M, Carleer R, Weyens N, Ameloot M, Ramos JL, Vangronsveld J (2014) Exploring the rhizospheric and endophytic bacterial communities of Acer pseudoplatanus growing on a TNT-contaminated soil: towards the development of a rhizocompetent TNT-detoxifying plant growth promoting consortium. Plant Soil 385:15–36

Tripathi RD, Srivastava S, Mishra S, Singh N, Tuli R, Gupta DK, Maathuis FJM (2007) Arsenic hazards: strategies for tolerance and remediation by plants. Trends Biotechnol 25:158–165

Turan K, Levent K, Muhittin S, Songul K (2008) Bacterial biodegradation of aldicarb and determination of bacterium which has the most biodegradative effect. Turk J Biochem 33:209–214

Ullah A, Heng S, Munis MF, Fahad S, Yang X (2015) Phytoremediation of heavy metals assisted by plant growth promoting (PGP) bacteria: a review. Environ Exp Bot 117:28–40

Vasileva-Tonkova E, Galabova D (2003) Hydrolytic enzymes and surfactants of bacterial isolates from lubricant-contaminated wastewater. Z Naturforsch 58:87–92

Vauterin L, Swings J (1997) Are classification and phytopathological diversity compatible in Xanthomonas. J Ind Microbiol Biotechnol 19:77–82

Vesely T, Neuberg M, Trakal L, Szakova J, Tlustoa P (2011) Water lettuce *Pistia stratiotes* L. response to lead toxicity. Water Air Soil Pollut 223:1847–1859

Viers J, Olivia P, Nonell A, Gelabert A, Sonke JE, Freydier R, Gainville R, Dupre B (2007) Evidence of Zn isotopic fractionation in a soil–plant system of a pristine tropical watershed (Nsimi, Cameroon). Chem Geol 239:124–137

Visioli G, D'Egidio S, Vamerali T, Mattarozzi M, Sanangelantoni AM (2014) Culturable endophytic bacteria enhance Ni translocation in the hyperaccumulator *Noccaea caerulescens*. Chemosphere 117:538–544

Visioli G, Vamerali T, Mattarozzi M, Dramis L, Sanangelantoni AM (2015) Combined endophytic inoculants enhance nickel phytoextraction from serpentine soil in the hyperaccumulator *Noccaea caerulescens*. Front Plant Sci 6:638

Wan Y, Luo S, Chen J, Xiao X, Chen L, Zeng G, Liu C, He Y (2012) Effect of endophyte-infection on growth parameters and Cd-induced phytotoxicity of Cd-hyperaccumulator *Solanum nigrum* L. Chemosphere 89(6):743–750

Wang W-D, Niu JL, Cui ZJ (2005) Biodegradation of pesticides: a review. J Heilongj Aug First Land Reclam Univ 17(2):18

Weckx J, Vangronsveld J, Clijster H (1993) Heavy metal induction of ethylene production and stress enzymes. I. Kinetics of response. In: Pech JC, Latche A, Balaque C (eds) Cellular and molecular aspects of the plant hormone ethylene. Kluwer Academic Publishers, Dordrecht, pp 238–239

Weyens N, Beckers B, Schellingen K, Ceulemans R, Croes S, Janssen J, Haenen S, Witters N, Vangronsveld J (2013) Plant-associated bacteria and their role in the success or failure of metal phytoextraction projects: first observations of a field-related experiment. Microb Biotechnol 6(3):288–299. https://doi.org/10.1111/1751-7915.12038

Wood TK (2008) Molecular approaches in bioremediation. Curr Opin Biotechnol 19:572–578

Xia XJ, Zhang Y, Wu JX, Wang JT, Zhou YH, Shi K, Yu YL, Yu JQ (2009) Brassinosteroids promote metabolism of pesticides in cucumber. J Agric Food Chem 57(18):8406–8413

Xinxian L, Xuemei C, Yagang C, Woon-Chung WJ, Zebin W, Qitang W (2011) Isolation and characterization endophytic bacteria from hyperaccumulator *Sedum alfredii* Hance and their potential to promote phytoextraction of zinc soil. World J Microbiol Biotechnol 27:1197–1207. https://doi.org/10.1007/s11274-010-0568-3

Yadav S, Prajapati R, Atri N (2016) Effects of UV-B and heavy metals on nitrogen and phosphorus metabolism in three cyanobacteria. J Basic Microbiol 56:2–13

Yuan M, He H, Xiao L, Zhong T, Liu H, Li S, Deng P, Ye Z, Jing Y (2014) Enhancement of Cd phytoextraction by two Amaranthus species with endophytic *Rahnella* sp. JN27. Chemosphere 103:99–104

Yuan L, Zhi W, Liu Y, Karyala S, Vikesland PJ, Chen X, Zhang H (2015) Lead toxicity to the performance, viability, and community composition of activated sludge microorganisms. Environ Sci Technol 49(2):824–830

Zhang X, Li C, Nan Z (2010) Effects of cadmium stress on growth and anti-oxidative systems in *Achnatherum inebrians* symbiotic with *Neotyphodium* gansuense. J Hazard Mater 175(1):703–709

Zhang WJ, Jiang FB, Ou JF (2011a) Global pesticide consumption and pollution: with China as a focus. Proc Int Acad Ecol Environ Sci 1(2):125–144

Zhang YF, He LY, Chen ZJ, Zhang WH, Wang QY, Qian M, Sheng XF (2011b) Characterization of lead-resistant and ACC deaminase-producing endophytic bacteria and their potential in promoting lead accumulation of rape. J Hazard Mater 186:1720–1725

Zhang YF, He LY, Chen ZJ, Wang QY, Qian M, Sheng XF (2011c) Characterization of ACC deaminase-producing endophytic bacteria isolated from copper-tolerant plants and their potential in promoting the growth and copper accumulation of *Brassica napus*. Chemosphere 83:57–62

Zhang X, Lin L, Zhu Z, Yang X, Wang Y, An Q (2013) Colonization and modulation of host growth and metal uptake by endophytic bacteria of *Sedum Alfredii*. Int J Phytoremediation 15:51–64

Zhang WH, He LY, Wang Q, Sheng XF (2015) Inoculation with endophytic *Bacillus* megaterium 1Y31 increases Mn accumulation and induces the growth and energy metabolism-related differentially-expressed proteome in Mn hyperaccumulator hybrid *pennisetum*. J Hazard Mater 300:513–521

Zhu LJ, Guan DX, Luo J, Rathinasabapathi B, Ma LQ (2014) Characterization of arsenic-resistant endophytic bacteria from hyperaccumuators *Pteris vittata* and *Pteris multifida*. Chemosphere 113:9–16

Chapter 15
Microbial Siderophores in Metal Detoxification and Therapeutics: Recent Prospective and Applications

Resham Sharma, Renu Bhardwaj, Vandana Gautam, Sukhmeen Kaur Kohli, Parminder Kaur, Ravinder Singh Bali, Poonam Saini, Ashwani Kumar Thukral, Saroj Arora, and Adarsh Pal Vig

Abstract Siderophores are small molecular weight metal scavengers which are released by plants, plant growth-promoting bacterial strains and fungi into the rhizosphere. These molecules have been widely reported as Fe^{3+} carriers under poor iron ion mobilization; however, recently they are being exposed for affinity towards other metal ions such as copper, zinc, etc. highlighting their phytoremedial potential. They are also effective anti-pathogenic agents, important signals towards oxidative stress and new age therapeutics. To understand the mechanism by which these moieties solubilize metal ions at both genetic and protein levels is the crux of our studies as these are extremely versatile molecules having myriad applications in the fields of agriculture, physiology, drug therapy, diagnosis, etc. Additionally, this paper also covers the biosynthesis and classification of microbial siderophores and their roles in plant and animal physiology.

Keywords Detoxification • Ferric ions • Metal scavengers • Siderophores • Medicinal applications

15.1 Introduction

Since the past six decades, various studies have been focused on some 500 small low-molecular-mass (≤10 kDa) molecules called the siderophores which are secreted by both plants and microbes into the rhizosphere (Hider and Kong 2010;

R. Sharma (✉) • R. Bhardwaj • V. Gautam • S.K. Kohli • P. Kaur
R.S. Bali • P. Saini • A.K. Thukral • S. Arora • A.P. Vig
Department of Botanical & Environmental Sciences, Guru Nanak Dev University, Amritsar, Punjab, India
e-mail: reshamksharma12@gmail.com

D. Egamberdieva, P. Ahmad (eds.), *Plant Microbiome: Stress Response*, Microorganisms for Sustainability 5, https://doi.org/10.1007/978-981-10-5514-0_15

Ahmed and Holmström 2014; Johnstone and Nolana 2015). The bacterial siderophores possess higher affinity for metal ions (especially ferric ions or Fe^{+3}) as compared to phytosiderophores (mainly mugineic acid) and are often present in lower concentrations (Kraemer 2004; Kraemer et al. 2006; Glick 2012). Many physiochemical factors such as ligand-binding sites onto metal ion, denticity, pH, redox, etc. govern the metal-binding ability of these molecules (Akafia et al. 2014). Their role as metal scavengers in the rhizosphere is very well known specially for iron ions (Aznar and Dellagi 2015). Other pertinent functions of siderophores include their antibiotic activity against many resistant bacterial strains and potential superbugs as sideromycins (Braun et al. 2009). They possess strong affinity for non-iron metal ions such as copper, manganese, molybdenum, vanadium, zinc, etc. (Hood and Skar 2012). Siderophores from bacterial strains bind to Zn to form zincophores or tsinkophores (Prentice et al. 2007), *Pseudomonad* strains show affinity for Mn (Harrington et al. 2012; Duckworth et al. 2014), and several other microbial siderophores attract Mo and Vn to form stable complexes (Deicke et al. 2013). Methanobactin is a copper-binding compound (CBC) or chalkophores (Kenney and Rosenzweig 2012), and other well-known copper siderophores include coproporphyrin and yersiniabactin (Chaturvedi et al. 2014). Enterobactin, yersiniabactin and aerobactin possess the capacity to form gold nanoparticles (Wyatt et al. 2014). Their active role as transporters of non-metal moieties such as boron and silicon and signalling molecules in plant defence mechanisms and oxidative stress is also well documented now (Chaturvedi and Henderson 2014; Butler and Theisen 2010; Nadal-Jimenez et al. 2012). Stable non-metal-siderophore complexes exist due to marine siderophores such as vibrioferrein from *Marinobacter* spp. (Amin et al. 2007). Citrate and catecholate siderophores interact and bind to boron to form strong signalling and sensing molecules (Sandy and Butler 2009). Additionally, they have gained relevance as therapeutic and diagnostic molecules in medical science (Ali and Vidhale 2013). Antioxidant and hormonal signalling cascades indicate the role of iron siderophores in various spheres of biology (Aznar et al. 2015). More recently, siderocalin, a mammalian siderophore-binding protein from the lipocalin family, specifically binding to actinide and lanthanide complexes, has been discovered (Allred et al. 2015). Despite the mind-boggling diversity on display across the microbial, plant and mammalian spheres, this review focuses on the most pertinent and applicative aspects of microbial siderophores in agriculture, therapeutics, etc.

15.2 Biosynthesis, Classification and Functional Diversity

Siderophores are synthesized by several bacteria and show significant variation in structures. These are mainly classified on the basis of characterization of functional or coordinating groups that bind with Fe^{3+} ions. The most important and commonly occurring groups include catecholates, hydroxamates and carboxylates (Ali and Vidhale 2013; Sah and Singh 2015; Gupta et al. 2015) (Fig. 15.1). A very small group of siderophores include pyoverdines, which are also termed as mixed ligands.

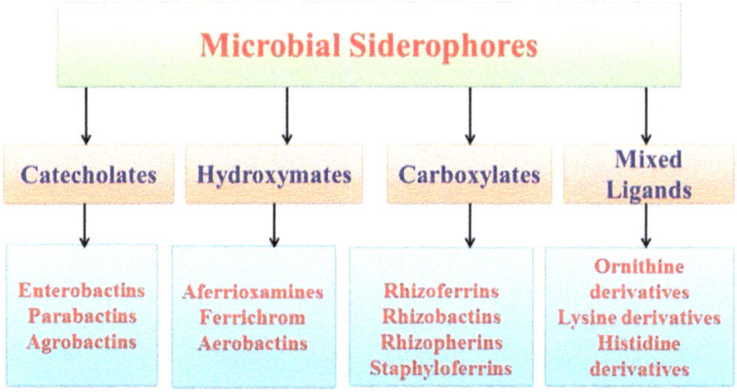

Fig. 15.1 Classification of microbial siderophores: an outline

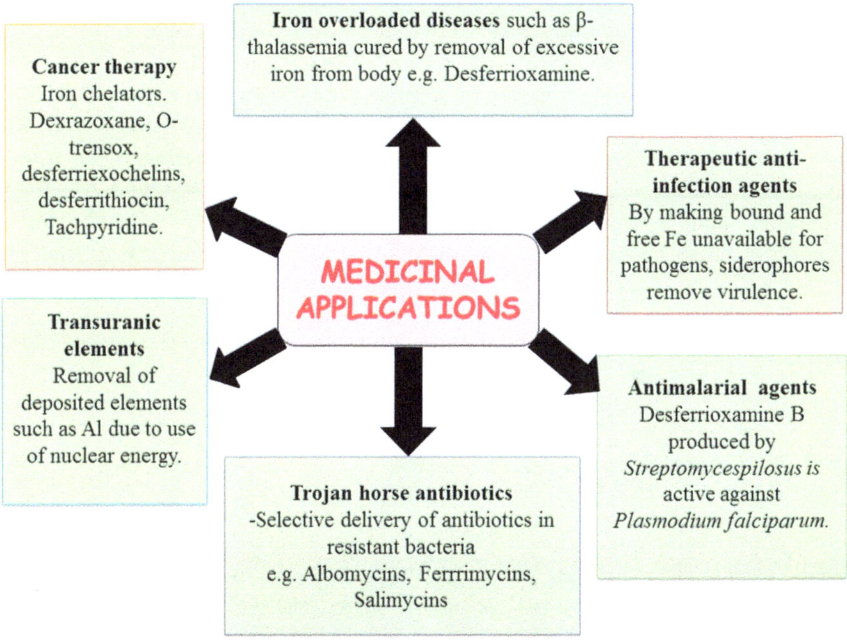

Fig. 15.2 Therapeutic approaches and applications

They constitute the fourth class of siderophores which have functional groups that are classified in chemically distinct classes. Numerous types of siderophores have been identified employing latest techniques of spectrophotometry, mass spectrometry, acid hydrolysis, electrophoretic mobility, proton NMR (nuclear magnetic resonance) spectroscopy and biological activities (Sah and Singh 2015; Kurth et al. 2016) (Fig. 15.2).

15.2.1 Biosynthesis

Microbial siderophore synthesis takes place through two pathways: non-ribosomal peptide synthetases (NRPSs) multienzyme dependent and NRPS independent (Sah and Singh 2015). NRPS-dependent biosynthesis involves the enzyme ATP pyrophosphate for the formation of hydroxamate siderophores (Lautru and Challis 2004). NRPSs are enzymes with large subunits which catalyse non-ribosomal peptide (NRP) synthesis by incorporating one amino acid per unit into the peptide chain. For instance, NRPSs synthesize the chromophores and the peptide chains of the microbial siderophore pyoverdine (Mossialos et al. 2002; Crosa and Walsh 2002).

Peptide synthetase is a multicomplex enzyme that produces peptide products without RNA template. 2, 3-dihydroxybenzoic acid (DHBA) is one of the precursor compounds of siderophores, which is synthesized from chorismate through the sequential action of a series of enzymes (Farrell et al. 1990). For instance, in anguibactin, the coordinating bonds are synthesized by molecular oxygen from various groups such as diphenoxylate group, hydroxamate group, imidazole group and thiazoline group. The structure of anguibactin is completed by two molecules of anguibactin, metal ion and solvent each. Anguibactin retrobiosynthesis, {(−N-hydroxy-N)- [2-(2,3-dihydroxyphenyl) thiazolin-4-yl] carboxyl} involving histamine, indicates the presence of 2,3-dihydroxybenzoic acid (DHBA), L-cysteine and N-hydroxy-histamine. In retrobiosynthesis of vibriobactin in *V.cholerae*, N1-(2,3-dihydroxybenzoyl)-N5,N9-bis[2- (2,3-dihydroxyphenyl)-5-methyloxazolinyl-4-carboxamido] norspermidine shows that it is comprised of DHBA, L-threonine and unusual polyamine norspermidine [bis (3-aminopropyl) amine] (Keating and Walsh 1999; Yamamoto et al. 1991).

15.2.2 Catecholate Siderophores

Siderophores belonging to the catecholate category have 2, 3-dihydroxybenzoate (DHB) or phenolate chelating groups as functional moieties (Table 15.1). They are also termed as pyrocatechols or 1, 2-dihydroxybenzene [$C_6H_4(OH)_2$] (Cornish and Page 1998; Wittmann et al. 2001). Every catecholate group bestows two oxygen atoms to chelate with Fe ions by forming bidentate ligand complexes. As a result of this, a hexadentate octahedral complex is formed (Ali and Vidhale 2013). Catecholates are naturally occurring colourless compounds and are present as trace amounts in environment. They are composed of three isomeric benzenediols which make them an orthoisomeric molecule. One of the most important catecholate widely characterized is enterobactin or enterochelin. It is a prototype of catecholate siderophore and has a cyclic trimester coordinating group (2,3-dyhydroxyserine). It has been reported to be produced by *Salmonella typhimurium* and *Klebsiella pneumoniae* (Ali and Vidhale 2013; Achard et al. 2013).

Table 15.1 Siderophores in plant roles and associated chemical moieties

Sr.no	Siderophore	Source of siderophore	Role in plants	References
1.	Catecholate (2,3-dihydroxybenzoate-DHB)	*Enterobacteriaceae*	Chelating agent for Fe^{3+} with agricultural applications	Wittmann et al. (2001), Gregory et al. (2012) and Sah and Singh (2015)
	(a) Enterobactin	*Enterobacteriaceae*	Aids in transportation and capture of ferric ions	Höfte (1993), Ward et al. (1999) and Sah and Singh (2015)
		Escherichia coli		
		Aerobacter aerogenes		
		Salmonella typhimurium	Nitrogen fixing	Ward et al. (1999)
		Klebsiella pneumoniae	Oxidative stress response	Raymond et al. (2003)
		Erwinia herbicola		
	(b) Agrobactin (linear catecholate)	*Agrobacterium tumefaciens*	Iron-chelating agent	Leong and Neilands (1982)
	(c) Parabactin (linear catecholate)	*Paracoccus denitrificans*	Iron-chelating agent	Leong and Neilands (1982)
2.	Rhizobactin (diaminopropane acytylated with citric acid via amine bonds to terminal carboxylate of citric acid)	Fungi, member of zygomycetes	Biotechnological applications	Gregory et al. (2012)
			Metal-chelating properties	Sah and Singh (2015)
	(a) Rhizoferrin	*Rhizobium meliloti*	Produces R, R-rhizoferrin	Ali and Vidhale (2013)
	(b) Staphyloferrin	*Staphylococcus hyicus*	Biodegradable in nature	Munzinger et al. (1999)
3.	Hydroxamate type (ferrioxamines)	*Frankia* strains (HsIi2, HsIi4 and CpI2)	Siderophores minimize the Mg^{2+}, Cu^{2+} and Zn^{2+} metal-induced growth inhibition in *Frankia* by acting as decontaminant agents in the moderate metal-contaminated soil	Singh et al. (2010)
4.	Hydroxamate type(aerobactin)	*Escherichia coli*, *Klebsiella pneumoniae*	Sequester iron in iron-poor environments	Gregory et al. (2012) and Buyer et al. (1991)
		Aerobacter aerogenes		
5.	Hydroxamate type (aerobactin)	*Pseudomonas*	Increased seed germination, root length, foliage and chlorophyll content	Manwar et al. (2001)

(continued)

Table 15.1 (continued)

Sr.no	Siderophore	Source of siderophore	Role in plants	References
6.	Hydroxamate type (ferrichromes)	*Pseudomonas* strain GRP3	When subjected to Fe deficiency, the siderophore-inoculated *Vigna radiata* plants showed reduced chlorotic symptoms and improved chlorophyll content as compared to uninoculated plant	Sharma et al. (2003)
7.	Hydroxamate type (ferrichromes)	*Pseudomonas* sp.	Supply iron to plants, immobilization of the heavy metals in soil so that microbial activity increases and soil fertility improves	Joshi et al. (2014) and Ahemad (2014)
8.	Mixed siderophores (pyoverdines)	*Pseudomonas fluorescens* C7	Improved plant growth in *Arabidopsis thaliana* plants because of uptake of Fe–pyoverdine complex and increase in iron inside plant tissues	Vansuyt et al. (2007)
9.	Mixed siderophores (pseudobactin)	*Fluorescent pseudomonas*, *Pseudomonas aeruginosa*	Protect plants from pathogenic fungi and other harmful organisms	Husen (2003) and Martin (2003)
10.	Mixed siderophores (pseudobactin)	*Pseudomonas putida*	Increases development and yield of various plants	Chaiharn et al. (2008)

15.2.3 Hydroxamate Siderophores

The most commonly occurring group of siderophores is the hydroxamate type, which is made up of C(=O) N-(OH) R. Here, R is an amino acid or its derivative that is primarily released by bacteria (Renshaw et al. 2002). Hydroxamate siderophores contain a fixed constancy ratio of 1:1 with Fe (III), which is in close proximity to that of the Fe (III)-EDTA complex (Mosa et al. 2016). On the basis of the side chain of the hydroxamate functional group, the hydroxamate siderophores are divided into three categories, i.e. ferrioxamines, ferrichrome and aerobactin (Winkelmann 2007). Ferrioxamines is linear in structure and its molecular formula is $C_{25}H_{48}N_6O$. The ferrichromes are cyclic in structure, made of two unpredictable amino acids (alanine, glycine or serine), three N-acyl-N-hydroxyl-L-ornithine and a glycine connected by peptide bonds (Ali et al. 2011). Aerobactin is the third type of hydroxamate siderophore with a molecular formula of $C_{22}H_{36}N_4O_{13}$ (Neilands 1995). It is found in *E. coli, Pseudomonas*, *K. pneumoniae*, *A. aerogenes* and other bacteria (Buyer et al. 1991).

15.2.4 Carboxylate Siderophores

A recent group of siderophores have been identified, whose members neither exhibit hydromate nor 2, 3-dihydroxybenzoate (DHB) chelating groups (Table 15.1.). The chelation in this category of siderophores is done by carboxylate or hydroxyl carboxylate groups (Sah and Singh 2015; Schwyn and Neilands 1987). One of the most important carboxylate siderophore was also isolated from *Rhizobium meliloti* strain DM4. Rhizobactin is an aminopoly (carboxylic acid) which has hydroxycarboxyl and ethylenediamine dicarboxyl moieties or coordinating groups (Bergeron et al. 2014). Another imperative member of carboxylate siderophores is staphyloferrin A which is synthesized by *Staphylococcus hyicus* DSM20459. This siderophore consists of two citric acid residues and one D-ornithine residue, which bind by two amide bonds (Ali and Vidhale 2013). Rhizoferrin is another carboxylate siderophore synthesized by fungi belonging to zygomycetes family (Holinsworth and Martin 2009; Al-Fakih 2014).

15.2.5 Mixed Siderophores

Mixed siderophores possess a minimum of two different Fe-binding ligands (Aznar and Dellagi 2015). Mixed ligands are those siderophores which are derived from ornithine (pyoverdines), lysine (mycobactin) and histamine (anguibactin) (Sah and Singh 2015). Pyoverdine is the ornithine derivative type of mixed siderophore which is also known as pseudobactin. It is actively produced by *Pseudomonas* species (Meneely and Lamb 2007). Mycobactin is the lysine derivative type of mixed siderophores. It is produced by *Mycobacterium tuberculosis* and *Mycobacterium smegmatis* (Varma and Podila 2005). Anguibactin is the histamine derivative type of mixed siderophore. It is synthesized by marine pathogen *Vibrio anguillarum* (Naka et al. 2013). These diverse classes have been elaborated with their applications in plants in the table given below.

15.3 Siderophore-Mediated Responses Against Various Abiotic Stresses

Phytoremediation is today acknowledged as the most accepted green technology which is an effective in situ method for removal/treatment of heavy metals (Gratão et al. 2005). Rhizosphere is the region of soil and root interface and has an important role in the phytoremediation of various pollutants most importantly the heavy metals. This rhizospheric region is an extremely microbial active region due to the presence of siderophore-producing bacteria (SPB) (Rajkumar et al. 2010). These bacteria are reported to improve the phytoremediation process by increasing the mobility and bioavailability of heavy metals through their various secretions such as chelating compounds, phosphate-solubilizing complexes, production of phytohormones, changing redox state, etc. (Ma et al. 2011). Most common metals like Cd, Ni, Cu, Pb and Zn and actinides like U(IV), Th(IV) and Pu(IV) are found to be highly solubilized and bioavailable in the presence of siderophores (Schalk et al. 2011). But, the ability of siderophores in increasing the phytoremediation mainly depends upon their ligand specificity or selectivity to form a stable metal-siderophore complex (Braud et al. 2006, 2007). The siderophore-producing bacteria which are resistant to metal play a vital role in growth and endurance of plants by providing necessary nutrients (e.g. iron) to plants which grow in contaminated soils. Increased growth in presence of siderophore-producing bacteria will further improve the efficiency of phytoremediation process (Braud et al. 2009; He and Yang 2007; Rajkumar et al. 2010).

Recent advances indicate incorporating the siderophore-producing genes from bacterial and fungal genomes into the plant genomes or direct application of isolated siderophores onto the plant. Many studies have come forward to support the active production of siderophores by root-dwelling bacterial strains to overcome metal stress. Iron-phytosiderophore complexes and their transporters were found to be present in high concentrations in the root extracts of transgenic *Petunia hybrida* plants grown in iron deficient highly alkaline soils through the electrospray ionization-Fourier transform-ion cyclotron resonance mass spectrometry (Murata et al. 2015). Plants possess the ability to produce multiple siderophores such as enterobactin, which further facilitates *E. coli* colonization and commensalism in inducing stress tolerance (Searle et al. 2015). cDNA of ferritin siderophores from chickpea plants exposed to extreme dehydration and high salt stress showed immense induction of stress signals, which suggested a strong iron buffering role in the soil medium (Parveen et al. 2016). Systematic DNA analysis of siderophore producing bacteria *Klebsiella* sp. D5A genome and identification of its genes contributing to plant growth and stress management resulted in an increase in salt tolerance and wide pH adaptability. It became evident that they had well-defined roles to play under extreme environmental conditions (Liu et al. 2016). Soil- borne Cd-resistant bacterium *Enterobacter* sp. strain EG16 was found to produce multiple siderophores and plant hormone indole-3-acetic acid (IAA), both of which promote plant growth. The isolated extracts from bacteria were applied to plants which showed 31% Cd accumulation as compared to controls which made the bacterial strain a very apt instrument for inducing assisted phytoremediation through (Chen et al. 2016).

15.4 Diagnostic and Therapeutic Values

Most of the siderophores are reported to have major role in virulence by acting as iron scavengers, and these ferrisiderophores reenter the bacterial cells by means of specific cell surface receptors (Lamont et al. 2002). Convergence of sensitive technologies leads to siderophore neutralization by mammals and their re-consumption by bacterial pathogens (Aznar et al. 2015). Similarly, the hosts have also developed certain important cell conversion and siderophore-based iron delivery methods which are of great interest for diagnostic and therapeutic studies. There are different possible methods for exploitation of iron requirement which ultimately effect multiplication of pathogens and development of virulence (Aznr and Dellagi 2015). In recent past, the usage of various natural and synthetic compounds for effective treatment of iron-dependent infections and others had become popular. However, the use of bacterial siderophores against pathogen inhibition, removal of transuranic elements and against malaria has also emerged as a potential strategy (Beneduzi et al. 2012). These siderophores can adopt different mechanisms by which they can cut the supply of iron which effects the pathogen development and multiplication by either acting as "Trojan horse" toxins or by inhibiting siderophore synthesis pathway through the formation of siderophore-antibiotic conjugates. Application of siderophores in conception of "Trojan horse" makes them to act as intermediates which assist the uptake of antibiotics in the cells. The other ways include either the depletion of iron by application of siderophores which cannot be consumed as a source of iron by the pathogens or inhibition of siderophore utilization endogenously (Miethke and Marahiel 2007; Ahmed and Holmström 2014). However, all of the three mechanisms act differently for different pathogens. Different studies on the role of siderophores in biocontrol methods of pathogen development are available, e.g. siderophore secreted by *Bacillus subtilis* effectively controls the growth in *Fusarium oxysporum*, which causes the *Fusarium* wilting of pepper (Yu et al. 2011). Similarly, siderophores produced from *Azadirachta indica* effectively chelates Fe (III) from the soil which later affects the growth of various fungal pathogens negatively (Verma et al. 2011). The siderophore-triggered immunity is regulated by *MYB72* gene, which imbalances the metal homeostasis and is required along with MYB10 to combat with deficiency of iron (Palmer et al. 2013). It was also reported that the dual function of NAGL (neutrophil gelatinase-associated lipocalin) can be used to kill cancer cells, by declining the supply of iron and increased efflux of iron leading to cell death due to inactivation of major oxidative enzymes (Tang et al. 2016). Similarly, the main causative agent of tuberculosis, i.e. *Mycobacterium tuberculosis,* secretes siderophores like mycobactin and carboxymycobactin. It was reported by Jones et al. (2014) that *M. tuberculosis* reuses its siderophores to effectively use the iron source. When this process is disordered, accumulation of siderophores in intracellular spaces was observed which later harms and detoxifies *M. tuberculosis*. These siderophores are poisonous and hamper the capacity of recycling of iron and the use of haeme as iron source. Thus, the enhancement of siderophore recycling can be used for development of one of the major pathogenic bacteria causing tuberculosis. The antibacterial property of siderophores was observed with the use of gallium to quench iron-scavenging siderophores in pathogenic bacteria *Pseudomonas aeruginosa*. It was observed that in gallium-mediated siderophore quenching is able to resist the bacterial growth and restrict virulence development (Ross-Gillespie et al. 2014).

15.5 Current Relevance and Future Prospects

Microbial siderophores synthesized and secreted by bacterial pathogenic strains such as *Aerobacter*, *E. coli*, *Enterobacter*, *Pseudomonas* sp. *Salmonella*, *Vibrio*, *Yersinia*, etc. acquire metal ions from the surrounding plant rhizospheric environment and end up generating several defence responses against fungal and bacterial pathogens and oxidative damage as well. However, more investigation is needed for getting a clear idea for the metal-siderophore interaction phenomenon. Metal scavenging is a competitive soil phenomenon for the diverse class of compounds, and many side benefits of prime agricultural and plant stress physiology regulation emerge. Not only this, these versatile agents have been studied as a special case of coordination chemistry in the living systems using techniques like NMR and X-ray crystallography. Advanced investigation has substantiated their role in phytoremediation, therapy against many contagious human diseases and improvising agents in imaging techniques such as MRI. In the coming times, the use of siderophores in immediate sensor-based technologies to curb spread of epidemics holds a lot of promise. Hence, microbial siderophores could be the next "wonder drugs" and "new age agricultural wizards" of our era. But, despite seeing siderophores in this new light of information and facts, we need to find out better ways of isolating them, applying them to living systems, inducing them in transgenic organisms and making all this cost-effective as well. A very interesting fact being that unlike other signalling molecules, siderophores are short-lived and don't persist after triggering plant immunity. It will be amazing to discover the involvement of the lipocalin family in siderophore activation and a siderocalin-like response system in plants as in the case of mammals. Thus, the question of the role of such proteins in siderophore-mediated immunity remains to be addressed. If we are able to ace up research at the genetic level and crack the molecular mechanisms that bestow precision and versatility to siderophores, this could lead to better crop management strategies and extensive bio-patenting of siderophores to be used as novel therapeutic agents for fortifying both plant and human health care.

References

Achard ME, Chen KW, Sweet MJ, Watts RE, Schroder K, Schembri MA (2013) An antioxidant role for catecholate siderophores in *Salmonella*. Biochem J 454(3):543–549

Ahemad M (2014) Remediation of metalliferous soils through the heavy metal resistant plant growth promoting bacteria: paradigms and prospects. Arab J Chem. https://doi.org/10.1016/j.arabjc.2014.11.020

Ahmed E, Holmström SJ (2014) Siderophores in environmental research: roles and applications. Micro Biotechnol 7(3):196–208

Akafia MM, Harrington JM, Bargar JR, Duckworth OW (2014) Metal oxyhydroxide dissolution as promoted by structurally diverse siderophores and oxalate. Geochim Cosmochim Acta 141:258–269

Al-Fakih AA (2014) Overview on the fungal metabolites involved in mycopathy. Open J Med Microbiol 4:38–63

Ali SS, Vidhale NN (2013) Bacterial siderophores and their application: a review. Int J Curr Microbiol App Sci 2(12):303–312

Ali T, Bylund D, Essén SA, Lundström US (2011) Liquid extraction of low molecular mass organic acids and hydroxamate siderophores from boreal forest soil. Soil Biol Biochem 43:2417–2422

Allred BE, Rupertb PB, Gaunya SS, Ana DD, Ralstonc CY, Hoehnea MS, Strongb RK, Abergela RJ (2015) Siderocalin-mediated recognition, sensitization, and cellular uptake of actinides. Proc Natl Acad Sci 112(33):10342–10347

Amin SA, Küpper FC, Green DH, Harris WR, Carrano CJ (2007) Boron binding by a siderophore isolated from marine bacteria associated with the toxic dinoflagellate Gymnodinium catenatum. J Am Chem Soc 129(3):478–479

Aznar A, Dellagi A (2015) New insights into the role of siderophores as triggers of plant immunity: what can we learn from animals? J Exp Bot 66(11):3001–3010

Aznar A, Chen NW, Thomine S et al (2015) Immunity to plant pathogens and iron homeostasis. Plant Sci 240:90–97

Beneduzi A, Ambrosini A, Passaglia LM (2012) Plant growth-promoting rhizobacteria (PGPR): their potential as antagonists and biocontrol agents. Genet Mol Biol 35(4):1044–1051

Bergeron RJ, Wiegand J, McManis JS, Bharti N (2014) Desferrithiocin: a search for clinically effective iron chelators. J Med Chem 57(22):9259–9291

Braud A, Jézéquel K, Vieille E, Tritter A, Lebeau T (2006) Changes in extractability of Cr and Pb in a polycontaminated soil after bioaugmentation with microbial producers of biosurfactants, organic acids and siderophores. Water Air Soil Pollut 6:261–279

Braud A, Jézéquel K, Lebeau T (2007) Impact of substrates and cell immobilization on siderophore activity by Pseudomonads in a Fe and/or Cr, Hg, Pb containing-medium. J Hazard Mater 144(1–2):229–239

Braud A, Jézéquel K, Bazot S et al (2009) Enhanced phytoextraction of an agricultural Cr-and Pb-contaminated soil by bioaugmentation with siderophore-producing bacteria. Chemosphere 74(2):280–286

Braun V, Pramanik A, Gwinner T, Köberle M, Bohn E (2009) Sideromycins: tools and antibiotics. Biometals 22(1):3–13

Butler A, Theisen RM (2010) Iron(III)-siderophore coordination chemistry: reactivity of marine siderophores. Coord Chem 254(3–4):288–296

Buyer JS, Lorenzo VDE, Neilands JB (1991) Production of the siderophore aerobactin by a halophilic *Pseudomonas*. Appl Environ Microbiol 57(8):2246–2250

Chaiharn M, Chunhaleuchanon S, Kozo A, Lumyong S (2008) Screening of rhizobacteria for their plant growth promoting activities. J KMITL Sci Tech 8:18–23

Chaturvedi KS, Henderson JP (2014) Pathogenic adaptations to host-derived antibacterial copper. Front Cell Infect Microbiol 4:3. https://doi.org/10.3389/fcimb.2014.00003

Chaturvedi KS, Hung CS, Giblin DE, Urushidani S, Austin AM, Dinauer MC, Henderson JP (2014) ACS Chem Biol 9:551–561. [PubMed: 24283977]

Chen Y, Chao Y, Li Y, Lin Q, Bai J, Tang L, Wang S, Ying R, Qiua R (2016) Survival strategies of the plant-associated bacterium Enterobacter sp. Strain EG16 under cadmium stress. Appl Environ Microbiol 82(6):1734–1744

Cornish AS, Page WJ (1998) The catecholate siderophores of *Azotobacter vinelandii*: their affinity for iron and role in oxygen stress management. Microbiology 144(7):1747–1754

Crosa JH, Walsh CT (2002) Genetics and assembly line enzymology of siderophore biosynthesis in bacteria. Microbiol Mol Biol Rev 66:223–249

Deicke M, Bellenger JP, Wichard TJ (2013) Direct quantification of bacterial molybdenum and iron metallophores with ultra-high-performance liquid chromatography coupled to time-of-flight mass spectrometry. Chromatogr A 1298:50–60

Duckworth OW, Akafia MM, Andrews MY, Bargar JR (2014) Siderophore-promoted dissolution of chromium from hydroxide minerals. Environ Sci Process Impact 16:1348–1359

Farrell DH, Mikesell P, Actis LA, Crosa JH (1990) A regulatory gene, angR, of the iron uptake system of *Vibrio anguillarum*: similarity with phage P 22 cro and regulation by iron. Gene 86:45–51

Glick BR (2012) Plant growth-promoting bacteria: mechanisms and applications. Hindawi Sci. https://doi.org/10.6064/2012/963401

Gratão PL, Prasad MNV, Cardoso PF et al (2005) Phytoremediation: green technology for the cleanup of toxic metals in the environment. Braz J Plant Physiol 17(1):53–64

Gregory JA, Li F, Tomosada LM, Cox CJ (2012) Topol AB, algae – produced Pfs25 elicits antibodies that inhibit malaria transmission. PLoS One 7(5):371–379

Gupta G, Parihar SS, Ahirwar NK, Snehi SK, Singh V (2015) Plant growth promoting rhizobacteria (PGPR): current and future prospects for development of sustainable agriculture. J Microb Biochem Technol 7:096–102

Harrington JM, Parker DL, Bargar JR et al (2012) Structural dependence of Mn complexation by siderophores: donor group dependence on complex stability and reactivity. Geochim Cosmochim Acta 88:106–119

He ZL, Yang XE (2007) Role of soil rhizobacteria in phytoremediation of heavy metal contaminated soils. J Zhejiang Univ Sci B 8(3):192–207

Hider RC, Kong X (2010) Chemistry and biology of siderophores. Nat Prod Rep 27(5):637–657

Höfte M (1993) Classes of microbial siderophores. In: Barton LL, Hemming BC (eds) Iron chelation in plants and soil microorganisms. Academic, San Diego, pp 3–26

Holinsworth B, Martin JD (2009) Siderophore production by marine-derived fungi. Biometals 22(4):625–632

Hood MI, Skaar EP (2012) Nutritional immunity: transition metals at the pathogen-host interface. Nat Rev Microbiol 10(8):525–537

Husen E (2003) Screening of soil bacteria for plant growth promotion activities in vitro. Indones J Agric Sci 4(1):27–31

Johnstone TC, Nolana EM (2015) Beyond iron: non-classical biological functions of bacterial siderophores. Dalton Trans 44(14):6320–6339

Jones CM, Wells RM, Madduri AV et al (2014) Self-poisoning of mycobacterium tuberculosis by interrupting siderophore recycling. Proc Natl Acad Sci U S A 111(5):1945–1950

Joshi H, Dave R, Venugopalan VP (2014) Pumping iron to keep fit: modulation of siderophore secretion helps efficient aromatic utilization in *Pseudomonas putida* KT2440. Microbiology 160:1393–1400

Keating T, Walsh C (1999) Initiation, elongation, and termination strategies in polyketide and polypeptide antibiotic biosynthesis. Curr Opin Chem Biol 3:598–606

Kenney GE, Rosenzweig AC (2012) Chemistry and biology of the copper chelator methanobactin. ACS Chem Biol 7:260–268

Kraemer SM (2004) Iron oxide dissolution and solubility in the presence of siderophores. Aquat Sci 66:3–18

Kraemer SM, Crowley D, Kretzschmar R (2006) Siderophores in plant iron acquisition: geochemical aspects. Adv Agron 91:1–46

Kurth C, Kage H, Nett (2016) Siderophores as molecular tools in medical and environmental applications. Org Biomol Chem 14:8212–8227

Lamont IL, Beare PA, Ochsner U et al (2002) Siderophore-mediated signaling regulates virulence factor production in *Pseudomonas aeruginosa*. Proc Natl Acad Sci U S A 99(10):7072–7077

Lautru S, Challis GL (2004) Substrate recognition by non-ribosomal peptide synthetase multi-enzymes. Microbiology 150:1629–1636

Leong SA, Neilands JB (1982) Siderophore production by phytopathogenic microbial species. Arch Biochem Biophys 281:351–359

Liu W, Wang Q, Hou J, Tu C, Luo Y, Christie P (2016) Whole genome analysis of halotolerant and alkalotolerant plant growth-promoting rhizobacterium Klebsiella sp.D5A. Sci Rep 6:26710. https://doi.org/10.1038/srep26710

Ma Y, Prasad MN, Rajkumar M et al (2011) Plant growth promoting rhizobacteria and endophytes accelerate phytoremediation of metalliferous soils. Biotechnol Adv 29(2):248–258

Manwar AV, Khandelwal SR, Chaudhari BL, Kothari RM, Chincholkar SB (2001) Generic technology for assured biocontrol of groundnut infections leading to its yield improvement. Chem Weekly 6(26):157–158

Meneely KM, Lamb AL (2007) Biochemical characterization of an FAD-dependent monooxygenase, the ornithine hydroxylase from *Pseudomonas aeruginosa*, suggests a novel reaction mechanism. Biochemistry 46:11930–11937

Miethke M, Marahiel MA (2007) Siderophore-based iron acquisition and pathogen control. Microbiol Mol Biol Rev 3:413–451

Mosa KA, Saadoun I, Kumar K, Helmy M, Dhankher OP (2016) Potential biotechnological strategies for the cleanup of heavy metals and metalloids. Front Plant Sci doi. https://doi.org/10.3389/fpls.2016.00303

Mossialos D, Ochsner U, Baysse C, Chablain P, Pirnay JP, Koedam N (2002) Identification of new, conserved, non-ribosomal peptide synthetases from fluorescent pseudomonads involved in the biosynthesis of the siderophore pyoverdine. Mol Microbiol 45(6):1673–1685

Munzinger M, Taraz K, Budzikiewicz H (1999) SS-rhizoferrin (enantio-rhizoferrin) – a siderophore of Ralstonia (Pseudomonas) pickettii DSM 6297 – the optical antipode of R, R-rhizoferrin isolated from fungi. Biometals 12:189–193

Murata Y, Itoh Y, Iwashita T, Namba K (2015) Transgenic petunia with the iron(III)-phytosiderophore transporter gene acquires tolerance to iron deficiency in alkaline environments. PLoS One 10(3):e0120227. https://doi.org/10.1371/journal.pone.0120227

Nadal-Jimenez P, Koch G, Thompson JA, Xavier KB, Cool RH, Quax WJ (2012) The multiple signaling systems regulating virulence in *Pseudomonas aeruginosa*. Microbiol Mol Biol 76(1):46–65

Naka H, Liu M, Actis LA, Crosa JH (2013) Plasmid- and chromosome-encoded siderophore anguibactin systems found in marine vibrios: biosynthesis, transport and evolution. Biometals 26(4):537–547

Neilands JB (1995) Siderophores: structure and function of microbial iron transport compounds. J Biol Chem 70(45):26723–26726

Palmer CM, Hindt MN, Schmidt H et al (2013) MYB10 and MYB72 are required for growth under iron-limiting conditions. PLoS Genet 9:e1003953

Parveen S, Gupta DB, Dass S, Kumar A, Pandey A, Chakraborty S, Chakraborty N (2016) Chickpea Ferritin CaFer1 participates in oxidative stress response, and promotes growth and development. Sci Rep 6:31218

Prentice AM, Ghattas H, Cox SE (2007) Host-pathogen interactions: can micronutrients tip the balance? J Nutr 137(5):1334–1337

Rajkumar M, Ae N, Prasad MN et al (2010) Potential of siderophore-producing bacteria for improving heavy metal phytoextraction. Trends Biochem Sci 28(3):142–149

Raymond KN, Emily AD, Sanggoo SK (2003) Enterobactin: an archetype for microbial iron transport. Proc Natl Acad Sci U S A 100(7):3584–3588

Renshaw JC, Robson GD, Trinci APJ, Wiebe MG, Livens FR, Collison D, Taylor RJ (2002) Fungal siderophores: structures, functions and applications. Mycol Res 106:1123–1142
Ross-Gillespie A, Weigert M, Brown SP et al (2014) Gallium-mediated siderophore quenching as an evolutionarily robust antibacterial treatment. Evol Med Public Health 1:18–29
Sah S, Singh R (2015) Siderophore: structural and functional characterisation – a comprehensive review. Agriculture (Poľnohospodárstvo) 61(3):97–114
Sandy M, Butler A (2009) Microbial iron acquisition: marine and terrestrial siderophores. Chem Rev 109(10):4580–4595
Schalk IJ, Hannauer M, Braud A (2011) New roles for bacterial siderophores in metal transport and tolerance. Environ Microbiol 13(11):2844–2854
Schwyn B, Neilands JB (1987) Universal chemical assay for the detection and determination of siderophores. Anal Biochem 160(1):47–56
Searle LJ, Méric G, Porcelli I, Sheppard SK, Lucchini S (2015) Variation in siderophore biosynthetic gene distribution and production across environmental and faecal populations of *Escherichia coli*. PLoS One 10(3):e0117906. https://doi.org/10.1371/journal.pone.0117906
Sharma A, Johri BN, Sharma AK, Glick BR (2003) Plant growth-promoting bacterium *Pseudomonas sp.* strain GRP3 influences iron acquisition in mung bean (*Vigna radiata* L. Wilzeck). Soil Biol Biochem 35(7):887–894
Singh A, Singh SS, Pandey PC, Mishra AK (2010) Attenuation of metal toxicity by frankial siderophores. Toxicol Environ Chem 92(7):1339–1346
Tang HC, Chang PC, Chen YC (2016) Iron depletion strategy for targeted cancer therapy: utilizing the dual roles of neutrophil gelatinase-associated lipocalin protein. J Mol Model 22(1):1–5
Vansuyt G, Robin A, Briat JF, Curie C, Lemanceau P (2007) Iron acquisition from Fe-pyoverdine by *Arabidopsis thaliana*. Mol Plant-Microbe Interact 20(4):441–447
Varma A, Podila GK (2005) Siderophore their biotechnological application. Biotech Appl Microbes:177–199
Verma VC, Singh SK, Prakash S (2011) Bio-control and plant growth promotion potential of siderophore producing endophytic Streptomyces from *Azadirachta indica* A Juss. J Basic Microbiol 51(5):550–556
Ward TR, Reas L, Serge P, Parel JE, Philipp G, Peter B, Chris O (1999) An iron-based molecular redox switch as a model for iron release from enterobactin via the salicylate binding mode. Inorg Chem 38(22):5007–5017
Winkelmann G (2007) Ecology of siderophores with special reference to the fungi. Biometals 20:379–392
Wittmann S, Heinisch L, Scherlitz-Hofmann INA, Stoiber T, Dorothe AF, Möllmann U (2001) Catecholates and mixed catecholate hydroxamates as artificial siderophores for mycobacteria. Biometals 17:53–64
Wyatt MA, Johnston CW, Magarvey NA (2014) Gold nanoparticle formation via microbial metallophore chemistries. J Nanopart 16:2212. https://doi.org/10.1007/s11051-013-2212-2
Yamamoto S, Chowdhury MAR, Kuroda M, Nakano T, Koumoto Y, Shinoda S (1991) Further study on polyamine compositions in Vibrionaceae. Can J Microbiol 37:148–153
Yu X, Ai C, Xin L, Zhou G (2011) The siderophore-producing bacterium, *Bacillus subtilis* CAS15, has a biocontrol effect on Fusarium wilt and promotes the growth of pepper. Eur J Soil Biol 47(2):138–145

Chapter 16
Linking Plant Nutritional Status to Plant-AMF Interactions

Amanda Azevedo Bertolazi, Muriel da Silva Folli-Pereira, Gustavo Caione, Lucas Zanchetta Passamani, Carlos Moacir Colodete, Sávio Bastos de Souza, Alessandro Coutinho Ramos, Nazima Rasool, Getúlio de Freitas Seben Júnior, and Evandro Luiz Schoninger

Abstract Soils with low fertility due to low availability of nutrients limit agricultural productivity, once crops will not possess the ideal conditions for development. Plants depend on their root systems to acquire the water and nutrients necessary for their survival in nature and for their yield and nutritional quality in agriculture. Thus, the use of plant growth-promoting microorganisms that form symbiosis with plant roots is an essential strategy to promote more efficient use of mineral resources in agriculture. Mycorrhizae are symbiotic associations between soil fungi and roots of vascular plants, which provide several benefits to the plant, like absorption and translocation of nutrients, mainly nitrogen and phosphate. The benefits of mycorrhizal symbiosis depend on the interaction between plant fungus and the environmental characteristics, such as availability of phosphorus and carbon supply to the symbiont. Mycorrhizal fungi can also protect plants from several biotic and abiotic

A.A. Bertolazi • S.B. de Souza
Plant Physiology Lab, UENF - Universidade Estadual do Norte Fluminense Darcy Ribeiro, Campos dos Goytacazes, RJ, Brazil

M. da Silva Folli-Pereira (✉) • G. Caione • G. de Freitas Seben Júnior • E.L. Schoninger
Universidade do Estado de Mato Grosso – Unemat, Alta Floresta, MT, Brazil
e-mail: muriel.folli@unemat.br

L.Z. Passamani
Integrative Biology Unit, Biotechnology Lab, Universidade Estadual do Norte Fluminense (UENF), Campos dos Goytacazes, RJ, Brazil

C.M. Colodete
Biotechnology and Environmental Microbiology Lab, Universidade Vila Velha (UVV), Vila Velha, ES, Brazil

A.C. Ramos
Laboratory of Physiology and Biochemistry of Microorganisms, Universidade Estadual do Norte Fluminense (UENF), Campos dos Goytacazes, RJ, Brazil

N. Rasool
Department of Botany, Satellite Campus University of Kashmir, Kargil, J & K, India

D. Egamberdieva, P. Ahmad (eds.), *Plant Microbiome: Stress Response*, Microorganisms for Sustainability 5, https://doi.org/10.1007/978-981-10-5514-0_16

stresses such as heavy metal contamination, salt and water stress, and pathogens, and they can influence soil properties like increase soil aggregation and stability. But, to assure the efficiency of these benefits, an adequate management of native fungi is necessary, which can be assured through the adoption of conservation practices related to soil preparation and management and which has been associated to the increased diversity of native mycorrhizal fungi. In these conditions, mycorrhizal fungi can represent an important technology to make agricultural production sustainable, both ecologically and economically, decreasing the costs with mineral fertilizers, irrigation, and pesticides.

Keywords Arbuscular mycorrhizal fungi • Nutrient uptake • Nutritional stress • Phosphorus • Rhizosphere

16.1 Introduction

Environmental degradation caused by conventional agricultural production systems has led to the depletion of natural resources. These issues have made society rethink the use of ecologically sustainable agricultural production systems. The knowledge about the biological processes of soil and its importance for the functioning of terrestrial ecosystems is necessary. Soil microorganisms can act in a beneficial or harmful way in the most diverse agricultural systems (Antoun 2012; Zachow et al. 2016). In this context, the main challenge is the efficient application of processes mediated by beneficial microorganisms such as arbuscular mycorrhizal fungi (AMF), to increase plant nutrition and tolerance to the most diverse environmental stress conditions to which agricultural crops are exposed (Wei et al. 2016).

Arbuscular mycorrhizae (AM) are symbiotic associations established between fungi of the phylum *Glomeromycota* and the roots of most plant species (INVAN 2016). The occurrence of AM has been observed in about 95% of plant species (Trappe 1987). This fact makes mycorrhizae a rule and not an exception in nature. Studies in recent years have shown that agricultural production, especially in tropical regions, can be enhanced by the use of sustainable technologies that are based on microbiological processes; thus, the management of native AMF has been shown to be an important technique for increasing the efficiency of soil phosphorus absorption by plants, when grown in soils that are deficient in this nutrient.

In soils that have low levels of fertility, AM may be fundamental for overcoming the nutritional stress to which plants are exposed. With the application of management techniques appropriate to the stimulation of native soil AMF communities, growth and production of agricultural crops can be sustained under conditions of nutritional deficiency, decreasing the costs associated with mineral fertilizers and irrigation of the plants in soils with little water availability. Agronomic research involving AM has been carried out, focusing mainly on the ability of different species of AMF and plant-AMF combinations to improve the nutritional status of plants undergoing such stress (Zhang et al. 2016). The application of AM in conditions of nutritional stress, to which most plant species from tropical regions are exposed, becomes even more relevant in view of high prices of industrialized fertilizers.

This chapter will focus on the general aspects of AM application in soils with low natural fertility, highlighting the benefits of the fungi to the partner plants under conditions of stress, especially nutritional, and their importance as a strategy available for sustainable agricultural production.

16.2 Plant Mineral Nutrition

Proper nutrition of plants depends on the availability of nutrients to be absorbed by the roots. In this context, knowledge on the factors that influence the uptake of nutrients and the physiological functions and nutritional disorders of each nutrient becomes very important for success in production. Before the process of nutrient uptake by the roots occurs, the contact of the root with the ion, either by the ion movement in the soil solution of the rhizosphere (diffusion or mass flow) or by root finding the ion during its growth (root interception), is necessary. Several inherent plant factors such as the anatomy of the root, genetic potential of plant, internal ionic state, and intensity of metabolism and transpiration can affect the absorption process. Other factors not related to the plant like soil pH, chemical form of the nutrient in the soil, aeration, moisture, organic matter, temperature, and presence of other ions and microorganisms also affect the nutrient uptake (Prado 2008). When the amount of nutrients in the soil is insufficient to fulfill the demand of plants, the symbiosis with mycorrhizal fungi improves the absorption of nutrients by the roots (Bücking and Kafle 2015). This happens because arbuscular mycorrhizal fungi (AMF) can act as an extension of the root system, thereby increasing nutrient uptake (Bücking and Kafle 2015). Moreover, AMF have been reported to induce expression of plant phosphate transporters (Harrison et al. 2002; Paszkowski et al. 2002; Nagy et al. 2005; Xie et al. 2013; Walder et al. 2015).

While positive effects of AMF in the absorption of P are commonly reported (Smith and Read 2008; Smith et al. 2011), the contribution of these fungi to the N nutrition of vegetables is still debated (Smith et al. 2011). Several studies have suggested that higher N content in plants with mycorrhizae is just an effect of increased absorption of P (Reynolds et al. 2005), because, due to the high mobility of inorganic forms of N in the soil (ammonium and nitrate), depletion of nutrients in the rhizosphere is less likely to occur (Bücking and Kafle 2015). However, recent studies have suggested that the ammonium absorption system of AMF has five times more affinity for NH^{4+} than typical absorption systems of plants and that these fungi are able to absorb ammonium from the soil even in low availability conditions (Pérez-Tienda et al. 2012). In addition to the absorption of P and N, S can also be transferred into plants through the AMF (Allen and Shachar-Hill 2009; Sieh et al. 2013). Recently, a sulfate charger was identified to be specifically involved in the absorption from the AMF (Giovannetti et al. 2014).

The importance of potassium (K) in plant nutrition is well known, but the contribution of the AMF in the K uptake is not very well studied (Garcia and Zimmermann 2014). However, accumulation of this element has been reported in spores (Pallon et al. 2007), hyphae (Olsson et al. 2008), and vesicles (Olsson et al. 2011). In addition to the macronutrients, AMF can also improve the absorption and accumulation of micronutrients in plants. Especially for Zn, Lehmann et al. (2014) concluded that symbiosis with AMF increased the concentration of this nutrient in various tissue culture experiments. Effects of AMF on Cu, Fe, and Mn accumulation were also reported by Lehmann and Rillig (2015), although under specific conditions. Currently, 17 chemical elements are considered plant nutrients. They are classified into organic (C, H, and O), primary mineral macronutrients (N, P, and K), secondary mineral macronutrients (Ca, Mg, and S), and mineral micronutrients (B, Cu, Fe, Mn, Mo, Ni, and Zn). It is noteworthy that Ni is an element that some researchers do not consider a plant nutrient.

16.2.1 Nitrogen

Nitrogen is the nutrient extracted and exported in higher amounts in plants. The atmosphere has approximately 78% of N as N_2. The N_2 is a natural gaseous source and nondirectly used by plants, because they need to process the prior combined forms, which are N-NH^{4+} (ammonium) and N-NO^{3-} (nitrate). The main processes responsible for the fixation of atmospheric N_2 into the combined forms are biological, industrial, and atmospheric nitrogen fixation. About 90% of plant nitrogen is in the organic form, and this is how this element performs its functions as a structural component of macromolecules and enzyme constituent. Nitrogen, before performing their functions in plants, undergoes "nitrate assimilation process", which requires reductase enzymes. The nitrate reduction occurs essentially in two stages: first there is the reduction of NO^{3-} to NO^{2-} (nitrite), and then the NO_2 is reduced to NH_3. Two enzymes are involved in the process, the nitrate reductase (RNO^{3-}) and nitrite reductase (RNO^{2-}) (Mengel and Kirkby 1987) resulting in the NH_3 form of nitrogen which can be assimilated by the higher plants. Nitrogen is present in amino acids and other important nitrogenous compounds such as nitrogen bases (purines and pyrimidines) and nucleic acids (DNA and RNA), which account for about 10% of the total nitrogen in the plant (Mengel and Kirkby 1987). In leaves, nitrogen is present in chloroplasts as a constituent of the chlorophyll molecule, wherein each magnesium atom is bonded to four nitrogen atoms; nitrogen is also involved in the synthesis of vitamins, hormones, coenzyme, alkaloids, hexosamine, and other compounds. Thus, nitrogen is a nutrient that is involved in almost all the important physiological processes in plants, such as photosynthesis, respiration, ion absorption of other nutrients, growth, and cellular and genetic differentiation.

Nitrogen deficiency initially causes yellowing (chlorosis) of older leaves followed by stunted growth. Because of its high redistribution, new leaves remain green, but if the problem is not corrected, the yellowing spreads to younger leaves as well. Excess of nitrogen causes appearance of leaves with dark-green color and overgrowth which can lead to plant lodging.

16.2.2 Phosphorus

Phosphorus is considered one of the most limiting nutrients for crop production in tropical soils due to low natural availability, high adsorption capacity, and precipitation in the soils. Therefore, it is difficult to maintain high levels of phosphorus in the soil solution for plant growth and development. Thus, to obtain satisfactory yields, the use of phosphate fertilizers to meet demands of this nutrient is highly necessary. For greater efficiency of phosphorus fertilization, some factors should be considered: liming before the phosphate fertilizer (except in the case of application of natural phosphates); the higher the content of clay and oxides of iron and aluminum, the greater will be the potential of the soil to adsorb phosphorus. Still, phosphorus has low mobility in the soils, especially in clay soils, which implies lower efficiency of fertilization without incorporation in soils with low content of the nutrient.

Phosphorus plays an important role in all metabolic processes of plants, such as photosynthesis, respiration, energy transfer, storage, cell division, growth, and nutrient uptake. This element is also important in transfer and storage of energy as a constituent of adenosine triphosphate (ATP), promoting rapid formation and growth of roots and improving the quality of fruits, vegetables, and grains. Phosphorus deficiency leads to the emergence of smaller seedlings (lower growth), old leaves with purple coloration (characteristic symptom in corn plants), and weak roots. Thus, the production is impaired.

16.2.3 Potassium

Potassium is the second element extracted in higher amounts by plants. The exchangeable potassium represents the fraction available to plants, although in some soils, non-exchangeable forms can also help to provide short-term supply of this nutrient. Potassium present in plant tissues is not incorporated into the organic fraction, so this nutrient does not have a structural function and remains as an ion. The main biochemical function of potassium is enzyme activation. More than 60 enzymes like synthetases, oxidoreductases, dehydrogenases, transferases, and kinases are dependent on potassium for their normal activity. Besides being a potent enzymatic activator, potassium exerts a physiological function essential to plants, which is the opening and closing of stomata. Due to the mobility of potassium in tissues, the deficiency symptoms occur first on older leaves with yellowing (chlorosis), followed by necrosis on the tips and margins of the leaves, and in the damaged regions there is putrescine accumulation.

16.2.4 Calcium

The major proportion of calcium is in the apoplast, where it is strongly retained in the cell wall structures. One of the main functions of calcium is in plant structure as an integrant of the cell wall, increasing the mechanical strength of the tissues and neutralizing organic acids in the cytosol. The cell wall is quantitatively the greatest "product" of plants, composing their real structure. Typically, when cells grow there is an increased contact surface between them enhancing the need of calcium supply to form pectin, providing for the elongation of the cell wall to reach its final size. In addition, the calcium in the form of calcium pectate is also part of the medium lamella which has the "cementing" function, connecting the neighboring cells. One of the key functions of calcium is to maintain the structural integrity of the membranes of various organelles. During calcium deficiency the membranes begin to leak, cellular compartmentalization is breached, and calcium binding to the pectin wall is affected. Moreover, root growth is smaller, and being immobile in phloem, calcium deficiency is initially observed in meristematic and younger leaves. These leaves appear distorted, because the margins don't expand fully due to poor cell wall formation. With the advance of symptoms of deficiency, growing points may die.

16.2.5 Magnesium

Magnesium plays structural and functional roles and is also an enzyme activator. Magnesium is the central atom in the chlorophyll molecule besides activator of a large number of enzymes. The main magnesium deficiency symptom is interveinal chlorosis starting in old leaves of plants due to high redistribution of magnesium.

16.2.6 Sulfur

Sulfur is required in considerable amounts by plants, in particular by legumes and brassics. Sulfur is important for the production of amino acids, proteins, and chlorophyll, and is a component of vitamins and plant hormones. It is found in plants in the organic forms of amino acids (cysteine, cystine, and methionine). Moreover, sulfur is a component of enzymes of the sulfhydryl group (−SH) and ferredoxin, participating in redox reactions. In conditions of sulfur deficiency, plants show yellowing of young leaves due to the low mobility of this nutrient in the phloem.

16.2.7 Boron

Boron exerts a structural as well as functional role. It is associated with the pectic substances associated with cell wall, the medium lamella, and plasmalemma. It is also an activator of various enzymes. Thus, boron is responsible for cell wall synthesis, cellular elongation, integrity of the membranes, transport of carbohydrates, and reproductive growth. Given the immobility of this nutrient in the plants, as evidenced by the low redistribution of boron from older tissues to the growth regions, the deficiency symptoms appear initially in the new parts and can result in death of growing points.

16.2.8 Chlorine

The main function of chlorine in plants is enzymatic, where this nutrient acts as an enzymatic cofactor in the photolysis of water during photosynthesis and tonoplast ATPases, stimulates the asparagine synthesis, and also acts in regulating the cellular osmotic pressure during opening and closing of stomata. Chlorine deficiency rarely occurs, because this element is present in all environments, and can be applied together with potassium as potassium chloride. Chlorine deficiency symptoms include wilting, chlorosis, leaf bronzing, and deformation.

16.2.9 Copper

Copper is a nutrient that easily transports electrons and, as such, is important in the physiological processes of redox reactions. The main enzymes found in higher plants that contain copper are plastocyanin, laccase, ascorbic acid oxidase, and cytochrome oxidase complex. A moderate copper deficiency causes only minor growth and reduction in harvest without characteristic symptoms, while more severe deficiencies can cause yellowing of younger leaves (or blue-green color), and leaves can wilt or have margins rolled up or even become larger than normal, and growing points of the branches may die.

16.2.10 Iron

Iron is another nutrient that plays structural as well as functional role participating in the composition of organic compounds, such as phytoferritin and Leg H6, and also has an enzymatic function. This nutrient is important in the biosynthesis of chlorophyll and is also a part of proteins and enzymatic constituents that transport electrons. Iron deficiency symptom is interveinal chlorosis, initially in younger leaves. With the aggravation of the deficiency, the green color disappears completely including in the main veins.

16.2.11 Manganese

Manganese plays an important role in redox processes in plants such as electron transport in photosynthesis and detoxification of free radicals O^{2-}. This nutrient is also involved in enzymatic systems of plants, either as a cofactor or an activator of enzymes, such as decarboxylase, hydrolases, and the transferring groups (phosphokinases and phosphotransferases). Manganese also participates in the Hill reaction (photolysis of water in photosystem II) of photosynthesis, chlorophyll formation, and proliferation and function of chloroplasts. Manganese deficiency causes chlorosis between the veins in young leaves.

16.2.12 Molybdenum

Molybdenum is a constituent of several enzymes, especially those operating in nitrogen metabolism as nitrate reductase and nitrogenase. It also acts in electron transfer, because it can change its oxidation state from Mo^{6+} (oxidized) to Mo^{5+} (reduced). Some species show the molybdenum deficiency symptoms in older leaves and others in the new leaves. General chlorosis, yellow-green spots on older leaves, and necrosis have been described. It was also reported a curving of the limb wilting the margins of new leaves, up (tomato) or down (coffee) the leaves. In the genus *Brassica*, a common symptom called "whiptail" wherein new leaves grow almost without limb and only main veins grow is very common.

16.2.13 Nickel

Nickel is the component of the enzyme urease that catalyzes breakdown of urea. The essentiality of nickel for plants is still debated. While some authors put this element on the list of micronutrients, others say that this element is only essential for some species of plants, and claims of its role in other plants need further studies. In severe nickel deficiency, urea accumulation occurs, causing necrotic lesions.

16.2.14 Zinc

Zinc is an activator of several enzymes, although it may be a constituent of some of them like tryptophan synthetase, RNase, RNA polymerase, and carbonic anhydrase. The micronutrient can affect important processes such as hormonal control of IAA: the absence of zinc reduces the synthesis of IAA, and therefore the cells get smaller, reducing the protein synthesis and nitrogen metabolism. Typically, zinc deficiency in several species is represented by the shortening of the internodes and the leaves becoming smaller; however, yellow tracks (or white ones) between the vein and edges of the leaves may also occur.

16.3 Arbuscular Mycorrhizal Fungi and Plant Mineral Nutrition

Arbuscular mycorrhiza (AM) is the most common and widespread plant symbiosis (Rasool 2012; Bonneau et al. 2013). The presence of arbuscular mycorrhizal fungi (AMF), the composition of the community, and the variety of species are affected by several factors including, chemical and physical properties of soils and the physiology of the host (Leal et al. 2013). These fungi are obligatory mutualistic, with over 250 known species belonging to the phylum *Glomeromycota* and genera *Gigaspora*, *Scutellospora*, *Glomus*, *Acaulospora*, and *Archaeospora* (Brundrett and Ashwath 2013).

The development of symbiosis between AMF and plants has been considered a successful strategy that allowed the colonization of terrestrial environments by the plants since the beginning of the evolutionary process. It has been demonstrated, by the analysis of fossil records, that AM have emerged over 400 million years ago (Reinhardt 2007; Smith and Read 2008). The success of this association is due to the existing benefits for both parties, whereas the fungus increases the area of host root exploration, because of mycelium expansion, which absorbs nutrients and water from the soil transferring these for the plant, receiving in return organic carbon (C) that is used in its development. All this exchange between the AMF and the plant is performed by a specialized structure of fungus called arbuscule. This structure is formed from the cell wall penetration by the hyphae with no penetration of the plant cell plasma membrane, thereby increasing the contact surface for the bidirectional exchange between the symbionts (Smith and Smith 2011).

Mycorrhizal plants can absorb nutrients from the soil by two main routes: the route of the plant, which involves the absorption of nutrients from the soil directly by the root epidermis and root hairs, and the route of the mycorrhizal fungus, which involves the uptake of nutrients from the mycelium and the transport of nutrients to the apoplastic interface of root cells, then being actually absorbed by plants. One reason that the symbiosis with the AM is so interesting is that the absorption of nutrients via plant is not efficient, especially in relation to the nutrients that have low mobility in soil. For example, phosphate (P) when available near the roots is rapidly taken up, and due to its low mobility, a depletion zone of this nutrient is formed near the root, so the uptake is performed more efficiently by the hyphae of the fungus, because they have a greater power of land exploration, reaching regions that are more distant from the roots (Schachtman et al. 1998). Thus, the main benefit of the symbiosis of plants with AMF is the major nutrient acquisition; however, several studies have shown that symbiosis can confer other benefits, such as increased tolerance to pathogenic microorganisms (Azcón-Aguilar and Barea 1996; Gange 2001), increased reproductive success and plant growth, tolerance to water stress, high temperatures, salinity, and soil acidity (Cavagnaro et al. 2001; Marulanda et al. 2003; Ruiz-Lozano et al. 2006; Smith et al. 2004; Pozo and Azcón-Aguilar 2007; Rasool 2012).

AMF is usually more important for the uptake of nutrients with low mobility in soil (like P and Zn) than nutrients with higher mobility. However, AMF have the ability to absorb N in amounts even higher than P, but it is believed that the plants do not require the AMF for their nitrogen nutrition, and because their root system is able to absorb it, this nutrient has high soil mobility (Gamper et al. 2004). Nevertheless, there is controversy regarding the importance of plant colonization by AMF for the N absorption. Nitrogen sources in its inorganic and organic form can be effectively absorbed by AMF and translocated to the host plant, representing a significant route for the absorption of N by plants (Jin et al. 2012). Although it may vary depending on the experimental conditions and the plant host, it has been shown that between 21 and 75% of all nitrogen absorbed by the roots originate from the extraradicular mycelium of the AMF (Tian et al. 2010). In addition to the nitrogen present in the soil, the fixation of N_2 by bacteria is another important source of nitrogen for the plants. There is evidence that these bacteria have increased ability of nitrogen fixing due to the phosphate supply and the micronutrients provided by the mycorrhizae, which is an indirect effect of mycorrhizal symbiosis on plant nitrogen nutrition (Bethlenfalvay and Lindcnnan 1992; Smith and Read 2008).

Arbuscular mycorrhizae are recognized for their ability to stimulate plant growth, mainly through the increase in the uptake of nutrients, especially phosphate. Ryan et al. (2003) identified high levels of nutrients in intra-radical hyphae, highlighting the fact that the phosphorus concentration was higher than that found in soil confirming the ability of the fungus in the uptake and accumulation of this element.

Phosphorous is a macronutrient present in the soil at low concentrations and is not very mobile. It is in these conditions that the AMF play a key role in the survival of several species that are unable to mobilize this element. P is a structural nutrient in the formation of nucleic acids, phospholipids, as well as several enzymes

(Lehninger et al. 1995). It is directly involved in the phosphorylation process and therefore in energy metabolism, in the signal transduction, and in the regulation of cellular activity. When missing, it can cause a significant decline in ATP (−74%) and ADP (−91%) content, as well as enzyme content (Duff et al. 1989). Thus, the maintenance of cellular homeostasis of this element is central to organisms in general and especially tropical plants in soils with low fertility. The rate of uptake and transport of inorganic phosphorus (Pi) by the roots is higher than its diffusion rate in the soil, so a depletion zone for this element is formed in the rhizosphere environment (Smith et al. 2011). The increase in the P absorption rate provided by AMF can be attributed to:

1. Increase in the soil volume explored by the extraradical hyphae of mycorrhizal fungi
2. Small diameter of the hyphae, which allows the exploration of soil spaces unachievable by the roots
3. Higher influx rate per unit area
4. Formation of polyphosphates, which are organic molecules synthesized by the AMF that are rich in P and lead to the reduction of Pi concentration within the hyphae, with concomitant accumulation of P in conditions of high availability in stress conditions, thereby allowing a continuous stream to the host
5. Production of enzymes such as phosphatases, which catalyze the release of organic complexes of P allowing their absorption by the plants in the arbuscular units (Marschner and Dell 1994; Smith et al. 2011)

The enhanced P nutrition in plants colonized with AMF will result in (a) increased growth and photosynthetic activity, (b) increase in carbohydrate transfer rate to the roots, and (c) increase in its efflux to apoplast and into the drain imposed by the mycorrhizal fungi (Bücking and Shachar-Hill 2005). Due to increased absorption of P (and to a lesser extent Zn), the pH of the rhizosphere is usually lower in the presence of AM which can lead to increased solubility of P in soil (Mohammad et al. 2004).

Like other nutrients, phosphate is absorbed selectively against an electrochemical potential gradient, from concentrations in the soil around 1 μmol L^{-1} to more than 1000 times this amount within the cell. This absorption process is, therefore, energetically dependent of P transporters (symport) and the activity of plasma membrane H^+-ATPases (Ramos et al. 2008, Wang et al. 2014). Studies performed by Smith (2003) and Smith et al. (2004) demonstrated that the transporters involved in phosphate absorption by plant roots are different from those involved in the absorption of colonized roots. This result suggests that there is a genetic regulation of Pi transport mechanisms in AM systems and this regulation is controlled directly by the fungus, because it is known that genes encoding these carriers are only expressed in the presence of symbiotic fungus (Karandashov and Bucher 2005), which is another indicative of the importance of mycorrhizal symbiosis in the uptake of this macronutrient.

16.4 Plant-AMF Interactions in the Rhizosphere Are Important for Plant Productivity

The symbiotic association between roots and mycorrhizal fungi is known for some time, and the benefit for both symbionts is well described in the literature. Among the main benefits for the plant are the increase in the capacity of nutrient uptake by the roots and a greater ability to respond to and survive under different biotic and abiotic stresses. These associations can result in higher productivity of plants of economic interest without the use of chemical fertilizers or pesticides (Smith and Read 2008).

In general, soils with low availability of Pi are favorable for the establishment of mycorrhizal symbiosis, a condition where the plant can get the greatest benefit of symbiosis. The benefits of mycorrhizal symbiosis depend on the interaction between plant, fungus and the environmental characteristics, such as availability of phosphorus and carbon supply to the symbiont (Berbara et al. 2006). This relationship appears to depend on the nutritional status of the plant, once high levels of Pi in the soil solution inhibit the arbuscular mycorrhizal colonization (Colozzi-Filho and Siqueira 1986). Previously, Fernández et al. (1987) and Pereira et al. (1996) showed an inverse linear relationship between the Pi concentration in the soil and AM symbiosis.

P is an important macronutrient for plants, because it is required in photosynthetic process, plant productivity, formation of nucleic acids and membrane phospholipids, substrate transporters (glucose phosphate/coenzymes), cell signaling (inositol trisphosphate), and modulation of proteins (Lynch and Deikman 1998; Fraùsto da Silva and Williams 1991; Birhane et al. 2012).

The improvement of the phosphate mineral nutrition in plants has been recognized as one of the greatest benefits of mycorrhizae, especially regarding plant productivity (Marschner and Dell 1994; Birhane et al. 2012; Caldeira et al. 1983; Van der Heijden et al. 2006). Several studies point out that the presence of mycorrhizae improved plant development, which was mainly measured by the production of biomass (dry weight) and increased nutrient content (25% nitrogen (N), 25% zinc (Zn), and 10% potassium (K)) (Marschner and Dell 1994; Van der Heijden et al. 2006).

Positive effects of mycorrhizal inoculation on plant growth and P uptake have been reported in *Brachiaria*, *Andropogon*, *Panicum*, and *Sorghum* genera (Sano 1984; Howeler et al. 1987). However, such responses are conditioned to the interrelations between soil characteristics, grass species, and AMF (Powell 1977). Caldeira et al. (1983) studying various species of plants such as *Coffea arabica*, *Citrus limonia*, and *Melinis minutiflora* inoculated with AMF, found significant gains in the dry matter and general nutritional content. The inoculation of coffee plants with the AMF *Acaulospora* sp. promoted 100% increase in dry matter (DM) of the shoot and increased levels of P, K, Ca, and Mg. *Citrus limonia* inoculated with *G. margarita* increased shoot DM by approximately 75% and accumulation of nutrients by 80%. *Melinis minutiflora* inoculated with *G. fasciculatus* increased shoot dry matter by 65%, P content by around 100%, and Mg content 80%, but there was no significant increase in Ca.

Costa and Paulino (1989) evaluated the effect of eight species of mycorrhiza: *Glomus mosseae*, *G. fasciculatum*, *G. macrocarpum*, *G. etunicatum*, *Gigaspora margarita*, *G. heterogama*, *Acaulospora laevis*, and *A. muricata,* on growth, concentration and uptake of P in grass *Andropogon gayanus*. According to the authors, there were differences in the efficiency of the AMF tested regarding growth and absorption of P by the grass. The highest yield increase in DM was observed after the inoculation with *G. heterogama*, *G. margarita*, and *G. macrocarpum* in the order of 201, 164, and 155%, respectively, compared to the control treatment. Regarding the P content, *G. fasciculatum*, *G. etunicatum*, *G. macrocarpum*, and *A. muricata* were the most effective AMF. Plants inoculated with *G. heterogama*, *G. margarita*, *G. macrocarpum*, and *G. fasciculatum* had the highest amounts of absorbed P. Paula et al. (1990) found that different AMF showed different absorptions of P, shoot DM, and grain yields of soybean. Ezeta and Carvalho (1982) revealed that cassava (*Manihot esculenta* Crantz.) inoculated with *Acaulospora* sp. had higher growth and P content than the non-inoculated plants.

Other studies indicate that AMF improve the acquisition of N (Hawkins et al. 2000; Mäder et al. 2000; Hodge 2001). It is known that the absorption of N by mycorrhizal plants is preferably in ammonium form (NH^{4+}) (George et al.1992) and it can be related to the mobility of the ammonium form of the N, which is considerably lower than that of nitrate (Cantarella 2007); thus, higher contribution from AMF in ammonium absorption is expected. The cationic form of N absorbed by mycorrhizal plants promotes the acidification of the hypha-soil interface, also called hyphosphere, and according to Liu et al. (2002), this process can contribute to the acquisition of P by the external mycelium of the fungus. Furthermore, due to the small diameter, hyphae penetrate more easily through the decomposing organic material, better absorbing the newly mineralized N (Hodge 2003). Fungal hyphae also help mycorrhizal plants to accumulate more K (George et al. 1992).

AMF influence not only the absorption of macronutrient but also micronutrient, especially the ones that are less mobile in the soil, for example, Zn, copper (Cu), and manganese (Mn) (Marschner and Dell 1994). Kothari et al. (1991), studied the mechanism of increased Zn absorption in maize plants (*Zea mays*) inoculated into two compartments with *G. mosseae* and a mixture of native AMF. These chambers were combined with different concentrations of P and micronutrients, containing sterilized limestone soil. Roots inoculated with *G. mosseae* had an average of 115% higher P and 164% higher Zn, when compared to non-mycorrhizal ones. These results were attributed to a highly efficient translocation of these elements via hyphal compartments to the plant. Roots colonized by *G. mosseae* recorded an increase in Cu concentration (135%) that was not observed in the shoot. In contrast, the Mn levels in roots of mycorrhizal plants were significantly lower, especially in plants inoculated with the native mixture of AMF. These results reveal the importance of hyphae in increasing the acquisition of P and Zn by mycorrhizal plants. Trindade et al. (2003), evaluating the effect of inoculation of *Gigaspora margarita* with banana seedlings in different substrates, found that in the absence of inoculation, different substrates did not promote any changes in Zn content; however, in inoculated seedlings, the concentration of this nutrient increased along with P. As for Cu, there was no significant effect for the inoculated seedlings. Other studies corroborate with the contribution of the AM association to the absorption of these micronutrients (Melloni and Cardoso 1999; Oliveira and Oliveira 2005). However, this response can vary depending on the species of plant and AM (Oliveira and Oliveira 2004). Other benefits provided by AMF, that are not essentially nutritional, have also been reported, such as improvement in the aggregation and structure of the soil, resistance against biotic and abiotic stress, and heavy metals (Rillig and Mummey 2006; Smith and Read 2008; Ruiz-Lozano and Aroca 2010; Yamato et al. 2008; Azcón and Barea 2010).

AMF benefit plants by altering the physical-chemical characteristics of the substrate, contributing to the formation and maintenance of soil structure, and aggregating the soil particles through the extraradicular hyphae and their exudates (Rillig 2004; Rillig and Mummey 2006). These fungi produce a glycoprotein known as glomalin, which plays a key role in the stability of the soil structure (Bedini et al. 2009). Glomalin contains about 60% of carbohydrate, has N binding to oligosaccharide, contains Fe, is insoluble in water, and has good hydrophobicity, which contributes to initiate aggregation. The amount of extractable glomalin from the soil has a high correlation with the stability of the soil aggregates (Rosier et al. 2006) and may influence, indirectly, the soil C storage by stabilizing the soil aggregates (Zhu and Miller 2003). Soil aggregate stability has a direct influence on the establishment (Van der Heijden 2004), growth (Jones and Smith 2004), and productivity of plants. Interestingly, Van der Heijden et al. (2006) studying 11 species of plants inoculated with mycorrhizal fungi, in different conditions and different cultivation periods, in pastures, did not observed an increase in overall plant productivity. However, there were responses temporally variable in the growth, increased biomass, increased acquisition of P, soil aggregation, and survival of various species of mycorrhizal plants.

The effect of abiotic stress (temperature, humidity, light, water, salinity, or heavy metals) varies according to its intensity and may directly affect plant growth and productivity (Schulze et al. 2002). Water stress can severely affect the plants (Gebrekirstos et al. 2006); however AMF can change the water relations of plants by improving their resistance to drought (Augé 2001; Smith and Read 2008; Ruiz-Lozano and Aroca 2010; Apple 2010). Several mechanisms have been proposed to explain the increased drought resistance by mycorrhizal plants. These mechanisms are partially nutritious (greater P absorption, but also K, N, Ca, Mg, Zn, and Cu) and partially non-nutritious which include hormonal effects (by abscisic acid), higher contact of soil hyphae, direct water uptake by hyphae, and increased photosynthesis (Augé 2001; Kaschuk et al. 2009; Smith et al. 2010). Augé (2004) showed that AMF also affects properties of soil moisture retention and stomatal conductance. Augé (2001) noted that the in situ effects of AMF increase in plant performance under drought conditions, are usually a combination of nutritional and non-nutritional effects. However, several studies have also reported a mycorrhiza-induced increase in water use efficiency in plants (Kaya et al. 2003; Ruiz-Lozano and Aroca 2010).

Birhane et al. (2012) tested the effects of mycorrhizal symbiosis in *Boswellia papyrifera,* in continuous precipitation system, with continuous watering to field capacity every 2 days for 4 months and uneven precipitation by pulsed irrigation during 4 months, with 15 days of irrigation followed by 15 days without water. These authors observed a positive effect of symbiosis on growth, DM, assimilation rate, water use efficiency, stomatal conductance, leaf area, and photosynthesis in *Boswellia papyrifera*, especially under conditions of pulsating water, confirming the crucial role of symbiosis in environments under stress. The positive mycorrhizal effect was correlated with improved P nutrition, which was evident by the largest P fractions in the shoots and roots of mycorrhizal plants, compared with non-mycorrhizal plants. The authors also argued that the increased stomatal conductance in mycorrhizal plants compared to non-mycorrhizal plants translates into increased photosynthesis. The beneficial effect of symbiosis was higher in irregular rain (or pulsed water) than in normal precipitation (continuous supply of water). Finally, they concluded that mycorrhizal plants in irregular rainfall conditions improve the photosynthetic rate, efficiency in water use, and growth.

Other studies have evaluated the role of AMF in protecting against salt stress (Wang et al. 2004), which is a serious problem and is consistently increasing in several parts of the world (Giri et al. 2003). AMF are known to occur naturally in saline environments (Yamato et al. 2008), and some studies have reported that arbuscular mycorrhiza symbiosis decreases salt stress via increased efficiency of water use and photosynthetic rate, suggesting that the reduction of this stress is due to a biochemical, physiological and nutritional combination (Aliasgharzadeh et al. 2001). Daei et al. (2009) also demonstrated the positive influence of different AMF on the growth of plants in saline soils. All AMF were beneficial to wheat plants; however *Glomus etunicatum* was more efficient under such conditions, compared with other AM species. This indicates the importance of selecting the right combination of AM species and host plant to make cultivation under salinity more likely.

Soils contaminated with heavy metals represent an unfavorable environment for plant growth (Paula et al. 2006) and AMF can improve plant growth in case of excess of heavy metals in soil (Canton et al. 2016). Klauberg-Filho (1999) evaluating the symbiotic efficiency of different AMF in promoting the shoot growth of *Panicum maximum* in soil polluted with heavy metals, observed that inoculation with *G. clarum* and *Scutellospora fulgida*, originated from areas contaminated with heavy metals, had higher efficiency in relation to the non-inoculated treatments, increasing the DM by 47% and 31%, respectively. Silva et al. (2006) found that inoculation with AMF with *Brachiaria decumbens* increased the metal phytoextraction capacity of this species up to 58%, 79%, 205%, and 930% for Cd, Zn, Pb, and Cu, respectively, when compared to control (non-inoculated). Mycorrhizal symbiosis improves plant health, through a higher protection, against biotic stress (e.g., pathogen attack) (Azcón and Barea 2010). During plant-pathogen interactions, plant proteins are released into the apoplastic space as a plant defense response. AM fungal colonization between the AMF and their hosts also increases release of these proteins (Suo and Leung 2002). The mechanisms of induction and suppression are associated with plant defense system, which plays an important role in colonization and compatibility between AMF and its host (Garcia-Garrido and Ocampo 2002). Thus, the mycorrhizal colonization stimulates the primary defense system of the plant to pathogen attack, thereby increasing plant tolerance to biotic stress caused by diseases (Elsen et al. 2008, Vos et al. 2012.). The bioprotector effect of AMF against plant pathogens may be related to the induction of resistance in a localized or systemic form (Pozo et al. 2002; Elsen et al. 2008). When colonized by AMF, plants promote biochemical, physiological, and molecular changes related to the plant's defense system (Garcia-Garrido and Ocampo 2002; Selosse et al. 2004).

16.4.1 Regulation of the Protein Profile During the Arbuscular Mycorrhizal Symbiosis Aiming Enhanced Plant Productivity

The benefits of mycorrhizal association are known for some time, and although they have been extensively investigated in recent years, the molecular mechanisms associated with these benefits remain poorly understood (Valot et al. 2005, 2006; Cangahuala-Inocente et al. 2011; Cosette Abdallah et al. 2014). One of the strategies of analysis of this association is through proteomics technique (Couto et al. 2013). This technique allows a comprehensive analysis of all the regulatory protein profile in both organisms during the symbiosis. The analysis of the main groups of regulated proteins during the association can reveal the molecular mechanisms of the main effects observed during the symbiosis. Based on the papers published in recent years, Table 16.1 shows the main proteins regulated during mycorrhizal association, especially highlighting the proteins involved in response to biotic and abiotic stress and associated with higher uptake of nutrients. As mentioned previously, higher uptake of nutrients in plants colonized by AMF is one of the main advantages of the association and is directly related to the increase in plant productivity. Mohanta and Bae (2015) reported all transporters involved in the absorption and transport of nutrients from the soil to the plant during the interaction with the AMF. Given the importance of these proteins to the successful association, some studies have been focused on plasma membrane protein expression during symbiotic association (Valot et al. 2005, 2006; Cosette Abdallah et al. 2014). Phosphate transporter (Javot et al. 2007), ammonium transporter (Kobae et al. 2010), and ABC transporter (Zhang et al. 2010; Gutjahr et al. 2012) were identified and regulated during plant-fungus association. All these proteins are known to be important in mycorrhizal interaction, especially because the increased expression of these proteins leads to a higher uptake of nutrients by the plant (Cosette Abdallah et al. 2014). In addition to increasing the expression of these transporters, the mycorrhizal interaction also promotes an increase in the expression of H^+-ATPase (Valot et al. 2005, 2006). The increased expression and activity of these enzymes are commonly observed during the mycorrhizal interaction, and it is suggested that the increased activity of this enzyme is responsible for the change in the electrochemical gradient, that provides the energy for the functioning of the aforementioned carries (Valot et al. 2006; Ramos et al. 2005).

Table 16.1 Proteins related to enhanced plant productivity regulated in the arbuscular mycorrhizal symbiosis

Description	Function	References
Phosphate transporter	Cellular response to phosphate starvation; inositol phosphate-mediated signaling; polyphosphate biosynthetic process	Javot et al. (2007)
Ammonium transporter	Ammonium transport; lateral root formation; organic cation transport	Kobae et al. (2010)
ABC transporter	Cellular iron ion homeostasis; ion transport; transmembrane transport	Zhang et al. (2010) and Gutjahr et al. (2012)
H^+-ATPase	ATP biosynthetic process; ATP hydrolysis coupled proton transport; regulation of intracellular pH	Valot et al. (2006) and Ramos et al. (2005)
Catalase	Cellular response to nitrogen, phosphate, and sulfate starvation; response to cadmium ion; response to oxidative stress	Cangahuala-Inocente et al. (2011)
Superoxide dismutase	Response to cadmium ion; response to copper ion; response to oxidative stress	Abdel and Lafet (2011)
Ascorbate peroxidase	Response to cadmium ion; response to heat; response to reactive oxygen species; response to salt stress	Abdel and Lafet (2013)
Peroxidase	Defense response; response to oxidative stress; hydrogen peroxide catabolic process	Abdel and Lafet (2013)
Chitinase	Cell wall macromolecule catabolic process; regulation of salicylic acid metabolic process; root epidermal cell differentiation; response to cytokinin	Pozo et al. (2002)
β-1,3-Glucanase	Cell wall organization; cortical microtubule organization; shoot system development; response to cytokinin	Pozo et al. (2002)

Improved tolerance to biotic and abiotic stresses is also directly related to the increase in productivity observed in mycorrhizal plants (Pozo and Azcón-Aguilar 2007). Proteins as catalase (Cangahuala-Inocente et al. 2011), superoxide dismutase (Abdel and Lafet 2011), ascorbate peroxidase (Abdel and Lafet 2013), and peroxidase (Abdel and Lafet 2013) are more expressed in mycorrhizal plants when subjected to conditions of abiotic stress. Among other functions, these proteins act in the capture and degradation of reactive oxygen species and decrease oxidative damage caused by abiotic stress (Noctor et al. 2014), alleviating the damage that stresses can cause and promoting the growth and development of plant even in unfavorable conditions. Plant-AMF interaction is also capable of increasing plant tolerance to biotic stress, especially reducing the incidence of plant diseases (Pozo and Azcón-Aguilar 2007). Pozo et al. (2002) showed that the plant-AMF interaction is responsible for increasing the expression of proteins as chitinase e β-1,3-glucanase. These proteins can act directly in the initial response to the attack of pathogens and increase the protection of plants. Thus, for both responses to biotic and abiotic stresses, the AM symbiosis was shown capable of increasing the expression of proteins important in defense to these conditions, alleviating the damage and keeping high productivity even in conditions adverse to development. Thus, the multifunctionality of the arbuscular mycorrhizal symbiosis for plants in general, as well as benefits in plant productivity (e.g., increased plant nutrition, photosynthesis, growth, and biomass), is emphasized. Consequently, other indirect benefits can be also be highlighted such as improvement in the aggregate structure of soil, resistance against biotic/abiotic stress and heavy metal stress.

16.5 Customized Adjustment of Soil AMF Communities

Arbuscular mycorrhizal fungi (AMF) are one of the most important components of soil biota in natural and agricultural systems, colonizing about 80% of the plant species of most plant families (INVAN 2016). In association with plant roots, these fungi establish one of the most specialized symbioses in nature: arbuscular mycorrhizae (AM). AM occur in most agricultural crops, such as the plants of the families Poaceae, Fabaceae, Solanaceae, and Cucurbitaceae, considered important botanical families for humans in terms of human nutrition (Carvalho 2015).

Mycorrhizae are complex symbioses formed by several components that determine the rate of colonization, the incidence of propagules, and the effects and functions of the symbiosis for plants and ecosystems. The main components of this system are the fungus, the plant, and the environment (soil), which have strong interrelationships and interdependence.

AMF rely on the photosynthates, produced by plants through photosynthesis, for their growth and reproduction, and, in return, they benefit the plants during this association through a series of improvements in their nutritional and physiological status, which have been pointed out as the main tolerance mechanism of these plants to the most diverse conditions of stress. AMF also provide the most evident benefit for the establishment of plants in the most diverse natural and agricultural systems, which correspond to the absorption, translocation, and availability of nutrients (mainly phosphorus) to the plant root, via extraradicular mycelium, in areas localized beyond those that the plant root system is capable of exploiting (Smith and Read 2008). Given all these benefits that the AMF provide to the plants, the management of native AMF communities can be considered an important tool for sustainable agricultural production, especially in soils that usually have low fertility.

Among the benefits of the arbuscular mycorrhizal association, better access to soil resources (Folli-Pereira et al. 2012), increased soil aggregation and stability (Wu et al. 2016), and increased plant tolerance to water stress (Rasool 2012; Dell Fabbro and Prati 2014) can be cited, thus contributing to agricultural productivity, sustainability, conservation, and functionality of natural ecosystems (Purin and Rillig 2007).

The arbuscular mycorrhizal symbiosis is present in all-natural ecosystems, even those affected by adverse environmental conditions, and can be defined as a specialized system for more efficient nutrient uptake and transfer, than in non-mycorrhized roots. However, the physiological role of AM is not limited to absorption and transfer of nutrients to the host plant. Several other beneficial effects to the host plant and to ecosystems have been described, including enhancement of water stress tolerance

(Amiri et al. 2015). It is currently accepted that the contribution of symbiosis between AMF and plants grown under drought conditions, results from water stress tolerance due to a combination of physical, nutritional, and cellular effects (Ruiz-Lozano 2003).

Although AMF can colonize the roots of the majority (about 95%) of the plant species, and considering that all species of AMF identified and cataloged until now, are far from overcoming the diversity of plant species in which they can associate (INVAN 2016), it is necessary to consider that these fungi do not show any specificity with the host plants in which AM is established. However, in 2006, Pouyu-Rojas and his collaborators demonstrated that the effects of a particular AMF species, may vary according to the host species in which the arbuscular mycorrhizal association is established, thus proving that, although it does not present host specificity, these fungi present certain functional specificity. In this sense, an important fact to be considered in relation to the productivity of agricultural and ecological systems, is the functional diversity of these fungi, considering that this functionality can determine the performance of the agricultural crops in the field, as well as determinants for the structure of plant communities (Malik et al. 2016).

Although AMF do not exhibit functional specificity in relation to their hosts, the occurrence of certain species and consequently the AMF communities may vary when considering different ecosystems, once these environmental conditions change a lot, which may interfere in the growth and sporulation of certain species. Thus, it is important to consider in the studies about AMF communities, in addition to their composition, their efficiency of the community as a whole, as well as each isolate that composes it. This efficiency can be determined by the percentage of colonized roots and the diversity of species of colonized plants, considering also the time spent by each isolate or a given AMF community to colonize the host plant. Regarding the efficiency of a given AMF community, the capacity of these fungi to favor the absorption of nutrients from the soil and transfer them to the plant, thus stimulating its growth, development, and production, should be considered (Gholamhoseini et al. 2013).

In addition to their effects on individual plants and at the plant community level, AMF can be mediators of competition between plants, thus influencing plant biodiversity (León et al. 2016) and the sustainability of terrestrial ecosystems. The taxonomic, genetic, and functional diversities are directly related to the processes that occur in plants and soil (Bouffaud et al. 2016), and therefore, there is an increasing interest in assessing the biodiversity and functions of AMF communities (Lekberg and Waller 2016). Although the biodiversity theme has been widely explored in terrestrial ecology in recent years, it has been largely ignored in terms of soil biota, mainly in tropical regions (Kivlin et al. 2011; Shi et al. 2016).

Adequate management of native AMF can be assured through the adoption of conservation practices related to soil preparation and management, which has been associated to the increased diversity of native AMF. This increased diversity, in turn, can ensure that a more efficient fungus species can be established in that community. For this efficiency to be determined, it is necessary to know the structure of the

AMF community of a given soil, as well as the functional diversity of these symbionts (Finlay 2004).

Although several studies have already been developed, the richness, diversity, and symbiotic potential of AMF population in the most diverse ecosystems are not yet sufficiently studied. The main results on the occurrence of AMF, encompass surveys in various crops and uncultivated ecosystems. They reveal a wealth of species, whereas several of them are not yet identified (about 20% of the species found), which allows to affirm that more studies on the diversity of AMF in the most diverse environments are necessary, since the diversity and composition of the fungal community exerts a great diversity of effects on the plant and the plant composition (Silva et al. 2015).

The soil cultivation and the imposition of environmental stress cause great modification of the structure of fungal communities, changing the distribution and the dominance of species. This happens due to biotic and abiotic changes in the soil environment, such as changes in vegetation (roots) and soil chemical properties, especially the components of acidity, availability of nutrients, water, salinity, and contamination by heavy metals. The propagules of these fungi are present in almost all soils, and the occurrence and the degree of root colonization are determined by the type of vegetation and the environment. AMF have a low occurrence or are absent in soils: fumigated, severely disturbed by erosion, mining soils, soils that go under long periods of fallow or under flood, and in those cultivated for long periods with non-host species and with high concentrations of environmental pollutants.

As mycorrhizae are compartmentalized biological systems, they suffer enormous influence from the environment and from numerous edaphic factors of each component, that directly or indirectly influence the formation, functioning, and occurrence of AM. The components and the controlling factors present constant and intense interaction, in a way that the alteration in any of them will exert influence on the mycorrhizae and propagules of AMF.

The AMF species differ markedly in their action to improve plant nutrition and health; however relatively little is known about the functional diversity at each taxonomic level (Carvalho 2015). The refinement and application of molecular identification methods in recent years, allowed to verify that the degree of functional specificity of some mycorrhizal fungi may be higher than predicted (Řezáčová et al. 2016), given the determination of occurrence of intraspecific differences in the species of AMF (Clapp et al. 2001).

In recent years, several studies have pointed out the arbuscular mycorrhizal association as the main responsible for the increase of yield of several agricultural crops in the field. However, these same studies have pointed out the adequate management of the native AMF of the soils under cultivation of these crops, as the guarantor of all the benefits promoted by these fungi. Johnson et al. (1991) adopted crop rotation as a conservation management technique aiming to optimize the native AMF communities of the soil. They verified an increase in yield of maize and soybean crops that could be associated to an increase in the diversity of AMF, caused by the rotation of these two agricultural crops, which can be explained by the different rates of sporulation and mycorrhizal colonization that the same species of AMF can present when considering different species of plants (Bever 2002). This last observation was confirmed by the work carried out by Carrenho et al. (2002), who studied the diversity and sporulation of several species of AMF in soil with cultivation of peanut, sorghum, and maize and found a higher sporulation related to the peanut crop in relation to sorghum and maize cultivation, although the sorghum culture had provided a higher diversity of AMF species.

The results obtained by these researches allow us to conclude that the optimization of the AMF communities that are most beneficial to a particular crop may be able to guarantee a series of "advantages" to these plants, especially regarding the increased absorption and availability of nutrients and the increased tolerance of these plants to biotic and abiotic stresses.

The genetic, physiological, and molecular basis of the control of AM formation by plants is unknown (Parniske 2004), but the mechanism by which the plant restricts the formation of AM seems to depend on both environmental and nutritional factors, but is most likely to occur an interaction between the two. It has been suggested that the ability of plants to regulate the partition and allocation of carbohydrates, plays a fundamental role in the control of the root colonization process. Thus, the suppression of AM formation under extreme environmental stress conditions, can be attributed to a limitation in the availability of photosynthates to AMF nutrition and/or activation of the plant defense system.

The ability of colonization (Avio et al. 2006) is used to describe the propagation capacity of AMF inside the plant roots. So, it should be treated as a steady-state measure, differing from the level of colonization observed in a particular root segment at any given time. A dynamic colonization process requires a continuous signal exchange during hyphae and root growth.

The establishment of AM, as well as the richness of AMF in the soils of the most varied ecosystems, seems to depend on both the host plant species, which compose a particular plant community (Silva et al. 2015), and the life strategy of the AMF species. The latter can be easily explained by the colonization rate of the host root by the fungi and the sporulation capacity of the microorganism, once other species have low colonization and sporulation rates (Husband et al. 2002). Another factor to consider is the type of inoculum that can greatly vary in relation to the colonization rate, wherein certain spores of one type of inoculum can result in slower responses in terms of colonization, because its germination requires ideal environmental conditions to occur (Carvalho 2015). Thus, it will take more time for the mycorrhizal association to occur. When the inoculum originates from the extraradicular mycelium network of native AMF of the soil, when established in this soil, the culture will be able to immediately have its roots colonized by this inoculum and will rapidly utilize the benefits provided by mycorrhizae, once the fungal hyphae are not damaged, which in turn will depend on adequate management through soil conservation practices, such as direct sowing (Carvalho 2015).

The inoculation of agricultural soils with AMF has not been possible, mainly due to the difficulty of multiplying these fungi on a commercial scale, since these are obligate biotrophic (Smith and Read 2008). In addition, the few commercial inoculants available on the market have not presented a number of spores with sufficient viability to ensure a good rate of root colonization. All these factors, in particular the insufficient technology available so far for large-scale AMF inoculum production, make the inoculation of agricultural soils unfeasible (Saggin Júnior and Lovato 1999). In this context, the most viable in terms of the use of AM in agriculture would be to adopt conservationist practices of soil management, that guarantee the maintenance of native AMF communities in these soils, so they may be sufficiently viable to guarantee maximum rates of root colonization of cultivated agricultural plants.

16.6 Role and Application of AMF in the Tolerance to the Nutritional Stress in Plants

The mineral nutrients available in the soil vary in their classification, depending on the absorption needs of each plant species (Zhang et al. 2015), and can, therefore, be classified according to the amount of macro- and micronutrients required by plants (Nicolodelli et al. 2016). Mineral nutrients are required in several vital metabolic processes to the plant development, and the nutritional stress in plants can occur when the stock of nutrients available in the soil is not sufficient to meet the nutritional demands of plants, or even when there is an excess of absorption of these nutrients (Wang et al. 2016). Nutritional stress in plants may lead to a decrease in plant growth rates, with severe consequences on agricultural production (Elmer and Datnoff 2014).

In this context, arbuscular mycorrhizae (AM) are an important mechanism of bioprotection for plants against the most diverse stress situations (García-Sánchez et al. 2014). This association gains more importance when plant communities or agricultural crops are inserted in soils with low concentrations of nutrients, mainly phosphorus, which is very common in tropical regions (Pankaj et al. 2016). In these conditions, AM can represent an important technology to make agricultural production sustainable, both ecologically and economically, because there would be a decrease in the costs with mineral fertilizers, irrigation, and pesticides. AM have been known since the early nineteenth century, but the benefits to plants began to be reported only a century later, when studies about how soil sterilization would influence on the ability of plant pathogens to infect plants, verified that in these soils plants showed symptoms of nutritional stress due to low nutrient absorption (Smith and Read 2008). Later, it was confirmed that plants inoculated with AMF extracted from the soil had higher growth rates and mineral nutrient contents in their tissues, the latter caused by greater absorption and transport through the fungal hyphae (Marschner and Dell 1994).

The application of AMF to enhance plant nutrition in low fertility soils and to increase the tolerance of plants subject to the most diverse biotic and abiotic stress conditions is indisputable. Several studies have pointed to the use of AM to increase nutrient uptake, especially phosphorus in low fertility soils (Folli-Pereira et al. 2012;

Zhang et al. 2016; Zhao et al. 2015), from sources of slightly soluble P (Bolan 1991). Although the first studies on the application of AMF are from the 1950s (Mosse 1957), confirmation of its beneficial effects on plant nutrition and increased interest in its application gained strength only from the 1970s, with the publication of Kleinschmidt and Gerdemann (1972) studies. The mechanisms that result in the beneficial effects of AMF on plants are not yet fully understood, but what has been more widely accepted is that this interaction directly affects the nutrition of plants by increasing the absorption of nutrients (especially phosphorus), or indirectly, through synergism with diazotrophic bacteria, as well as through physiological changes in roots and rhizospheric changes.

Plants may present different responses to AMF in terms of nutrient absorption, because each plant species has certain particularities to be considered, like their nutritional requirement. Also, the capacity and speed of mycorrhizal colonization of AMF can be variable between plants and can be strongly influenced by P levels in the soil, because higher levels of P in soil will lead to lower velocity and rate of root colonization by the fungi (Wei et al. 2016). The main role of AMF is to help plants to overcome nutritional stress, and this happens due to the increased nutrient uptake, especially phosphorus, absorbed by the extraradicular mycelium of the AMF, which act as true extensions of the root system of plants, absorbing nutrients beyond the rhizosphere and transporting nutrients to the root cortex cells (Smith et al. 2015). This improvement in the nutritional status increases the plant resistance to other stressful conditions besides nutritional stress (Heidari and Karami 2014).

In terms of plant nutrition, mycorrhizae can also contribute to the increase of biological nitrogen fixation (BNF) by diazotrophic bacteria, since mycorrhizal plants have higher P content, which is one of the main requirements for BNF (Artursson et al. 2006). The synergism of diazotrophic bacteria and mycorrhizal fungi has been proven in different species and pointed out as responsible for conferring benefits, both in nutrition and in the protection of seedlings (Miyauchi et al. 2008; Barea et al. 2005).

In relation to increased nutrient uptake provided by mycorrhizae, in addition to the facts already discussed, it should also be considered the increase of the contact surface with the mineral particles provided by the fungal mycelium, once their hyphae have a diameter ranging from 2 to 20 μm (Baláz and Vosátka 2001), which are, on average, ten times smaller than the root diameter, which makes hyphae much more efficient in absorbing nutrients than root hairs. In addition to the mean diameter of fungal hyphae, their length should also be considered, since they can reach distances much higher than those reached by the roots (Sylvia 1992), thus considerably increasing the influx of P in the soil (Pedersen and Sylvia 1996).

In general, AMF increase the uptake of P by up to 80% (Marschner and Dell 1994). In addition to high phosphorus absorption efficiency, the hyphae of AMF can act beyond the nutrient depletion zone by increasing the uptake of several other nutrients by up to 25%, which makes AMF promising to be used as a natural biofertilizer in sustainable agricultural systems. The composition of rhizosphere microorganisms may influence the establishment and nutritional role played by the AMF in plants. When there is little supply of nutrients in the soil, the other microorganisms can compete with the AMF for its greater absorption, or when the supply of nutrients is enough so that the competition does not occur, the other microorganisms can work together with the AMF, expanding the range of benefits offered to the plants (Smith and Read 2008). This synergism between AMF and other soil microorganisms has been reported in the literature, mainly with growth-promoting rhizobacteria, where it has been observed that some species are able to increase the germination of the spores used as inoculum and consequently the rates of mycorrhizal colonization, which in turn will cause increased plant growth (Barea et al. 1997).

Although there are studies showing that high levels of several soil nutrients, such as N, Zn, and Cu (Schweiger 2016; Bąba et al. 2016), are able to inhibit the establishment of AM, P has been the most studied and pointed out as the nutrient that mostly influences the formation of this association (Moreira and Siqueira 2002). To the majority of host genotypes and soil types, it has been highlighted that in high levels of phosphorus in the soil, the rates of mycorrhizal colonization are lower and the formation of AM is compromised, but at low levels of P in the soil, the germination of spores increases, and consequently, the rates of root colonization by AMF are also enhanced (Cobb et al. 2016). The responses of AMF may vary in relation to the P uptake (Urcoviche et al. 2015), and the host plants may also vary in this response, and in terms of the nutritional benefits of AM, the entire soil-plant-FMA system should be considered. Unlike ectomycorrhizal fungi, AMF are not able to alter the morphology of the roots of the plants, but, like the first ones, they are able to cause significant physiological alterations, which in turn alter the growth rate and the vigor of the roots (Folli-Pereira et al. 2012). Usually, plants inoculated with AMF present a root system that grows better and is more branched, which may contribute to an increased area of soil that is exploited by them (Qiang-Sheng et al. 2016).

AMF can also act on nutrient availability to the plant (Marschner 1998), through the release of organic acids from the hyphae, also increasing the cation uptake by plants. These two factors mentioned above may contribute, together, to a decrease in soil pH. AMF can also stimulate the growth of microorganisms that solubilize phosphate (Muthukumar et al. 2001). From the 1990s, some compounds secreted by plant roots during nutritional stress, capable of stimulating the development of AM, were recognized (Nair et al. 1991; Paula and Siqueira 1990). The recognition of these compounds was an important milestone for the development of technologies that aimed the production of commercial inoculums of AMF, since several organic compounds, like the isoflavonoid formononetine, are capable of stimulating the formation of AM.

Although the use of commercial inoculum of AMF in agriculture is already possible, the guarantee of success in the establishment of AM does not seem to depend only on the quantity and viability of the spores found in the commercial inoculum. In this context, the success of mycorrhization will depend primarily on the symbiotic plant being compatible with a wide range of native AMF and on the soil used for the commercial inoculum to present a reasonable number of viable spores. In addition, the amount of P levels in the soil must be capable of guaranteeing that some degree of nutritional stress is imposed on the plants, since high levels of P can inhibit the establishment of AM. The decision to use commercial AMF inoculants, seems to depend strongly on the benefits that they can bring in terms of productivity and reduced use of chemical fertilizers. When all the benefits outweigh all the risks and difficulties involved, the use of this commercial inoculum becomes feasible. Once all the risks and difficulties involved in the production of commercial inoculum of AMF are overcome, their large-scale application may contribute to a reduction in the use of agrochemicals, reduce crop losses caused by various stresses, increase production, and promote environmental conservation. Therefore, AMF are important components of agricultural production and, if properly managed, can contribute substantially to the sustainability of agrosystems.

16.7 Conclusions and Future Perspectives

AM represent an important and highly ecologically sustainable resource for guaranteeing productivity in soils with low P availability for plants, which has its biotechnological potential reinforced by the relative absence of host specificity. The limitation for the large-scale use of AMF in agriculture is related to the difficulty of producing quality inoculum on a commercial scale, due to the fact that these fungi are biotrophic obligatory, making only viable the inoculation of plants in the nursery stage, when it is the case. Data on the diversity and performance of AMF are not yet sufficient to provide all the information necessary for its large-scale application. Thus, the management of native species, represents the most viable and promising alternative for increasing the productivity of agricultural crops in soil with nutrient shortages. Based on the functional character of the AMF, it is possible to stimulate the most efficient fungi species to improve the nutrient uptake by plants and increase their availability.

The major challenge in the studies regarding the role of AMF in overcoming the plant nutritional stress, is to understand the complex relationship established among the partners involved in this association, especially when dealing with agricultural soils of tropical regions, because these soils present serious nutritional limitations, and also because most of the work has focused on soils from temperate regions. These advances in AMF studies could reveal certain peculiarities of this symbiosis and could be relevant to the selection of fungal communities that are most beneficial to the crops to be established.

References

Abdel A, Lafet HA (2013) Growth and some physiological activities of pepper (*Capsicum annuum* L.) in response to cadmium stress and mycorrhizal symbiosis. J Agric Sci Technol 15:1437–1448

Abdel A, Lafet HA (2011) Influence of arbuscular mycorrhizal fungi and copper on growth, accumulation of osmolyte, mineral nutrition and antioxidant enzyme activity of pepper (*Capsicum annuum* L.) Mycorrhiza 21:495–503

Aliasgharzadeh N, Saleh RN, Towfighi H, Alizadeh A (2001) Occurrence of arbuscular mycorrhizal fungi in saline soils of the Tabriz plain of Iran in relation to some physical and chemical properties of soil. Mycorrhiza 11:119–122

Allen JW, Shachar-Hill Y (2009) Sulfur transfer through an arbuscular mycorrhiza. Plant Physiol 149:549–560

Amiri R, Nikbakht A, Etemadi N (2015) Alleviation of drought stress on rose geranium [*Pelargonium graveolens* (L.) Herit.] in terms of antioxidant activity and secondary metabolites by mycorrhizal inoculation. Sci Hortic 197:373–380

Antoun H (2012) Beneficial microorganisms for the sustainable use of phosphates in agriculture. Procedia Eng 46:62–67

Apple ME (2010) Aspects of mycorrhizae in desert plants. In: Ramawat KG (ed) Desert plants. Springer, Berlin, pp 121–134

Artursson V, Finlay RD, Jansson JK (2006) Interactions between arbuscular mycorrhizal fungi and bacteria and their potential for stimulating plant growth. Environ Microbiol 8:1–10

Augé RM (2001) Water relations, drought, and vesicular-arbuscular mycorrhizal symbiosis. Mycorrhiza 11:3–42

Augé RM (2004) Arbuscular mycorrhizae and soil/plant water relations. Can J Soil Sci 84:373–381

Avio L, Pellegrino E, Bonari E, Giovannetti M (2006) Functional diversity of arbuscular mycorrhizal fungal isolates in relation to extraradical mycelial networks. New Phytol 172:347–357

Azcón-Aguilar C, Barea JM (1996) Arbuscular mycorrhizal and biological control of soil-borne plant pathogens-an overview of the mechanisms involved. Mycorrhiza 6:457–464

Azcón R, Barea JM (2010) Mycorrhizosphere interactions for legume improvement. In: Khan MS, Zaidi A, Musarrat J (eds) Microbes for legume improvement. Springer, Vienna, pp 237–271

Bąba W, Błońska A, Kompała-Bąba A, Małkowski Ł, Ziemer B, Sierka E, Nowak T, Woźniak G, Besenyei L (2016) Arbuscular mycorrhizal fungi (AMF) root colonization dynamics of *Molinia caerulea* (L.) Moench. In grasslands and post-industrial sites. Ecol Eng 95:817–827

Baláz M, Vosátka M (2001) A novel inserted membrane technique for studies of mycorrhizal extraradical mycelium. Mycorrhiza 11:291–296

Barea JM, Azcón-Aguilar C, Azcón R (1997) Interactions between mycorrhizal fungi and rhizosphere microorganisms within the context of sustainable soil-plant systems. In: Gange AC, Brown VK (eds) Multitrophic interactions in terrestrial systems. Blackwell Science, Oxford, pp 65–77

Barea JM, Pozo MJ, Azcón R, Azcón-Aguilar C (2005) Microbial cooperation in the rhizosphere. J Exp Bot 56:1761–1778

Bedini S, Pellegrino E, Avio L, Pellegrini S, Bazzoffi P, Argese E, Giovannetti M (2009) Changes in soil aggregation and glomalin-related soil protein content as affected by the arbuscular mycorrhizal fungal species *Glomus mosseae* and *Glomus intraradices*. Soil Biol Biochem 41:1491–1496

Berbara RLL, De Souza FA, Fonseca HMAC (2006) Fungos Micorrízicos arbusculares: Muito além da nutrição. In: Fernandes MS (ed) Nutrição mineral de Plantas. Sociedade Brasileira de Ciência do Solo, Viçosa, pp 53–88. [in portuguese]

Bethlenfalvay GJ, Lindcnnan RG (1992) Mycorrhizae in sustainable agriculture, Special publication: 54. American Society of Agronomy, Madison

Bever JD (2002) Host-specificity of AM fungal population growth rates can generate feedback on plant growth. Plant Soil 244:281–290

Birhane E, Sterck FJ, Fetene M, Bongers F, Kuyper TW (2012) Arbuscular mycorrhizal fungi enhance photosynthesis, water use efficiency, and growth of frankincense seedlings under pulsed water availability conditions. Oecologia 169:895–904

Bolan NS (1991) A critical review on the role of mycorrhizal fungi in the uptake of phosphorus by plants. Plant Soil 134:189–207

Bonneau L, Huguet S, Wipf D, Pauly N, Truong HN (2013) Combined phosphate and nitrogen limitation generates a nutrient stress transcriptome favorable for arbuscular mycorrhizal symbiosis in *Medicago truncatula*. New Phytol 199:188–202

Bouffaud M-L, Creamer RE, Stone D, Plassart P, van Tuinen D, Lemanceau P, Wipf D, Redecker D (2016) Indicator species and co-occurrence in communities of arbuscular mycorrhizal fungi at the European scale. Soil Biol Biochem 103:464–470

Brundrett MC, Ashwath N (2013) Glomeromycotan mycorrhizal fungi from tropical Australia III. Measuring diversity in natural and disturbed habitats. Plant Soil 370:419–433

Bücking H, Kafle A (2015) Role of arbuscular mycorrhizal fungi in the nitrogen uptake of plants: current knowledge and research gaps. Agronomy 5:587–612

Bücking H, Shachar-Hill Y (2005) Phosphate uptake, transport and transfer by the arbuscular mycorrhizal fungus Glomus intraradices is stimulated by increased carbohydrate availability. New Phytologist 165:899–912

Caldeira SF, Chaves GM, Zambolim L (1983) Observações de micorriza vesicular-arbuscular em diferentes espécies de plantas. Rev Cer 30:19–24. [in portuguese]

Cangahuala-Inocente GC, Da Silva MF, Johnson JM, Anicet Manga A, Van Tuinen D, Henry C, Lovato PE, Dumas-Gaudot E (2011) Arbuscular mycorrhizal symbiosis elicits proteome responses opposite of P-starvation in SO_4 grapevine rootstock upon root colonization with two Glomus species. Mycorrhiza 21:473–493

Cantarella H (2007) Nitrogênio. In: Novais RF, Alvarez V, Barros NF, Fontes RLF, Cantarutti RB, Neves JCL (eds) Fertilidade do solo, 2nd edn. Sociedada Brasileira de Ciência do Solo, Viçosa, pp 375–470. [in portuguese]

Canton GC, Bertolazi AA, Cogo AJD, Eutrópio FJ, Melo J, Souza SB, Krohling CA, Campostrini E, Silva AG, Façanha AR, Sepúlveda N, Cruz C, Ramos AC (2016) Biochemical and ecophysiological responses to manganese stress by ectomycorrhizal fungus Pisolithus tinctorius and in association with Eucalyptus Grandis. Mycorrhiza 26:475–487

Carrenho R, Trufem SFB, Bononi VLR (2002) Effects of using different host plants on the detected biodiversity of arbuscular mycorrhizal fungi from an agroecosystem. Rev Bras Bot 25:93–101

Carvalho DJVM (2015) Functional diversity of native mycorrhizal fungi on wheat protection against the toxicity of Mn. Master thesis. University of Évora, Évora, Portugal

Cavagnaro TR, Gao LL, Smith SE, Smith P (2001) Morphology of arbuscular mycorrhizas is influenced by fungal identity. New Phytol 151:469–475

Colozzi-Filho A, Siqueira JO (1986) Micorrizas vesículoarbuscular em mudas de cafeeiro.1 -Efeitos da Gigaspora Margarita e adubação fosfatada no crescimento e nutrição. Rev Bras Ci Solo 10:119–205. [in portuguese]

Costa LN, Paulino VT (1989) Efeito de micorrizas vesiculo-arbusculares sobre o crescimento e absorção de fosforo em *Andropogon gayanus* cv. Planaltina. Ciên Agron 20:21–24. [in portuguese]

Clapp JP, Rodriguez A, Dodd JC (2001) Inter- and intra-isolate rRNA large subunit variation in *Glomus coronatum* spores. New Phytol 149:539–554

Cobb AB, Wilson GWT, Goad CL, Bean SR, Kaufman RC, Herald TJ, Wilson JD (2016) The role of arbuscular mycorrhizal fungi in grain production and nutrition of sorghum genotypes: enhancing sustainability through plant-microbial. Agric Ecosyst Environ 233:432–440

Cosette Abdallah C, Valot B, Guillier C, Mounier A, Balliau T, Zivy M, Van Tuinen D, Renaut J, Wipf D, Dumas-Gaudot E, Recorbet G (2014) The membrane proteome of *Medicago truncatula* roots displays qualitative and quantitative changes in response to arbuscular mycorrhizal symbiosis. J Prot 108:354–368

Couto MSR, Lovato PE, Wipf D, Dumas-Gaudot E (2013) Proteomic studies of arbuscular mycorrhizal associations. Adv Biol Chem 3:48–58

Daei G, Ardekani MR, Rejali F, Teimuri S, Miransari M (2009) Alleviation of salinity stress on wheat yield, yield components, and nutrient uptake using arbuscular mycorrhizal fungi under field conditions. J Plant Physiol 166:617–625

Dell Fabbro C, Prati D (2014) Early responses of wild plant seedlings to arbuscular mycorrhizal fungi and pathogens. Basic Appl Ecol 15:534–542

Duff SMG, Moorhead CGB, Lefebvre DD, Plaxton WC (1989) Phosphate starvation inducible 'bypasses' of adenylate and phosphate dependent glycolytic enzymes in *Brassica nigra* suspension cells. Plant Physiol 90:1275–1278

Elmer WH, Datnoff LE (2014) Mineral nutrition and suppression of plant disease. Renew Agric Food Syst 4:231–244

Elsen A, Gervasio D, Swennen R, De Waele D (2008) AMF-induced biocontrol against plant parasitic nematodes in *Musa* sp.: a systemic effect. Mycorrhiza 18:251–256

Ezeta FN, Carvalho PCL (1982) Influência da endomicorriza na absorção de P e K e no crescimento da mandioca. Rev Bras Ci Solo 6:25–28. [in portuguese]

Fernández AB, Siqueira JO, Menezes MAL, Guedes GAA (1987) Efeito diferenciado do fósforo sobre o estabelecimento e efetividade da simbiose endomicorrízica em milho e soja. Rev Bras Ci Solo 11:101–108. [in portuguese]

Finlay RD (2004) Mycorrhizal fungi and their multifunctional roles. Mycologist 18:91–96

Folli-Pereira MS, Meira-Haddad LS, Rassol N, Otoni WC, Kasuya MCM (2012) Development of mycorrhized vitroplants of *Jatropha curcas* L. at different rooting stages. Plant Biotech Rep. https://doi.org/10.1007/s11816–012-0232–5

Fraùsto da Silva JJR, Williams RJP (1991) The biological chemistry of the elements. Oxford University Press, Oxford, pp 467–496

Gamper H, Peter M, Jansa J, Lüscher A, Hartwig UA, Leuchtmann A (2004) Arbuscular mycorrhizal fungi benefit from 7 years of free air CO_2 enrichment in well-fertilized grass and legume monocultures. Glob Chang Biol 10:189–199

Gange AC (2001) Species-specific responses of a root-and shoot-feeding insect to arbuscular mycorrhizal colonization of its host plant. New Phytol 150:611–618

Garcia-Garrido JM, Ocampo JA (2002) Regulation of the plant defence response in arbuscular mycorrhizal symbiosis. J Exp Bot 53:1377–1386

Garcia K, Zimmermann SD (2014) The role of mycorrhizal associations in plant potassium nutrition. Front Plant Sci 5:337. https://doi.org/10.3389/fpls.2014.00337

García-Sánchez M, Palma JM, Ocampo JA, García-Romera I, Aranda E (2014) Arbuscular mycorrhizal fungi alleviate oxidative stress induced by ADOR and enhance antioxidant responses of tomato plants. J Plant Physiol 171:421–428

Gebrekirstos A, Teketay D, Fetene M, Mithloehner R (2006) Adaptation of Wve co-occurring tree and shrub species to water stress and its implication in restoration of degraded lands. For Ecol Manag 229:259–267

George E, Häussler KU, Vetterlein D, Gorgus E, Marschner H (1992) Water and nutrient translocation by hyphae of *Glomus mosseae*. Can J Bot 70:2130–2137

Gholamhoseini M, Ghalavand A, Dolatabadian A, Jamshidi E, Khodaei-Joghan A (2013) Effects of arbuscular mycorrhizal inoculation on growth, yield, nutrient uptake and irrigation water productivity of sunflowers grown under drought stress. Agric Water Manage 117:106–114

Giovannetti M, Tolosano M, Volpe V, Kopriva S, Bonfante P (2014) Identification and functional characterization of a sulfate transporter induced by both sulfur starvation and mycorrhiza formation in *Lotus japonicus*. New Phytol 204:609–619

Giri B, Kapoor R, Mukerji KG (2003) Influence of arbuscular mycorrhizal fungi and salinity on growth, biomass and mineral nutrition of *Acacia auriculiformis*. Biol Fertil Soils 38:170–175

Gutjahr C, Radovanovic D, Geoffroy J, Zhang Q, Siegler H, Chiapello M, Casieri L, An K, An G, Guiderdoni E, Kumar CS, Sundaresan V, Harrison MJ, Paszkowski U (2012) The half-size ABC transporters STR1 and STR2 are indispensable for mycorrhizal arbuscule formation in rice. Plant J 69:906–920

Harrison MJ, Dewbre GR, Liu J (2002) A phosphate transporter from *Medicago truncatula* involved in the acquisition of phosphate released by arbuscular mycorrhizal fungi. Plant Cell 14:2413–2429

Hawkins HJ, Johansen A, George E (2000) Uptake and transport of organic and inorganic nitrogen by arbuscular mycorrhizal fungi. Plant Soil 226:275–285

Heidari M, Karami V (2014) Effects of different mycorrhiza species on grain yield, nutrient uptake and oil content of sunflower under water stress. J Saudi Soc Agric Sci 13:9–13
Hodge A (2001) Arbuscular Mycorrhizal fungi influence decomposition of but not plant nutrient capture from, glycine patches in soil. New Phytol 151:725–734
Hodge A (2003) Plant nitrogen capture from organic matter as affected by dispersion, interspecific competition and mycorrhizal colonization. New Phytol 157:303–314
Howeler RH, Sieverding E, Saif S (1987) Practical aspects of mycorrhizal technology in some tropical crops and pastures. Plant Soil 100:249–283
Husband R, Herre EA, Turner SL, Gallery R, Young JPW (2002) Molecular diversity of arbuscular mycorrhizal fungi and patterns of host association over time and space in a tropical forest. Mol Ecol 11:2669–2678
International culture collection of vesicular arbuscular mycorrhizal fungi – INVAN (2016) http://invam.caf.wvu.edu
Javot H, Penmetsa RV, Terzaghi N, Cook DR, Harrison MJ (2007) A *Medicago truncatula* phosphate transporter indispensable for the arbuscular mycorrhizal symbiosis. Proc Natl Acad Sci U S A 104:1720–1725
Jin H, Liu J, Liu J, Huang X (2012) Forms of nitrogen uptake, translocation, and transfer via arbuscular mycorrhizal fungi: a review. Sci China Life Sci 55:474–482
Johnson NC, Pfleger FL, Crookston RK, Simmons SR, Copeland PJ (1991) Vesicular arbuscular mycorrhizas respond to corn and soybean cropping history. New Phytol 117:657–663
Jones MD, Smith SE (2004) Exploring functional definitions of mycorrhizas: are mycorrhizas always mutualisms? Can J Bot 82:1089–1109
Karandashov V, Bucher M (2005) Symbiotic phosphate transport in arbuscular mycorrhizas. Trends Plant Sci 10:22–29
Kaschuk G, Kuyper TW, LeVelaar PA, Hungria M, Giller KE (2009) Are the rates of photosynthesis stimulated by the carbon sink strength of rhizobial and arbuscular mycorrhizal symbioses? Soil Biol Biochem 41:1233–1244
Kaya C, Higgs D, Kirnak H, Tas I (2003) Mycorrhizal colonization improves fruit yield and water use efficiency in watermelon (*Citrullus lanatus* Thunb) grown under well-watered and water stressed conditions. Plant Soil 253:287–292
Klauberg-Filho O (1999) Ecologia e atividade de fungos micorrízicos arbusculares em solo poluído com metais pesados 161p Tese (Doutorado)—Universidade Federal de Lavras. [in portuguese]
Kleinschmidt GD, Gerdemann JW (1972) Stunting of citrus seedlings in fumigated nursery soils related to the absence of endomycorrhizae. Phytopathology 64:1447–1453
Kivlin SN, Hawkes CV, Treseder KK (2011) Global diversity and distribution of arbuscular mycorrhizal fungi. Soil Biol Biochem 43:2294–2303
Kobae Y, Tamura Y, Takai S, Banba M, Hata S (2010) Localized expression of arbuscular mycorrhiza-inducible ammonium transporters in soybean. Plant Cell Physiol 51:1411–1415
Kothari SK, Marschner H, Römheld V (1991) Effect of a vesicular-arbuscular mycorrhizal fungus and rhizosphere micro-organisms on manganese reduction in the rhizosphere and manganese concentrations in maize (*Zea mays* L.) New Phytol 117:649–655
Leal PL, Siqueira JO, Stürmer SL (2013) Switch of tropical Amazon forest to pasture affects taxonomic composition but not species abundance and diversity of arbuscular mycorrhizal fungal community. Appl Soil Ecol 71:72–80
Lehmann A, Rillig MC (2015) Arbuscular mycorrhizal contribution to copper, manganese and iron nutrient concentrations in crops – a meta analysis. Soil Biol Biochem 81:147–158
Lehmann A, Veresoglou SD, Leifheit EF, Andrillig MC (2014) Arbuscular mycorrhizal influence on zinc nutrition in crop plants – a meta-analysis. Soil Biol Biochem 69:123–131
Lehninger AL, David LN, Michael MC (1995) Lehninger principles of biochemistry. Worth Publishers, New York
Lekberg Y, Waller LP (2016) What drives differences in arbuscular mycorrhizal fungal communities among plant species? Fungal Ecol. https://doi.org/10.1016/j.funeco.2016.05.012

León DG, Moora M, Öpik M, Jairus T, Neuenkamp L, Vasar M, Bueno G, Gerz M, Davison J, Zobel M (2016) Dispersal of arbuscular mycorrhizal fungi and plants during succession. Acta Oecol 77:128–135

Liu A, Hamel C, Elmi A, Costa C, Ma BL, Smith DL (2002) Concentrations of K, Ca and mg in maize colonized by arbuscular mycorrhizal fungi under field conditions. Can J Soil Sci 82:271–278

Lynch J, Deikman J (1998) Phosphorus in plant biology: regulatory roles in molecular, cellular, organismic, and ecosystem processes, vol 19. ASPBR, Rockville

Malik RJ, Dixon MH, Bever JD (2016) Mycorrhizal composition can predict foliar pathogen colonization in soybean. Biol Control 103:46–53

Mäder P, Vierheilig H, Streitwolf-Engel R, Boller T, Frey B, Christie P (2000) Transport of N-15 from a soil compartment separated by a polytetrafluoroethylene membrane to plant roots via the hyphae of arbuscular mycorrhizal fungi. New Phytol 146:155–161

Marschner H (1998) Role of root growth, arbuscular mycorrhiza, and root exudates for the efficiency in nutrient acquisition. Field Crop Res 56:203–207

Marschner H, Dell B (1994) Nutrient-uptake in mycorrhizal symbiosis. Plant Soil 159:89–102

Marulanda A, Azcon R, Ruiz-Lozano JM (2003) Contribution of six arbuscular mycorrhizal fungal isolates to water uptake by *Lactuca sativa* plants under drought stress. Physiol Plant 119:526–533

Melloni R, Cardoso EJBN (1999) Quantificação de micélio extrarradicular de fungos micorrízicos arbusculares de plantas cítricas. II. Comparação entre diferentes espécies cítricas e endófitos. Rev Bras Ci Solo 23:59–67. [in portuguese]

Mengel K, Kirkby EA (1987) Principles of plant nutrition. International Potash Institute, Bern. 687p

Mohammad A, Mitra B, Khan AG (2004) Effects of sheared-root inoculum of *Glomus intraradices* on wheat grown at different phosphorus levels in the field. Agric Ecosyst Environ 103:245–249

Mohanta TK, Bae H (2015) Functional genomics and signaling events in mycorrhizal symbiosis. J Plant Interact 10:21–40

Moreira FMS, Siqueira JO (2002) Microbiologia e Bioquímica do solo. Editora UFLA, Lavras. [in portuguese]

Mosse B (1957) Growth and chemical composition of mycorrhizal and nonmycorrhizal apples. Nature 179:922–924

Muthukumar T, Udayan K, Rajeshkannan V (2001) Response of neem (Azadirachta Indica a. Juss) to indigenous arbuscular mycorrhizal fungi, phosphate-solubilizing and asymbiotic nitrogen-fixing bacteria under tropical nursery conditions. Biol Fertil Soils 34:417–426

Miyauchi MYH, Lima DS, Nogueira MA, Lovato GM, Murate LS, Cruz MF, Ferreira JM, Zangaro W, Andrade G (2008) Interactions between diazotrophic bacteria and mycorrhizal fungus in maize genotypes. Sci Agric 65:525–531

Nagy R, Karandashov V, Chague V, Kalinkevich K, Tamasloukht M, Xu G et al (2005) The characterization of novel mycorrhiza-specific phosphate transporters from *Lycopersicon esculentum* and *Solanum tuberosum* uncovers functional redundancy in symbiotic phosphate transport in solanaceous species. Plant J 42:236–250

Nair MG, Safir GN, Siqueira JO (1991) Isolation and identification of vesicular-arbuscular mycorrhiza-stimulatory compounds from clover (*Trifolium repens*) roots. Appl Environ Microbiol 57:434–439

Nicolodelli G, Senesi GS, Perazzoli ILO, Marangoni BS, Benites VM, Pereira DMBM (2016) Double pulse laser induced breakdown spectroscopy: a potential tool for the analysis of contaminants and macro/micronutrients in organic mineral fertilizers. Sci Total Environ 565:1116–1123

Noctor G, Mhamdi A, Foyer CH (2014) The roles of reactive oxygen metabolism in drought: not so cut and dried. Plant Physiol 164:1636–1648

Oliveira AN, Oliveira LA (2004) Associação micorrízica e teores de nutrientes nas folhas de cupuazeiro (*Theobroma grandiflorum*) e guaranazeiro (*Paullinia cupana*) de um sistema agroflorestal de Manaus Amazonas. Rev Bras Ci Solo 28:1063–1068. [in portuguese]

Oliveira NA, Oliveira LA (2005) Colonização por fungos micorrízicos arbusculares e teores de nutrientes em cinco cultivares de bananeiras em um latossolo da amazônia. Rev Bras Ci Solo 29:481–488. [in portuguese]

Olsson PA, Hammer EC, Pallon J, Van Aarle IM, Wallander H (2011) Elemental composition in vesicles of an arbuscular mycorrhizal fungus, as revealed by PIXE analysis. Fungal Biol 115:643–648

Olsson PA, Hammer EC, Wallander H, Pallon J (2008) Phosphorus availability influences elemental uptake in the mycorrhizal fungus *Glomus intraradices*, as revealed by particle-induced X-ray emission analysis. Appl Environ Microbiol 74:4144–4148

Pallon J, Wallander H, Hammer E, Arteaga Marrero N, Auzelyte V, Elfman M et al (2007) Symbiotic fungi that are essential for plant nutrient uptake investigated with NMP. Nucl Instrum Methods Phys Res Sect B 260:149–152

Pankaj U, Verma SK, Semwal M, Verma RK (2016) Assessment of natural mycorrhizal colonization and soil fertility status of lemongrass [(Cymbopogon Flexuosus, Nees ex Steud) W. Watson] crop in subtropical India. J Appl Res Med Arom Plants. https://doi.org/10.1016/j.jarmap.2016.10.002

Parniske M (2004) Molecular genetics of the arbuscular mycorrhizal symbiosis. Curr Opin Plant Biol 7:414–421

Paszkowski U, Kroken S, Roux C, Briggs SP (2002) Rice phosphate transporters include an evolutionarily divergent gene specifically activated in arbuscular mycorrhizal symbiosis. Proc Natl Acad Sci U S A 99:13324–13329

Paula MA, Siqueira JO (1990) Stimulation of hyphal growth of the VA mycorrhizal fungus Gigaspora Margarita by suspension-cultured *Pueraria phaseoloides* cells and products. New Phytol 115:69–75

Paula MA, Siqueira JO, Hoshika E (1990) Crescimento, nutrição e produção de soja inoculada com populações de fungos micorrízicos vesículo-arbusculares. Rev Bras Ci Solo 14:151–156. [in portuguese]

Paula MA, Soares CRFS, Siqueira JO (2006) Biomassa, atividade microbiana e fungos micorrízicos arbusculares em solos "*landfarming*" de resíduos petroquímicos. Rev Bras Eng Agric Amb 10(2):448–455. [in portuguese]

Pedersen CT, Sylvia DM (1996) Mycorrhiza: ecological implications for plant interactions. In: Mukerji KG (ed) Concepts in mycorrhizal research. Kluwer, Dordrecht, pp 195–222

Pereira EGJO, Siqueira N, Curi FMS, Moreira AAC (1996) Efeitos da micorriza e do suprimento de fósforo na atividade enzimática e na resposta de espécies arbóreas ao nitrogênio. Rev Bras Fis Veg 11:59–65. [in portuguese]

Pérez-Tienda J, Valderas A, Camañes G, García-Agustín P, Ferrol N (2012) Kinetics of NH_4^+ uptake by the arbuscular mycorrhizal fungus *Rhizophagus irregularis*. Mycorrhiza 22:485–491

Powell CL (1977) Mycorrhizas in hill-country soils I. Spore bearing mycorrhizal fungi in thirty-seven soils. J Agric Res 20:53–57

Pozo MJ, Azcón-Aguilar C (2007) Unraveling mycorrhiza-induced resistance. Curr Opin Plant Biol 10:393–398

Pozo MJ, Cordier C, Dumas-Gaudot E, Gianinazzi S, Barea JM, Azcón-Aguilar C (2002) Localized versus systemic effect of arbuscular mycorrhizal fungi on defence responses to *Phytophthora* infection in tomato plants. J Exp Bot 53:525–534

Prado RM (2008) Nutrição de plantas. UNESP, Jaboticabal. 407 p. [in portuguese]

Purin S, Rillig MC (2007) The arbuscular mycorrhizal fungal protein glomalin: limitations, progress, and new hypothesis for its function. Pedobiologia 51:123–130

Qiang-Sheng W, Ming-Qin C, Ying-Ning Z, Chu W, Xin-Hua H (2016) Mycorrhizal colonization represents functional equilibrium on root morphology and carbon distribution of trifoliate orange grown in a split-root system. Sci Hortic-Amsterdam 199:95–102

Ramos AC, Façanha AR, Feijó JA (2008) Proton (H^+) flux signature for the presymbiotic development of the arbuscular mycorrhizal fungi. New Phytol 178:177–188
Ramos AC, Martins MA, Façanha AR (2005) Atividade ATPásica e pirofosfatásica em microssomos de raízes de milho colonizadas com fungos micorrízicos arbusculares. Rev Bras Ciênc Solo 29:207–213. [in portuguese]
Rasool N (2012) Effect of some alien invasive plant species on soil microbial structure and function. Ph. D thesis, Department of Botany, University of Kkashmir, Srinagar, J & K, India
Raimam MP, Albino U, Cruz MF, Lovato GM, Spago F, Ferracin TP, Lima DS, Goulart T, Bernardi CM, Miyauchi M, Nogueira MA, Andrade G (2007) Interaction among free-living N-fixing bacteria isolated from *Drosera villosa* var. villosa and AM fungi (*Glomus clarum*) in rice (*Oryza sativa*). Appl Soil Ecol 35:25–34
Reinhardt D (2007) Programming good relations—development of the arbuscular mycorrhizal symbiosis. Curr Opin Plant Biol 10:98–105
Reynolds HL, Hartley AE, Vogelsang KM, Bever JD, Schultz PA (2005) Arbuscular mycorrhizal fungi do not enhance nitrogen acquisition and growth of old-field perennials under low nitrogen supply in glasshouse culture. New Phytol 167:869–880
Řezáčová V, Gryndler M, Bukovská P, Šmilauer P, Jansa J (2016) Molecular community analysis of arbuscular mycorrhizal fungi—contributions of PCR primer and host plant selectivity to the detected community profiles. Pedobiologia 59:179–187
Rillig MC (2004) Arbuscular mycorrhizae, glomalin and soil quality. Can J Soil Sci 84:355–363
Rillig MC, Mummey DL (2006) Mycorrhizas and soil structure. New Phytol 171:41–53
Rosier CL, Hoye AT, Rillig MC (2006) Glomalin-related soil protein: assessment of current detection and quantification tools. Soil Biol Biochem 38:2205–2211
Ruiz-Lozano JM (2003) Arbuscular mycorrhizal symbiosis and alleviation of osmotic stress. New perspectives for molecular studies. Mycorrhiza 13:309–317
Ruiz-Lozano JM, Aroca R (2010) Host response to osmotic stresses: stomatal behaviour and water use efficiency of arbuscular mycorrhizal plants. In: Koltai H, Kapulnik Y (eds) Arbuscular mycorrhizas: physiology and function. Springer, Berlin, pp 239–256
Ruiz-Lozano JM, Porcel R, Aroca R (2006) Does the enhanced tolerance of arbuscular mycorrhizal plants to water deficit involve modulation of drought-induced plant genes? New Phytol 171:693–698
Ryan MH, Mccully ME, Huang CX (2003) Location and quantification of phosphorus and other elements in fully hydrated, soil-grown arbuscular mycorrhizas: a cryo-analytical scanning electron microscopy study. New Phytol 160:429–441
Saggin Júnior OJ, Lovato PE (1999) Aplicação de micorrizas arbusculares na produção de mudas e plantas micropropagadas. In: Siqueira JO, Moreira FMS, Lopes SA, Guilherme LR, Faquin V, Fuitinni AE, Carvalho JG (eds) Inter-relação fertilidade, biologia do solo e nutrição de plantas. SBCS, Viçosa, pp 725–773
Sano SM (1984) Influência de endomicorrizas nativas do cerrado no crescimento de plantas. Rev Bras Ci Solo 8:25–29. [in portuguese]
Schachtman DP, Reid RJ, Ayling SM (1998) Phosphorus uptake by plants: from soil to cell. Plant Physiol 116:447–453
Schulze E, Beck E, Müller-Hohenstein K (2002) Plant ecology. Springer, Berlin, p 702
Schweiger PF (2016) Nitrogen isotope fractionation during N uptake via arbuscular mycorrhizal and ectomycorrhizal fungi into grey alder. J Plant Physiol 205:84–92
Selosse MA, Baudoin E, Vandenkoornhuyse P (2004) Symbiotic microorganisms, a key for ecological success and protection of plants. C R Biol 327:639–648
Shi N-N, Gao C, Zheng Y, Guo L-D (2016) Arbuscular mycorrhizal fungus identity and diversity influence subtropical tree competition. Fungal Ecol 20:115–123
Sieh D, Watanabe M, Devers EA, Brueckner F, Hoefgen R, Krajinski F (2013) The arbuscular mycorrhizal symbiosis influences sulfur starvation responses of *Medicago truncatula*. New Phytol 197:606–616

Silva DKA, Souza RG, Velez BAA, Silva GA, Oehl F, Maia LC (2015) Communities of arbuscular mycorrhizal fungi on a vegetation gradient in tropical coastal dunes. Appl Soil Ecol 96:7–17
Silva S, Siqueira JO, Soares CRFS (2006) Fungos micorrízicos no crescimento e extração de metais pesados pela Braquiária em solo contaminado. Pesq Agropec Bras 12:1749–1757. [in portuguese]
Smith SE (2003) Mycorrhizal fungi can dominate phosphate supply to plants irrespective of growth responses. Plant Physiology 133:16–20
Smith SE, Smith FA, Jakobsen I (2004) Functional diversity in arbuscular mycorrhizal (AM) symbioses: the contribution of the mycorrhizal P uptake pathway is not correlated with mycorrhizal responses in growth or total P uptake. New Phytologist 162:511–524
Smith SE, Facelli E, Pope S, Smith FA (2010) Plant performance in stressful environments: interpreting new and established knowledge of the roles of arbuscular mycorrhizas. Plant Soil 326:3–20
Smith SE, Jakobsen I, Gronlund M, Smith FA (2011) Roles of arbuscular mycorrhizas in plant phosphorus nutrition: interactions between pathways of phosphorus uptake in arbuscular mycorrhizal roots have important implications for understanding and manipulating plant phosphorus acquisition. Plant Physiol 156:1050–1057
Smith SE, Read DJ (2008) Mycorrhizal Symbiosis, 3rd edn. Academic, London
Smith SE, Smith FA (2011) Roles of arbuscular mycorrhizas in plant nutrition and growth: new paradigms from cellular to ecosystem scales. Annu Rev Plant Biol 62:227–250
Smith SE, Stonor MMR, McNeill A, Andrew Smith F (2015) Indigenous arbuscular mycorrhizal (AM) fungi contribute to wheat phosphate uptake in a semi-arid field environment, shown by tracking with radioactive phosphorus. Appl Soil Ecol 96:68–74
Suo Y, Leung DWM (2002) Accumulation of extracellular pathogenesis-related proteins in rose leaves following inoculation of *in vitro* shoots with *Diplocarpon rosae*. Sci Hortic 93:167–178
Sylvia DM (1992) Quantification of external hyphae of vesicular-arbuscular mycorrhizal fungi. In: Norris JR, Read DJ, Varma AK (eds) Methods in microbiology, vol 24. Academic, San Diego, pp 53–66
Tian C, Kasiborski B, Koul R, Lammers PJ, Bücking H, Shachar-Hill Y (2010) Regulation of the nitrogen transfer pathway in the arbuscular mycorrhizal symbiosis: gene characterization and the coordination of expression with nitrogen flux. Plant Physiol 153:1175–1187
Trappe JM (1987) Phylogenetic and ecological aspects of mycotrophy in the angiosperms from an evolutionary standpoint. In: Safir G (ed) Ecophysiology of VA mycorrhizal plants. CRC Press, Boca Raton, pp 5–25
Trindade AV, Salqueiro JLL, Santos TMC (2003) Crescimento e produtividade do mamoeiro a partir de mudas micorrizadas, sob diferentes esquemas de adubação. In: Papaya Brasil: Qualidade do Mamão para o Mercado Interno. Incaper, Vitória. 728 p. [in portuguese]
Urcoviche RC, Gazim ZC, Dragunski DC, Barcellos FG, Alberton O (2015) Plant growth and essential oil content of Mentha Crispa inoculated with arbuscular mycorrhizal fungi under different levels of phosphorus. Ind Crop Prod 67:103–107
Valot B, Dieu M, Recorbet G, Raes M, Gianinazzi S, Dumas-Gaudot E (2005) Identification of membrane-associated proteins regulated by the arbuscular mycorrhizal symbiosis. Plant Mol Biol 59:565–580
Valot B, Negroni L, Zivy M, Gianinazzi S, Dumas-Gaudot E (2006) A mass spectrometric approach to identify arbuscular mycorrhiza-related proteins in root plasma membrane fractions. Proteomics 6:S145–S155
Van der Heijden MGA, Streitwolf-Engel R, Riedl R, Siegrist S, Neudecker A, Ineichen K, Boller T, Wiemken A, Sanders IR (2006) The mycorrhizal contribution to plant productivity, plant nutrition and soil structure in experimental grassland. New Phytologist 172:739–752
Van der Heijden MGA (2004) Arbuscular mycorrhizal fungi as support systems for seedling establishment in grassland. Ecol Lett 7:293–303
Vos C, Broucke DVD, Lombi VM, Waele DD, Elsen A (2012) Mycorrhiza-induced resistance in banana acts on nematode host location and penetration. Soil Biol Biochem 10:1–7

Walder F, Brulé D, Koegel S, Wiemken A, Boller T, Courty PE (2015) Plant phosphorus acquisition in a common mycorrhizal network: regulation of phosphate transporter genes of the Pht1 family in sorghum and flax. New Phytol 205:1632–1645

Wang E, Yu N, Bano SA, Liu C, Miller AJ, Cousins D, Zhang X, Ratet P, Tadege M, Mysore KS, Downie JA, Murray JD, Oldroyd GED, Schultze M (2014) A H -ATPase that energizes nutrient uptake during mycorrhizal symbioses in rice and Medicago truncatula. The Plant Cell 26:1818–1830

Wang FY, Liu RJ, Lin XG, Zhou JM (2004) Arbuscular mycorrhizal status of wild plants in saline-alkaline soils of the Yellow River Delta. Mycorrhiza 14:133–137

Wang R, Dungait JAJ, Buss HL, Yang S, Zhang Y, Xu Z, Jiang Y (2016) Base cations and micronutrients in soil aggregates as affected by enhanced nitrogen and water inputs in a semi-arid steppe grassland. Sci Total Environ. https://doi.org/10.1016/j.scitotenv.2016.09.018

Wei Y, Su Q, Sun ZJ, Shen YQ, Li JN, Zhu XL, Hou H, Chen ZP, Wu FC (2016) The role of arbuscular mycorrhizal fungi in plant uptake, fractions, and speciation of antimony. Appl Soil Ecol 107:244–250

Wu Q-S, Wang S, Srivastava AK (2016) Mycorrhizal hyphal disruption induces changes in plant growth, glomalin-related soil protein and soil aggregation of trifoliate orange in a core system. Soil Tillage Res 160:82–91

Xie X, Huang W, Liu F, Tang N, Liu Y, Lin H et al (2013) Functional analysis of the novel mycorrhiza-specific phosphate transporter AsPT1 and PHT1 family from *Astragalus sinicus* during the arbuscular mycorrhizal symbiosis. New Phytol 198:836–852

Yamato M, Ikeda S, Iwase K (2008) Community of arbuscular mycorrhizal fungi in coastal vegetation on Okinawa Island and effect of the isolated fungi on growth of sorghum under salt-treated conditions. Mycorrhiza 18:241–249

Zachow C, Berg C, Müller H, Monk J, Berg G (2016) Endemic plants harbour specific Trichoderma communities with an exceptional potential for biocontrol of phytopathogens. J Biotechnol 235:162–170

Zhang F-F, Gao S, Zhao Y-Y, Zhao X-L, Liu X-M, Xiao K (2015) Growth traits and nitrogen assimilation-associated physiological parameters of wheat (*Triticum aestivum* L.) under low and high N conditions. J Integr Agric 14:1295–1308

Zhang Q, Blaylock LA, Harrison MJ (2010) Two *Medicago truncatula* half-ABC transporters are essential for arbuscule development in arbuscular mycorrhizal symbiosis. Plant Cell 22:1483–1497

Zhang S, Wang L, Ma F, Zhang X, Fu D (2016) Arbuscular mycorrhiza improved phosphorus efficiency in paddy fields. Ecol Eng 95:64–72

Zhao R, Guo W, Bi N, Guo J, Wang L, Zhao J, Zhang J (2015) Arbuscular mycorrhizal fungi affect the growth, nutrient uptake and water status of maize (*Zea mays* L.) grown in two types of coal mine spoils under drought stress. Appl Soil Ecol 88:41–49

Zhu YG, Miller RM (2003) Carbon cycling by arbuscular mycorrhizal fungi in soil-plant systems. Trends Plant Sci 8:407–409

Printed by Printforce, the Netherlands